Applied Life Sciences

Applied Life Sciences

Edited by **Misha Bob**

Callisto Reference

New York

Published by Callisto Reference,
106 Park Avenue, Suite 200,
New York, NY 10016, USA
www.callistoreference.com

Applied Life Sciences
Edited by Misha Bob

International Standard Book Number: 978-1-63239-077-6 (Hardback)

Printed in China.

Contents

Preface

Significant advancements are continually being made in the study of life sciences and their diverse applications for sustainable development and ecosystem preservation. This book presents discourses and discussions on various issues in applied life sciences that were presented and conferred upon in an International Conference on Applied Life Sciences. It was organized by the International Society for Applied Life Sciences (ISALS). ISALS assists experts to publish their studies on discoveries and advancements in its science journals.

After months of intensive research and writing, this book is the end result of all who devoted their time and efforts in the initiation and progress of this book. It will surely be a source of reference in enhancing the required knowledge of the new developments in the area. During the course of developing this book, certain measures such as accuracy, authenticity and research focused analytical studies were given preference in order to produce a comprehensive book in the area of study.

This book would not have been possible without the efforts of the authors and the publisher. I extend my sincere thanks to them. Secondly, I express my gratitude to my family and well-wishers. And most importantly, I thank my students for constantly expressing their willingness and curiosity in enhancing their knowledge in the field, which encourages me to take up further research projects for the advancement of the area.

Editor

1

Effect of Environmental and Socioeconomical Change on Agricultural Production in Konya Region

Sukru Dursun*, Serpil Onder, Ramazan Acar, Mithat Direk, Osman Mucehver

Selcuk University, Konya, Turkey;

*Correrspónding author

Abstract

Recently environmental problems are increasing all over the world with increasing population and industrialisation which is the increasing human requirements. In the same way, there are many environmental problems in the Konya Closed Basin, such as scarcity of water supplies in the region for realization of the unplanned agricultural production with related irrigations, since water storage difficulties owing to geologic and topographic characteristics of the area and drainage problems of irrigated agricultural areas. Climatic conditions are shaped large flat steppe areas in Konya basin. There will be important water shortage problems in the basin and irrigation water demand should be planned. Water losses should be kept to a minimum so that the amount of water obtained from underground water resources can be reduced and eventually the possible biological and ecological adverse impacts of the project will decrease and the diversity of species of flora and fauna will be preserved and endemic species will be protected. With this investigation, we aimed to put forward source of problems come from the future, appeared recently and continue in the future. Our effort may help to reduce critical problem or save our environment for our children.

Keywords: Konya Close Basin, Environmental problem, Agriculture related to water, Climate change,

1. Introduction

Turkey straddles Europe and Asia with an area of 779,452 km2, across the Marmara Sea with Istanbul and Canakkale Straits. Turkey's border is on the northwest by Bulgaria and Greece, on the east by Georgia, Armenia, Azerbaijan and Iran, and on the south by Iraq and Syria (Figure 1). Turkey's 8,333 km coastline extends along the Black Sea, the Marmara Sea, the Aegean and the Mediterranean Sea. Turkey is one of the fortunate countries in which all four seasons and various climate types are present. Large plains, highlands, mountains and deep valleys mainly characterize its geography. Turkey lies between the temperate zone and the sub-tropical zone. These peculiarities are causing different seasons to be lived at the same time, over the different climatic regions of the country. Turkey divides into five basic climate regions: Mediterranean Climate, Black Sea Climate, Semi-humid Marmara Climate, Steppe Climate and Continental Eastern Anatolia Climate. Average altitude is 1,132 m and only 10% of the country is less than 250 m above the sea level. High mountains are concentrated in central and eastern Anatolia. The physical landscape is closely linked to environmental conditions.

Erosion is one of the most severe rural environmental problems affecting 81% of the total land surface in varying levels of severity. About 73% of the cultivated land and 68% of the prime agricultural land (Classes I-IV) is prone to erosion. As a result, about one billion tons of soil is carried away each year. Turkey has about 120 natural lakes, including small lakes in the mountains. The largest and deepest lake is Lake Van with a surface area of 3,712 km^2 and an altitude of 1,646 m from sea level. There are four main regions where lakes are intensively dispersed: The "Lakes District" (Eğirdir, Burdur, Beyşehir, Acıgöl Lakes), Southern Marmara (Sapanca, İznik, Ulubat, Kuş Lakes), Lake Van and its environs, and Lake Tuz and its environs. Turkey has 555 large dam reservoirs. The names and surface areas (km^2) of the large ones are Atatürk (817), Keban (675), Karakaya (268), Hirfanlı (263), Altınkaya (118), Kurtboğazı (6). Turkey is rich for streams and rivers which many of them meet with seas within Turkey border. However, Turkey is not enough water sources for their requirements. Many lakes are under drying risk that needs to take measure for this problem (Table 1).

2. General View of Turkey

A large part of the Konya province is coincided on the high plains of Central Anatolia. Southern and western parts of Konya area are included the southern Mediterranean. Konya takes geographically in between 36° 41′ and 39° 16′ north latitude and 31° 14′ and 34° 26′ east longitude respectively (Figure 1). Surface area (excluding lakes) is 38,257 km^2. This area is the largest face in cities area of Turkey (Anonymous 2008a). The average height of the city from the sea level is 1,024

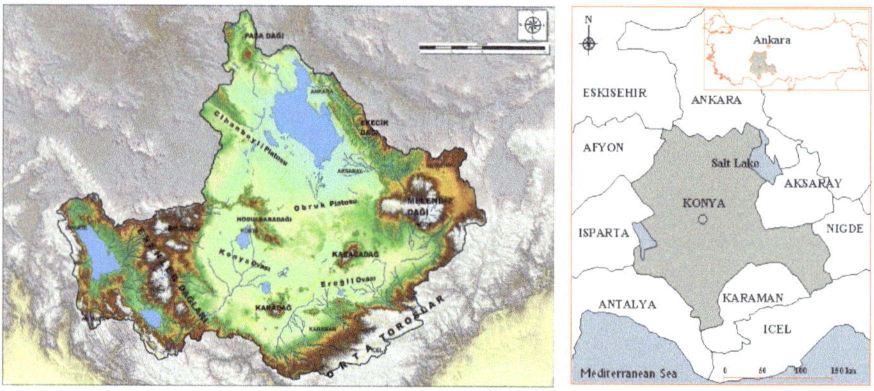

meters. Average altitude is 1,016 m in Konya (Anonymous, 1998; Anonim, 1997).

Figure 1. Location maps of Konya City and Konya Closed Basin

Konya Basin Wetlands	Original Scale*	Protection Status	Recent Situation
Samsam Lake	830 ha	SPA* (1992)	Drying activity for opening Agricultural. Lake area decrease under 400 ha
Kozanli Lake	650 ha	SPA (1996)	Lake area decreased to 50 ha
Kulu Lake	860 ha	SPA (1992)	Lake was dried with usage of underground water for irrigation of agricultural area. Kulu town wastewater and solid waste discharged via Kulu stream. Lake decrease % 40.11.
Eregli reedy (Akgöl)	37,000 ha	SPA (1992), Nature reserve (6,787 ha)	The reedy dried during Eşmekaya Barrage construction. It lost Natural SPA status in 2005. Surface area decrease 92.85%.
Esmekaya reedy	11,250 ha	SPA (1992), Permanent wildlife reserve (4,500 ha)	Area nearly dried with barrage building on water source for irrigation system construction. Lake area decrease 41.76 %. Single water source of area wastewater swage channel of Eregli town.
Beysehir Lake	73,000 ha	SPA (1988/91)	Extreme water pulled for irrigation Konya-Cumra plain. Wastes of around settlement and industry discharged into lake. Lake was decreased 2.96%.
Sugla Lake	16,500 ha	-	Irrigation of 14,600 ha agricultural area in Konya was planned with building of Sugla barrage in Seydişehir Town in. Lake was turn to water reservoir with decreasing surface area to 2,500 ha.
Hotamis reedy	16,500 ha	SPA (1992)	Lake area was decreased 27.85%.
Bolluk Lake	1,100 ha	SPA (1992)	It dried in 2007 year.
Tersakan Lake	6,400 ha	SPA (1992)	It was decreased 81.24%
Tuz Lake	92,562 ha	SPA (1992), national site status (71,44 km^2)	Huge amount of untreated domestic and industrial waste water of neighbour city and town was discharged. Lake area 62.84%.
Meke Lake	40 ha, depth 12 m	Meke crater (493 ha) accepted as Ramsar side in 2005.	Lake depth was decrease to 1 m with extreme irrigation water usage.
Karapınar reedy, Sultaniye reedy	6000 ha	-	Surface was decrease % 43.43

Table 1. Wetlands and Lakes in Konya Closed Basin (Anonymous, 2008b; Durduran, 2008)

SPA: strict protected area.*

2.1. Population

The population of the city was about 56,462 in 1940, which increased slightly to 119,841 in 1960 and finally showed a high rate of increase and reached 980,953 in 2008. Estimations for 2020 showed that population will be 1,311,200 in 2020 year. The density of the population is 57 people /km^2. Konya is the 2nd most crowded city centre in the Interior Anatolian region and 7th overall in Turkey (Anonim, 1994; Anonym, 2010; Anonim, 1996; DİE, (2003)).

2.2. Soil

Different big soil groups were originated due to differences climate, topography and a variety of major soil main components in Konya. In addition, some of the land types without soil cover can be seen. Konya plain is covered with young geological formations. Plain towards the centre is covered the young segmented edges, increasing the thickness of 500 m. 4/5 of the province is flat and the rest is mountainous. Konya is reputation as a wheat warehouse for Turkey, there are other plains Akşehir and Eregli (Anonim, 1997; Anonim, 2000a; Anonim, 2003a).

Land Use Main Class	Land Use Sub-Class	Area (ha)	%
Artificial areas	City building	75,265	1.5
	Industry, commerce and Transport area	12,464	0.3
	Mine, wasteland and Construction area	11,284	0.2
	Artificial Non-agricultural green areas	497	0.0
Agricultural Areas	Agricultural suitable Areas	2,000,966	40.2
	Continual products	39,608	0.8
	pasture Areas	281,841	5.7
	Mixed Agricultural Areas	443,531	8.9
Forest and semi natural areas	Forest areas	72,511	1.5
	scrubs and grass plant areas	863,707	17.3
	nude and low plant areas	927,316	18.6
Wetlands	Terrestrial Wetlands	61,408	1.2
	Coastal Wetlands	4,655	0.1
Water surface	Terrestrial Water	184,663	3.7
	Total	4,980,534	100

Table 2. Land use value in Konya Closed Basin (*CORINE 2nd level classification; ÇOB, 2009*)
Source: TUBITAK CBS; ÇOB, 2009

Territory of the Konya province, flat or slightly undulating topography formed on the old lake and marine sediments and volcanic rocks. The Provence soil quality is middle degree despite moderate forces the yield limiting factors such as drainage and erosion composed of fine-grained components of this land with the water thickness. The most important limiting factor is water. Unconsciously irrigation activity in provinces areas made unavailable inefficient agricultural. Floods were caused serious damage in the past few years with more rainfall. The ground water problems flooding and soil salting in a large area will be reduced with followed the completion of "Konya Plain Drainage Project" (Tables 2, 3).

Town	Total area	Planting Land Area		Fallowing area		Vegetables Garden Area		Fruit area	
	ha	ha	%	ha	%	ha	%	ha	%
Karatay	165.592	87.675	52,9	76.956	46,5	695	0,4	266	0,2
Meram	58.945	34.059	57,8	19.653	33,3	4.581	7,8	653	1,1
Selçuklu	87.406	40.514	46,4	46.291	53,0	420	0,5	181	0,2
Ahırlı	5.282	3.022	57,2	1.417	26,8	14	0,3	830	15,7
Akören	21.334	10.409	48,8	10.806	50,7	30	0,1	88	0,4
Altınekin	74.683	63.479	85,0	10.900	14,6	265	0,4	39	0,1
Beyşehir	59.634	43.672	73,2	13.847	23,2	1.302	2,2	814	1,4
Bozkır	19.614	6.031	30,7	9.695	49,4	59	0,3	3.830	19,5
Cihanbeyli	206.793	121.522	58,8	85.247	41,2	24	0,0	0	0,0
Çumra	140.676	90.263	64,2	46.716	33,2	3.374	2,4	323	0,2
Derbent	15.131	10.661	70,5	4.151	27,4	176	1,2	144	1,0
Derebucak	2.668	1.326	49,7	1.264	47,4	7	0,3	70	2,6
Emirgazi	73.509	18.871	25,7	54.630	74,3	0	0,0	8	0,0
Ereğli	125.866	57.529	45,7	58.574	46,5	4.916	3,9	4.847	3,9
Güneysınır	27.962	10.656	38,1	15.437	55,2	650	2,3	1.219	4,4
Halkapınar	7.318	2.455	33,5	4.048	55,3	2	0,0	813	11,1
Hüyük	19.733	9.738	49,4	9.363	47,4	171	0,9	461	2,3
Karapınar	136.426	94.652	69,4	40.437	29,6	1.121	0,8	216	0,2
Kulu	133.094	55.301	41,5	77.766	58,4	4	0,0	24	0,0
Sarayönü	108.819	71.499	65,7	37.093	34,1	123	0,1	105	0,1
Seydişehir	35.957	33.177	92,3	819	2,3	1.482	4,1	478	1,3
Tuzlukçu	49.725	23.387	47,0	25.833	52,0	170	0,3	335	0,7
Yalıhüyük	3.769	2.760	73,2	660	17,5	6	0,2	342	9,1
İL TOPLAMI	**1.579.933**	**892.656**	**56,5**	**651.603**	**41,2**	**19.591**	**1,2**	**16.084**	**1,0**

Table 3. Agricultural land usage status in Konya closed basin (*TUBITAK CBS; ÇOB, 2009;* Anonim, 2000/b Anonim, 2003/b*)*

2.3. Vegetation

Steppe vegetation is dominant in both Konya and its vicinity. The flora of the providence consists of Irano-Turanien species and Anatolian endemic species (Çetik, 1985; Akman, 1990).

2.4. Vegetation

The climate of Konya is typical terrestrial climate which is mostly known as hot, semiarid summers with cold, rainy and snowy winters. The annual average of total precipitation is 319.2 mm/year. Konya has located in the lowest precipitation receiving region of Turkey. The average value of relative humidity is %59 (ACIR, 2009; Anonim, 1998). There has not been a study about how much is used the consumed net annually total amount of solar energy in the province of Konya. But the solar energy is used in provinces residences during summer. Konya city centre Average monthly sun irradiation level and Figure 4. Monthly Rainfall (mm) levels of Konya city centre are given below (Figures 2 – 4; Table 4).

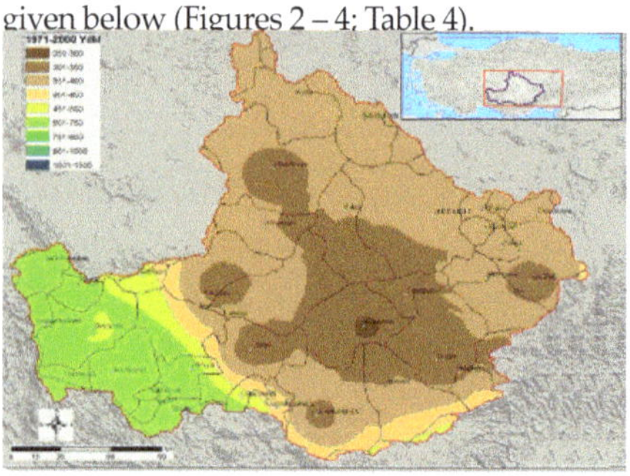

Figure 2. Yearly total precipitation distribution map of Konya closed basin (DMI, 2010 DMI, 2006)

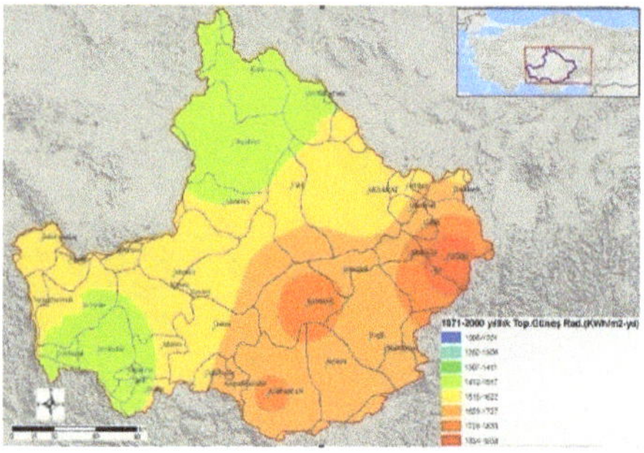

Figure 3. Yearly total sun radiation distribution map of Konya closed basin (DMI, 2010)

Average monthly sun irradiation of Konya city centre		Konya city centre Average monthly sun irradiation level of	
Months	**Average period (h/min.)**	**Months**	**Average level (cal/cm^2min)**
January	3:2	January	186.36
February	4:6	February	287.80
March	6:1	March	381.84
April	7:0	April	465.09
May	8:6	May	576.70
June	10:4	June	652.92
July	11:3	July	649.94
August	11:0	August	600.03
September	9:6	September	478.69
October	7:2	October	344.93
November	5:0	November	247.99
December	3:0	December	164.47
Annual Mean	7:25	Annual Mean	419.73

Table 4. Konya city centre Average monthly sun irradiation level (ACIR 2009; Anonim, 1998; Çiftçi, 1991)

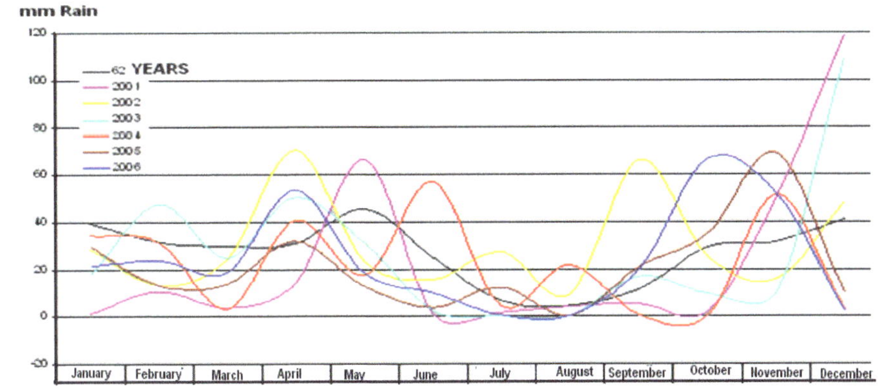

	January	February	March	April	May	June	July	August	Septem.	October	Novem.	Decem.
62year	39.3	31.4	29.8	31	45.5	25	6.5	4.4	11.4	29.3	31.4	40.8
2001	1	10.6	3.7	14.1	66	0.7	1.3	4.1	5.1	1.9	50.1	118.4
2002	27.8	12.9	24.2	70	22.9	15.3	27.	8.7	65.8	24.6	15.3	48
2003	17.6	47.5	24.6	50.2	30.9	2.3	0	0	16.6	9.5	9.8	106.6
2004	34.1	31.1	3.1	40.6	17.2	56.9	4	21.4	0	0	51.3	2.8
2005	29.5	12.9	13.8	31.8	12.5	3.5	12.2	0.1	20.9	34.7	68.8	9.8
2006	21.2	23.8	18.4	53.4	17.9	9.9	0.3	0	20	66.1	51.9	2

Figure 4. Monthly Rainfall (mm) levels of Konya city centre (Meteorology region directorate-2009)

2.5. Water Resource

Drinking Water Resources and Dams: There are more than 200 drinking water wells in the Konya city centre (General directory of KOSKI, 2008; Anonym, 2008). Their outputs change from 10 to 50 L/sec and average deeps are 150 m. Water taking levels were less than 100 m about 20 years ago but it increased up to 250 m in recent years. That means high amount of water sucking from the underground water, decreasing ground water levels with increasing city centre population (about on millions in 2010). There are four dams west of Konya to protect the Konya land from the flood, with the intent of water use and irrigation to ensure. These are Apa, Altınapa, May, Sille dams. Also Beyşehir Lake is used as a source of drinking water for the centre of the County of Beyşehir.

Ground Water Resources: Determining the situation of ground waters in the Konya province will be discussed separately within province and independent Konya-Çumra-Karapınar and; Cihanbeyli-Yeniceoba-Kulu Plains. Studying area, the geological formations in the plains, which those transport to water (aquifers), the depth of these formations in groundwater, its flow direction, purpose, status and location distribution of the number of wells will be discussed (Anonym, 2008).

Groundwater Depth and Flow Direction: Konya-Çumra-Karapınar plain is the most important one of the subdivided plains in the province. Water-bearing formations are varied. Therefore, these formations also vary depending on the state of the ground water. General groundwater flow direction is from southwest to northeast. There is largely groundwater recharge From Neogene age limestone in the south and marls west, in the Kreatase age limestone to plains. DSI, Rural Services and well as many private companies have opened number of wells in order to meet drinking and irrigation water requirements in Konya-Çumra- Karapınar plains since 1956. The total area irrigated by these wells are approximately 76 000 ha. Depths ranging between 23 m and 409 m wells have different flow rates MTA, 2006a,b).

Konya closed basin, especially wetland area keeps the widest and virgin halophytic steps of Turkey and also the world contains special fauna and flora. WWF International declared the Konya closed basin in the 200 special ecological regions with its 16 important bird area and 6 special plant containing place. WWF-Turkey (National Life Protection Foundation) began the project "Intelligent Use of Konya Closed Basin" September, 2003. But most of the wet lands were lost with wrong and unconscious applications (*Anonymous, 2008b; Table 1).*

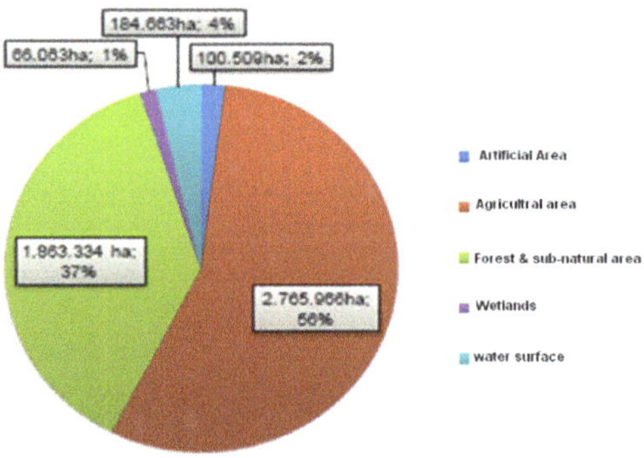

Land Use	Artificial areas	Agricultural Areas	Forest and semi natural areas	Wet-lands	Water surface	Total
Area (ha)	100,509	2,765,966	1,863,334	66,063	184,663	4,980,534
%	2.02	55.54	37.41	1.33	3.71	100

Table 5. Land use value in Konya Closed Basin (CORINE 1st level classification; Anonim, 1997; Günay, 1985)

Source ÇOB, 2009; TUBITAK CBS

Water potentials: Konya is the 7th city of Turkey having a high population. Water sources of the city are limited and annual precipitation (319.2 mm/a), for this reason, economical usage of irrigation water for green areas of Konya city centre is more important. In this study, the importance of irrigation for continuing usage of city centre green areas was emphasised, and water conservation measurements were explained for Konya city centre green area sample.

Wetlands: There are 12 important wetland area in the Konya closed Basin. A project used Landsat satellite images showed that most of the wetlands were dried or lost, as it has been wetland characteristics in last 20 years (Durduran 2008; Aygen, 1967).

Rivers: Çarşamba Brook is originated from west of the Hacımer Mountain, under the name of Sırıstad from 2200 m elevation. Meram River located near to the village Başarakavak is discharged from sources northwest of Konya, Akdağ hillsides. Other rivers and streams are Ivriz stream, Zanapa River, Gümüşler creek. İnsuyu Creek is the most important river in Cihanbeyli - Kulu region (Table 5). Table 6. Water storage areas in Konya closed basin, Table 7. Land use value in Konya Closed Basin (CORINE 1st level classification). Protected land and areas in Konya closed basin is given in Table 7.

2.6. Industry and Technology

City Industry Development, Site Selection Process and These Influencing Factors

Development of industry in the province of Konya

Economically active population is 9.1% in the Konya who work in the industrial sector, industrial investments, and significant investments began in 1950s with the sugar factory and realized after 1960. In 1963 the first large scale factory production is cement production plant in the 1960s. After then the chrome-magnesite brick factory was started production in 1968. The biggest aluminium plant in Turkey 'was started build in Seydişehir, 1969. Aluminium Factory was launched in 1970,

LAKE	LOCATION	RIVER	PURPOSE	AREA (m^2)
Güzelyurt Göleti	Aksaray	Selindiz	Sulama	442.538
Helvadere Göleti	Aksaray	Taşlıgöl	Sulama	400.498
Dokuzyol Göleti	Karaman	Kurudere	Sulama	198.427
Akören Göleti	Konya	Bayındır	Sulama	888.234
Aydoğmuş Göleti	Konya	Boğaz	Sulama	331.009
Aydoğmuş Göleti	Konya	Boğaz	Sulama	331.009
Başhüyük Göleti	Konya	Kurudere	Sulama	296.177
Bostandere Göleti	Konya	Kalaycı	Sulama	405.092
Cihanbeyli Göleti	Konya	İnsuyu	Sulama	1.574.576
Çağlayan Göleti	Konya	Yayla	Sulama	889.403
Çavuş Göleti	Konya	İlmen	Sulama	276.139
Çiftliközü Göleti	Konya	Karakaya	Sulama	356.115
Çukurçimen Göleti	Konya	Çökük	Sulama	165.530
Derbent Göleti	Konya	Belbaşı	Sulama	151.639
Erenkaya Göleti	Konya	Çarşak	Sulama	919.179
Evliyatekke Göleti	Konya	Arkil	Sulama	268.759
Güneydere Göleti	Konya	Gavur Deresi	Sulama	2.354.084
Kızılören Göleti	Konya	Yayla Deresi	Sulama	145.670
Malas Göleti	Konya	Uludere	Sulama ve Kullanma	235.065
May-Kayasu Göleti	Konya	Peynirli	Sulama	159.458
Sefaköy Göleti	Konya	Kavakdere	Sulama	140.738

Table 6. Lakes and characteristics in Konya closed basin kaynak gösterilmeli

which provides great contributions to the province and the country's economy, that Turkey's only plant producing raw aluminium. Tümosan diesel engine factory began production in 1981, has an annual production capacity of 25,000. 1970s, most large-scale enterprises in Konya belongs to public sector consists of businesses and establishments belonging to the private sector, the business demands of the region are not geared towards mass production for a production that consists of small and medium-sized businesses. These establishments are farm equipment, engines and spare parts and food industry engaged in the production of small places. Current industry state of Konya is listed below.

storage areas	location	river	area (m^2)
Hotamış Depolaması	Konya		52.138.774
Suğla Depolaması	Konya	Suberte, Irmak Çayı, BSA	44.359.924

Table 7. Water storage areas in Konya closed basin

Protected Area	Characteristic	Area (ha)
Antalya Cevizli Gidengelmez Dağı	Yaban Hayatı Geliştirme Sahası	5.925
Konya Bozdağ	Yaban Hayatı Geliştirme Sahası	59.308
Mersin Çamlıyayla Cehennemderesi	Yaban Hayatı Geliştirme Sahası	17
Kocakoru Ormanı	Tabiat Parkı	331
Akgöl(Ereğli Sazlığı)	Tabiatı Koruma Alanı	6.681
Beyşehir Gölü	Sulak Alan	90.671
Col Gölü	Sulak Alan	4.744
Uyuz Gölü	Sulak Alan	112
Kulu Gölü	Sulak Alan	2.206
Kozanlı Gökgöl	Sulak Alan	967
Tuz Gölü	Sulak Alan	328.347
Tersakan Gölü	Sulak Alan	9.511
Bolluk Gölü	Sulak Alan	9.697
Meke Maari	Sulak Alan	339
Konya Acıgöl	Sulak Alan	267
Kizoren Obruğu	Sulak Alan	281
Samsam Gölü	Sulak Alan	2.218
Ereğli Sazlıkları	Sulak Alan	22.263
Cirali Obruk	Sulak Alan	337
Meyil Obruğu	Sulak Alan	101
Tuz Gölü	Sulak Alan	5
Tersakan Gölü	Sulak Alan	5
Beyşehir Gölü	Milli Park	86.833
Kızıldağ	Milli Park	54.718
Beyşehir Gölü	Milli Park	0,07
Kızıldağ	Milli Park	0,07

Table 8. Protected land and areas in Konya closed basin

Small Industrial Sites

Small industrial estates, small and medium-scale industrialists and artists visited the existing infrastructure, education and social facilities of the Ministry to ensure that the credit support the work of a healthy workplace or directly carried out by own resources of entrepreneurs. There are a total of 17 units with Small Industrial Estate with a capacity of 4409 business launched in Konya (Table 9, 10).

Industry Region (KOSB)		
Sector	Firm nu.	Ratio%
Automotive side Ind.	27	18
Farm tool & Machine	16	11
Machine & spare part Ind.	13	9
Smelting Ind.	11	7
Paper & Packing Ind.	4	3
Sounding, Pipe & Irrigation sys.	13	9
Plastic, Paint & Chemistry Ind.	12	8
Construction Materials Ind.	4	3
Leader & Textile Ind.	5	3
Metal Property & hardware Ind.	1	1
Mill machine Ind.	6	4
Metal Ind. (No iron)	5	3
Other	33	22
Total	150	100

Table 9. Industrial sector distribution in Konya 1st Organised

Industry Region (KOSB)		
Sector	Firm nu.	Ratio %
Automotive spare Ind.	57	18
Smelting Ind.	30	10
Machine Ind.	27	9
Farm Machine & tools Ind.	23	7
Construction Ind.	21	7
Plastic Ind.	19	6
Food Ind.	13	4
Packing Ind.	13	4
Lorry box dumper Ind.	8	3
Mill machine Ind.	8	3
Textile Ind.	7	2
Rubber Ind.	7	2

Table 10. Industrial sector distribution in Konya Organised

3. Environmental Problems in Konya

3.1. Soil Pollution

An investigation on contamination of the Konya province on the territory by metals and microbial, the main discharge channel of Konya Plain is known to cause Salt Lake pollution, the water withdrawn for agricultural irrigation and soil be contaminated with the decrease in the efficiency of agricultural land indirectly. (Water samples taken at different times of Main Discharge Canal to results of analyze; and used in accordance with Food Ingredients Regulation this should be regarded as dangerous waters). According to the results of an other project, Konya Closed Basin province, the identification and elimination of pollution in the Salt Lake; Main Discharge Canal reached the Salt Lake and agricultural areas due to agricultural use for irrigation purposes, the necessary precautions are not taken, otherwise soil will be contaminated so that is not irreversible situation.

3.2. Atmospheric Pollution

There have been done some measurements in the city on the atmospheric SO_2, and fluorinated compounds, hydrogen sulphur, magnesite powder, carbon dust and heavy metals in the soil as it will cause accumulation of particulate matter.

3.3. Pollution from the waste

There is a little investigation on hazardous matter contamination on soil. There are some heavy metal accumulations in waste water watered agricultural lands.

3.4. Land Property

The raising quality classification of land use is taking into consideration the degree of conformity to the classification system of cultivars, land pasture, forest and home animal husbandry. Processing show considerable differences from the territory of the province in terms of features for this classification needs of the land, limiting factors (soil, topography, drainage failure, etc.) (Table 12).

3.5. Unplanned urbanization

Konya city population was rapid increased as a result of migration from rural to city centre after 1950 years due to the establishment of factories in, to be come mechanized, increasing the agriculture production of the province, connection of Konya to highway network, development of industry. However, Konya is one of the provinces of rare non shanty. When we look at the transportation, infrastructure and social service standard development pattern is a concern that the city is public improvements. However, the existing environmental impacts are not considered in the development of the city. Air pollution is to reach a large size especially during the winter months due to Meteorological and geomorphologic structure of the city. Establishment of industrial zones in the northern part of the city has important effect on air pollution problem, especially stainless steel sector.

Usage form	Land Use Capability Classes Forms (ha)										TOTAL
	I	II	III	IV	Total	V	VI	VII	Total	VIII	
Dry Farm (fallowing)	494,017	384,132	439,779	291,568	1,609,496		154,973	78,272	233,245		1,842,741
Irrigated farm	367,249	310,025	372,948	276,276	1,326,498		126,698	56,358	183,056		1,509,544
Less irrigated farm	7,931	5,941	9,512	1,971	25,355		84		84		248,081
Vineyard (dry)	374	663	1,512	2,655	5,241		9,919	21,536	31,455		36,696
Garden (irrigated)	77		215		292						292
Garden (dry)	400	783	928	367	2476		232	206	438		2,916
Vineyard (irrigated)	11,121	6,565	1,566	339	19,591		120	52	172		19,763
Meadow-pasture	39165	38252	103419	131641	312477	17041	214170	516660	747871		1,060,348
Meadow land		1994	6248	7146	15388	16635	5269	56668	78572		93960
pasture land	39165	38252	103419	131641	131477	17041	214170	516660	747871		1,060,348
Forest-shrub		1994	6348	7146	15388	16635	5269	56668	78572		93,960
Forest-land	39165	36258	971711	124495	297089	406	208901	459992	669299		966,288
Shrub land	176	322	1927	3588	6013		16987	109203	126190		132,203
Non-farming land	13119	3949	2774	1317	20250		2471	1715	4186	354´0	59,846
Residential (Heavy)	202	93			295			218	218	354´0	35,923
Residential (rare)	123353	2391	1446	1067	17257		484	815	1299		18,556
Industry land	564	93	271	235	1163		774	198	972		2,135
Army land		463	1057		1520		12131	463	1678		3,196
National Park				15	15			21	21		36
Other lands										160037	160,037
Water surface										189621	189,621
TOTAL	546477	426151	549756	436266	1958650	17041	414481	1050450	1481982	385068	3,825,700

Table 11. Land Use Capability Classes Forms of Konya Breakdown 2008 (Konya Province Land Property Inventory)

3.6. Reduction of Green Areas

Vegetation in the Konya province is shaped according to climatic conditions. Steppes are covered with large flat areas. Weeds will sprout with spring rains of the plains in spring season, but a variety of greenery lasts a short life would not be long. Herbs are immediately dry and roasted starting high summer temperatures. Wide plains take the steppe form. There are shrub and forest areas in the mountainous of province. Forested area is 559 759 hectares in the province. Percentage of forest in the province was 14.6%. Reforestation initiatives are supported by people and companies in Konya province. A major work is done for the proliferation of green areas. Reforestation campaign is underway.

3.7. Land Problems

It can be grouped in seven levels.

1. Large scale and fragmented lands: Land fragmentation directly affects on the soil entire work the in the agricultural sector based.

a. Effects of land use areas

1.Increased border losses

2.Losses caused by excess land routes

3.Losses small and far parcels leaving empty

b. Impact of working time and labour costs

Work and labour costs by shrinking has negative impact on non-economic forms with increasing fragmentation in acquiring parcels losses caused by excess land routes

1. Increased routes

2.Preparation of work duration increase

3.Returning time losses increase

c. Fragmentation effects on mechanization

Partly or messy parcels subdivided from the land, more of the time spends cultivation period and going each small part. Machine efficiency and usage trend is going lower. The efficiency reduces with time going to land and returns, preparation, going from a parcel to other parcel, increasing machinery work.

d. Effects of fragmentation on Culture Technical Services

Service surface area per unit area for common small land parcels is increases with road and irrigation, drainage length of the area and therefore production efficiency decreases accordingly. Also construction costs and service spends are increase.

e. Effects of fragmentation from Operating Activities

There is effect on planting time and harvest time, on alternation business organization, on the agricultural struggle, on transportation efficiency. Taking in all of these failures;

Increase the agricultural production, agricultural settlements in order to develop the agricultural sector and enterprises belonging the people and more than a small surplus divided into parcels scattered in various positions in the Management School or impractical, according to the principles of modern agriculture and land ways, shaped the development of the culture of technical services in accordance with the format, design and arrangement need a new or amended land consolidation.

2. Unconscious or More Soil Manipulating Activity: Manipulating the soil, the structure of the soil, namely sand, clay, silt, organic ratios are damage. Soil tillage must be suitable with rainfall, climatic conditions and plant species and necessary cultivation in accordance with the increasing costs balanced unnecessary and excessive soil tillage and soil compaction and underlying supporting structure, forming harm base stone. The is important effect of forming harm base stone on infiltration and surface of the water in the soil prevents the occurrence of leakage may be caused by water erosion in the region. High soil tillage has effect on such as fuel and labour costs, increased more and more crumbling supporting structure, and paves the way wind erosion.

3. Irrigation with Wastewater: Irrigation with wastewater is common in Konya city. Chemical and medical wastes in wastewater become leading as soon as the soil desertification and insolubility of soil available useful organic and inorganic substances.

Changing property of soil these pollution effect take negative rolls on ability of as needed chemical fertilizers by the plant in the soil. If the sewage and detergents, chemicals and medical waste speared and sewage water after treatment can be used safely to be applied to the irrigation of landscape areas and irrigation of agricultural land.

4. Erosion: Wind erosion is dangerous for some parts of Konya city. Erosion is the most important factor at L organic matter poor unproductive soils in sierozem lands at around Karapınar town.

5. Alternation application as need: Konya city is in continental climate zone. Product range is limited due to lack of sufficient irrigation ability. According to the available range of products with a suitable climate and soil characteristics, useful production pattern of land must be used. Applying rotation, production increase and as well as effective land use would be useful. Different plant groups have different nutrients uptake and left from the soil. For this reason, applying the necessary alternation rule, trying to get sufficient efficiency in the land in the region, and we must maintain the structure of the land and must start applying right functioning.

6. Unconscious Fertilization: Fertilizer applied in Konya city without taking account to soil and plant needs. Thus, the chemical properties of soil deterioration spoil and can not be get requested advantage from the fertilizer. Fertilizer should be applied after analysis of soil at amount and sort need of land.

7. Carelessness in application building industrial plants region: Stage of the establishment of industrial area must be kale in consideration of the predominant wind direction and the land will be established in the industry, not be agricultural land, regardless of whether the city centre in particular, the factory chimneys of particles and chemical compounds into the atmosphere, due to the prevailing wind hovering over farmland productivity is negatively affected. As a result of the establishment of industrial plants in productive agricultural lands located within the boundaries of city centre disposed directly or indirectly.

4. General Discussion and Recommendations

Water Economising Measures in the City Centre Green Areas Irrigation: Irrigation was described as watering the plant for their necessary in a controlled and an artificial way (Kara, 1983). Active way of usage the irrigation water is to give enough water to grow plants necessary. Water amount must be estimated concerning the plant type in the water poor places and used in control and order. Barış (2008) has recommended to followings for active and economical water usage in the green areas;

-Suitable planning and conceiving should be done. Present natural conditions should be considered while planning green areas.

-Soil should be well presented and soil quality must be improvement.

-Regional plant types should be selected for local climate.

-Simple growing grassing area and dry climate resistance grass type should be selected.

- Selection of the irrigation system should be suitable with green area wideness.

-Water carriage of the soil and soil temperature should be controlled. The mulching must be presented against to erosion.

-Necessary attention should be followed after construction of green areas (pruning, wild plant control, pest control, irrigation etc.).

The natural factors have effect on the environmental which urban data form climate, topography, rivers, flood areas, geomorphologic structure, vegetation and soil ability. The formation of urbanisation plans for the future is not determined for natural form, the macroeconomic data, but it played a role in limiting and conditioner direction. There is an interaction between artificial and natural environmental processes. This interaction may be contradictory as that may be in harmony with the environment but also damaging both give rise to negative results in some processes, jobs and products. In general, balance degradation such as environmental pollution, ecological processes are this kind. These events can be seen in Konya.

Water is the foundation of life and is a driving force for economic and social developments and for the poverty eradication. But drought and water crisis have been a global problem since last years of 21st century. For this reason saving water usage is the most important necessity in the solution of this problem.

Necessary precautions for economical water usage in irrigation of urban green areas may be explained as the followings:

-The existent irrigation system planning should take into consideration of plant pattern, water amount, water quality, soil characteristics, and a project may be presented in relation with selecting low water requiring plants and techniques.

-Before presenting green areas, soil characteristics should be known and prepared for activity, and necessary measures (quality improvement etc.) should be done in planning period.

-Natural and endemic plant species may be preferred for green area formation. Endemic and natural species have more resistance to ecological difficulties and they may be effected less from local climatic conditions if they were grown in right position.

-Selecting the plants for green areas other than natural plant species should have resistance to dryness and water requirements must be lower.

-Specially, presentation of wide scale green areas (park and recreation area, city centre and city sides, gardens of public institutions and university campus etc.) should be preferred xerophytic plants, brier, which have a large shelter over soil surface instead of grass and plants desire high water.

-Municipality water system is not preferred for irrigation and an alternative water sources should be constituted. For this matter, precipitation water in suitable reservoirs may be sawed during rainy period. These systems save the underground water source in dry period.

-The orderly and careful maintenance (irrigation, pruning, struggle with undesirable plant and animals etc.) will be followed after presentation of green areas.

It was attempted to give some examples which were chosen from the hundreds of them showing energy wasting in our country. As it could be seen by examples, unfortunately waste its own energy and energy source fast by using them ineffectively. This makes us vey upset as being as Turkish scientist. This paper have been presented to worn the people and want them, to take necessary precautions as soon as possible.

5. References

[1] ACIR, 2007. Air Conditioning Industry Report. The research report on Konya's air conditioner industry. Konya.

[2] Akman, Y. 1990. İklim ve Biyoiklim, Palme Yayınları, 319s, Ankara.

[3] Anonim, 1994, "1990 Genel Nüfus Sayımı", Nüfusun Sosyal ve Ekonomik Nitelikleri, 42: Konya, DİE Yayın No: 1641, Ankara.

[4] Anonim, 1996, "İstatistik Göstergeler", T.C. Başbakanlık Devlet İstatistik Enstitüsü, Ankara.

[5] Anonim, 1997, "Ülkemiz Arazi Varlığı Kullanım Analizi", Ziraat Mühendisliği Dergisi No:303,Sayfa:19, Ankara.

[6] Anonim, 1998. Cumhuriyetin 75. Yılında Konya. Konya İl Yıllığı, Konya Valiliği, Konya.

[7] Anonim, 2000/a, Tarım ve Köyişleri Bakanlığı, Konya İl Müdürlüğü Kayıtları, Konya.

[8] Anonim, 2000/b, DİE Tarım Şubesi BİM Kayıtları, Ankara.

[9] Anonim, 2003/a, "Konya Tarım Master Planı", T.C. Tarım ve Köyişleri Bakanlığı, Konya Tarım İl Müdürlüğü, İl Tarım ve Kırsal Kalkınma Master Planlarının Hazırlanmasına Destek Projesi, Konya.

[10] Anonim, 2003/b, http://www.die.gov.tr

[11] Anonim, 2010. Türkiye İstatistik Yıllığı. www.tuik.gov.tr (Erisme tarihi: 16.09.2011)

[12] Anonym, 2008, Devlet Su İşleri Genel Müdürlüğü Ankara http://www.dsi.gov.tr

[13] Anonym, 2008. Konya Valiliği KOP Tarım Eylem Planı, Konya.

[14] Anonymous 1998. Konya Büyükşehir Çevre Düzeni Revizyon Nazım İmar Planı Raporu. Konya City Municipality, Konya.

[15] Anonymous 2008a. Turkish Republic Konya Governorship. Available online at: www.konya.gov.tr

[16] Anonymous, 2008b. www.wwf.org.tr/wwf-tuerkiye-hakkinda/nerede-calisiyoruz/konya-kapali-havzasi/18k Erişim:Aralık 2008.

[17] Aygen, T, 1967, Manavgat-Oymapınar (Homa) Kemer Barajı ile Beyşehir-Suğla Gölü Manavgat Çayı Havzasının Hidrojeolojik ve Karstik Etüdü, DSİ Raporu, Ankara,.

[18] Barış ME, 2008. Sarıya Bezenen Kentlerimizi Kimler ve Nasıl Yeniden Yeşertebilir. Available online at: http://www.peyzajmimoda.org.tr/genel/bizden_detay.php? kod=1173&tipi=2&sube=0

[19] Çetik, R., 1985. İnterior Anatolia vegetation and ecology. S.Ü.Publish No: 7, Konya.

[20] Çiftçi, N. , 1991. Konya-Çumra Ovasında Sulamaya Açılmakta Olan Bazı Arazilerin Drenaj Sorunlarıyla Dren Derinliklerinin Belirlenmesi. Selçuk Üniv. Zir. Fak. Dergisi, Cilt:1, Sayı:2, Konya.

[21] DİE, (2003) Genel Nüfus Sayımı 2000, TC. Devlet İstatistik Enstitüsü Yay. No:2759, Ankara.

[22] DMI, 2006. Meteoroloji İşleri Müdürlüğü Çalışma Alanının Coğrafi Durumunu Gösteren 1 /250000 Ölçekli Harita, Konya.

[23] Durduran S. 2008. Göllerimiz Kuruyor. Yeni İpek Yolu Dergisi. Konya Ticaret Odası Yayını, Sayı:125, Konya.

[24] Günay, Y., Durusu, A. ve Tabuman, F.C., 1985, Isparta ve Yöresindeki Göllerde Su Kalitesi, Evsel ve Endüstriyel Atıklarla İlgili Parametreler; TÜBİTAK Çevre Araştırma Grubu. Proje No: ÇAĞ-47/C, 14-24, 48, Ankara.

[25] Kara M, 2005, Sulama ve Sulama Tesisleri . S.Ü. Ziraat Fakültesi Ders kitabı, ISBN 975-448-177-6 Konya.

[26] MTA Genel Müdürlüğü 2006a. Jeoloji Etütleri Dairesi, Araştırma Alanının Jeolojik Haritası,Konya.

2

Global Energy Consumption Paradigm: Future Trends and Trajectory

Vrushal T. Ghoble

Jai Hind College, India

Abstract

The energy debate and its discourse have changed drastically in the 21st century. The paradigm will go on changing with the coming time. This phenomenon will be in context of the consumers and producers; as well as the environment alike. Unlike the earlier times, today energy stands critical not just for the west but also for other consumer markets. This paper intends to address the energy consumption pattern and its changing paradigm. The new parameters of the global growth and development has been impacting the levels of energy consumption, thus, the changing statistics and fluctuating markets will see a tremendous increase and shift in the consumption pattern. The central theme of the paper is that, with the globalization process, the high income pockets have become largely vulnerable. The paper analyses the increased energy consumption and its future speculation. The paper also discusses the long term consumption trajectory.

Keywords: energy consumption; globalization; changing paradigm

1. Changing energy paradigm

Energy industry today has become very versatile and is expanding as time passes. However, fossil fuels will continue to be an important source of energy for the world. "Oil and natural gas alone will still make up almost 60 per cent of global energy supplies in 2030. The fastest - growing fossil fuel will be natural gas, because it is abundant, affordable and the cleanest - burning. By 2030, global demand for natural gas will be more than 55 per cent higher than in 2005"1. "With the United Nations predicting world population growth from 6.6 billion in 2007 to 8.2 billion by 2030….. The global electricity demand is increasing twice as fast as overall energy use and is likely to rise 76 per cent to 2030. Nuclear power provides about 14 per cent of the world's electricity, almost 24 per cent of electricity in OECD countries, and 34 per cent in the EU. Nuclear power generation is an established part of the world's electricity mix providing in 2007 some 15 per cent of the world's electricity (cf. coal 42 per cent, oil 6 per cent, natural gas 21 per cent and hydro & other 18 per cent)"2. As the renewable and alternative sources of energy are escalating their influence at the global level, the world will remain dependent on oil and gas for most of its usage.

The chief factors that are driving the global energy demand are expanding globalization process, where today, the people are connected globally through inter – connected roads and highways. This has also led to intense Industrialization process, where factories and manufacturing businesses require a huge amount of energy to run. At the same time, the rise of the middle class globally, which "..... has also raised the needs of the people, aspiring for a better and luxurious life"[4].

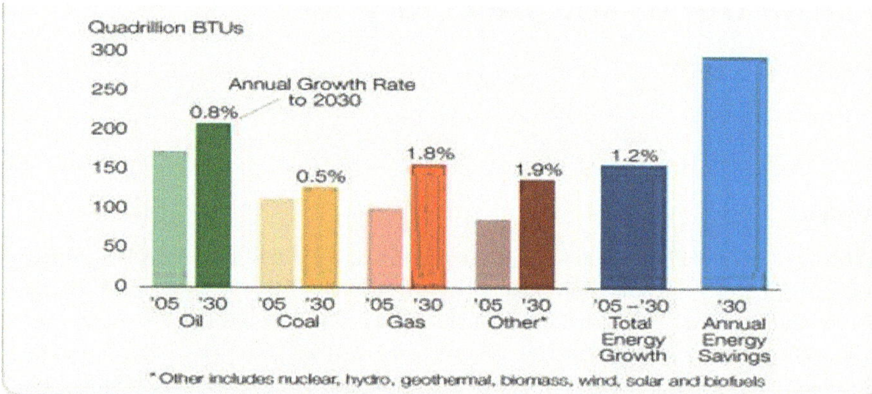

Graph 1. ExxonMobil, "Energy Demand".

2. New consumption markets

Worldwide energy markets have witnessed increasing activity by the energy consumers. Now, this presence could be divided into old and new players. In addition to the old players in the market, the new stakeholders are Asia. The new players are clearly defined as the energy consumers, which have experienced a drastic increase in its energy consumption due to the higher economic growth. Thus, the 21st century sees a new competition from the new energy rivals. As time passes, these energy consumers do see a stiff competition for hydrocarbon resources. Gradually, the competition is also carried out in other areas where there are huge reserves of oil and gas such as, Europe & Eurasia and Africa. With this, the competitors also face the challenge to contribute for regional security and development of these societies.

Post - Second World War, saw the emergence of hydrocarbon as a basic factor in the development of the world. Thus, the major oil producer i.e., the Middle East became the centre of attraction for the global powers. Oil, therefore became the basic factor for the powers involvement in many energy rich countries of the world. The Middle East especially, where there are the largest reserves of oil in the world, became the centre stage. Many of the energy rich states today are majorly dependent on income from oil exports. Some countries are dependent upto 90 per cent. This clearly identifies the space energy holds in the economic development of a country. However, the oil exports made the producers completely dependent on the revenues, with no development whatsoever. This made the producers a rentier economy. Although, the enormous wealth that was incurred as oil revenues, made the producing countries rich and powerful, and helped them in the development of their country. Thus, over a period of time, the intensity of competition has increased and become more complex.

In the last few years, the global energy demand has increased mainly due to the economic surge and the rise of global middle classes. The energy demand of the producers has also increased.

"Energy demand in the Gulf has more than doubled in the past 10 years and is forecast to increase by 85 per cent by 2030 compared with 2008 levels, a Wood Mackenzie report said. An energy demand surge in the Arabian Peninsula will be largely met by oil - fired generation, removing about 1.5 million barrels of oil equivalent per day (boepd) otherwise available for export....."[6]. The rise of new energy suppliers outside OPEC have also impacted on the Middle East exports. Due to the instability in the region, the consumers are looking for other sources, which happen to be Europe & Eurasia and Africa. Thus, the consumer market has shifted to new energy supplies, however, the oil finds are cheaper in the Middle East, which will be a major attraction for the consuming countries.

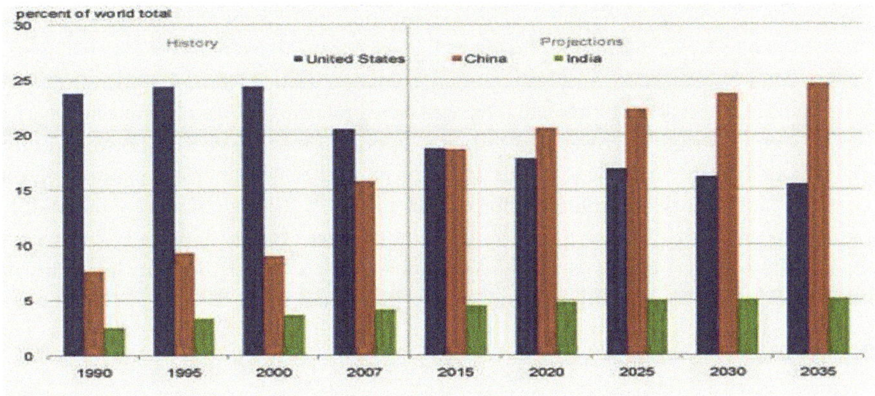

Graph 2. Energy Information Administration (EIA), "International Energy Outlook 2010".

2.1. China

China is the second largest oil consumer after US. From the net exporter of oil in the 90s, China has emerged a net importer of oil, and also consumes a large amount of natural gas. China imported its first shipment of LNG in the summer 2006, and the country has quickly ramped up imports since then, importing about 730 MMcf/d in 2009 and 1,120 MMcf/d in the first half of 2010[7]. Natural gas plays an important role and its usage for domestic purpose has increased lately, also due to the fact that China plans its growth in all sections such as construction business, transportation sector, industries, agriculture, etc. the overall demand has grown up. "Coal supplied the vast majority (71 per cent) of China's total energy consumption of 85 quadrillion British thermal units (Btu) in 2008. Oil is the second - largest source, accounting for 19 per cent of the country's total energy consumption. While China has made an effort to diversify its energy supplies, hydroelectric sources (6 per cent), natural gas (3 per cent), nuclear power (1 per cent), and other renewables (0.2 per cent) account for relatively small amounts of China's energy consumption mix"[8]. China's energy strategy is a complexity of hydrocarbon constraints and environment friendly energies[9]. Apparently, the Chinese energy demand exceeds supply bringing other stakeholders (producers) into picture.

2.2. India

In 2009, India was the fourth largest oil consumer in the world, after the United States, China, and Japan. "Oil accounts for nearly 24 per cent of total energy consumption, natural gas six per cent, hydroelectric power almost 2 per cent, nuclear nearly 1 per cent, and other renewables less than 0.5 per cent. Although nuclear power comprises a very small per centage of total energy consumption at this time, it is expected to increase in light of international civil nuclear energy co - operation deals"[10].

A safe and secure supply of energy resources to the end consumer is a major challenge for both the producers and the consumers. Today's consumers face tremendous challenges in terms of a safe passage of oil and gas through pipelines or tankers. "About two – thirds of the world's oil trade (crude oil & refined products) moves by tanker. Oil transported by sea flows through fixed routes, the most important being the Strait of Hormuz and Strait of Malacca. External threats like piracy and terrorism have threatened the secure transit of energy supplies in the straits"[11]. Any threat to these supplies would create a worst scenario for the global economy. "....., the worst hit would be the Gulf states, unable to export their oil and their main source of income would come to an end"[12]. The significance of the Middle Eastern countries such as, Saudi Arabia, producer of spare capacity of oil should be recognized[13]. Thus, today's energy business has become more complex, where any instability in the market will affect the producers and the consumers alike.

3. Speculative Trajectory

As the global energy demand increases, the energy rich areas specifically the countries which are benefitted by the oil wealth such as Saudi Arabia and UAE, their domestic demand has also taken an upswing. To meet the increased consumption, the Middle Eastern states need huge proportion of energy. Countries like Saudi Arabia, Qatar and UAE are known to be the economic powerhouses of the region. "Saudi Arabia is the largest consumer of petroleum in the Middle East, particularly in the area of transportation fuels. Domestic consumption growth has been spurred by the economic boom due to historically high oil prices and large fuel subsidies. In 2006, Saudi Arabia was the 15th largest consumer of total primary energy, of which 60 per cent was petroleum - based. The remainder was made up of natural gas, the growth of which has been limited by supply constraints. In 2008, Saudi Arabia consumed approximately 2.4 million bbl/d of oil, up 50 percent since 2000, due to strong economic and industrial growth and subsidized prices. According to independent analysis quoted in industry reports, demand is expected to rise by eight to 10 percent through 2010, mostly in the area of electricity and NGLs for petrochemical production. Saudi Arabia also does direct burn of crude oil for power generation during summer months"[14].

Energy today, has become a indispensable support- system of development and growth, for the western and the developing nations alike. Apparently, energy being the backbone of the global expansion and progress; it is fundamental in the onward industrial movement. With the energy supplies in other regions getting scarce, the demand has become ever more critical.

4. Conclusion

Looking at the current scenario and the future trajectories, the global competition and militarization of these resources is inevitable. The rapid growth of the 'rising' Asian economies will set its own repercussions for the world. China and India's rapid growth will facilitate the future growth of the hydrocarbon industry as its demand rises. Under these circumstances, it becomes vital for Asia to look out for alternative / renewable sources of energy, which are costly, unsafe and will take some time to be produced on a commercial scale, while simultaneously, developing the indigenous resources. Serious competition for the resources, will lead to lesser shares for individual stakeholders, hence the hike in the oil prices, also affecting the taxation policy of the countries. Apparently, the process of globalization has been efficient in integrating the global economies and connecting the producers and the consumers. With the oil politics expanding, the stakeholders involved will have to redefine their interest.

5. References

[1] ExxonMobil, "Global Energy Demand", Accessed 19th December 2010, See, http://www.exxonmobil.com/corporate/energy_issues_energydemand.aspx

[2] World Nuclear Association (December 2010), "World Energy Needs and Nuclear Power", Accessed 19th December 2010, See, http://www.world-nuclear.org/info/inf16.html

[3] Ibid.

[4] Kothari, Raj Kumar (2010), "India's Foreign Policy in the New Millennium", Academic Excellence, Delhi, p. 38.

[5] Energy Information Administration (EIA), "International Energy Outlook 2010", Accessed 20th December 2010, See, http://www.eia.doe.gov/oiaf/ieo/world.html

[6] "Gulf energy demand to 'soar 85pc by 2030" (20th July 2010), Accessed 18th December 2010, see, http://www.tradearabia.com/news/ogn_183277.html

[7] Energy Information Administration (EIA) (November 2010), "Country Analysis Briefs: China", p. 14, Accessed 19th December 2010, See, http://www.eia.doe.gov/emeu/cabs/China/pdf.pdf

[8] Ibid, p. 2.

[9] Ghoble, Vrushal (April – June 2010), "China and the Global Energy", Dialogue, 11 (4): 152 – 163.

[10] Energy Information Administration (EIA) (August 2010), "Country Analysis Briefs: India", p. 1, Accessed 19th December 2010, See, http://www.eia.doe.gov/emeu/cabs/India/pdf.pdf

[11] Ghoble, Vrushal (October – December 2009), "Maritime Energy Security and Hormuz and Malacca Straits", Think India Quarterly, 12 (4): 96.

[12] Ibid, p. 101.

[13] Saudi Arabia increased its production during the Iraq war in 2003 to compensate for the loss of Iraqi oil and made it unnecessary to release oil from the strategic stocks.

[14] Energy Information Administration (EIA) (November 2009), "Country Analysis Briefs: Saudi Arabia", p. 1 & 2, Accessed 19th December 2010, See, http://www.eia.doe.gov/emeu/cabs/Saudi_Arabia/pdf.pdf

3

An Analysis of Iran's Cities Distributions in Relation to Earthquake Hazard

Safar Ghaedrahmati, Mohammad Reza Rezaei

University of Yazd, Iran

Abstract

Natural disasters such as earthquakes often result in extensive casualties and damage. The location of Iran and many other developing countries in active tectonic regions of the world emphasizes the necessity to develop a comprehensive disaster management system. Iran is located in the Alpine-Himalayan seismic belt which is one of the most active tectonic regions of the world. Iran is a country with about 75 million people living and the history of the region indicates strong earthquakes. Population centralization in urban area and metropolitans with environmental disaster, especially in developing countries such as Iran increase metropolitan's vulnerability against earthquake hazard. Iran is a vulnerable against earthquake hazard, because of distribution of cities and population centralization in its metropolitan. But the amount of damage and injured in an earthquake in Iran's cities is different. Population growth with increase of cities number cause to sustain a loss of people and appurtenances. This research tries to help to planning for earthquake crisis. Type of research is applied and method of data collection is documentary and methods of analysis are; population analysis with urban system analysis and urban distribution system in related to earthquake hazard. In beginning the situation of urban system and locations of cities are studied, than the relationship between distribution of cities and earthquake hazard studied. This paper examines the spatial distribution of the population and focusing on urban system patterns.

Keywords: Iran's Cities, Distributions of Cities, Earthquake Hazard, Urban System.

1. Introduction

A disaster is a perceived tragedy, being either a natural calamity or man-made catastrophe. It is a hazard which has come to fruition. A hazard, in turn, is a situation which poses a level of threat to life, health, property, or that may deleteriously affect society or an environment. Planning for crisis, especially urban crisis in variable angles must be analysis and review. This discussion is interested in some sciences such as urban planning, environment science and crisis management. This topic is very important, because of related to life and property of people. Basically, researches can important role in decrease of manmade and environmental hazard. Developing countries suffer the greatest costs when a disaster hits – more than 95 percent of all deaths caused by disasters occur in developing countries. A disaster can be defined as any tragic event with great loss stemming from events such as earthquakes.

Various disasters like earthquake are natural hazard that kill thousands of people and destroy billions of dollars of habitat and property each year. The rapid growth of the Iran's population and its increased concentration often in hazardous environment has escalated both the frequency and severity of natural disasters. Among various natural hazards, an earthquake is the major disasters adversely affecting very large areas and population in the Iran. For few decades, the population of cities in developing countries, including Iran had a higher growth rate than the total growth rate of countries' population. The cities in the developing countries have become areas of very high vulnerability to natural hazards, where effectively, more than 40% of the urban population is directly or indirectly threatened [1].

Controlling the development of land on or close to active faults is a Resource Management Act 1991 issue. The guidelines provide direction on land use planning approaches for land on or close to active faults. They aim to assist planners, emergency managers, earth scientists, and people in building industry to avoid or mitigate the fault rupture hazard[2]. Many large cities in developing countries are subject to natural hazards .At global scale, large natural hazards are associated with unplanned and poor cities [3]. In fact the main pressure which threats our lives is because of hasty urbanization without analyzing the hazard risk [4].

While in developed countries, the development of the big cities began centuries ago and generally allows for controlled and planned urban expansion, the opposite is the case in developing countries, where rapid urbanization's process is characterized by an unplanned urban expansion. The developing metropolitan cities have enormous difficulties in coping both with the natural population increase and the urban physical expansion. In fact, urbanization process increases the vulnerability through centralization of human and property. Because of unplanned urbanization, growth of land in vulnerable areas with a high level of hazard risk, inadequate urban management and unsuitable construction measures, third world cities transform into vulnerable centers. If cities grow rapidly, without any plan and attention to the observance of urbanism regulation and factors resistance, it will cause an increase in urban vulnerability. This case deteriorates when the metropolitan cities extend on or close to active faults. On the other hand, urbanization programs must be based on natural hazard knowledge. Suitable location for residence and logical development of cities has a basic role in the decrease of earthquake damage [5].

2. Seism tectonic and seismicity of Iran

The Iranian plateau can be characterized by active faults, recentvolcanic and high surface elevation along the Alpine earthquake belt. Tectonicstudies indicate that the Iranian plateau has a very high density of active andrecent faults. Earthquake data of Iran show that most activity is concentratedalong the Zagros fold thrust belt while less activity is observed in central andeastern Iran. Thus, several regions are vulnerable to destructive earthquakesThe preparation of an earthquake hazard map is the delineation of theseism tectonic province and the assessment of the associated maximum earthquakepotentials[figure 1]. The boundaries of the provinces are established through analysis ofseismic history, relocated epicenter for the past several decades; tectonicenvironments, active faults, regional geomorphology, and plate boundaries.

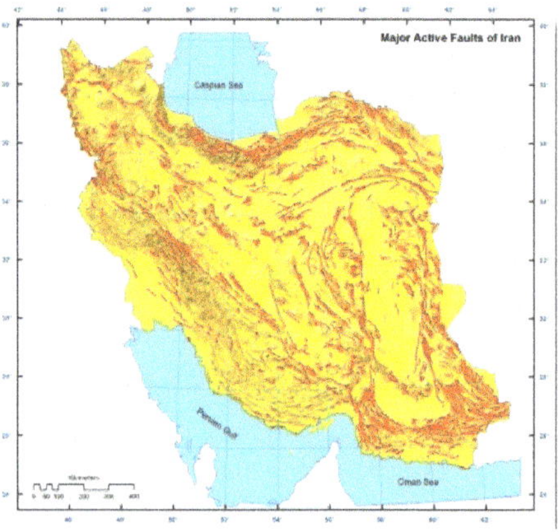

Figure 1. Major Active Fault of Iran

3. Population Analysis

Iranian cities are usually the heart of economic and cultural changes that have occurred after the Islamic revolution in 1979. These cities are increasingly having impacts via political–economical arrangement and chiefly by urban management structures. Structural features have led to the increasing growth of cities and urbanization (in number, population and physical frame) and the main problems in them. For instance, housing, water supply, traffic jams, social welfare, security, safety and health, etc. The beginning and continuation of the war with Iraq accelerated these circumstances. After the war, population movements resulted in a number of urban changes, such as

Local people moved from war-stricken regions to others, especially to cities; Young men moved from rural areas to towns and then to cities; Foreign migrants moved from such countries as Afghanistan and Iraq to Iran. On the other hand, because of the lack of birth control policies and also because of the deceptive attractions of cities, particularly big cities, the birth rate has shot up, something which has occurred mainly in rural regions and small cities.

The population of Iran has increased rapidly since 1956. The 1956 and 1966 decennial censuses counted the population of Iran at 18.9 million and 25.7 million, respectively, with a 3.1% annual growth rate during the 1956–1966 period. The 1976 and 1986 decennial censuses counted Iran's population at 33.7 and 49.4 million, respectively, a 2.7% and 3.9% annual growth rate during the 1966–1976 and 1976–1986 periods. The 1996 count put Iran's population at 60 million, a 1.96% annual growth rate from 1986–1996 and the 2007 count put Iran population at 72 million. The reasons of the recent decline in the annual population growth rate are attributed partially to the government's family planning efforts since 1989 and the dismal economic conditions and general decline in living standards for the average Iranian household. The urban population of Iran has nonetheless been rapidly growing from 1956 to 1986. It has expanded about 2.6 times in the period 1956–1976, and doubled again (2.52 times) from 1979 to 1996. The number of cities in Iran has gone from 199 to 393 and 614 and 900 and 1012 during the periods 1956–1976, and 1996–2007 [6].

4. Cities and urbanization in Iran

Rural to urban migrations are the major reason for Iran's rapid urbanization. The employment opportunities and declining living conditions forced the rural population to migrate. In 1956 only one Iranian city (namely Tehran), had a 500,000 population, whereas 9 of the nation's 199 cities had a 100,000 population or more; about 53% of the total urban population lived in cities with 100,000 or more. In 1976, only 23 (6.2%) of 373 cities in Iran had population of 100,000 or more, and, only four cities had population of 500,000 or more. In this year 28.6% of the urban population of Iran lived in Tehran.

By 1996, nearly 10% of 614 cities in Iran had a population of 100,000 or more and nine cities had a population of 500,000 or more. 36.8 million People lived in urban areas, 8.4% lived in Tehran, 41.8% in the nine largest cities with populations of 500,000 and more. Thus, the country's largest city (Tehran) had a very rapid population growth. In 1956, 1976 and 1996 its population had reached 1.5, 4.5 and 6.7 millions. Similarly, the population of the Greater Tehran Metropolitan area had reached about 10.5 millions from 1956 to 1996. Several conclusions can be drawn from the urbanization analyses in Iran during 1956–1976 and 1976–1996. First, Iran's urban population increased by 5.9 million to 15.9 and then 36.8 millions in a period of 40 years. Second, Iran's hierarchy of urban settlements had only one city in 1956, four in 1976 and nine cities in 1996 with a population of half a million or more. This represents a major gap in the hierarchy of urban settlements. Third, the number of large cities (defined as those with a population of at least 250,000) went from 3 to 8 and then to 23. Their share of the urban population increased from 31.6% to 47% and then to 61.3% of Iran's total urban population. Fourth, the number or medium-sized cities (with a population of 100–250,000) grew from 15 to 36, while their share of total urban population decreased from 42.5% to 15.4% to 13.9% during 1956–1996. This indicates that rural and urban migrants generally did not consider the medium-sized cities as their destinations and moved to large cities to take advantage of greater opportunities there. Fifth, the number of small cities (<100,000) increased from 190 to 533, while their share of the total urban population decreased from 49.1% to 31.3% [7].

5. Primacy City Index (PCI) in Iran

Almost all censuses after the revolution have revealed a continuation of large-scale urbanization and an increasing tendency towards the concentration of urban population in a few big cities. The proportion of urban population to the total population of the country in 1976 reached to 46.1% while in 1996 it increased to 61% and increased to 72% in 2010. Both the increase in the number of urban places and population increase in cities have contributed to the process of urbanization [8]. After the revolution, the number and population of cities continued to increase as before. These changes are shown in[Table1]. A recent major policy of urban economic and industrial decentralization is a persistent program of the government. The policy has been identified as a result of the massive growth of Tehran in recent years, up to 9 million by 2010. Part of the growth of the capital resulted from the lack of economic opportunities elsewhere and in order to redress the developing primacy of Tehran and the internal pressures which it is undergoing, the policy of decentralization is to be implemented as quickly as possible [9].

Year	1959	1966	1976	1986	1996	2007	2010
Total Population	18500000	25789000	33709000	49445000	60055000	70495000	74733230
Urban Population	5997000	9794000	15855000	26845000	36700000	48259000	53637652
Number of Cities	199	271	373	496	617	1012	1025
Urban Population (%)	31.4	38.7	46.1	54.3	61.3	68.4	72.77

Table 1. Trends and the number of urban population in Iran 1976–2007

Source: [8]

The characteristics of the urban system of Iran can be enumerated as follows:

1. High concentration of economic and commercial investment in several big cities, especially in Tehran, and the lack of control over it. The process of urbanization in the country has been accompanied by an overconcentration of productive activities and economic forces in big cities;

2. High concentration of social, cultural, educational and welfare facilities in the above-mentioned cities, which is mainly resulting from the lack of equal distribution of the capital and equipment in the totality of regions and cities.

3. Physical and spatial expansion of big cities and their irregular growth; in spite of the recent reforms (such as renovation plans), in the old cities or some of the sections of ancient cities, cities often have had not proper physical form yet and in the regional level they had not harmonious distribution spatially. This is affected by two factors: the first is the climatic and natural situation; the second is the national policy and planning.

4. Processes of urbanization lead to urban primacy, regional inequalities, centralization of political and economic power within cities and intra-urban ecological segregation and environmental crisis; this process has intensified because there is no accord on the necessary principles of urban sustainability within development [10].

Year	Population	Changes		Changes index 1957=100	Changes in proportion to base year	Annual growth (%)
		Absolute	Proportional			
1957	18954704	-	-	100	-	-
1967	25788722	6834018	36.1	136.1	36.1	3.13
1977	33708744	7920022	30.7	166.8	66.8	2.71
1987	49445010	15736266	46.7	213.5	113.5	3.91
1997	60055488	10610478	21.5	235	135	1.96
2007	70495782	10440294	17.3	252.3	152.3	1.62

Table 2. Population changes in Iran 1957- 2007

Source : [8].

Iranian cities are not often 'self-contained', and cannot act independently from their nation and region. Dependent urbanism not only leads to uneven urban hierarchies and high levels of 'intra-urban inequality', but also creates cities that are more likely to be economically 'parasitic' on the surrounding region than 'generative'. Therefore, today we are witnessing an increase in the gap between regions and cities and the creation and spread of squatter settlements or poor housing, especially in the suburbs. Certainly, several projects have been developed in order to reduce the inequalities of urban development, but none of them have been executed completely[Table2].

6. Cities at Risk

The beginning of changes in the process of urbanization in Iran goes back to the Qajarieh period (1906) the offices, embassies, new buildings, theaters, movies, shops and hotels were built according to new styles but without planning and discipline. In the first Pahlavi's period (1921), the changes had been without proper planning and mostly on the basis of modernization, which included the advent of machines, new building, offices, ministries, embassies, new urban laws, changing streets, etc., and also changes in economic, social and cultural ways, not only in Tehran but in the other cities of Iran. But up to 1948, no national, regional or urban planning had been undertaken in Iran.

In the first decades of second Pahlavi era, the Iranian government implemented five national development plans (1948–1978). Together, these plans reflected a gradual shift from agriculture to industry and from subsistence agriculture to large scale capital intensive commercial farming [11].

In the post-revolutionary period, the two national development plans were prepared in the same way as prior to 1979. The authorities were concerned with planning, war, and expansion of the universities, relying on oil income. The urban development exceeded the pre-revolutionary period; the peripheries of large cities expanded and the urban indices of centralization increased [12]. The traditional building forms of the cities and the tendency toward modernization had been increased[figure 2].

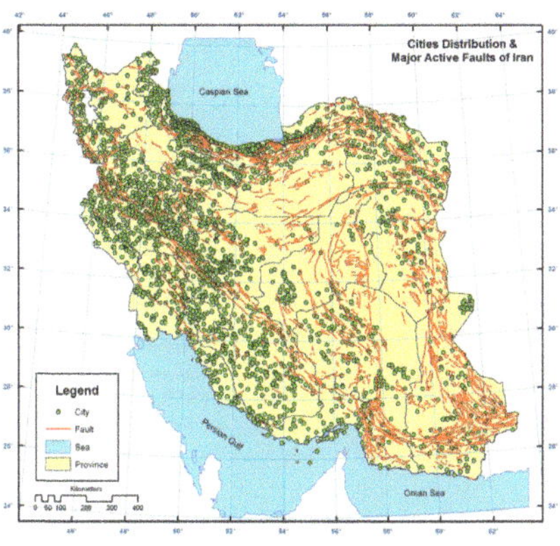

Figure 2. Cities Distribution and Major Active Fault of Iran

7. Conclusion

The sustainable urban expansion has become a key point in the context of urban studies. The management of the urban growth on the safe lands is the main issue in developing countries where the urban areas experienced rapid urban physical growth. In this study, we examined the relationship between urban extent and active faults around the cities. In general, the results of this paper are summarized as follows:

Demographic changes have become a main reason in transforming the urban dynamics. One of these dynamics is urban growth that causes uncontrolled and rapid expansion. The rapid growth of the Iran's population and its increased concentration often in hazardous cities has escalated both the frequency and severity of natural disasters.An earthquake is the major disasters adversely affecting very large cities and population in the Iran. For few decades, the population of cities in Iran had a higher growth rate than the total growth rate of countries' population. The cities in the Iran have become areas of very high vulnerability to earthquake hazards.

8. References

[1] Chardon, A. C. (1999). A geographic approach of the global vulnerability in urban area: case of Manizales, Colombian Andes. Geo Journal, No.49, pp. 197 - 212.

[2] Kerr, Janine, Nathan, Simon (2004). Planning for development of land on or close to active faults, Wellington, New Zealand. pp.1-68

[3] Batisani, N. and Yarnal B. (2009), urban expansion in center county, Pennsylvania: Spatial dynamics and landscape transformations. Applied Geography, 29, 235-249.

[4] Giddens, Anthony (2000).Runaway World. New York: Rout ledge.

[5] Nateghi, F. (2000) Disaster mitigation strategies in Teheran. Disaster Prevention and Management, 9, 205–211.

[6] Statistical Center of Iran (2003) Population Data. SCI Publication, Tehran.

[7] Javan, J (2002) Population Geography of Iran. Jahad Danesgahi, Tehran (Persian).

[8] Demographia world areas: population projection: edition 6.1(2010.07)

[9] Plan & Budget Organization (PBO) (1997) Report on the Performance of National Development Plan, Tehran.

[10] Drakakis-Smith, (199) Third World Cities: Sustainable Urban Development II—Population, Labour and Poverty, Urban Studies May 1996 33: 673-701.

[11] Sharbutoghlie, A (1991) Urbanization and Regional Disparities in Post Revolutionary Iran. West View Press, Boulder, CO.

[12] Ziari, Keramatollah,(2006). The planning and functioning of new towns in Iran, Cities, Vol. 23, No. 6, p. 412–422, 2006

4

MTA Genel Müdürlüğü, 2006b. Araştırma Alanının Topoğrafik Haritası, Konya. Feasibility of Ecotourism Absorption in Desert Zones

(Case study: Tezerjan Telecabin)

H. Zarei Mahmoodabady[1], M.Yazdi [2]

1 Department of Environmental Engineering, Islamic Azad University, Maybod Branch, Yazd, Iran

2 Department of Agricultural Engineering ,Payam E Noor University, yazd Center,Yazd, Iran

Abstract

Ecotourism is a modern phenomenon in which the tourists' principle motive is observation and pleasure of nature and phenomena and natural-cultural landscapes. We can consider it as one of the new source of income in the line of sustainable development of our province. The main object of this research is the examination and evaluation of natural phenomena which is attractive for tourism in order to recognize ecotourism fields and opportunities. Yazd province with an area of 131,551 square kilometers (the third province from an area view point) and with a population of about 880.000 is situated in the central plateau limits. The average yearly rainfall is 107 mm, the average yearly temperature is 18 degrees centigrade and the rate of proportional humidity is between 30 and 35 degrees. Yazd, as one of the regions with dry climate, has natural, historical and cultural attractions with tourist attraction potential. The city of Taft with the different climatic conditions in comparison with other regions (the average rainfall of 250mm) and with mountains and glacier is considered as tourism axis of province. With regard to ecological capability evaluation, ecotourism potential is obtained with GIS software and with tourism ecological model (Makhdum) in the Tezerjan region. In the mentioned region, areas with centralized outdoor recreation potential of grade 1 have very limited area that arising from limitation of soil, water, ground cover, and climate of this region. In the sites 1 and 2, they have a vast outdoor recreational potential of grade 2 and suggested sites 3 and 4 have grade 1 for centralized outdoor recreation.

Keywords: Ecotourism, Yazd, dry areas, industrial sustainable development

1. Introduction

In today world, optimized utilization of potential and existing facilities of any country has become one of the major local, national and international concerns within the framework of sustained development goals. This subject has more importance in dry and semi-dry regions of world and Iran, particularly in Yazd Province of Iran due to the vulnerability of the bio and live world resources and fragility of ecosystems.

Tourism is considered as one of the largest and most variable industry of world in modern time and many countries use this dynamic industry as a main source of income, job creation and development of their substructure. Ecotourism is a responsible trip to natural environments that contributes in protecting the environment as well as sustaining local people's economy. Tourism industry is the source of 5.1 percent of national income of world [10]. More than 50 percent of job difficulties in developing countries could be solved through tourism industry. In 2004, tourism has brought 75 billion dollars income for developing countries and 21 billion dollars of it; 28 percent, is the share of ecotourism [9].

Iran is in the first five countries with highest climatic varieties in world [7].There are more than 160 mammal species, 500 species of birds, 270 species of fish and more than 800 species of plants, giving this country an actual capability to attract ecotourism. On the other hand, Iran receives only one hundred percent of foreign tourists in the world while in a twenty years perspective, it should raise tourism income to 25 billion dollars [8]. In 2004, Iran had the last rank among other countries in world tourism organization in terms of tourists and incomes [1].

Yazd province, as a significant example of dry and semi-dry regions of the world has special characteristics in ecologic, environmental...terms; therefore, in order to achieve sustained development, any interference and actions of human groups in natural environment such as agriculture, industry, services...must be performed based on full and comprehensive knowledge on the ecologic tolerance and ability of the zone. On the other hand, due to the richness of natural phenomena in the province, such as significant desert characteristics in dry plains, Kevir and beautiful mountain foots or natural views with variety of animal and plant species, in addition to having recreational aspect, it has the opportunity to be utilized in developing local communities, remove poverty and create job along with maintaining and protecting the ecosystems of the region. Assessment of environment power means assessing possible use of man from land for agriculture, range, forestry and park management (tourism protection), engineering affairs urban development, industrial and rural improvement [4].

In Iran, development of preparations land started at State Forests and Ranges Organization in 1956. The plans were initiated in 1960 by executing forestry plan in Visar woods (south Noshahr) [3]. Eventually, the plan was executed in range administration, urban development, park management, cattle breeding, fishery management...The goal of this research is to study and assess natural phenomena that attract tourism for the purpose of identification and introduction of ecotourism grounds and opportunities in Yazd Province. The assumption is that agricultural and industrial and mining activities, with respect to the environmental and ecological conditions of the zone could not meet the present and future demands. The abundance of natural phenomena and special climatic, geology, wild life...have provided necessary potential to benefit from potential and existing chances to improve tourism. Using those potentials, by expanding and improving tourism facilities would attract ecotourism and cause economic flourish of the region. Therefore, providing telecabin in Tezerjan zone is an example of increasing potentials for attracting tourism that could peruse the goal of introducing the attractiveness of zone for executing the plan in order to improve job creation for local forces and preventing migration of rural population to cities.

2. Materials and Methods

The research method is a combination of field, research and literature review studies. After evaluation of ecology power of the limit subject of study, the systemic analysis methods. This method is based on full identification of needed resources and map development. First, the entire resources and parameters were identified, their maps were prepared and then, by analyzing data that included table processing, a map and model in GIS is used to process information layers, overlap them and in next stage, make environmental units.By using ecologic model that is developed through conformity with specifications of study zone, the units are measured in power terms for centralized and expanded recreation usage. To prepare the unit map of land figure, the map of slope layers, height layers and geographic direction layers are combined together. The placement of layers is done by using ILWIS software and the code of land shape units was extracted from the three combination formula.

E= [j3(j1 (1-1) + ja)-1]+j13

J3= Total number of map layers

J1= Total number of layers of height

Jii= Number f floors of height map

I= Number of floor on slope map

J13= Number of floor of map

The Range Subject of Study

Province Yazd has 131,551 square kilometers is in the center of Iran and its neighbors are Kerman, Isfahan, Fars and Khorasan provinces. The zone subject of study is located between 31 northern degree and 54 eastern degree. The maximum height 3750 meters and minimum 2066 meters, covering an area of 9,903,500 square meters, approximately 10 square kilometres [11]. The situation of the region in the province is shown in Figure 1.

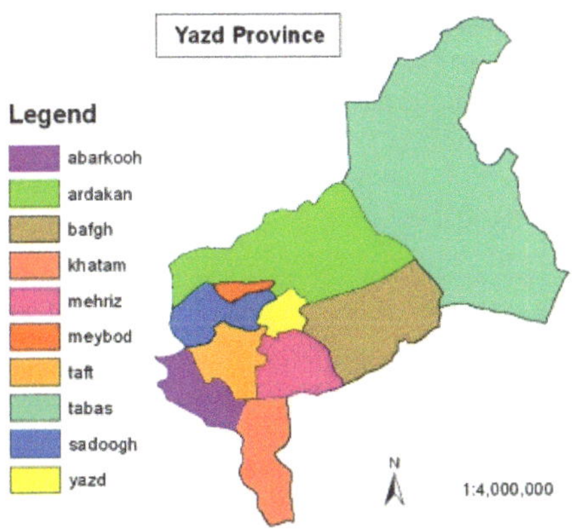

Figure 1. Political divisions of Yazd

3. Results

Natural attractions of the province

The geographic study of Yazd province shows that the existing situation of natural resources and environment of man is the product of long historical process. Yazd-Ardakan plain is one of the driest regions of the region as much that average rainfall of province is around 100 mm. There are several large and small desert pits around Yazd province, the most famous of them are Tafeh Taghestan, Abarkouh, Marvast and Daranjir [2].Those deserts start from west part of the province that covers its east south and north and makes one of the ecotourism attractions of province. Those zones are visited by local and foreign eco-tourists.

In terms of hills and high lands, Yazd province is very variable. The heights of different spots vary from 850 meters from sea in Tabas to 4070 meters in Shir Kouh Mountains. This range of height leads to emergence of mountain foots and plains in Shirkouh range, among which, one might note Manshad, Dehbala, Tazarjan, Sanich…green valley [5].

There are several springs in the region, the most important of them are Gharbal Biz spring in 40 Klm of Yazd, Tamehr Spring in 6 kilometers south of Taft city.Touran Post boiling lime spring is one of the unique springs in Iran where hot water boils out of land along with gypsum minerals. In the lapse of time, the boiled water has created laminated salina hills.

Existence of many caves in the province, some of them containing remains of cavemen are among natural attractions. There are thirty caves in the province, including three groups of epic, historical and geological caves that are visited by many tourists each year. Province Yazd with its considerable area and special desert capabilities has high environmental value. Despite undesirable climatic conditions and limited resources of water, it is the wild life habitation of wild animals such as ewe, panther, deer, bustard and partridge, they could play role in tourists attraction.

3.1. Climate

The climate subject of study, according to Amperage division has dry and cold climate. A considerable part of rainfall is snow fall. Average annual rainfall is 205 mm and mean average annual temperature is 11.4 centigrade degree. The temperature and rainfall maps are as follows.

3.2. Hydrology and water resources

The hydrology and water resources of subject of study are located in Yazd-Ardakan hydrological unit. The water resources of the region consist of springs, subterranean canals and wells and due to the existence of Snow Mountains in the surface and underground water resources in spring, the amount of water shed is very high. Tezerjan underground water beds are of mountain beds. Among significant specifications of those beds one may point out the low thickness of alluvium, large grain alluvium deposits, high penetration and relatively low reserve of the bed. The floor rocks in all regions are Shirkouh granite and the direction of underground water flows are along topography slope

3.3. Slope and aspect

To prepare the drawings of land shape, height layers, slope and direction maps are needed. To prepare the map of height layers, the numerical models of height was classified as per table one. To assess the area between height lines, Planimeter and Ilwis software was used and the relevant curves were drawn in Excel software environment. With respect to mountain morphology of the mountain, the maps of height layers have been prepared in three floors.

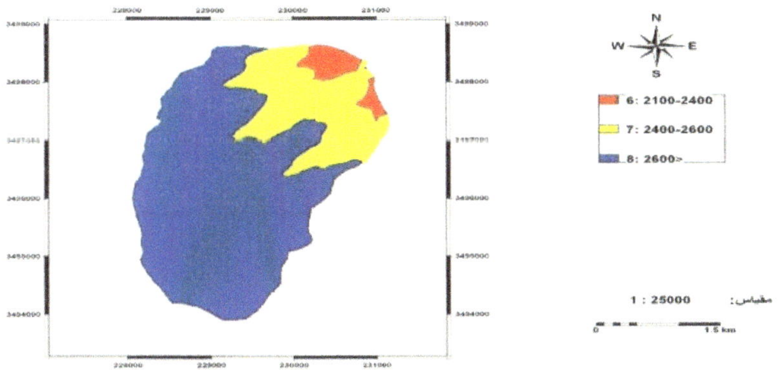

Figure 2. The height classes

Slope and aspect classification

Preparing slope classification is significant due to the importance of this parameter in using the land. The number of layers of slope of basin subject of study was determined with respect to the goal and type of plan usage. The geographic direction is effective on amount of receiving light and amount of light affects on the evaporation, perspiration and photosynthesis. Therefore, it is considered as an ecologic factor that could affect on frequency of plant society.

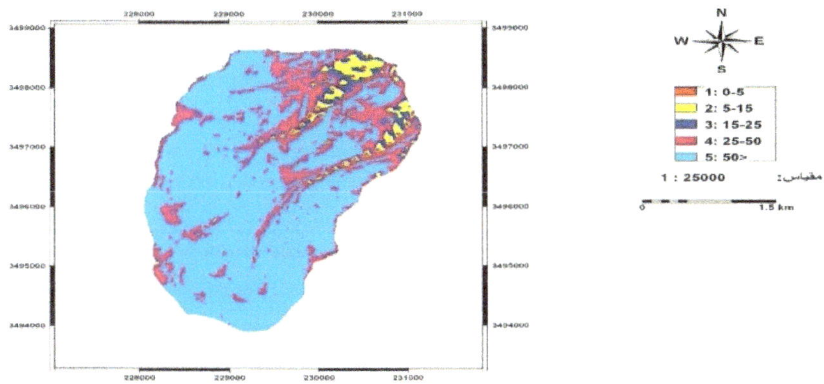

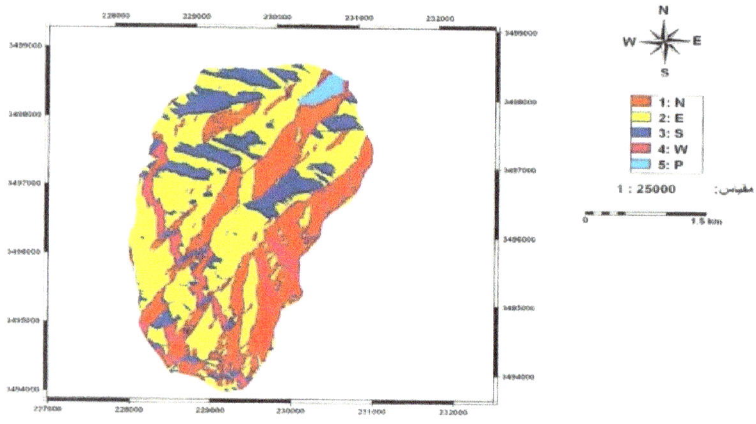

Figure 3. Classes for aspect and The slope of area

3.4. Geology

The geology and soil science show that in terms of texture of geology layers, the range of variety is not high. In large part of the area, Shirkouh granite is noticeable. In some parts, sand stone and cretaceous limes are located on Shirkouh granite. In addition, soil science studies show that there is no fertile and cultivable soil and the region is more consist of Rocky Mountains without soil. Most soils have light texture with pebbles.

3.5. Vegetation

Variety and compactness of natural vegetation is much limited in the area subject of study. Despite relatively good rainfalls, mountainous region and sharp slopes, particularly granite stone texture, the plants coverage development has been limited. The dominant vegetation of the region is mountain wormseed in +2400 meters height and plain wormseeds in less than 2400 meters height. In addition, there are different types of goat's thorn in the region. The vegetation coverage is less than 10%.

4. Discussion and Conclusion

4.1. Evaluation of environmental power for tourism application

In order to prepare a unit map of land, the maps of slope layers, height and geographic direction were mixed. The overlap is performed by using Ilwis software and the code and units of shape of land were given to the polygons through the three-combine formula, the results are in accordance with following table. The land units were prepared by GIS software. The unit map of shape of land and soil science map, slope, vegetation and plants coverage were overlapped and ultimately, the map of environmental units were specified. Ultimately, the overlap steps, preparing environmental unit maps and specifications of each unit were compared with the ecologic models for that scope and its power for vast recreation and centralized applications were determined.

As shown in table number 1, in the scope subject of study, the range with centralized recreation power with first class desirability has limited area. This factor is due to the limitations of soil, water, vegetation and climate. The complex of those factors has lowered the ability of the zone for centralized recreation and risks its success. With respect to the principle of multi-use, the application of expanded recreation and centralized recreation usage are compatible. Thus, to implement this plan in the zone, it is sufficient to perform local zoning for selecting proposed sites of the origin and destination stations. With respect to the base map, decisions and studies, the suggested sites have been specified for centralized recreation. Sites 1 and 2 have expanded recreation power with quality degree 2 and have closest distance to the access roads. Therefore, it could be discussed as a suggested site for origin stations. Sites 3 and 4 have the advantage of centralized recreation with quality degree 2 and closeness to Tezerjan snow peak as well as beautiful view.

Type	Area	Limits the number in each class
Intensive recreation class 1	3.918/462	5
Intensive recreation class 2	1.147.370/881	187
Extensive recreation class 1	4.034.100/933	850
Extensive recreation class 2	3.885.293/13	333

Table 1. Tvansnjy range of model produced using

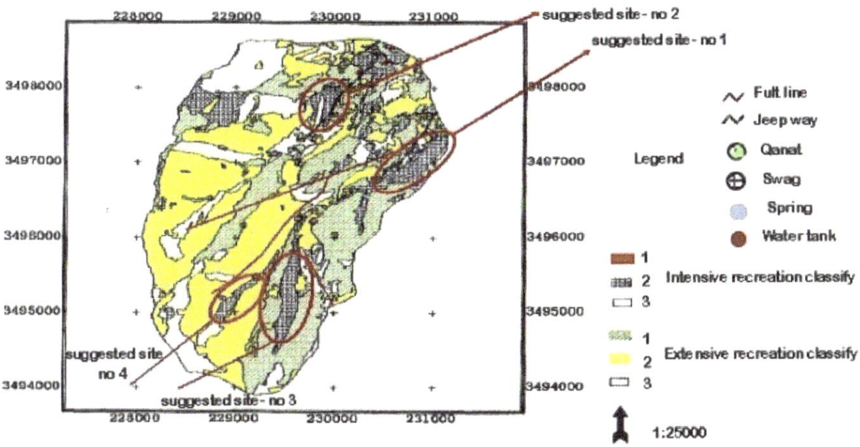

Figure 4. Map based decision making

5. References

[1] Hosseini, H, 2004.Pact of Environmental Development of Yazd Province, Environment Protection dept. of Yazd Province. p. 67

[2] Ghoabadian, A, 1982. Natural Figure of Yazd Province in connection with Desert Issues, first Ed. Mashhad .p. 41-54.

[3] Gharenejad, H, 1996. Tourism Economy and Policymaking, Mani Publications. p. 23.

[4] Makhdoum, M, 1999. Foundations of Land Preparation, Tehran University Press. p. 241-242.

[5] Mehrshahi, D, 1989. A View on Geo-morphology of Yazd province, Journal of Geographic Researches, Research Foundations of Razavi Shrine, Vols. 15 .p. 54

[6] Nasirzadeh, H, 2002. Tourism and Ecotourism Studies and their Impacts on Environmental and Development, Publications of Yazd Governor General Office.

[7] Parsaian, A. Arabi, M, 1998. Tourism in Comprehensive Perspective, Publication of Cultural Researches

[8] Rezaei, M, 2003. Ecotourism, Journal of Travel, 12th Year, No. 2.

[9] Saghaei, M, 2007. Feasibility of Ecotourism Development in Iran, Regional Tourism and Sustained Development Congress

[10] Sheybani, V, 2005.Tourism in Iran, Tehran, Research Center of Islamic Legislative Assembly. p. 191-193.

[11] Zare, A, 2004. Comprehensive Plan of Economic, Social and Cultural Development of Yazd Province, Yazd Governor General Publication. p. 161-164.

5

2 Rangeland Dynamics Monitoring Using Remotely-Sensed Data, in Dehdez Area, Iran

Ali A Torahi

Department of Forestry, Behbehan University of Technology, Behbehan, Iran,

Abstract

The land-cover dynamics has been quite conspicuous over the last three decades in Dehdez area, Iran. Therefore, the present study was undertaken in the Dehdez area to assess the trends of rangelands dynamics in the study area during the period 1990-2006. Two clear, cloud-free Landsat and one ASTER images were selected to classify the study area. All images were rectified to UTM zone 39, WGS84 using at least 25 well distributed ground control points and nearest neighbor resampling. Land-use/cover mapping is achieved through interpretation of Landsat TM satellite images of 1990, 1998 and ASTER image of 2006. Fieldwork carried out to collect data for training and validating land-use/cover interpretation from satellite image of 2006, and for qualitative description of the characteristics of each land-use/cover class. In order to create a testing sample set, first of all a set of testing points is selected randomly. A supervised classification technique with Maximum Likelihood Algorithm was applied based on 48 training samples for the image of 2006, and 42 samples for the images of 1990 and 1998 and the land-use/cover maps were produced. Error matrices were used to assess classification accuracy. The results showed, rangeland covers about 30.8%, 36.7% and 45% of the total geographical area of the Dehdez area in 1990, 1998 and 2006, respectively. Overall accuracies of land-use/cover classification for 1990, 1998 and 2006 were 89.37%, 75.24% and 71.14%, respectively. Kappa values obtained were of 78.71%, 55.61% and 51.41% of accuracy for the 1990, 1998 and 2006, respectively. During 16 years span period (1990-2006) about 1738.4 ha, 383.7 ha, 32.8 ha and 890.1 ha of rangelands were converted to forest, agriculture, water and settlement. The total rich rangelands in the area, accounts for 38.5%, 44% and 42.2% in 1990, 1998 and 2006, respectively. The total poor rangeland in the area, accounts for 61.5%, 56% and 57.8% in 1990, 1998 and 2006, respectively. Satellite Remote Sensing enabled the generation of a detailed rangeland map and the separation of grazing intensity levels in rangelands can be generated with the relatively little effort in areas that are difficult to access.

Keywords: Rangeland, Dynamics, Remote sensing, Dehdez, Iran

1. Introduction

On an area basis, rangelands contribute significantly to the world's land surface and an important share is devoted to grazing. Management of these areas alters drastically their natural characteristics. With increasing livestock, pressure on natural and semi-natural rangelands is also increased. But not all range types have the same production potential. Destruction of rangeland is mainly related to overpopulation. To enhance food production, grasslands possessing fertile soils have been ploughed and converted to agricultural lands. Three syndromes, i.e., desertification, woody encroachment, and deforestation are inherent to global grazing. These syndromes have widespread and differential effects on the structure, biochemistry, hydrology and biosphere-atmosphere exchange of grazed ecosystems as well as represent a major component of global environmental change [1]. Managing the rangelands represents a major shift in thinking and practice. In some parts of the world leading many to believe this is the required stimuli to develop a sustainable rangelands management approach [2]. There is an increasing need for better management of rangelands in developing countries in view of the alarming depletion causing by population pressure, agricultural expansion, and misuse of rangelands. Basic to the implementation of any rangelands management strategy, whether it is for assessment and allocation to sustainable uses or for rehabilitation of rangelands and denudate lands, is a clear understanding of how much natural resources there are, where they are located, and their present condition. Knowledge of the rangelands and its geographical environment are essential for proper planning of sustainable rangelands management. To meet those requirements, precise and up-to-date information regarding the status of the rangelands and potential of rangelands rehabilitation is important to upgrade and to design proper management for future improvement of the rangelands. Several studies reported successful mapping of rangelands in arid and semi-arid environments as well as in temperate areas based on remotely sensed data [3, 4]. In tropical and subtropical areas, attempts to classify land-cover have been performed [5]. In the study area, raring cattle are a very important economic activity and continuous grazing all year round is possible at almost all sites. Hence, precise land-cover information and a quantification of the study area are required. The present study was undertaken in the Dehdez area to assess the trends of rangelands dynamics in the study area during the period 1990-2006.

2. Study Area

Dehdez area is located in the south-eastern part of the state (50° 12′ 14″ to 50° 33′ 25″ E and 31° 35′ 05″ to 31° 58′ 12″ N). It has an elevational range of 580-3000 m a.s.l. with a total area of 513.12 km^2. The population of the area was 23745 in 2006 with an average density of 47 person per km^2. The structural geology of the Dehdez area consists of limestone, calcareous marl, and shale stone in Asmari and Pabdeh formations. Most of the soils were alkaline (pH ranged from 7.5 to 9) across the study area. Average annual precipitation was 620 mm, mean annual maximum temperature ranged from 17-34ºC and mean annual minimum temperature from 4-14ºC at the study sites.

3. Methodology

3.1. Image Pre-processing

Two clear, cloud-free Landsat and one ASTER images were selected to classify the study area: June 17, 1990; May 18, 1998 and June 21, 2006. The Dehdez area is entirely contained within Landsat and ASTER path 164, row 38. All images were rectified to UTM zone 39, WGS84 using at least 25 well distributed ground control points and nearest neighbor resampling. The root mean square errors were less than 0.25 pixel (7.5 m) for each of the three images. Image processing was performed using ENVI 4.5. Land-use/cover mapping is achieved through interpretation of Landsat TM satellite images of 1990, 1998 and ASTER image of 2006.

3.2. Image Classification

3.2.1. Training

Classification scheme was based on the land cover and land use classification system developed by [6] for interpretation of remote sensor data at various scales and resolutions. Based on the Anderson land-use/cover classification system, the land-use and land-covers are classified as forest land, rangeland, water bodies, agricultural land and residential land. The unsupervised image classification method carried out prior to field visit, in order to determine strata for ground truth. Fieldwork carried out to collect data for training and validating land-use/cover interpretation from satellite image of 2006, and for qualitative description of the characteristics of each land-use/cover class. Also, it is necessary to collect other ancillary data and historical data required for classification of 1990 and 1998 images. In order to create a testing sample set, first of all a set of testing points is selected randomly. However, reaching all those random points in practice is infeasible because study area is a very complex with very steep slope, and very difficult to access, especially areas which are very far from the road/path. So, a modification is made in the field, whereby 95 randomly points used and all cover classes, which were mapped in the vicinity of these points were checked. For classification of images of 1990 and 1998 simple random sampling is applied [7].

3.2.2. Allocation

The image classification is carried out in ENVI software. A supervised classification technique with Maximum Likelihood Algorithm was applied. The classification was based on 48 training samples for the image of 2006, and 42 samples for the images of 1990 and 1998. The land-use/cover maps of 1990, 1998 and 2006 were produced by using supervised image classification technique based on the Maximum Likelihood Classifier (MLC) and 132 training samples [8]. Finally, a 3*3 majority filter was applied to each classification to recode isolated pixels classified differently than the majority class of the window.

3.2.3. Testing

An independent sample of an average of 95 polygons, with about 100 pixels for each selected polygon, was randomly selected from each classification to assess classification accuracies. Error matrices as cross-tabulations of the mapped class vs. the reference class were used to assess classification accuracy [9]. Overall accuracy, user's and producer's accuracies, and the Kappa statistic were then derived from the error matrices.

4. Summary and Conclusion

The land-use/cover pattern in the area as a whole showed that the rangeland covers about 30.8%, 36.7% and 45% of the total geographical area of the Dehdez area in 1990, 1998 and 2006, respectively. According to the confusion matrix, the overall accuracies of land-use/cover classification for 1990, 1998 and 2006 were 89.37%, 75.24% and 71.14%, respectively. The Kappa statistic was made for verifying the accuracy and trusty of land-use/cover maps. The Kappa values obtained were of 78.71%, 55.61% and 51.41% of accuracy for the 1990, 1998 and 2006, respectively. Monitoring of land-use/cover reflected changes were greater in extent over the span of 16 years in the land under different categories. During 16 years span period (1990-2006) about 1738.4 ha, 383.7 ha, 32.8 ha and 890.1 ha of rangelands were converted to forest, agriculture, water and settlement. Overall, because of conversion of other land covers in to rangeland, the total rangeland increased by 14.2% during 1990-2006. The rate of rangeland increment was approximately 1.77% per year.

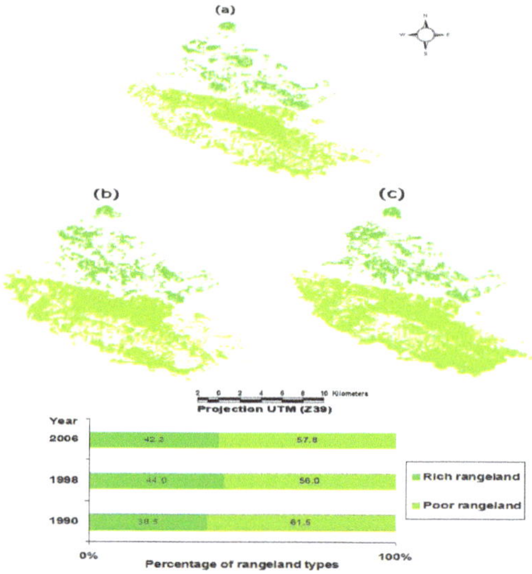

Figure 1. Rangeland in Dehdez area: (a) 1990, (b) 1998 and (c) 2006 and comparison of respective extents of rangeland types by percentage in the area

There are two types of rangelands in Dehdez area viz., rich rangeland and poor rangeland. Spatial distribution revealed that the northern and some eastern parts of the area dominated by the rich rangelands (Fig. 1). The total rich rangelands in the area, accounts for 38.5%, 44% and 42.2% in 1990, 1998 and 2006, respectively (Table 1). Because of steep slops and inaccessibility of these areas the rangelands have suitable condition and more species diversity than the other parts. Spatial distribution shows that the poor rangelands were dominated in southern and eastern parts of the area (Fig. 1). The total poor rangeland in the area, accounts for 61.5%, 56% and 57.8% in 1990, 1998 and 2006, respectively (Table 1). These rangelands are used to be as pasture lands and most of them are degraded by the livestock overgrazing. People used to graze their livestock wherever the accessible poor rangelands they found. Basically, the poor rangelands close to the settlement seems under intensive grazing. High degradation is visible in the poor rangelands of the area and species that are better able to withstand grazing have crowded out original species in many areas.

Rangeland type \ Year	1990		1998		2006	
	Area		Area		Area	
	(ha)	(%)	(ha)	(%)	(ha)	(%)
Rich rangeland	6094	38.5	8294	44.0	9762	42.2
Poor rangeland	9720	61.5	10545	56.0	13350	57.8
Total	15814	100	18839	100	23112	100

Table 1. Area under different rangeland types in Dehdez area

Satellite Remote Sensing enabled the generation of a detailed rangeland map and the separation of grazing intensity levels in rangelands can be generated with the relatively little effort in areas that are difficult to access. It is also presents a more realistic appraisal of livestock grazing pressure in the area and set a framework for future management plan. About 36.6% of Dehdez's area of rangeland showed mismanagement features and almost 26% was classified as intensively or overgrazed. Based on land-cover dynamic map, the area of rangelands is increased during the period of study, but because of the high pressure of human activity the quality of these rangelands seriously affected. Therefore, many changes have been occurred in the rangeland, because of over grazing. Certain factors are responsible for the changes are; human activities, domestic animals and ecological conditions. Still, human activity is the primary factor to affect the rangelands in this area. Animal husbandry is the factor of subsidiary kind as people cause a lot of consumption of grass and frequent pasturing by domestic animals. The rangelands got ruined by domestic animal, agricultural activities like deep plaguing and decreasing in the efficiency of soil in the area. The rangelands are exploited for medicinal plants as well as fuel and its wood quality, make cause for the demolition. This process produces further even more harmful conditions for the rangelands while quality of vegetation cover is reduced so much even for food and fodder. The damaging union of these factors is adversely affecting the ecological cycle in further. Natural rangelands are consequently changing into the agricultural and less natural vegetation cover. Quality of vegetation cover is decreased because of the unbalanced conditions like the unequal ratio of vegetation cover and animals. The lack of plant cover due to overgrazing causes soil erosion by water and wind. In conclusion, efforts should be made to prevent degradation and deterioration of rangelands irrespective of scale and geographic location.

5. References

[1] Asner, G. P., Elmore, A.G. Olander, L.P., Martin, R. E., and Harris, A. T. (2004). Grazing systems, ecosystem responses, and global change. Annual Review of Environment and Resources 29:261-299.

[2] Cook, F. M. (1999). The challenge of sustainable forests. Forest resources policy in Malaysia, 1970-1995. University of Hawaii Press, Honolulu.

[3] Laliberte, A. S., Rango, A., Havstad, K. M., Paris, J. F., Beck, R. F. and McNeely, R. (2004). Object-oriented image analysis for mapping shrub encroachment from 1937 to 2003 in southern New Mexico. Remote Sensing of Environment, 93, pp. 198-210.

[4] Evans, J. and Geerken, R. (2006). Classifying rangeland vegetation type and coverage using a Fourier component based similarity measure.pdf. Remote Sensing of Environment, Vol. 105, 1, 1-8.

[5] Sedano, F., Gong, P. and Ferrao, M. (2005). Land cover assessment with MODIS imagery in southern African Miombo ecosystems. Remote Sensing of Environment 98:p 429-441.

[6] Anderson, A. (1976). Land Use and Land Cover Classification System, Geological survey professional paper 964, Wash., D.C.

[7] Jensen, J. R. (1996). Introductory digital image processing. New Jersay: Trentice Hall.

[8] Richards, J. A. (1993). Remote Sensing Digital Image Analysis, an Introduction, second edition, Springer-Velarg.

[9] Congalton, R. G., & Green, K. (1999). Assessing the accuracy of remotely sensed data: Principles and practices (pp. 43–64). Boca Rotan, Florida' Lewis Publishers.

6

A Simple Method for Purification of Low Levels of Beluga (*Huso huso*) Vitellogenin

Mahdi Paktinat[1], Mahvash Khodabandeh[2], Bagher Mojazi Amiri[1], Hamid Farahmand[1]

1 Department of Fisheries, Faculty of Natural Resources, University of Tehran, Karaj, Iran

2 Industrial and Environmental Biotechnology Department, National Institute of Genetic Engineering and Biotechnology (NIGEB), Tehran, Iran

Abstract

The beluga sturgeon (*Huso huso*) is one of the most valuable commercial fish species in the Caspian Sea and has become highly endangered due to overfishing, loss of natural habitat and water pollution [2]. Vitellogein as a major beluga sturgeon yolk precursor, can be used as a suitable marker to monitor the contamination of the aquatic environments. In this study, we tried to compare some methods for purification of low levels of beluga vitellogenin. The efficiency of selective precipitation (EDTA-Mg+2), anion exchange chromatography and preparative electrophoresis in purification of low vitellogenin levels were compared. SDS electrophoresis analysis of the obtained fractions showed that DEAE-based chromatography and selective precipitation were unable to purify low levels of beluga vitellogenin. Preparative electrophoresis showed an appropriate result and low level of vitellogenin was purified efficiently by this method. These data suggest that preparative electrophoresis is a simple and efficient procedure for the purification of low levels of beluga Vtg.

Keywords: Beluga, vitellogenin, purification, endocrine disruptor

1. Introduction

Endocrine disrupting chemicals (EDCs) are considered to be a major source of *pollutants* in the aquatic environments and may interfere with the endocrine system and produce adverse developmental, reproductive, neurological, and immune effects. Fish, because of their constant and direct exposure to EDCs present in the aquatic environment are the most affected by EDCs [4]. This long-life exposure can influence reproduction functions including reduced fertility in both male and female, reduced fecundity, change in mating behaviors, partial sex reversal and skewed population sex ratios [10]. Certainly in this case, beluga sturgeon because of long reproductive cycle is a susceptible specie in contaminated environments. So regarding to the above mentioned issues, study on the effects of the endocrine disruptors on the reproductive functions in such endangered species is necessary. Vitellogenin (Vtg) is a dimeric phosphoglycolipoprotein with high molecular weight and the major yolk precursor protein synthesized by the liver in response to circulating estrogen, released into the bloodstream and taken up by growing oocytes via receptor-mediated endocytosis, where it undergoes proteolytic cleavage to form yolk proteins [8].

Vtg synthesis in normal conditions is confined to reproductive females, but it can be induced in either sex or every age by exposure to estrogens. Due to this trait, Vtg is a suitable marker to monitor the contamination of the aquatic environments with estrogenic compounds and often used as a biomarker of endocrine disruption in fish [1]. With regard to the significant importance of Vtg in screening and testing for endocrine disruption in fish, many studies have focused on the purification of this protein and then use of it as antigen in the production of antibody. So, setting up the appropriate methods for Vtg purification is important. Several purification methods such as ultracentrifugation, precipitation by dimethylformamide, selective precipitation using EDTA-Mg+2, ion exchange chromatography and gel filtration based on the physicochemical properties of Vtg have been developed [3]. Despite the efficiency of these methods in isolation of Vtg from most of the oviparous animals, some methods may be less efficient for purification of fish Vtg, due to high lipid content or less phosphorylation of fish Vtg [6]. Moreover, it's necessary to consider that Vtg obtained from various species respond to these methods differently. Therefore, optimization of Vtg purification methods is considered as an essential part of research about this protein. Accordingly, this study was designed to optimize the beluga Vtg purification methods and compare the performance of these methods for the purification of beluga Vtg.

2. Materials and methods

2.1. Plasma preparation

Blood plasma was obtained from naturally vitellogenic females and males reared at the Rajaee Sturgeon Hatchery Center (Sari, Mazandaran, Iran). Blood samples were collected from the caudal vein of the anesthetized fish using heparinized tubes. Plasma was then separated by centrifugation at 4000 rpm (4 °C) for 10 minutes. Aprotinin (Sigma #A6279) was added to the plasma samples (20 μL, 0.2 TIU aprotinin/1000 μL plasma), and all samples were stored at -70 °C until assay.

2.2. Purification of Vtg by DEAE-Sepharose, anion-exchanger resin

Anion exchanger resin (DEAE Sepharose CL-6B, Pharmacia) packed into 10 × 1.2 cm column and purification carried out according to the procedure described by Shi et al. [11]. Briefly, 0.5 ml of plasma sample were diluted with equal volume of buffer A (Tris-Cl 20 mM, pH: 9.0, NaCl 0 M) and centrifuged at 13000 rpm (4 °C) for 15 min. The obtained supernatant was loaded on the column equilibrated with the buffer A. Purification was performed at the room temperature with a flow rate of 36 ml h^{-1}. Unbound proteins were removed by an additional washing step with 2 column volumes of the buffer A. Bound proteins were eluted using a gradient of 0-0.8 M NaCl in 20 mM Tris-Cl (pH: 9.0), within 15 column volumes. Eluted fractions were collected at a volume of 4 ml and the elution profile was monitored at 280 nm. Fractions containing significant amounts of protein were evaluated for the presence of Vtg by SDS-PAGE.

2.3. Purification of Vtg by selective precipitation (EDTA-Mg^{+2})

Vtg precipitated from the plasma samples by method of Wiley et al. with some modifications [13]. Briefly, 0.5 ml of plasma was mixed with 2 ml of 20 mM EDTA. Then, precipitation was performed with adding 0.1 ml of 0.5 M $MgCl_2$ in this mixture. The precipitate was collected by centrifugation at 5000 rpm for 15 minutes at 4 °C and the supernatant was discarded. The obtained precipitate was re-dissolved in 1 ml of the buffer containing 1 M NaCl, 50 mM Tris-Cl and centrifuged at 13000 rpm for 30 minutes at 4 °C to remove any insoluble materials. The precipitate was discarded and supernatant fraction containing purified Vtg used for following assay. In order to determine the purity of Vtg, the obtained supernatant was subjected to the SDS-polyacrylamide gel electrophoresis.

2.4. Purification of Vtg using preparative electrophoresis

Selective precipitated samples were dissolved in the 125 µl sample buffer containing SDS and β-mercaptoethanol and loaded on a preparative gel (5 % stacking gel over 7.5 % resolving gel) with high molecular weight marker and electrophoresed at 120 V. After electrophoresis, the section related to the marker was cut and bands were visualized with staining. After comparing the stained section with the main section, the Vtg and contiguous slices were excised from the gel and liquefied in a glass-glass tissue grinder in 3 ml of Tris-Cl 50 mM (pH: 7.5) and shook overnight to extract proteins from the gel slices. Obtained fractions were centrifuged at 5000 rpm (4 °C) for 20 minutes and the supernatant were evaluated for the presence of Vtg by SDS-PAGE.

2.5. SDS-Polyacrylamide Gel Electrophoresis (SDS-PAGE)

SDS-PAGE performed with a 5% stacking gel and 7.5% separating gel according to Lämmli [5]. Gels were stained with 0.1% Coomassie Brilliant Blue G-250 for the visualization of proteins and the molecular mass of the individual protein bands was estimated using high molecular weight marker (Pharmacia Biotech, Piscataway, NJ).

2.6. Determination of Vtg concentration in purified samples

The absorbance of the each purified sample was measured at 280 and 260 nm and the following equation was used for calculating the Vtg concentration in the purified samples.

Vtg concentration = (1.55 × O.D. 280 nm) – (0.76 × O.D. 260 nm)

3. Results

3.1. Purification of Vtg by anion-exchange chromatography

Anion-exchange chromatography on DEAE-Sepharose was carried out for the purification of beluga Vtg. Result presented in fig. 1.A show that this column *was not able* to separate Vtg from plasma proteins. Accordingly, we tried to optimize chromatography conditions to purify Vtg (e.g. binding and elution buffers concentration, pH and volume). Fig. 1.B *shows the best result* of anion-exchange chromatography. Eluted Fractions containing significant amounts of protein were evaluated for the presence of Vtg by denaturing gel electrophoresis (fig. 1.B). As shown in fig. 1.B this anion-exchanger resin was not able to purify low plasma Vtg levels.

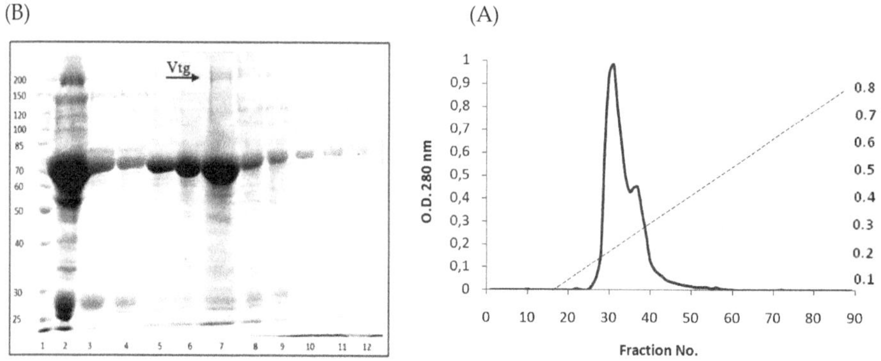

Fig 1. A: Elution profiles of plasma proteins from vitellogenic beluga on a DEAE-Sepharose chromatography column. Proteins bound to the column were eluted with a gradient of 0-0.8 M of NaCl in 20 mM Tris–HCl buffer. B: SDS-PAGE analysis of elution peak fractions in 7.5% reducing gel stained with coomassie brilliant blue. Lanes 1- molecular weight markers; 2- vitellogenic beluga plasma sample; lanes 3-12 are different elution fractions [24-42] from the column.

3.2. Purification of Vtg by selective precipitation (EDTA-Mg^{+2})

We observed that the addition of bivalent cations to plasma samples containing Vtg in the presence of EDTA was caused in the precipitation of the plasma Vtg (Fig. 2). The *SDS-PAGE results show* that Vtg has a molecular weight of about 200 kDa. A visible impurity of albumin with a molecular weight of approximately 75 kDa was present. Therefore, presence of impurities in the purified samples represents the inefficiency of this method for the purification of beluga Vtg. Nevertheless, this method could be a suitable way to identify Vtg among the plasma proteins.

3.3. Purification of Vtg by preparative electrophoresis

After termination of the preparative gel electrophoresis, slices related Vtg were excised from the gel. Proteins were extracted from the gel slices and electrophoresed for evaluating of the Vtg presence. As shown in fig. 3, purified Vtg was appeared as single major band of 200 kDa, along with some minor bands with lower apparent mass (fig. 3). It seems to be the degraded products of Vtg. With respect to protein patterns obtained from SDS-PAGE, this method has considerable ability in the purification of the beluga Vtg. To validate this method, we determined the concentration of protein in the purified sample that contained the appropriate levels of Vtg (232 µg/ml).

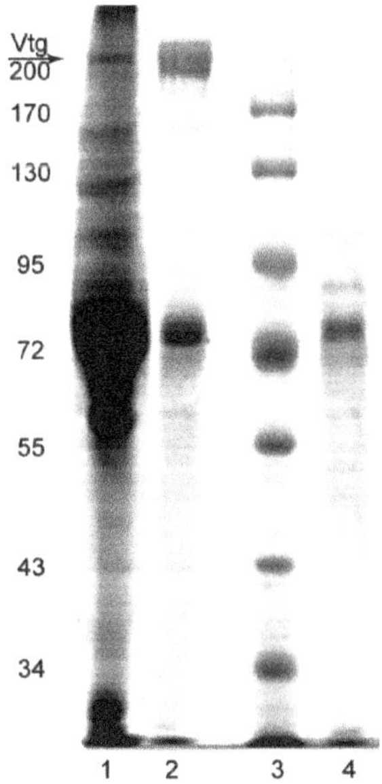

Fig 2. SDS-PAGE (7.5%) of male and female beluga Plasma. 1- beluga female plasma; 2- selective precipitated (EDTA-Mg^{+2}) beluga female plasma; 3- High molecular weight markers; 4- selective precipitated (EDTA-Mg^{+2}) beluga male plasma.

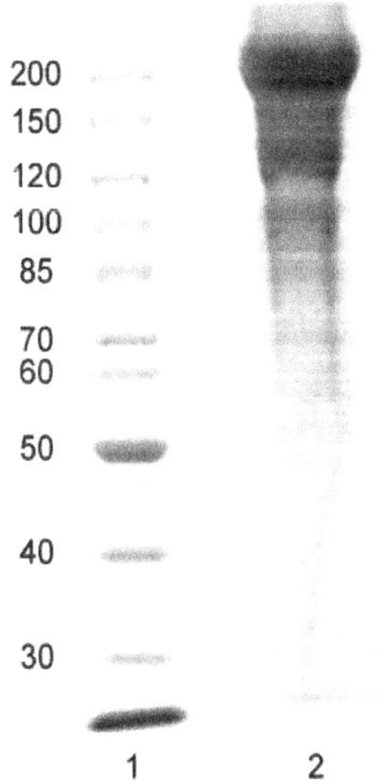

Fig 3. SDS PAGE (7.5%) of purified Vtg. 1- high molecular weight markers; 2- purified beluga Vtg.

4. Discussion

In the current study we investigated the efficiency of the anion-exchange chromatography, selective precipitation and preparative electrophoresis for the purification of the low levels of beluga Vtg.

The anion-exchange chromatography was the first method which used for the purification of beluga Vtg. *As represented in the results (fig. 1), DEAE based column was completely* unable to purify low Vtg levels and modification of some operational features couldn't to change the results. Various chromatographic methods have been used for the isolation of the fish Vtg [7] and some have used DEAE-Sepharose for purification of fish Vtg [9]. Low levels of Vtg in loading samples and low performance of DEAE-Sepharose for isolation of this levels of Vtg may be the *main reasons* why this column was *not* very *effective* in purification of beluga Vtg. As presented in fig. 2, selective precipitation was caused to the significant precipitation of the plasma Vtg. But structural similarity between Vtg and albumin in calcium binding properties was caused to the precipitation of the plasma albumin. Wiley H. S. [13] applied the selective precipitation procedure for purification of Xenopus laevis Vtg. Low plasma Vtg levels and less phosphorylation of fish Vtg might be the main reseems of the current results.

This method is specific for calcium binding proteins such as Vtg that carries the large amounts of the calcium ions. So, applying this method to the male and female plasma can help with the identification of Vtg.

Purification of beluga Vtg using preparative electrophoresis was the last procedure which was used in this study. According to our results (fig. 3), preparative electrophoresis was an appropriate procedure for purification of beluga Vtg. In this procedure, selective precipitation (EDTA-Mg^{+2}) showed good efficiency with regard to the precipitation of Vtg from beluga plasma samples, helping to acquire highly pure Vtg from the preparative electrophoresis. Inoculation of Vtg prepared from preparative electrophoresis has some advantages and disadvantages. Disadvantages of this procedure are partial degradation and the loss of secondary and tertiary structures of Vtg [12]. On the other hand, the Vtg purification method that has been developed in this study has clear advantages with regard to speed, low cost and simplicity when compared to current purification methods.

In conclusion, the efficiency of chromatographic methods are depend on the using of an appropriate resin that can be costly. In addition, they presented a disadvantage of being time consuming and generally are inefficient for the purification of low levels of plasma Vtg. So, such procedures need to stimulate of the Vtg production by an estrogen treatment. Selective precipitation of Vtg needs to the very high concentrations of plasma Vtg and essentially is efficient in purification of Vtg with high contents of the phosphate groups. Finally, preparative electrophoresis is a suitable method for the purification of the low levels of fish Vtg, especially when combined with selective precipitation (EDTA-Mg^{+2}). However, the loss of native structure of Vtg is the main disadvantage of this method.

5. References

[1] Cheek, A.O., King, V.W., Burse, J.R., Borton, D.L., Sullivan, C.V., 2004. Blue-gill (Lepomis macrochirus) vitellogenin: purification and enzyme-linked immunosorbent assay for detection of endocrine disruption by papermill effluent. Comparative Biochemistry and Physiology. C: 137, 249-260.

[2] Falahatkar, B., Soltani, M., Abtahi, B., Kalbassi, M.R., Pour kazemi, M., 2006. Effects of dietary vitamin C supplementation on performance, tissue chemical composition and alkaline phosphatase activity in great sturgeon (Huso huso). J. Appl. Ichth. 22, 283-286.

[3] Hara, A, Sullivan, C.V and Dickhoff, W.W., 1993. Isolation and some characterization of vitellogenin and its related egg yolk proteins from Coho Salmon (Oncorhynchus kisutch). Zool. Sci. 10, 245-256.

[4] Hutchinson, T.H., 2002. Reproductive and developmental effects of endocrine disrupters in invertebrates: in vitro and in vivo approaches. Toxicology Letters. 131 (1-2), 75-81.

[5] Lämmli, U.K., 1970. Cleavage of Structural Proteins During Assembly of the Head of Bacteriophage T4. Nature. 227, 680-685.

[6] Magalhaes, I., M.L. Ledrich, J.C. Pihan, and J. Falla., 2004. One-step, non-denaturing purification method of carp (Cyprinus carpio) vitellogenin. J. Chromatogr. (B) 799, 87-93.

[7] Maltais, D. and R.L. Roy., 2009. Purification and partial characterization of vitellogenin from shorthead redhorse (Moxostoma macrolepidotum) and copper redhorse (Moxostoma hubbsi) and detection in plasma and mucus with a heterologous antibody. Fish Physiol. Biochem. 35, 241-254.

[8] Mommsen, T.P., Walsh, P.J., 1988. Vitellogenesis and oocyte assembly. In: Hoar, W.S., Randall, D.J. (eds) Fish physiology, vol XIA. Academic Press, San Diego, p: 347.

[9] Mosconi, G.O., Carnevali, R., Carletta, M., Nabissi and A.M. Polzonetti-Magni., 1998. Gilthead Seabream (Sparus aurata) vitellogenin: Purification, partial characterization and validation of an enzyme-linked immunosorbent assay (ELISA). Gen. Comp. Endocrinol. 110, 252-261.

[10] Purdom, C.E., Hardiman, P.A., Bye, V.J., Eno, N.C., Tyler, C., AND Sumpter, J.P., 1994. Estrogenic effects of effluent from sewage treatment works. Chemistry and Ecology. 8, 275-285.

[11] Shi, G.Q., Shao, J., Jiang, G.B., Wang, Q.X., Lu, Y.Q., Liu, J.F., Liu, J.M., 2003. Membrane chromatographic method for the rapid purification of vitellogenin from fish plasma. J. Chromatogr., B. 785: 361-368.

[12] Watts, M., Pankhurst, N.W., Pryce, A., Sun, B., 2003. Vitellogenin isolation, purification and antigenic cross-reactivity in three teleost species. Comp. Biochem. Physiol., B. 134, 467-476.

[13] Wiley, H.S., Opresko, L., Wallace, R.A., 1978. New Methods for the Purification of Vertebrate Vitellogenin. Analytical Biochemistry. 97, 145-152.

7

Artificial Neural Networks: A Non-Linear Tool for Air Quality Modeling and Monitoring

Amirsasha Bnanankhah*, Farhad Nejadkoorki

Department of Environmental Engineering, Yazd University, Yazd, Iran

*Corresponding author

Abstract

Environmental problems including air pollution have increased in recent years. Due to the complexity and nonlinear nature of these phenomena and problems, this paper was prepared to explore the application of Artificial Neural Networks as a nonlinear model effective to tackle these problems. This paper applied main neural networks to atmospheric science, creation process, examples, limitations and advantages. Advantages of the model are its effective capacity to relate underlying relations between input and output variables and high tolerance to errors of input variables. Some disadvantages are that the model's success is dependent on both the quality and quantity of data.

Keywords: Artificial neural networks, air quality, predict, unique approach

1. Introduction

Artificial neural networks, as a branch of artificial intelligence, can model highly non-liner functions. Neural networks have been shown to be an effective alternative to more traditional techniques of statistical analysis when the complexity of a problem increases and theoretical understanding decreases. Neural networks approximate highly non-linear functions between input features and output features and require no prior knowledge of the nature of that relationship [1]. The basic element of a neural network is the neuron; several neurons are organized into layers; input, hidden and output. Each neuron has a simple structure that mimics the functionality of neurons found in a human brain. All connections between neurons are weighted and these interconnections are the basic parameters of a model and due to the difference between the target and the model output, they are adjusted during the learning or training process. Artificial neural networks can be divided into several groups according to their topology. Multilayer perceptron artificial Neural Network (MPNN) and Kohonen neural network (KNN) are the main artificial

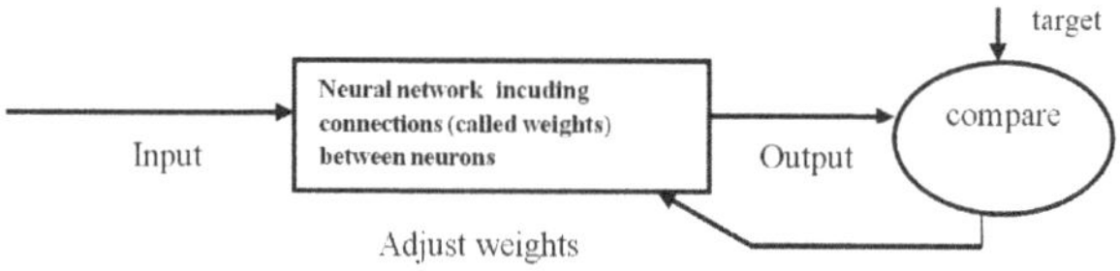

Figure 1. Tipical ANN structure

neural networks that can cover a huge variety of air pollution and meteorological modeling applications [8].

2. Multilayer perceptron And Kohonen neural networks

A multilayer perceptron consists of a system of simple interconnected neurons. A multilayer perceptron is able to learn during the training process. Training requires a set of training data, which consists of a series of input and associated output features. There are many algorithms that can be used to train a multilayer perceptron. The back propagation algorithm is the most computationally straightforward algorithm for training this network. When a network is trained with suitably representative training data, the multilayer perceptron can generalize to new, unseen input data. Therefore it has three main applications in atmospheric and meteorological sciences; prediction, function approximation and pat tern classification.

- **Prediction:** as respects the relationship between meteorology and pollution is complex and potentially multiscale in nature, the multilayer perceptron can be applied to quality prediction.
- **Function approximation**: unlike prediction, function approximation aims to use multilayer perceptron to better and more fully model relationships between data.
- **Pattern classification:** multilayer perceptron can be used to work in pattern classification in order to better distinguish data [1,3].

Kohonen neural network (KNN) significantly differs from multilayer perceptron and the main purpose of KNN is to sort multivariable patterns into groups of similar patterns. It is important that the grouping criteria need not be known, therefore this is unsupervised learning. So the kohonen neural network is a very practical and effective tool to establish groups of similar patterns in data sets [8].

3. Model creation process

The first stage is feature determination in order to define a model's domain. It is done to enable incorporation of all the important information; to simplify multilayer perceptron and therefore achieve more effective learning and to reduce the number of learning patterns needed. Then, the database of measurements is divided into several sets for construction of the model. The model includes a training set that is used to adjust the interconnection weights of the multilayer perceptron neural networks: a testing set that is used to periodically during the learning process to test a model's generalizing capacity and its is optimization during learning; a production set that is used for model verification to determine expected error, when the model has been trained; it can be used on patterns with unknown output values. This set of patterns is the on line set. In the next step, pattern selection can be used to sort patterns into groups to show which ones are the more important. These patterns contain all the information about the studied phenomenon. Network topology determination is another step; whereby numbers of neurons in the input, hidden, and output layers are determined and it is from the number of features and number of patterns that better performance of the model is achieved. Then the training and testing process is carried out periodically on the testing set and the training algorithm is used to determine the model's interconnection weights for the best results on the optimization and testing set. Finally when the model is trained it is validated by the production set to determine the expected error in further on line use. Feature determination and pattern selection are therefore the most crucial steps in a model's construction and usually determine a model's ability [8].

4. Examples of studies by artificial neural networks in the field of atmospheric science

4.1. Application of backpropagation neural network in predicting emissions from a palm oil mill

In Malaysia, the palm oil industry is one of the major main industries and it occupies a large sector of the country's economy. The industry produces air pollution from emissions that can contribute to health problems in nearby communities especially when fuels are mixed disproportionately. It is therefore necessary to reduce air polluting emissions either by using air pollutant removing devices or by improving the combustion efficiency of boilers used in the industry by detecting effective parameters to emission pollutants. Input variables were fiber flow, shell flow, steam capacity, feed water, steam pressure, power output, main pressure, flue gas temperature and output variables were carbon monoxide, nitrogen oxide, sulphur dioxide and particulate matters. Finally the trained data by NN agrees well with the measured data almost within 8 % error for emission pollutants [4].

4.2. Forecasting extreme PM_{10} concentrations using artificial neural networks

Particulate matter has a major effect on public health because of air pollution, this has been a major concern in Tehran for recent years and the city has been suffering from PM_{10}: an artificial neural network was used to forecast estimates for maximum PM_{10} concentrations 24 hour ahead in Tehran from meteorological and gaseous pollutants. Input features were date, day of week, month of year, mean of solar radiation, mean and max temperatures, mean wind direction and speed, mean CO, mean NO and mean and max of PM_{10} from the day before and mean solar radiation, mean temperature, mean wind direction, and mean of wind speed for the next day. Results showed that forecasting PM_{10} was promising with an index of agreement of up to 83%[5] .

4.3. Measurement and prediction of ozone levels around a heavy industrialized area: a neural network approach

It is well known that ozone is formed from the complex chemical interaction of primary pollutants and a presence of solar radiation. Therefore a neural network was applied to forecast ozone levels near an industrial area in Kuwait. The inputs to the network were wind speed and direction, relative humidity, temperature, solar intensity and concentrations of methane, carbon monoxide, carbon dioxide, nitrogen oxide, nitrogen dioxide, sulfur dioxide, non-methane hydrocarbons and dust. It was found that those precursors having the most effect on predicted ozone concentrations and temperature played an important role. In addition neural networks were compared against linear and non-linear regression models and it was found that the neural network model provided superior predictions [6].

4.4. An online air pollution forecasting system using artificial neural networks

Urban air quality management and information systems are required to predict the next days air pollution levels to implement the appropriate action and control strategy, therefore neural networking was applied to develop an online air pollution forecasting system for the greater Istanbul area. The system predicts three air pollution indicator levels for the next three days. The inputs were general condition, wind direction, pressure, day temperature, night temperature, relative humidity and wind speed. Output parameters were sulfur dioxide, particulate matter and carbon monoxide. The results showed that quite accurate predictions of air pollutant indicators are possible with a simple neural network and further optimization of the model can be achieved using different input parameters and better forecasts are observed using day of week as an input parameter [7].

5. Limitations and vantage of neural networks

5.1. Limitations

One of the limits of artificial neural networks in practice is that they are difficult to implement and interpret. The success of an artificial neural network depends on both the quality and quantity of data. Deciding on a network structure or architecture, determining a number of layers and neurons is another problem and there are no rules to help in this process. The lack of physical attributes and relations is another limitation. The inability to explain in a comprehensible form the process through which a given decision was made by the neural networks. Neural networks are not a miracle to all real world problems; therefore other traditional techniques are powerful in their own way [1,2].

5.2. Advantages

The ability of artificial neural networks is to gain an understanding of underlying relations between input variables and output variables and to solve complex and non linear problems and high tolerance to data containing noise and measurement errors due to distributed processing within a network [2].

6. Summary and conclusion

This paper has briefly reviewed techniques of artificial neural networks. According to performance of the networks in this study the results demonstrate that artificial neural networks can be useful a tool to model hidden phenomenon, particularly in atmospheric science and there are many freely available software packages to implement neural networks easily. Therefore the current approach can be extended by further research to model environmental and air pollution problems.

7. References

[1] Gardner, M. W. Dorling, S. R. ,Artificial neural networks(The multilayer perceptron) A review of applications in the atmospheric sciences. Atmospheric Environment.1998,32: pp. 2627- 2636.

[2] Basheer, I. A. Hajmeer, M. Artificial neural networks: fundamentals, computing, design, and application. Journal Microbiological Methods . 2000. 43: pp. 3-31.

[3] Chan, C. W. Huang, G. H. Artificial intelligence for management and control of pollution minimization and mitigation processes. Engineering Application of Artificial Intelligence. 2003,16: pp. 75-90.

[4] Azid, I. A. Yusoff, A. R. Seetharamu, K. N. Application backpropagation neural network in predicting palm oil mill emission. AJSTD. 2003, 20: pp. 71-86.

[5] Nejadkoorki, F. Baroutian, S. Forecasting extreme PM_{10} concentrations using artificial neural networks. International Journal Of Environmental Research,.2012, 6(1):pp. 277-284.

[6] Elkamal, A. Abdul-Wahab, S. Bouhamra, W. Alper, E. Measurement and prediction of ozone levels around a heavily industrialized area : a neural network approach. Advances in Environmental Research. 2001, 5: pp. 47-59.

[7] Kurt, A. Gulbagci, B. Karaca, F. Alagha, O. An online air pollution forecasting system using neural networks. Environmental International. 2008, 34:pp. 592-598.

[8] F. Nejadkoorki. Advanced Air Pollution.InTech.2011.

8

Evaluation of the Effects of Climate Change on Temperature, Precipitation and Evapotranspiration in Iran

M.T. Dastorani* and S. Poormohammadi

Faculty of Natural Resources, Yazd University, Iran

*Corresponding author

Abstract

The purpose of this research is a comprehensive evaluation of climate change effects on temperature, precipitation and evapotranspiration over the country of Iran in next time periods of 2010-2039, 2040-2069 and 2070-2099, and under scenarios of A2 and B2. After preparation of measured temperature and precipitation data and calculation of evapotranspiration for the base time period of 1960- 1990 in 46 meteorological stations (with a nationwide distribution), it was tried to have initial zoning of these three parameters over the country. Then after receiving of maximum and minimum temperature and the values of precipitation from HadCM3 model under scenarios A2 and B2 for the mentioned next three time periods, these data were downscaled. By evaluation of uncertainties, most critical parts of the country that would be affected by climate change were specified. Results show that the highest increase in temperature occurs in west parts of the country, but the highest increase of evapotranspiration belongs to central regions of Iran. However, variation of precipitation is different in different parts of the country depending on the used scenarios and the selected time periods.

Keywords: Climate Change, Iran, Precipitation, Temperature, Evapotranspiration.

1. Introduction

The climate change phenomenon is caused by greenhouse gasses, which affect the greenhouse properties of the earth's atmosphere. Emissions of greenhouse gasses have been increased since industrialization in the 1900s, due to increasing in fossil fuel burning. These gasses allow solar radiation to travel from the sun to the ground but prevent the reflected heat from the surface into the space. This causes to rise the earth's temperature, gradually (Takara et al., 2009)

It is expected that climate change would strongly affect the hydrologic cycle of water in future decades (Gedney et al, 2006, Milly et al, 2005). It will also have significant impacts on the availability, as well as the quality and quantity of water. Among the climatic variables, precipitation (P) and evapotranspiration (ET) have the great importance in long term changes of water resources (Piao et al, 2006). In this regard, many researchers predicted that climate change accelerates water cycles with more ET and increased precipitation (Betts et al, 2007, Oki and Kanae, 2006). But increased precipitation does not necessarily lead to sustainable water resources because less frequent but heavier precipitation may lead to extremely flood or drought occurrence (Andreadis and Lettenmaier, 2006). Therefore, it should be emphasized that to monitor and assess the impact of climate change on drought occurrence, ET and P should be considered together as two major climatic variables. Kirono et al. (2011) study is an example of RDI application for characterization of Australian droughts under enhanced greenhouse conditions. In their study, RDI was applied to simulate climate variables from 14 GCMs performed for the IPCC 4th assessment report. The results showed a general increase in drought areal extent and frequency for most regions of the country. Karamouz et al.(2009) simulated the flood flow under climate change phenomenon using GCM models in Kajoo river basin located in arid and semi-arid region of south-east Iran.

The purpose of present research is the evaluation and mapping of the impacts of climate change on the parameters of precipitation, temperature and evapotranspiration in Iran under the scenaries of A2 and B2 for the periods of 2010-2039, 2040-2069 and 2070-2099. Awareness about the type and the amount of the impacts would help the authorities and planners to take more optimized and effective management strategies on water resources to cope with the expected condition.

2. Materials and methods

2.1. Study area

The study area of the research is the country of Iran. Climate condition varies considerably over the country especially from north to the south. In a narrow strip in north of Iran annual precipitation is over 1000 mm and in some parts of it reachs over 1700 mm covered by dense forests. However the most parts of Iran especially central and south east regions are very warm hypper arid areas with less than 100 mm annual precipitation and over 3500 mm potential evapotranspiration. This considerable variations in climate condition causes wide range of biodiversity in animal and plant communities. In this study it was tried to use data from different parts of Iran to cover these variations.

Table 1 presents general information of 46 meteorological stations used in this research. As it is seen from the table, the highest mean annual precipitation belongs to Anzali station with 1780 mm, while Zabol station recievs only 54 mm per year as the lowest value through the selected sites. The warmest site is Bandarabbas with annual average temperature of 27.4 ºC while the value of this parameter is Saghez as the coldest site in northwest is 11.5 ºC.

Station	Lat.	Long.	P (mm)	T (ºC)	Station	Lat.	Long.	P(mm)	T (ºC)
Abadan	30.37	48.25	128	25.15	Saghez	36.25	46.27	422	11.55
Ahvaz	31.33	48.67	196	24.8	Sanandaj	35.33	47.00	470	13.55
Anzali	37.47	49.47	1780	16	Semnan	35.55	53.38	105	17.65
Arak	34.10	49.40	354	13.95	Shahrekord	32.32	50.85	285	11.95
Babulsar	36.72	52.65	813	16.7	Shahroud	36.42	55.03	135	14.3
Bakhtaran	34.27	47.12	443	14.05	Shiraz	29.53	52.58	323	17.15

Station	Lat.	Long.	P (mm)	T (°C)	Station	Lat.	Long.	P(mm)	T (°C)
Bam	29.10	58.40	67	22.3	Tabas	33.60	56.90	74	21.05
Bandarabbas	27.22	56.37	139	27.4	Tabriz	38.08	46.28	222	11.85
Bandar lengeh	26.58	54.83	81	26.1	Tehran	35.68	51.32	226	16.65
Birjand	32.87	59.20	161	16.95	Torbat- Hey.	35.27	59.22	237	14.45
Bushehr	28.98	50.83	256	24.25	Varamin	35.35	51.68	156	16.5
Chabahar	25.42	60.75	87	26.1	Yazd	31.90	54.40	57	18.85
Dezful	32.40	48.38	366	24.35	Zabol	31.33	61.48	54	21.75
Esfahan	32.62	51.07	110	15.8	Zahedan	29.47	60.88	108	18.25
Fasa	28.97	53.68	219	19.25	Zanjan	36.23	48.48	320	11.45
Garmsar	35.25	52.17	100	17.55	Khoramabad	33.50	48.30	516	17.95
Ghazvin	36.25	50.00	285	14.5	Khoy	38.55	44.97	269	12.5
Gorgan	36.82	54.47	655	17.8	Mashhad	36.27	59.63	239	13.6
Iranshahr	27.20	60.70	81	26.6	Nowjeh	35.20	48.72	343	11.5
Jask	25.63	57.77	152	26.7	Orumiyeh	37.53	45.08	367	12.3
Kashafrud	35.98	60.83	284	17.15	Ramsar	36.90	50.67	1234	15.9
Kashan	33.98	51.45	134	19.5	Rasht	37.25	49.60	1278	15.6
Kerman	30.25	56.97	164	15.9	Sabzevar	36.22	57.67	155	16.5

Table 1. Main properties of stations used in this research

2.2. Methodology

In this study three main sources of data were used which are as follows:

1-Historical daily temperature and precipitation data of the selected meteorological stations from 1961 to 1990 (T_{min}, T_{max} and P).

2-Projected monthly data of HadCM3 for projected period of 2010 to 2039 (T_{min}, T_{max} and P), 2040-2069 (T_{min}, T_{max} and P), 2070-2099 (T_{min}, T_{max} and P) that were resulted from GCM-runs for the Third Assessment Report (TAR) based on the IPCC-SRES scenario of A2.

3-Projected monthly data of HadCM3 from 2010 to 2039 (T_{min}, T_{max} and P), 2040 to 2069 (T_{min}, T_{max} and P), 2070 to 2099 (T_{min}, T_{max} and P), based on scenario of B2.

4- Calculated evapotranspiration for each time period using T_{min} and T_{max}.

Figure 1 illustrates the procedure for study the impact of climate change on temperature, Precipitation and Evapotranspiration in this research. After downscaling of the temperature and precipitation data for three time periods of 2010-2039, 2040-2069 and 2070-2099 in all selected sites, the values of reference evaporanspiration were calculated for the base time period as well as the mentioned future time periods. Then the nationwide maps of mean temperature, precipitation and the evapotranspiration of the mentioned future time periods were prepared. Based on these maps the effects of climate change phenomenon on the studied parameters (T,P and ETO) has been analysed.

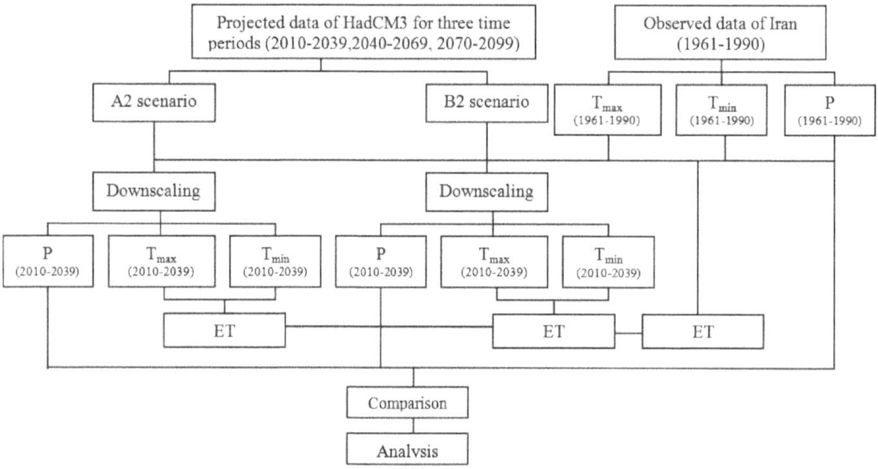

Figure 1. Proposed methodology for study of climate change impacts on temperature, precipitation and evapotranspiration in this research.

3. Results and discussion

Scenario A2: As a sample, figure 2 shows the maps of evapotranspiration rate of Iran in the base time period of 1961-1990 and the next periods of 2010-2039, 2040-2069 and 2070-2099 under the scenario A2. It must be mentioned that several maps for studied parameters including temperature, precipitation and evapotranspiration unber both scenarios were prepared and interpreted, which is not possible to show here due to page limitationjh.Comparing the map 11-B and 11-A, it is understood that evaporanspiration increased in all time periods of 2010-2039 2040-2069 and 2070-2099 and almost in all stations.

Results show that under scenario A2, stations located in north parts of Iran would experience the highest increase of precipitation. However, under scenario B2 Bushehr and other stations located in south coast show the highest increase in precipitation during next decades. Therefore, north and south coasts are the regions experiencing highest increase in precipitation under both scenarios and therefore, management of water resulted from extra precipitation in agriculture, natural resources, storage and hydropower plants are important. In the other hand, increase in precipitation in these regions would increase the risk of flooding, soil erosion as well as land slides that needs affective management and planning strategies. In this regard investigation on seasonal variation of precipitation is of importance, as seasonal variation of precipitation affects

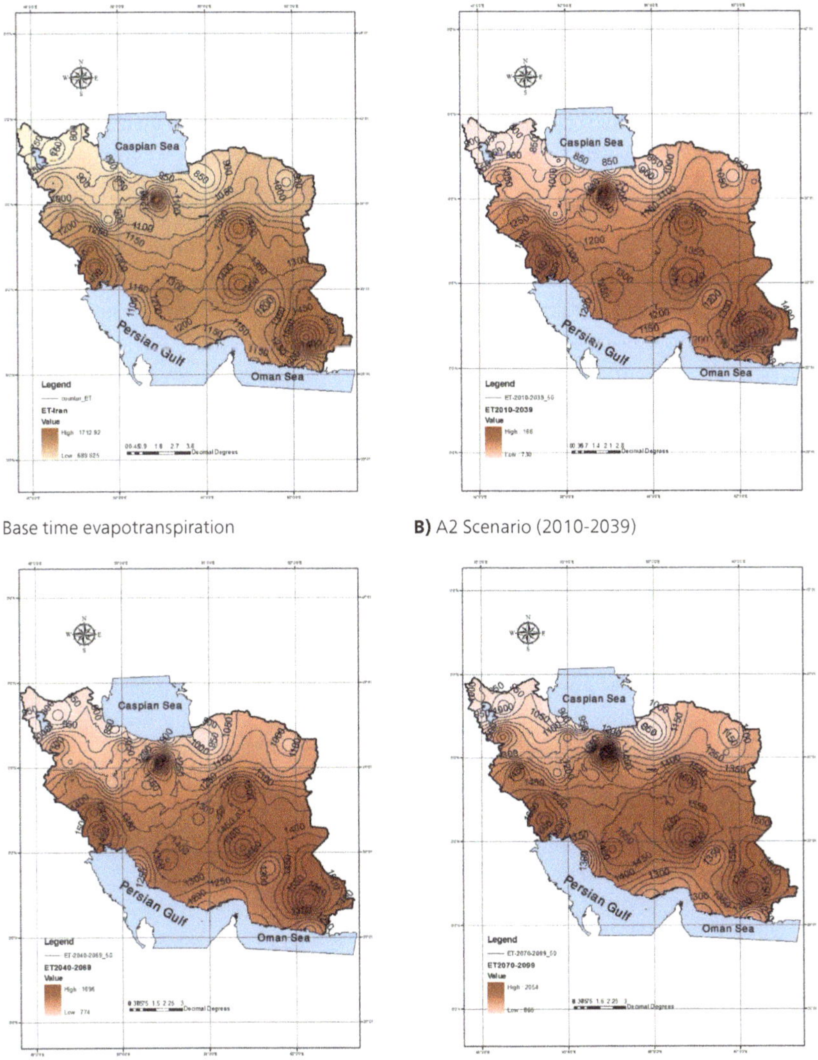

A) Base time evapotranspiration

B) A2 Scenario (2010-2039)

C) A2 Scenario (2040-2069)

D) A2 Scenario (2070-2099)

Figure 2. Maps of evapotranspiration values in base and the future time periods under scenario A2.

the type of compatible species and growing period for crops especially in rainfed farming. Results also show that under both scenarios highest decrease in precipitation amount occurs in west and northwest parts of the country. Therefore, according to the results of this research, west and northwest parts of Iran will be affected by precipitation decrease and water scarcity in time period of 2010-2039. This necessitates specific planning and management of water resources for these regions to overcome the problem of water shortage especially in drought periods when water availability would be more limited due to climate change effects. West and northwest regions of Iran are the main parts of the country for rainfed agriculture (especially for rainfed wheat production) which will be considerably affected by precipitation decrease caused by climate change. Estimation and analysis of seasonal variation of precipitation under climate change scenarios is of importance for sustainable agricultural planning in these parts of the country. Results also show that under both scenarios, in 2010-2039 the highest increase of temperature would be occur in west half of the country, where is mostly mountainous and cold. Temperature increase in these regions will considerably affect agricultural and natural resources, and due to population density that are mostly dependent on agricultural activities in this regions, the effects of climate change on precipitation decrease and temperature increase (in 2010-2039) on people life cannot be ignored.

4. Conclusions

Climate change modeling is usually with many uncertainties. As was seen in this research, the results of model for three parameters of temperature, precipitation and evapotranspiration under two scenarios of A2 and B2 were different in different areas over the Iran. The extreme values produced by scenarios A2 were higher than those of B2. Although in all stations, and for all three time periods of 2010-2039, 2040-2069 and 2070-2099 under both scenarios, temperature is increasing but the rate of increase under scenario A2 is higher than B2. Evapotranspiration is another parameter that is increasing in all three studied time periods under both scenaeios and in all stations (except few stations in time period 2010-2039), but the rate of increase under scenario A2 is again higher than scenario B2. Therefore, it can be said that duing next decades and under both scenarios, temperature and evapotranspiration are increased, and this increase is internsified from 2010-2039 toward 2070-2099. This condition will limit water availability and increase the demand for water in different sectors. However, about precipitation, the results do not follow a clear asending or desending trend like what was seen about temperature and evapotranspiration. For example, under both scenarios in time period of 2010-2039, some stations show increase but in some other decrease in precipitation occurs. From 2010-2039 toward 2070-2099 more stations show decrease in precipitation, and in 2070-2099 except few of them, in all other stations decrease of precipitation will occur comparing to the base time period. Variation of precipitation under scenario A2 is hgher than B2. Athough results produced under scenarios A2 and B2 are different but the general trends for all three parameters of temperature, precipitation and evapotranspiration for both scenarios are almost the same. This indicates that Iran will be strongly affected by global warming and climate change during next decades. As the main parts of Iran are hyper arid, arid and semi-arid regions, water shortage is even now a big problem against social and economical development. Mean annual precipitation over the country is about 250 mm where mean annual potential evapotranspiration is over 2100mm, showing high sensitivity of the state to water related problems. Climate change will definitly intensify this condition over next decades. Therefore, to be able able to cope with the next expected hard condition, more research as well as efficient management and planning strategies are required especially for risky zones where the higher effects are expected to take place.

5. References

[1] Andreadis K. M. and D. P. Lettenmaier, 2006, Trends in 20th century drought over the continental United States, Geophysical Research Letters, Vol.33, L10403

[2] Betts R. A., O. Boucher, M. Collins, P.M. Cox, P.D. Falloon, N. Gedney, D. L. Hemming, C. Huntingford, C. D. Jones, D. M. H. Sexton, and M. J. Webb, 2007, Projected increase in continental runoff due to plant responses to increasing carbon dioxide, Nature, Vol.448, pp. 1037-1041

[3] Gedney N., P. M. Cox, R. A. Betts, O. Boucher, C. Huntingford, and P. A. Stott, 2006, Detection of a direct carbon dioxide effect in continental river runoff records, Nature, 439, pp. 835-838.

[4] Karamouz M., M. Fallahi, S. Nazif and A. Ahmadi, 2009, Long lead flood simulation using downscaled GCM data in arid and semi-arid regions: A case study, Journal of Desert, V. 14 No. 2.

[5] Kirono D. G. C., D.M. Kent, K.J. Hennessy, F. Mpelasoka, 2011, Characteristics of Australian droughts under enhanced greenhouse conditions: Results from 14 global climate models, Journal of Arid Environments 75 (2011), pp. 566-575.

[6] Milly P.C.D., K.A. Dunne, and A.V. Vecchia, 2005, Global pattern of trends in streamflow and water availability in a changing climate, Nature, 438, pp. 347-350.

[7] Oki T. and S. Kanae, 2006, Global hydrological cycles and world water resources, Science, Vol.313, pp. 1068-1072.

[8] Piao S., P. Friedlingstein, P. Ciais, N. de Noblet-Ducoudre, D. Labat, and S. Zaehle, 2006, Changes in climate and land use have a larger direct impact than rising CO2 on global river runoff trends, Proceedings of the National Academy of Sciences, Vol.104, No.39, pp. 15242-15247

[9] Takara K., S. Kim, Y. Tachikawa and E. Nakakita, 2009, Assessing Climate Change Impact on Water resources in the Tone River Basin, Japan, Using Super-High-resolution Atmospheric Model Output, Journal of Disaster Research, Vol.4, No.1.

9

The Study of the Urban Environment in Old and Modern Forms (Case Study of Yazd, Iran)

Elham Yousefi*, Farhad Nejadkoorki

Department of Environmental Engineering, Yazd University, Yazd, Iran

*Corresponding author

Abstract

From a sustainable point of view, social justice and environmental parameters should be given similar attention to economical aspects. However, similar to many developing countries there have been less attention to these compared to economical part in Iranian cities. Iranian cities have mainly two forms, old and modern. While the city limits are expending continuously, the old and historical forms are vacated and in this way a great amount of public capital is wasted. The growth model of Yazd, a city in center of Iran is not environmentally friendly. In the second half of the last century there were changes that led to cultural and historical inconsistency. This paper studies the urban environment in both old and new forms of Yazd , as well as horizontal growth , noise pollution, architecture and tourism problem

Key words: urban from, Yazd, environment, tourism, architecture.

1. Introduction

Historical, cultural, economical, social, and environmental factors are all important in architecture. Iran contemporary architecture is influenced continuously by modernization, industry. Technology has changed the people's life and their environment. It can be said that modernization has been started about one century ago, and it is keeping on now[1]. The contemporary architecture of Iran has followed modern architecture, such as in expanding wide and straight streets, suburban and intra urban highways, high buildings, urban chess forms, but in many cases the rules and norms have not been considered. Segmentation in Iranian cities especially ones with old forms has left the country with massive destruction. The modern architecture fails to design environmentally friendly constructions. In other words some of the valuable conceptions that the modernization used for rejection the simplicity and strengthen the life were fainted and changed into disvalues [2]. In order to meet a sustainable city, one should not only take into account economical efficiency but also social justice and environmental protect. Today it is not regarded in Iran cities especially Yazd. Moreover, much supplication of land and wrong using has caused inactive lands with the most suitable services and installations. While city limits are expanding continuously, old and historical forms vacated and consequently a great amount of public capital is wasted[3]. Horizontal growth of the city is considered as urban dispersion. Furthermore, an imbalance between urban ecology and room lead to a fragile environment with less people in some area and more people in somewhere else and unmanaged lands. [3].In Iran before urbanization, there were a consistency within urban areas and the city growth was controlled by people' social, economical and safety condition needs [4,5].

2. Horizontal expansion in modern form and forgetting old form.

Yazd has two kinds of growth, organic and inorganic. The former changes both urban elements and internal population [6]. The later led to cultural and historical destruction. Though Yazd compared to other Iranian provinces has been less affected, economical and social scanning of land reform, rapid modernization and capital relationship expansion in Iran have some effects on Yazd. The physical expansion of Yazd in this period was because of the changes in population and life style. In the way that external city growth has stimulated and the signs of inorganic growth are increasing rapidly. Most recently formation of different districts especially in south, west and east of the city has decreased the distance of between the city and the other villages and connect them to the city and increase urban growth which leads to more inorganic growth. Therefore, several small towns were made around the city by different organizations.

3. Noise pollution as one of the environmental issues

In one study [7] sampling noise levels campaign was carried out, in 20 areas (10 for historical zone and 10 for modern zone) during the work days of the week using a 2260 Bruel & Kjaer between August and September 2010.The instrument was installed on streets line adjacent and 80 cm above of the road and was calibrated prior to taking measurement using its corresponding calibrator, 4231B8K. A wind screen was also used, although no sensible and strong blow during sampling days was observed. Two traffic peak periods, morning peak (7:00 – 9:00) and evening peak (19:00 – 21:00) was considered. The coordinates of study stations were defined by GPS and transferred to GIS. Figure 1 shows the place of stations in the GIS platform.

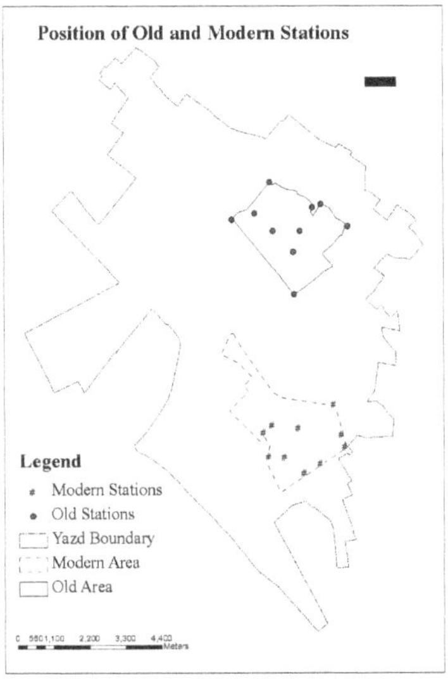

Fig 1. Sampling stations in two zones in the study area

In order to compare traffic noise levels, the average of traffic noise levels in the morning and evening in each street were considered. Then using SPSS and one sample T-test comparison between street noise levels was carried out. In one study [8] considering related researches about noise pollution and its effects on people in work places, a questionnaire was provided . For examining noise effects on people in commercial - official zones, ten street that their L_{eq} was measured in 2008, were chosen [9].This questionnaire was distributed between 500 persons in commercial places randomly .Table 1,2 show the results of study noise levels of 20 street in Yazd (old and new from) with standard SPSS.

One-Sample statistics

	N	Mean	Std. Deviation	Std. Error Mean
leq	20	72.6450	2.85228	0.63779

Table 1. Standard Deviation, standard Error, Mean of noise levels in sample stations

The output of one sample T-test is made up of two parts. In the first table (one-Sample Statistics) the Number, Mean, Standard Deviation and Standard Error Mean has been reported. It is seen that the mean of street noise levels has been 72.64 dB. In the second table (One Sample -Test), t-statistic and degree of freedom (df) has been reported. For One-Sample T-test the df equals to one subtracted from the sum of case numbers (n-1 that is 20-1=19 in here).

One-Sample Test

	Test Value = 65					
	t	df	Sig. (2-tailed)	Mean Difference	95% Confidence Interval of Difference	
					Lower	Upper
leq	11.987	19	0.00	7.64500	6.31001	8.9799

Table 2. Comparing noise levels of streets studied with standard amount (65dB)

In performing the Iranian law made in (19.3, 1387), the permissible limitation of noise in commercial zone in the day is 65dB. As we see P-value is 0 that is less than 0/05. So it can be said that that noise levels with standard amount has meaning full difference and is more than standard. Fig 2 shows the relationship between noise level and the number of people.

In all sampling days and different seasons, the noise levels were higher than the standard levels that can be dangerous for Yazd citizens in a long time period. It is because of the increasing number of vehicle such as motorcycles. According to a research done in 2008-2009 the result was that People in commercial – official zones are annoyed because of the noise pollution traffic. Reaction of the persons to noise pollution was confusion, fear, anxiety, anger, headache and problems in sleeping that among them anger was the most importance of all. The environmental pollution such as soil and water because physical illnesses, while the noise pollution is important because not only its physical impacts but

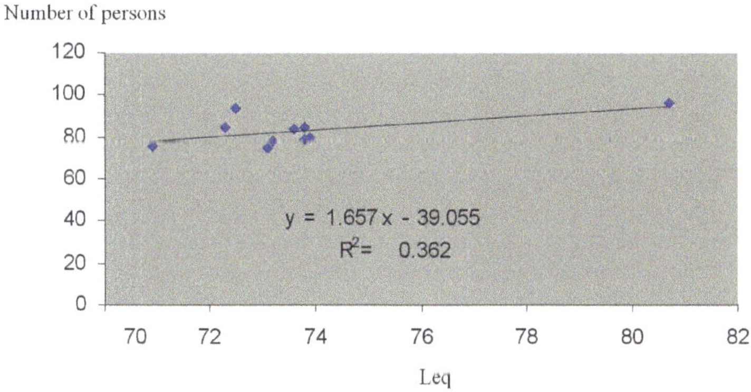

Figure 2 . The relationship between noise and annoyed people

mental consequences. A great number of motor vehicles are added to the transportation navy each year. Most of them in cities and for city traffic are used. City construction in Iran, especially Yazd, has a traditional system that is related to geographical, cultural and religious conditions. These two means that increasing of vehicles and traditional system are not compatible, so the condition planning should be in the way that the society can enjoy both of them in the best.

4. Tourism industry

Of the Iranian cities, Yazd is the most important one because of the historical structure in old form, with valuable architecture and unique features [10]. Although the old form has some problems such as destruction, the less wide of pavements and migration of people, it has important points that have caused its reduction [11]. The old form regarding to the cultural and historical structure has attracted a lot of tourists. Diplomacy and programming for new and better construction as well as old form reconstruction protect historical and cultural heritage and paying attention to their suitability and fitness with environmental and cultural structure causes constant expansion of cities. Old sections of Yazd need a controlled movement for continuing the life [12]. Figure 3 shows one of the tourist programs in Yazd old form (Walking Tour).

5. Architecture

One example of architecture elements in Yazd is Air trap. As its name shows it is an important part of the buildings in hot and dry or hot and humid areas, that causes wind blow and using pure energy in mild temperature in the houses to a comfortable degree[13]. The other examples are old houses in Yazd, which are in the direction of Kiblah. With this direction summer space and winter rooms are placed around the central yard. The central yard is the main space in Yazd houses. There is a pool in the middle of the yard with small gardens around it, full of trees needing a little water such as: pomegranate, grape, fig and pistachio. They provide a green space, shadow and cool weather [14]. Unfortunately, despite this conscious architecture, a new architecture is appearing that is inconsistent with the climate of area. There are line houses madding up of concretes, irons and bricks. Their yard with short walls from one hand unable to make a suitable shadow, protect the buildings against strong winds and sunlight and on the other hand, using thin walls, ceilings and using bitumen on the roofs that because of darkness cause increasing temperature in summer and decreasing temperature in winter and finally using heater with fossil fuels and cooling machines used in different seasons. It can be said that the new architecture with unsuitable structure causes the residents' discomfort and with unfit construction materials it makes instability [15]. Figure 4 shows an old house in Yazd with an Air trap.

Fig 3. Walking Tour

Fig 4. An old house in Yazd with an Air trap

6. Summer and result

Exploring the city forms and its platform for different aspects of economic, social, environmental highlighted that we need a holistic approach to assess. New development should take into account problems such as negative growth, lack of facilities, increasing ruin spaces, sport and medical needing. The old from should be cared because it contains monuments, historical and cultural works and it is important in attracting tourists. It can be suggested that local authorities consider: land preparation, arid land situation, reduction of destroyed forms and transferring unsuitable using with their unequal distribution all over the city and also the differences between requiring lands in various districts. Finally we suggest further research on relationship the city, climate conditions, and old from.

7. References

[1] Ebrahimi , So , Eslami , Gh , (2010). architecture and city building during GoZar period . Journal of Shahr Hoviat .

[2] Islami, S. Gh. R., (1998). "Endogenous Development: A Model for the Process of Man-environment Transaction", unpublished PhD Thesis, Heriot-Watt University , Faculty of Environmental Studies, Edinburgh College of Art, Department of Architecture, Edinburgh, U.K.

[3] Tagh vaei , M., Sarei , M. H ,(2006).The horizontal expansion and current capacity of land (case study of Yazd) Geografical researches, 152, 55- 133 ,

[4] Majed , Hamid, (1997 0. land the main problem of city expansion Abadi magazine , number 33, sudy and search center of city bulding and Iran architecture .

[5] Athari , Kamal ,(2000) .Toward government interfering efficiency in city land market , home economy journal ,30, Home & land organization .

[6] Nazarian , Asghar ,(1995). Iran urban Geography , Peyam nor university

[7] Yosefi , E, (2011),Considering the effects of urban form on noise pollution in Yazd using GIS,Masters thesis,Tehran university.

[8] Yousefi , E, (2010).The study of noise pollution effects on urban communities the effects of noise –Traffic pollution of Yazd on persons in work places , Tehran,The 4th conference & exhibition on environmental engineering,30 October-03 November 2010,Tehran,Iran.

[9] Nejad koorki,F.,E,Yousefi.,F,Naseri,(2008).Analyzing street traffic noise pollution in the city of Yazd. Journal of Environmental Health Science & Engineering, 53-62.

[10] Aslani , A.(2004).old form unique and unrepetitive . Shahr dariha magazine . 6th year , 70 number .

[11] Zeyari, k. Bernvand ,M .Ebrahimi. s,(2008).,presentation of new and Better construction of urban old forms(Yazd). The first conference of regeneration of urban distressed areas, Mashhad,10-11 December 2008 .

[12] Zeyari , M, Beranvand zadeh ,M,Alizadeh ,Y.Ebrahimi , S(2009). examination and evaluation Better and new construction of urban old forms , Yazd , 2 ,85 -99

[13] Mahmudi , M . Mofidi ,(2008)Analysis of architectural form of form Yazd air traps and finding . Best Honarhay ziba journal , 36 , 27-36 .

[14] [14] http://fa.wikipedia.org

[15] khoda Bakhshshi , sh., Mofidi , M,(2001) constant construction in relationship with iran traditional architecture , 3rd National Energy Congress, Tehran,Iran.

10

Economic Impact of Lake Edku Pollution

El-Tatawy Nashwa

Economics and Agribusiness Dept., Faculty of Agricalture, Alexandria University, Egypt

Abstract

Edku lake lay on the west branch of rashid, 40 km from Alexandria governorate, the lake relate to northern side of the western Mediterranean Sea through Almadia Bogaz. As for the sources of water supply is from salt and fresh water. This paper aims to shed light on the effect of Edku lake pollution and compute the effect on lake pollution on income and Maximum sustainable yield and the expected production under these conditions in 2015.

Keywords: Edko lake-fish production-impact of pollution- Maximum sustainable yield-Egypt

1. Introduction

In the last two decades several studied has been carried out on the lake Edko pollution. Most of these studied concerned with the technical component of that pollution i.e. toxicity, water quality and effect on biological conditions. Nevertheless a very few studies has concerned with the economic aspects of the fishery pollution. None of the published studies had concerned with the social effects of the lake pollution.

As the welfare of both economy-and individuals is the end purpose of the sustainable development, Socio-economic aspects should be considered by every plan for the development and rehabilitate of the lake. The objectives of this study are:

1. To investigate the effect of the pollution on production and productivity.
2. To define and compute the prevailing output-input relationships under pollution.
3. To define and measure the economic impact of the lake pollution on income.
4. expected production of lake Edko under the pollutio

2. Material and methods

Data required was collected from the fish statistical reports of the national institute of oceanography and fisheries, central agency for public Mobilization and statistics and from published papers and studies on Lake Edko pollution. Data collected the period 1985-2010.

Descriptive statistical analysis was carried out to identify the pollution effect on production, Labor and capital. The Econometric analysis was utilized to identify and compute the output-input relationship. The ordinary least squares (OLS) was the method of parameters estimation. The study has utilized the descriptive as well as the quantitative methods in the analysis. The simple as well as the multiple regressions have been used to estimate the functions. Different functional forms have been estimated using static as well Double Exponential Smoothing.

3. The impact of pollution on production

The historical data on production is graphically represented in Figure (1). The data can be distinguished into three time periods. The first from 1985 to 1992 reflects a slow decreasing trend of production. The second from 1993 to 2003 reflecting an increasing trend and the period from 2004 to 2010 this reflects a dramatically decreasing production trend.

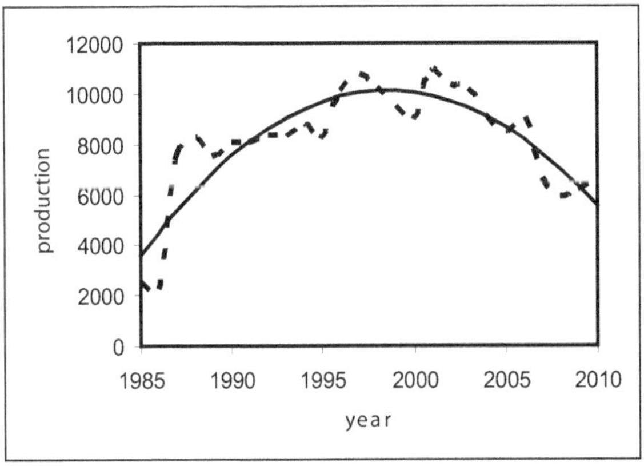

Fig 1. Production trend of lake Edko

Table (1) shoes the descriptive statistics of the previous three periods. The statistics reflects the features of every period. The overall trend of production during the hall period (1985-2010) can be represented by the following equation:

$$Y = 2627.72 + 1028.19\,t - 35.24\,t2$$

(8.67) ***, (8.27) ***

R^{-2}=0.77, F=37.75***

Where: Y: production T: time ( ) ***: t values significant at (0.01) level.

Main statistics	1985-92	1993-2003	2004-2010
Minimum (ton)	2210	8209	5886
Maximum (ton)	8300	10910	9056
Mean (ton)	6758	9808	7395
Standard deviation	2488	908	1388
Coefficient of variation	36.82	9.25	18.77

Table 1. Main statistics of the three time periods of production

Source: computed from: Central Agency for public Mobilization and statistics (CAPMAS) annual fishery statistics.

4. The development of productivity

Productivity is a measurement of efficiency which affects profitability. Total factor productivity refers the amount of production to all factors of production (Land, Area, Capital...). Partial factor productivity refers production to only one of the factors of production. Due to lack on proper data needed for computing total factor productivity, only partial productivities are computed. Table (2) summarizes the descriptive statistics of this productiveness. it is clear to notice the declining of the three partial productivities particularly from 1997 to 2010.

	Labor Productivity	Capital Productivity	Hectare Productivity
Minimum (ton)	2.13	6.35	1.21
Maximum (ton)	3.67	10.99	1.93
Mean (ton)	2.68	8.14	1.55
Standard deviation	0.42	1.19	0.23
Coefficient of variation	15.68	14.62	14.84

Table 2. Descriptive Statistics for productiveness

Source: computed from: CAPMAS, annual fishery statistics, and National institute of oceanography and fisheries, Annual fishery estimation reports.

The Maximum Labor productivity (i.e. per capita production) amounted to 3.67 ton in 2001. The minimum was 2.13 in 1998. The Maximum per boat production (capital productivity) amounted to 10.99 ton in the same year 2001 and only 6.35 ton in 1998. The maximum production per Hectare.

Of the lake reached 1.93 ton in 2001 and only 1.21 ton in 2008. The previous result confirms the detraction of the lake productivity due to the increasing pollution particularly in the last decade.

5. The relationship between production, Labor and Capital

In spite of declining of production and productivity the fishery, particularly in the last ten years, the labor and capital (the number of boats) in the fishery has taken a rapidly increasing trend.

	The equation	R-2	F
1	$\log Y = 8.35 + 0.73 \log L$ (2.01)*	0.19	4.04*
2	$\log Y = 7.22 + 0.88 \log C$ (2.4)**	0.27	5.75**
3	$\log Y = 6.60 + 1.443 \log L - 0.55 \text{ Lon } C$ (1.20) (-0. 49)	0.22	2.83*

Table 3. The relationship between production, Labor and Capital

Number between bracts are the t values

* significant at the (0.1) probability level. * *significant at the (0.05) probability level

Technique was utilized to estimate the quantitative relationships between production & labor and capital for the period 1997-2010. Results or shown in Table (3).

Equation (1) and (2) characterize one input-output relationship. The coefficients of labor (L) and No. of boats (C) are negative. This means that the marginal product as well as the elasticity is negative. In another word, the value of marginal product is actually less than the opportune cost of labor and capital. It can be concluded that under prevailing conditions of the fishery if labor and capital is reduced the production would not decrease. Equation (3) is a two input-output relationship. Although the equation represents positive Marginal product of labor, however it is not statistically significant, therefore it was interpreted.

$$Y = 10.73 - 0.003\ F$$

$$(-2.17)^{**}$$

$$R^{-2}=0.18\ ,\ F=4.72^{**}$$

Where: Y : production (catch)F : effort(number of boats) ( )** : t values significant at (0.05) level.

Maximum Sustainable yield(MSY) will be 9594 ton and it was not happened in this period ,and Maximum Effort which make this Sustainable yield will be 2104 boats.

The study has been estimated fish production using static as well Double Exponential Smoothing as seen in table (4).

Year	Production Ton	No. of boats
2012	5188	690
2013	4748	640
2014	4308	590
2015	3868	541

Table 4. expectation of fish Production , and the number of boats in lake edko

Source: computed from: CAPMAS, annual fishery statistics, and National institute of oceanography and fisheries, Annual fishery estimation reports.

6. Economic impact of the lake pollution

The losses due to pollution can be measured the difference between the maximum value of production before pollution and the yearly values after pollution.

The Maximum production during the period 1997-2010 was 10784 ton in 1997. The minimum o the other hand was 5886 tons in 2008. Therefore 1997 were considered as a base year (minimum pollution). Table (4) represents the economy losses due pollution during 1997-2010. The economy losses amounted to 6.70 Million dollar in 2007. The economy losses reached 6.70 Million dollar at current prices in 2007. The whole losses during the period 1997-2010 reached 42 Million dollar. The expected economy losses will reach 6916 ton, the value of this loss will reach9.31 Million dollar at current prices in 2010. The whole losses during the period 2012-2015 will reach18.11 thousand ton, the value of this loss will reach 244 Million dollar at current prices in 2010.

7. Summary and conclusion

Edku lake was one of the richest lakes fish in Egypt ,especially marine fish , accounted for its contribution to Egyptian fish production about 3.5%during 1985,as a result of exposure to the pollution problem and constrains in production ,this ratio reached about 1%in 2010, so that this paper aims to shed light on the effect of Edku lake pollution and compute the effect on lake pollution on income and Maximum sustainable yield and the expected production under this conditions in 2015 to identify the constrains and problems that led to this decline and to development of fish production of Edko lake.

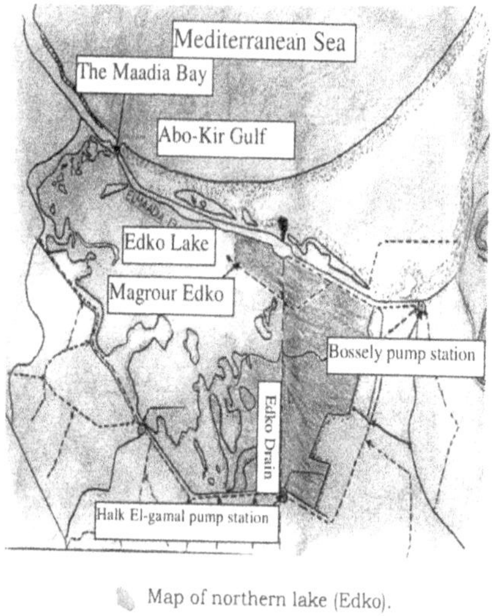

Fig 1. Map of northern lake (Edko)

Year	Production (ton)	Deviation from base year (ton)	Economy Losses Value (1000$)*
1997	10784	-	-
1998	10280	504	949
1999	9494	1290	2457
2000	8922	1862	4173
2001	10510	174	428
2002	10336	448	780
2003	10230	554	710
2004	9056	1728	2191
2005	8490	2294	3019
2006	8986	1798	2886
2007	6645	4139	6709
2008	5886	4898	6020
2009	6206	4578	6308
2010	6493	4291	5776

Table 5. National welfare losses due to lake pollution

*Deviation from base year X average price per ton

Source: Computed from: CAPMAS, annual fishery statistics, and National institute of oceanography and fisheries, Annual fishery estimation reports.

8. References

[1] Central Agency for public Mobilization and Statistics, Annual Fishery Statistics.

[2] David. L. (1988). Production Economics, Mac Millan, New York.

[3] El-Caryony , F., (1986), Bioeconomic Analysis of fishery Management in lake Mariut (Egypt) Ph.D. thesis, Olsztyn, Poland.

[4] El-Mallah,G.,S.Fattouh and Nahwa El-Tatawy, (April,1998) Economic impact of lake Marut pollution, Mansoura,J.Agric.sciences,Vol.19,No.(4), Mansoura University..

[5] El-Sharkawy, F. (1978). Final report on the study of: Pollution of lake Mariot Due To Sewage and Industrial Wastes, The High Institute of Public Health and Academy if Scientific Research and Technology, Alexandria.

[6] Gamil, H. and El-Karyony. L. (1994) Phosphor, Nitrogen, Loading and trend of fish catch as index of Lake Mariot Eutrophication.

[7] Hanaw, A. and Rustemyer, F. (1975). Productivity: Definition, Measurement, use, Agrarwirtshaft, Hanover (in German).

[8] H.M. Nagy, S.A. Kantoush and K. Watanabe, (2003). Simulation for flood routing and tidal propagation in coastal drains, Alexandria Engineering Journal, Vol., 42.

[9] National Institute of Oceanography and Fisheries, Annual Fishery Estimation Reports.

[10] Shimaa Ameen, (2009). An economic study for the production and marketing of fish in Lake Edco, the Degree of Master of Science In Agricultural economics, dept. of agric. Alex. University, Alex. Egypt.

[11] Tomas,G.H,El-Caryony,I.A ,Fisheries of the Egyptian Northern Delta Lakes during (19621989),Alex,J.Agric.Res.,Vol.40,No.(1),Alexandria Journal of Agricultural reseaech,Alex. University,April,1995.

11

Biodegradation of Hydrocarbons (Oil Fuels) by *Pseudomonas aeroginosa, Candida sp* and *Aspergillus terreus* by Isolated from the Coast Line of Arzew – Oran-Algeria

Kheira Hammadi, Mahjouba Aznouz, Miloud Halbouche

Laboratory of PAA Faculty SNV, Abdelhamid Ben Badis Mostaganem University Algeria

Abstract

Among many refineries are located along the coast of Algeria, the refinery of Arzew in the north-west of Algeria is the subject of our study. Since always the sea was the universal receptacle of pollution by hydrocarbons negatively modified the natural balance of the aquatic environment and can give many problems for the environment. Our study aimed on the biodegradation as a natural elimination of these pollutants and used as control of this pollution. The aim of this work is the study of marine pollution by physical; chemical and biological methods. The species of *Pseudomonas aeruginosa, :Candida petrolium* and *Aspergillus terreus* isolated from the sea water of three stations port from Hyproc, fishing port and Marsa el Hajadj showed their capacity of adaptation and assimilation of strong concentration of the hydrocarbons oil Arabian light and crude oil of Hassi Messoud 10% in a natural environment and 3% in a synthetic medium , their roles of transformation and degradation of the crude oil of Hassi Messoud and the petrol of the Arabian light.

Keywords: Bioterioration, crude oil;light arab oil; Pseudomonas aeruginosa; Candida sp, Aspergillus terreus

1. Introduction

The rejection of hydrocarbons (HC) of oil-bearing origin in the environment constitutes one of the most alarming phenomena of pollution in the sense that these HC are toxic for the man, fauna and flora (Belhaj and *al.*, 2000).

The elimination of the oil in marine environment requires the intervention of the various biotic and abiotic factors. Among these factors; the biological breakdown by the micro-organisms and in particular the bacteria is the natural process most important in depletion of maritime environment. Consequently, mechanisms of the biological breakdown of the substances tankers (linear alkanes, phénylalcanes, cycloalcanes, hydrocarbons polycyclic and polyaromatic) by the marine bacteria (Soltani, 2010).

Metabolic reactions of the bacteria and other micro-organisms which are naturally present in the seamen circles are usually called mechanisms of biological breakdown.

According to several authors, metabolic ways of degradation by stocks of Pseudomonas sp. were the first studied ways and are very known (Sutherland and *al.*, 1995).

The objective of this study is the insulation of the micro-organisms marinades able to eliminate the oil substances or in the event of waste from an industrial complex (Andrade, 2001; Cardoso da Silva. and *al.*, 2003).

2. Material and methods

To appreciate the phenomenon of the biodegradation of hydrocarbons in sea water we prepared a culture medium by natural sea water for that a source of carbon was selected as well as a microbial population.

2.1. Sampling sources

It is natural sea water taken in a not polluted zone. A quantity of one liter is filtered on Whatman paper. At summer then added of ammonium chloride (2 g/l) as source of nitrogen and sodium phosphate (0.1 g/l) as source of phosphorus. To agitate this medium magnetically. To preserve at 4° C with the darkness for one month. The pH is adjusted to 8 (Boutefnoucht and *al.*, 2009).

According to Boutefnoucht and *al.* (2009) the source of carbon added in the middle of culture is a derivative of the crude oil (Arabian light) of Arzew "Oran".

3. Determination the microbial biodegradable

The source of carbon is a light fraction oil, a bacterial;yeast and fungi species would be able with it to only degrade this source of carbon in the Oil "Arabian Light", it is what directed us with the insulation and the purification of various stocks starting from our studies microbiological of the 3 stations and to test them on the oil crude.

The bacteria;yeast and fungi used for the inoculation of our test come from our insulation and identification with tests bacteriological first part of our experimental of 3 stations.

4. Experimental device

Technique used in our experimental and based known the manometer technique of the apparatus of Warburg . For each culture to be tested one needs 14 bottle of Warburg for pipe side, clean and dry: for each of the 12 substrates, like endogenous witness and the last like barometric witness thermo. Each bottle of Warburg with pipe has three compartments: the principal compartment, a compartment with pipe and a central tank.

To measure the reagents and to introduce them into the compartments of each bottle, as follows:

Bottles of test	Principal compartment	Central flask
	1.4 ml of oil where oil and 1.4 ml of calibrated cellular suspension (microorganisms)	**0.1 ml of solution of KOH with 20%α**
Endogenous pilot bottle	1.9 ml of oil plug where oil and 1.0 ml of calibrated cellular suspension (bacteria)	0.1 ml of solution of KOH with 20%α
Pilot bottle thermo barometric	3.0 ml of plug of oil where oil	20%α (p/v) of potassium hydroxide in distilled water -

5. Analytical method

5.1. Potential of degradation

The potential of degradation is given thanks to the analyzes the rate of CO2 release in the medium by the relationship between the quantity of substrate consumed in the tests and that presents in the abiotic witnesses in each 4 days of the incubation period 20 days.

5.2. Determination the rate of mineralization (CO_2)

The output of mineralization is the relationship between the numbers of moles of carbons released in the form of CO2.

Measurement was taken each 4 days for one 20 days period, and this technique based on the method of Warburg and calculates the carbonization gas rate of them according to Waes (1971) the formula was used for calculations.

$$KCO_2 = \frac{\frac{Vg273}{T} + Vfa}{Po}$$

X	Representing the quantity of gas in µl (0° C, 760 m Hg)
h	Representing the modification in mm of the open arms of the pressure gauge
KCO_2	Representing the constant of the bottle
Vf	Representing the quantity in µl, of liquid in the bottle
Vg	Representing the difference, in µl, between the total volume of the pressure gauge and the bottle and the number of µl of liquid of the bottle
T	Representing 273 + the temperature of operation (27° C)
α	Representing the solubility of CO2 in the solutions, in ul CO2/ul solution
P_0	Representing the standard pressure expressed according to the manometer solution

The value used was the value of CO_2 in the water, which is of 0.759 with 25° C; the manometer solution was the known solution of Brodie with density 1.033, so that P_0:

$$Po = 760 \times \frac{13,6}{1,033} = 10000$$

6. Results

The produced CO_2 rate is deferred in the Graph .On notices after the incubation period, that the production of CO_2 is increasing according to time for each pure population of the microorganisms, the results showed an increase in the CO_2 rate which is in direct contribution with the reduction in the rate of the oil crude and light arab oil.

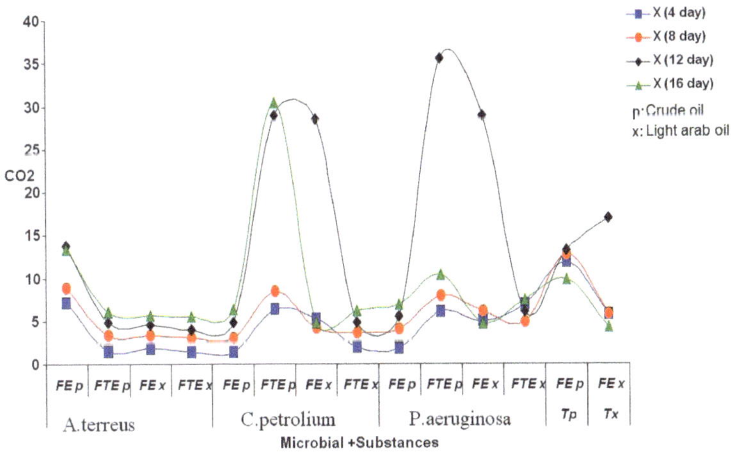

Fig 1. The concentration of CO2 (µl) released by *P.aeruginosa*; *C.petrolium* and *A.terreus*

7. The biodegradability test of DCPIP indicator

The biodegradability of the microorganisms was verified using the technique based on the rodox indicator 2,6-dichlorophenol indophenol (DCPIP) (Hanson et al 1993).

The principal of this technique is that during the microbial oxidation of the carbon source ,electrons are transferred to electron acceptors by incorporating an electron acceptor such DCPIP to the culture medium; the ability of the microorganisms to utilize the substrate by observing the color change of DCPIP from blue (oxidized) to colorless (reduced) This technique was used by pirollo et al 2008

The time to decolorization of the DCPIP indicator was registered of each microorganism Pseudomonas aerugenosa was 8hours .Candida petrolium 12 hours and Aspergillus terreus in 17 hours. we have noticed during the experiment no decolorization of the substrate controles (Without inoculums) or of the inoculums controls(without oils)was observed, similar results were found by junior et al 2009.

8. Discussion

From the graph we noticed a difference in rate of mineralization both have substance by the rate of CO_2 to release by *Pseudomonas aerugenosa .Candida petrolium* and *Aspergillus terreus,* it reaches 35.62 µl for *Aspergillus terreus.* after 12 days of incubation; according to Cerniglia (1992) that the metabolic way of degradation of Naphthalene by *Aspergillus terreus* and utilizes a dioxygenase which oxidizes one of the benzene cycles to form a cis-dihydrodiol. Clear mineralization is regular positive in the suspensions after one 12 days and 16 days period; the metabolic ways of degradation by stocks of *Aspergillus terreus* . Were the first studied ways and are much known. The contribution of oil biological breakdown causes significant increase in the rate of mineralization of carbon by report the, which rises with (28.99 µl crude oil; 28.6 µl for

Arabian light oil /12 days) for *Candida petrolium*, and (35.62 μl Oil; 28.99 μl Arabian light oil /12 days) for , and weak for *Pseudomonas aerugenosa* with (13.78 μl crude oil; 5.72 μl arabiant light/12 days) and this quantity CO_2 to release decreases with time before 12 days for *Aspergillus terreus* The oil carbon very quickly mineralizes by report Arabian light according to our results for the different microorganisms.In the 16 days have observed one followed by increase in the rate of CO_2 release by to 30.55 μl for (*Pseudomonas aeurogenosa*), and an absence of the release of CO2 for (*Aspergillus terreus* and *Candida petrolium*).

9. References

[1] Andrade, M.L., Reyzabal, M.L., Marcet, P., Montero, J.M., 2001. Industrial Impact on Marsh Soils at the Bahia Blanca Ria, J. Envir. Quality, 31, 532.

[2] Belhaj A., Elmerich C., Ismaili M., Zouhdi M., Hajjam Z., Assafi M., Alaoui M.A., 2000. Bacilles à gram positif : identification et potentiel de biodégradation des hydrocarbures. Biologie Infectiologie, VI (1).

[3] Boutefnouchet N., Bouzerna N. et Chettibi H., 2009. Biodégradation des hydrocarbures en eau de mer : Cas de la Naphta B. Scientific Study & Research, Vol VII : 1582-540X

[4] Cardoso da Silva J.R., G., Carlos, E., de Carvalho Lange, I., 2003. Hydrogeology of study of mangrove area around Guanabara bay, Rio de Janeiro, Brazil, Unuaro do Instito de Geociencias, 2003,26, 92.

[5] Cerniglia, C.E. (1992). Biodegradation of polycyclic aromatic hydrocarbons. *Biodeg.* 3:351-368.

[6] Hanson KG. Desal JD.Desai, AJ (1993) :Arapid and simple screening technique for a potential of crude oil degrading microorganisms ; Biothnol;Techniques; 7.745-746

[7] Junior,SJ ;Adreano,pm and Dejaniora(2009): Biodegradation of biodiesiel by C.viswanathii;African journal of biothechnology Vol 8(12);pp2774-2778.

[8] PirolloMPS.Mariano AP.Lovagio RB ;Costa SGVAO.Walter V,Housanmann R Contiero J (2008):Biosulfactant synthes by P. aeruginosa LBI isolatedfrom a hydrocarbon contaminated siteJ. Appl.Microbiol.106;1484-1490

[9] Soltani M., 2010. Distribution lipidique et voies métaboliques chez quatre bactéries Gram-négatives hydrocarbonoclastes. Variation en fonction de la source de carbone. These de doctorat université de paris 6.

[10] Sutherland, J.B; Raffi, F; Khan, A. A. and Cerniglia, C.E. (1995). Mechanisms of Polycyclic Aromatic Hydrocarbon Degradation. In Microbial Transformation and degradation of toxic organic chemicals. Edited by Young, L.L and C.E. Cerniglia. Wiley-Liss. New York.

[11] Waes G., 1971. La production d'acide carbonique par les ferments lactiques. Le Lait, 51, 123.

12

To Review Climate Change Effects on Basic Resources (A Case Study of These Effects on Zagros Forests)

Majid Loghmanpour[1], Zhirair Vardanian[2],
Hadi Kiadaliri[3], Mohsen Elahi[4]

1 Yerevan State Agrinean University, Armenia and Expert of Forest, Range & Watershed Management Organization

2 Dept. of Yerevan State Agrinean University, Armenia

3 Science & Research Unit of Azad University, Tehran, Iran

4 Tarbiat Modares University of Noor-Mazandarn, Iran

Abstract

Human civilization was accompanied with the pollution, destruction of natural resources and biodiversity. The era of industrialization begins with emissions of carbon dioxide and other gases into the atmosphere. As researchers believe average temperatures of earth has increased over 74 %. Global warming causes droughts, rising of sea water level , penetrating saline water into freshwater resources, melting of polar ice, increasing of desertification trend and tropical diseases outbreak. Loss of biodiversity, reduction in forest production, changing of the border of farmlands and forests on high-latitude and wasting of semi-persistent of forest species are other consequences of the global warming. The effects of this phenomenon can be clearly seen in Zagros forests. According to studies done,Zagros region warming has lead to prolonged droughts period in the region. Besides, dusty storms and their sediments on the leaves can result in trees' tension and therefore their physiologic weakness which at least cause trees being attacked by wood eater beetles. Using superseded energies such as sun , wind and water , protection of forests and forest plantation with species having high potential of carbon sequestration as well as standardizing the transportation vehicles for economizing energy are some effective strategies recommended for reducing the global warming effects.

Keywords: Global warming, carbon dioxide, biodiversity loss, deforestation

1. Introduction

The reduction of forest areas and their converting to deserts were the obvious impacts of human long settlement on the earth. This reduction was accelerated when men were living together and began to exploit the resources around him more and more. Nowadays, since this exploitation exceeds self- recovery of nature over the time, thus we witness more unbalanced situation than before. (Shamekhi, 2009). Continued environmental degradation resulted in more pollution and biodiversity loss On the other hand; public awareness leads to a movement for changing the current critical situation of natural resources and the issue of legislation and policy making to achieve more applied management regarding natural resources. Stockholm conference convened in June 5th1972 in Stockholm- the capital of Sweden – and attended by 1400 representatives from 113 countries offered new initiatives for protection of the environment. (Shamekhi, 2009). Twenty years later on June 5th 1992 and in Stockholm conference anniversary, a conference titled Environment and United Nations Development convened in Rio-de-Janeiro –Brasilia which addressed again the issue of environment protection. Simultaneously and immediately after the Rio Conference several international Conventions including the Convention of Climate Change were ratified. The convention was signed by majority of member countries participating in the conference, including Iran (Shamekhi 2009).

2. Methods and Procedures

The effects of climate change have been reviewed in this article, using library and documentary studies as well as descriptive method. Excel software was used for analyzing the statistics and preparing diagrams. For obtaining metrological dates in this research, height from sea level and meteorological data of all stations of western provinces were studied at first and then the station which its height was near the mean height of all stations and had most meteorological data was selected and analyzed.

2.1. Forest ecosystems and their impacts on global warming:

Warming will cause forests of northern hemisphere move ahead to north pole which this phenomenon will cause people resort to farming of premature species and breed their animals in lands which were forests before , so changing of border of agriculture and forest affects the conservation policies because of undesirable conditions of land. It is obviously that desirable situations of high latitudes can't neutralize the effects of malfunction of middle latitudes.(Koochaki, Sharifi , Zand

Year	Burned forests(ha)	The number of forest fires
1985	2233	75
1986	8426	79
1987	5407	10
1988	611	73
1989	407	116
1990	1133	16
1991	288	146
1992	3923	100
1993	33379	192
1994	6119	143
1995	1977	722

Table 1. Sever forest fires of IRAN

1998). The effects of global warming in lower latitudes in which temperature changes is less than the higher one , these effects occur in the form of the amount of humidity and its distribution. Changing of the place of natural growth of resistant forest species and loss of semi-resistant species are other impacts of global warming on forest ecosystems which result in abnormal growth of forest plants and reduction of production of primary and secondary forest products. Sea level rising will also result in degradation of mangrove.(Nezhat, Erfanifard 2009) . Meanwhile increasing of land temperature will lead early spring in the region and untimed melting of ices as well as drying of forest regions. Dried forest regions also increases contingency of forest fires. As it has been showed in table No. 1, the incidence of forest fires has been increased in Iran in 1985-1995.

2.2. Warming and increasing trend of desertification:

Another important and devastating effect of global warming is increasing of the desertification process. Although it is difficult to specify the accurate contribution of climate change in desertification , but undoubtedly its role in intensifying severity of desertification which taken place through deduction of vegetation cover and agricultural productions as well as soil degradation and underground waters falling makes difficult the accessibility of sustainable development. (Akbari, Nasseri and AshgarTousi, 2009).

3. A noticeable example of the effects of warming in Iran

Increasing of the temperature in late winter and early spring in Saudi Arabia Peninsula caused to increasing of the temperature of the air near the soil surface. This situation will result in turbulence and wind blowing in lower layer of atmosphere and therefore dust entrance in to it(Abbassi ,RafieiEmam and Roohipour 2008). Satellite images of Middle East reveal that Rub'Al Khali the vast sandy desert of the world is the origin of this phenomenon. Most violent sandstorms which originated from this desert pass the Persian Gulf and smash west provinces of the country. The risen dust covers the leaves of quarks sp. and its combination with humidity arisen from transpiration of the foliages makes them as a firm material which stick hardly to the surface of the leaves in such a way that even wind can't remove them. Besides the sand storm phenomenon ,rainfall reduction and droughts, conscious and unconscious man made degradation such as subsoil plough aiming at dry farming which cuts the roots and forfeit the existent insufficient moisture of the region as well as animal grazing which prevents oaks' regeneration are the main factors of the weakness of Zagros Forests. At present, Zagros Forests with 5500 antiquity and 5 million hectares area as the main source of the water of the country have been threatened by Wood Eater Beetles species (Azizkhani 2010). This is because drought tension results in the increasing of amino acid density of the plant and therefore pests uprising. As a general, water tensions have stunts the growth of the plant and therefore they are very sensitive to pests uprising. Moreover, the surface temperature of the plants under water tension usually is 2 °c to 4° c more than other plants and this itself leads to increasing of growth speed of the insects (Koochaki , Sharifi and Zand 1998). As mentioned before, flooding, water shortages, dusty storms, aging and persistency, lack of revitalization ,incomplete age pyramid are the elements of incidence of pests uprising and its intensification there to. (Azizkhani, 2010). In which, of course, climate change is the first chain of all mentioned factors of a cycle that man is the main reason himself.

Province	Temperature °C		Forests areas	
	Mean temp.till 80 decade	Mean temp. in 80 decade	Forest areas before 70 decade	Forest areas in 70 decade
Lorestan	16/8	17/18	880000	585537
Kohkilooyeh & Boyerahmad	15/12	15/56	992000	427044

Table 2. The increasing of mean temperature and reduction of forests areas in two western provinces of the country

4. Conclusion

- Today global warming is a world issues which its effects and a consequence affects all living creatures on the earth.
- The area of the world ′ forests is estimated to be about 9 / 3 billion hectares of which over 94 million hectares have been degraded during 10 years , 90% of this destruction in tropical regions.Forests degradation in Iran had also similar situation.
- According to studies increasing of temperature in Lorestan and Kohkilooyeh and Boyerahmad provinces has caused the reduction in forest areas.
- As it has been shown in table No. 2 the incidence of forest fires have been also increased abnormally in the country like other regions of the world in recent years.

5. Some strategies for declining global warming:

1. Minimizing the fossil fuels usage and replacing of other energies such as solar energy, hydrogen, waves, wind and nuclear energy;
2. Forest plantation and preventing the degradation of remaining forests;
3. To put more taxes on fossil fuels;
4. To use public transportation, standardizing the vehicles for saving energy;
5. To use utensils with more efficient energy consumption;
6. To absorb carbon dioxide in power stations for producing energy which is an effective method for decreasing global warming if buried correctly and don't add pollution in to the environment.

6. References

[1] A. Partovi. Iran's role in global warming . Jame Jam newspaper report,2010.

[2] Gh.AbdiNejad .T0 review the effective factors on desertification and drought .JangalvaMarta′ a quarterly published newsletter.2008, **78** : 8-10.

[3] H. Ahmadi , Surveying the effective factors on desertification. JangalvaMarta' a quarterly published newsletter. 2004 , **62** : 66 -70.

[4] H.R.Abbasi , A.RafieiEmam , and H. Roohipour .Analyzing the origin of dusts in Boushehr and Khozistan provinces using satellite images . JangalvaMarta' a quarterly published newsletter. 2008,**78**: 18-51.

[5] Akbari , and M.Naseri Ashgar Tousi . To review the effects of climate change on desertification. JangalvaMarta' a quarterly published newsletter. 2009, **83** : 48 – 52.

[6] M.R.Ardakani .Ecology, Tehran University Press, 2004.

[7] M. R,Arefipour .To review the reasons of drought of oaks forests of Zagros. Research Institute of Forest and Range. 2009.

[8] N.Rahimi.Climate Change and its Environmental Effects .Akhavan Press center, 2004.

[9] T.Shamekhi .Rules and Regulations and Management of Natural Resources (Forests and Ranges). Tehran University Press, 2008.

[10] T. Shamekh.Natural Resources Policies . Dr. Javanshir Education Center of Natural resources, Karaj, Iran, 2008.

13

Nitrate Removal from Water Using Synthesis Nanoscale Zero-Valent Iron (NZVI)

Shima Ziajahromi[1,*], Ali Daryabeigi Zand[1] and Meysam Khanizadeh[2]

1 Environmental Sciences Research Institute, Shahid Beheshti University, G.C., Tehran, Iran

2 Faculty of Nuclear Engineering and Physics, Amirkabir University, Tehran, Iran

* Corresponding author

Abstract

This study was conducted to investigate chemical reduction efficiency of nitrate by synthesis nanoscale zero-valent iron (NZVI) in aqueous solution, under aerobic condition. TEM image shows synthesis nano zero-valent iron has a size in the range of 40–150nm. Experimental results suggest that the reduction efficiency of nitrate decreased quickly with increasing initial pH value from 4 to 10 increased considerably with the increasing dosage of nanoscale zero-valent iron from 0.25 to $1gl^{-1}$ and did not vary much with initial nitrate concentration changing from 30 to 50 mg l^{-1}(NO_3^--N). With reductive denitrification of nitrate by nano zero-valent iron, the removal rate of nitrate reached 80% in 60 min with nano zero-valent dosage of $1.0gl^{-1}$ and pH_{in} 4, in room temperature.

Keywords: nitrate, water, iron nanoparticles.

1. Introduction

Nowadays, regarding to increasing demand on safe drinking water, removal of widespread pollutants such as nitrate is creating a significant challenge in water treatment industry. Anthropogenic sources such as nitrogen fertilizer, nitrogen pesticides and industrial waste effluent discharge account for most nitrate contamination of ground and surface waters [1]. Elevated nitrate concentrations in drinking water supplies present a potential risk to public health. In infants NO_3 is reduced to NO_2, which combine with hemoglobin in the blood to form methemoglobin leading to cyanosis in babies under six month old [2]. A research conducted by Mayo Clinic Center in Minnesota also showed that drinking tap water with a high concentration of nitrate would have a higher risk of causing bladder cancer and ovary cancer [3]. Therefore many countries have regulated the concentration of nitrate in drinking water. In the US, EPA established a maximum contaminant level (MCL) of 10mg/L NO_3-N for drinking water [4]. In Iran the regulatory thresholds for NO_3-N in drinking water sources are set as 10mg/L which is equivalent to 44.82 mg/L NO_3. Current technology to remove nitrate from water include ion exchange, reverse osmosis, biological denitrification and chemical reduction [3]. Among different water treatment methods, using nano materials such as nano zero-valent iron as a new method has a good potential for removal of nitrate. Researchers have studied the use of zero-valent iron in halogenated organics, azoaromatic nitroaromatics and the treatment of different kinds of compounds such as inorganic compounds like heavy metals [5]. In recent years, there has been a growing interest in the use of zero-valent iron in the treatment of water containing nitrate. Previous studies have demonstrated that nitrate could be completely reduced by metallic iron under anoxic and aerobic conditions; furthermore, the major reduction product was ammonia [6,7]. Nano zero-valent iron in contrast with iron powder have some advantages of high specific surface area, high active surface, easily being sca- tered and so on, which lead to the increasing denitrification rate of nitrate [5]. In this paper, we studied the parameters which affect on the effectiveness of nitrate removal by synthesis iron nano particles. These parameters include pH, dosage of nanoscale Fe° and nitrate concentration. The Purpose of this study is to improve our understanding of denitrifcation condition by nanoscale Fe.

2. Materials and Methods

2.1. Chemicals and materials

The following chemicals were purchased from Merck: $NaBH_4$ (for synthesis), $FeSO_4.7H_2O$ (98%), Methanol (99%), NaOH (99%), H_2SO_4 (98%), KNO_3 (98%). Nitrate reagent (Nitraver 5) was obtained from Hack company.

2.2. Method for nanoscale Fe^0 synthesis

In a typical synthesis of Fe° Nanoparticles by borohydride reduction, 4.0g of $FeSO_4{\cdot}7H2O$ was dissolved in 200mL of 30% methanol and 70% de-ionized water (v/v). The pH was adjusted to about 6.8 by 3.8M NaOH. Then 1.5 g of $NaBH_4$ powder was dissolved in 10mL de-ionized water and the solution was added incrementally to the mixture in ultrasonic shaker at 25ºC temperature for 45 min After addition of all of the $NaBH_4$ solution the mixture was stirred in jar test for another 45 min and then centrifuged for 15 min at 5000 rpm. The solid particles were washed at least three times with methanol and then dried for 4hr under vacuum condition, and then broken up with a spatula to form a fine black powder and immediately added to the aqueous solution to react with nitrate. The ferrous iron was reduced to zero-valent iron according to the following reaction Eq(1):

$$Fe\,(H_2O)_6 + 3BH_4^- + 3H_2O \rightarrow Fe^{\circ} + 3B(OH)_3 + 10.5\,H_2 \qquad (1)$$

2.3. Preparation of aqueous nitrate solution

Different concentrations of nitrate in aqueous solution were prepared by dissolving desired quantities of KNO_3 in de-ionized water. An initial concentration of 30mg/l NO_3-N (133 mg/l NO_3) was used for studying the effects of pH and dosage of iron nanoparticles, whereas 50mg/l NO_3-N (222 mg/l NO_3) also was used for studying the effect of nitrate concentration on removal effectiveness.

2.4. Experiments for chemical reduction of nitrate by iron

Five hundred milliliter of aqueous nitrate solution of a selected concentration was first put in each of glass beakers for each set of experiments. Freshly prepared nanosized ZVI at arbitrary concentration (1, 0.5 or 0.2g/L) was added to each glass beaker. Chemical reduction of nitrate by nanosized ZVI at ambient temperature, and desired pH(4,7,10) and/or desired nitrate concentrations (30, 50 mg/l) were simultaneously conducted in various glass beakers using a jar test apparatus at a mixing rate of 300 rpm. In each set of experiments the reaction vessels were removed one by one from the jar test apparatus at intervals of 10 min during the 60-min reaction. Periodically, 20ml of the aqueous solution passed through a 0.45 μm membrane filter to separate iron nanoparticles. The concentrations of unreacted nitrate (NO3) were determined by VIS spectrophotometer (VIS DR2800, HACH). The pH value was measured using pH meter (satorius-150).

2.5. Mechanism of denitrification by nanoscale Fe^0

In aqueous systems, zero-valent iron (Fe°) is readily oxidized to ferrous ion (Fe^{+2}) by many substances. under anaerobic condition, H^+ is the only electron acceptor that will be involved in the reaction. Therefore, the overall process of corrosion in anaerobic Fe°–H_2O system is described by the following reaction [6], Eq(2). But Under aerobic conditions dissolved oxygen would play a role of the electron acceptor in the cathodic half-reaction. In this case, the primary reaction yields only OH^- and not H_2 [8], Eq(3):

$$Fe^{\circ} + 2H_2O \rightarrow Fe^{+2} + H_2 + 2OH^- \quad (2)$$

$$2Fe^{\circ} + O_2 + 2H_2O \rightarrow 2Fe^{+2} + 2OH^- \quad (3)$$

The reaction and its mechanisms between nitrate and ZVI is a true redox reaction (Yang and Lee, 2005). Several studies have indicated the final products of chemical reduction of nitrate by ZVI could be N_2 or NH_3 depending on the experimental conditions [6,7,8]. But certainly, the main product of this reaction is ammonium [7], (Eq(4)):

$$4Fe^{\circ} + NO_3 + 7H_2O \rightarrow 4Fe^{+2} + NH_4^+ + 10OH^- \quad (4)$$

3. Results and Discussion

3.1. Characterization of nanoscale Fe^0

The particle size are determined by PHILIPS (EM208 S, the Netherlands), transmission electron microscopy (TEM) at 100 kV of acceleration voltage. Figure 1, shows TEM image of the synthesized nanoscale $Fe^{\underline{o}}$. The particles are spherical with the size ranging from 60 to 120 nm in diameter.

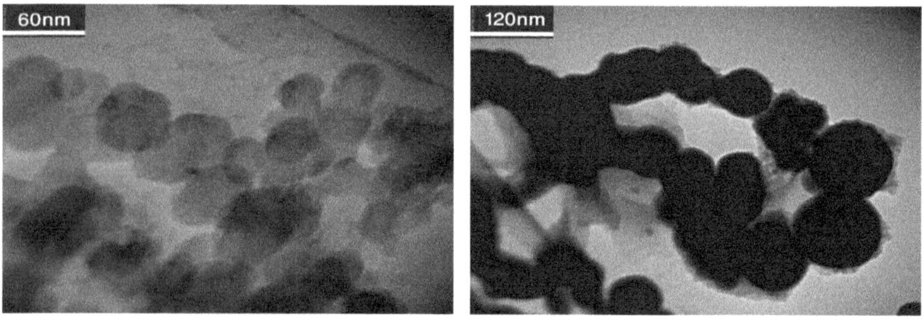

Fig 1. TEM images of synthesis nanoscale Fe° particles

3.2. Effect of Fe^0 dosage on nitrate reduction by nanoscale Fe^0

Fe° dosage is a significant variable parameter in nitrate reduction by nanoscale Fe°. Since the denitrifcation of nitrate by Fe° involves reaction at the metal surface, it was anticipated that the quantity of metal surface area should strongly influence the efficiency of nitrate reduction. In this study we used three different dosage of nanoscale Fe° (0.2, 0.5, 1 g/L). As shown in figure 2, with the dosage increasing, the removal efficiency of nitrate become higher and higher, In 0.2g/L Fe° dosage, after 60 min nitrate removal efficiency reached 57%, then increasing Fe dosage of 0.2 to 0.5, cause increasing efficiency to 70% ,and finally in 1g/L dosage after 60min, nitrate removal reached near 80% . Therefore with increasing Fe° concentration, metal surface area also increased, and efficiency of nitrate reduction will be increased [2,7,9].

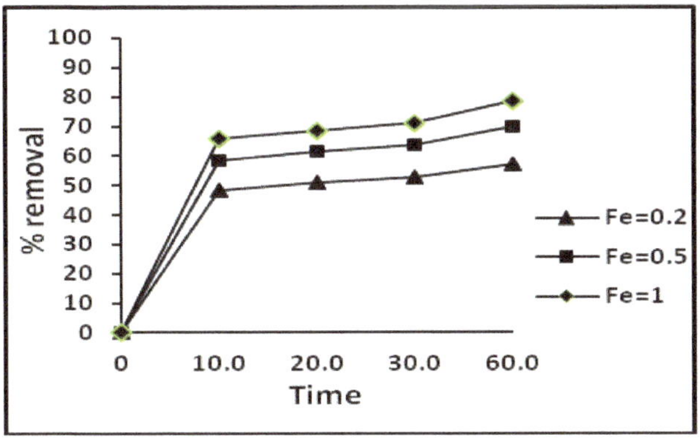

Fig 2. Effect of Fe° dosage on nitrate reduction by nanoscale Fe°, *T*=20°C, pHin= 4, C_0=30mg/l NO_3-N, stirring at 300rpm

3.3. Effect of pH value on the reduction of nitrate by nanoscale Fe^0

In this study three different pH value (4, 7, 10) were employed to study the effects of pH on nitrate reduction efficiency by nanoscale Fe°. Figure 3, shows the effect of initial pH value on the reduction of nitrate. The removal efficiencies of nitrate decreased with the increasing initial pH value. When the initial pH was 4, about 80% of nitrate was reduced in 60 min, while the removal efficiencies decreased to 70% and 64%, when the initial pH values were 7 and 10, respectively. This suggests that the reduction of nitrate could be well performed in acidic conditions. In fact, the reduction of nitrate proceeded on the surface of iron particles, lowing pH, would dissolve away ferrous hydroxide and other protective layers at the surface of nanosized Fe° yielding more fresh reactive sites for chemical reduction of nitrate. Therefore, Based on the experimental results obtained, in general, the efficiency of nitrate removal by nanosized ZVI increases as the system pH decreases. This finding is in accord with that of reported by Yang et al , Chen et al and Zhang et al [2,5,10].

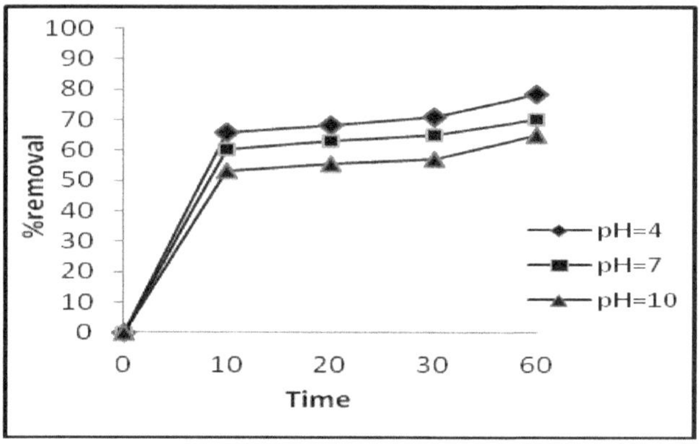

Fig 3. Effect of initial pH on nitrate reduction by nanoscale Fe°, T=20°C, Fe°= 1g/L, C_0=30mg/l NO_3-N, stirring at 300rpm

3.4. Effect of NO_3– concentration on nitrate reduction by nanoscale Fe^0

Two different initial nitrate concentrations (30, 50 mg/L NO_3-N) were employed to study the denitrification efficiency by nanoscale Fe^0. Figure 4, shows the effect of nitrate concentrations on the nitrate removal by nanoscale Fe^0 at initial acidic pH. The removal efficiency of nitrate for 30, 50 mg/L NO_3-N reached 78.3% and 79.98 respectively after60 min. Though the final removal efficiencies of nitrate for different initial nitrate concentrations were close to each other, the final efficiency increased with the increase of the initial nitrate concentration. This finding is in accord with Alowitz et al and Liou et al [11,12].

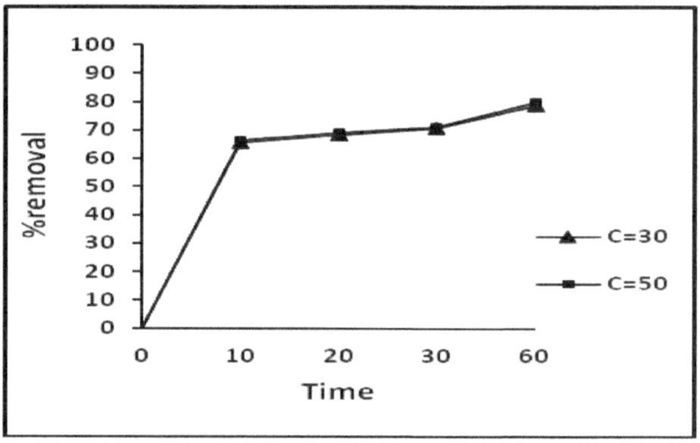

Fig 4. Effect of NO3–N initial concentration on nitrate reduction by nanoscale Fe°, T=20°C, Fe°= 1g/L, pH_{in}=4, stirring at 300rpm

4. Acknowledgements

This work was supported by the Environmental laboratory of Sciences Research Institute, Shahid Beheshti University. The author cordially appreciates the extensive and constructive comments made by Dr. Kassaee at Department of Chemistry, Tarbiat Modares University.

5. References

[1] A. Agrawal, and P.G. Tratnyek. Reduction of nitro aromatic compounds by zero-valent iron metal. Environ. Sci. Technol. 1996, **30**: 153–160.

[2] M. J. Alowitz, and M. M. Scherer. Kinetics of nitrate, and Cr (VI) reduction by iron metal. Environmental Science and Technology. 2002, **36**: 299–306.

[3] Y. Choe, Y. Young-Yang, and K.Y. Hwang. Kinetics of reductive denitrification by nanoscale zero-valent iron. Chemosphere .2000, **41**:1307–1311.

[4] Y. M. Chen, C.W. Li, and S. Sh. Chen. Fluidized zero valent iron bed reactor for nitrate removal. Chemosphere.2005, **59**:753–759.

[5] H-Y. Hu, N. Goto, and K. Fugie. Effect of pH on the reduction of nitrite in water by metallic iron. Water research.2001, **35**:2789–2793.

[6] C-W. Li, Y-M. Chen, and W-S. Yen. Pressurized CO2/zero valent iron system for nitrate removal. Chemosphere.2007, **68**: 310–316.

[7] Y. H. Liou, S. L. Lu, C. J. Lin , W. H. Kuan, and S. C Weng. Chemical reduction of an unbuffered nitrate solution using catalyzed and uncatalyzed nanoscale iron particles .Hazard Mater. 2005, **127**: 102-110.

[8] L. Lemaignen, C. Tong, V. Begon, R. Burch, and D. Chadwick. Catalytic denitrification of water with palladium-based catalysts supported on activated carbons. Catal Today. 2002, **75**: 43-48.

[9] C. Su, and R.W. Puls. Nitrate reduction by zerovalent iron: effects of formate, oxalate, citrate, chloride, sulfate borate, and phosphate. Environmental Science Technology.2004, **38** : 2715- 2720.

[10] G. C. C. Yang, and H. L. Lee. Chemical reduction of nitrate by nanosized iron: kinetics and pathways. Water Research.2005, **39** : 884-894.

[11] W.X. Zhang. Nanoscale iron particles for environmental remediation: an overview. Nanoparticle Resource. 2003, **5**: 323-332.

[12] J. Zhang, Z. Hao, Z. Zhang, Y. Yang, and X. Xu. Kinetics of nitrate reductive denitrification by nanoscale zero-valent iron. Process Safety and Environmental Protection. 2010, **150**: 7-12.

14

Rangeland Degredation and Its Impacts on Water Quality in Zayandehroud River Basin

Fatemeh Bateni, Sima Fakheran and Alireza Soffianian

Department of Natural Resources, Isfahan University of Technology, Isfahan, Iran

Abstract

The quality of water in many regions of world are threatened by overuse, misuse and pollution, and it is increasingly recognized that water quality of rivers, streams and wetlands are strongly influenced by landscape characteristics of their watershed including landscape composition (i.e. land use/land cover types and their fractions) in uplands and the spatial configuration of these land use/land cover types. This study focuses on the effects of land cover changes on the water quality of Zayandehroud River. The main goal of this study was to quantify the change in rangelands and forests in Zayandehroud river basin, which suffered intense human interference, in a period of eleven years (1997–2008) and to evaluate how landscape patterns (including Number of Patches, Edge Density, Percentage of Rangelands and Forests) influence on the water quality indices (including BOD5, EC, NO_3, P and TDS) measured in 10 stations along the Zayandehroud river. The results indicated that water quality were significantly correlated with both the proportions and configuration of Rangeland areas. Total edge of range land area had positive effects on water quality, especially on BOD_5 and EC. The proportion of rangeland was negatively correlated with water quality variables. Also PLAND and LPI metrics of range land had positive effect on decreasing nutrient (NO_3, PO_4) of water in this river. However, there was no significant correlation between water quality variables and proportion of Forest in Zayandehroud basin. Because Zayandehroud basin is located in a semi-arid area, where forests are very limited and are occurred only in small patches with low density. It was shown that degradation of range land lead in to degradation of water quality which highlights the importance of rangeland conservation in water quality management at landscape scale.

Keywords: Land cover change, Rangeland, Forest, Landscape metrics, Water quality, Zayanderoud River

1. Introduction

The quality of water in many regions of world are threatened by overuse, misuse and pollution, and it is increasingly recognized that water quality of rivers, streams and wetlands are strongly influenced by landscape characteristics of the watersheds including landscape composition (i.e. land use/land cover types and their fractions) in uplands and the spatial configuration of these land use/land cover types. Many studies have shown that composition and spatial arrangement of landscapes within watersheds can account for the variability of nutrient concentration in streams [2,4] Using landscape metrics for quantitative analysis of landscape pattern structure and its change have been widely adopted by landscape ecology researchers.

The main goal of this study was to quantify the change in rangelands and forest areas in Zayandehroud river basin in Isfahan province, Iran, in a period of eleven years (1997–2008) and to evaluate how landscape patterns influence on the Zayandehroud river water quality. Human activities such as urban development and intensification of agriculture, played an important role in the drastic change of rangeland and forest areas in recent decades, particularly in semi-arid areas in Iran. The rapidly increasing of water demand and water pollutions due to population growth, industrial and agricultural development around Zayandehroud river causing Zayanderoud water quality to severely downgrade over the past decades. Therefore, monitoring of Zayandehroud water quality is a critical issue, especially due to the concern that freshwater is a scare resource in this region of Iran. In this study we examined 1)whether there is a significant relationships between land cover changes and surface water quality in Zayandehroud watershed, 2) Whether landscape metrics are good indicators for predicting impacts of landscape structure on surface water quality, 3) which metrics can be more accountable in predicting water quality in the study area.

2. Methods

2.1. Study Area

The Zayandehroud River is the most important river in central Iran which stretches over a length of 400 km, originating in the Zardkouh Mountain and ending in the Gavkhooni swamp after passing through the city of Isfahan. The Zayandehroud River basin has an area of 41,500 square kilometres, and an average rain fall of 130 millimeters. There are 2,700 square kilometres of irrigated land in the Zayandehroud River basin, with water derived from the nine main hydraulic units of the Zayandehroud River, wells, qanat and springs in lateral valleys. The location of the study area is shown in Figure.1 [7].

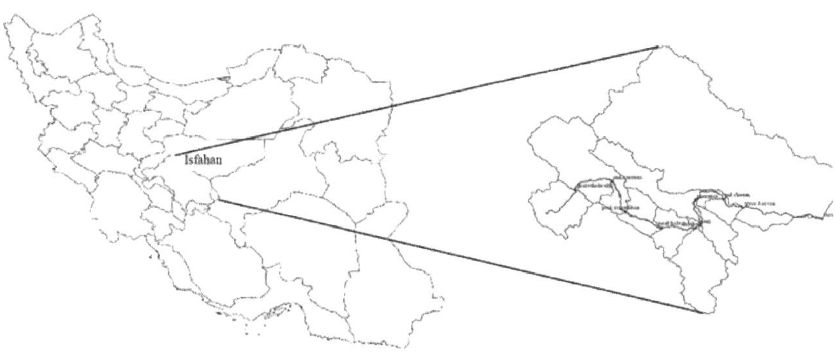

Figure 1. The location of Zayandehroud Basin in Iran

2.2. Data collection and preparation

Water quality such as BOD5, EC, NO_3^- and PO_4 data for 10 sampling stations along Zayanderour river in Ghale shahrokh, Tanzimi Dam, Zama khan bridge, Kalle bridge, Dizicheh, Lenj, Mousiyan, Choum bridge, Ziyar bridge and Varzaneh. For 1997 and 2008 were obtained from theWater Organization and Environment organization of Iran.

Maps of forests and range land for study area were prepared using hybrid classification of multi-temporal Landsat5 (ETM) images taken in September 1997 and 2008.

2.3. Watershed delineation

Identification and quantification of nonpoint source pollution in a large basin like Zayanderoud basin are logistically challenging. Therefore, the Zayanderoud basin was divided into 10 distinct sub basins based on elevation and available hydrographical data, using Arc-SWAT extension in ArcGIS 9.3 [1,8].

2.4. Quantifying landscape pattern changes

Changes of landscape pattern can be detected and measured by landscape metrics which quantified and categorized complex landscapes into identifiable patterns. Various metrics, including: Edge Density (ED), Largest Patch Index (LPI) and Percentage of Landscape (PLAND) were calculated using Fragstat 3.3 [9] to quantify the landscape patterns changes in 1997–2008.

2.5. Statistical analysis

Pearson correlation test and regression analysis was applied to assess the relationship between landscape indices and water quality parameters in R 2.7.12(R Development Core Team 2007).

3. Results

Table 1 shows the changes in landscape pattern which is calculated by landscape metrics in two census dates (1997 and 2008).

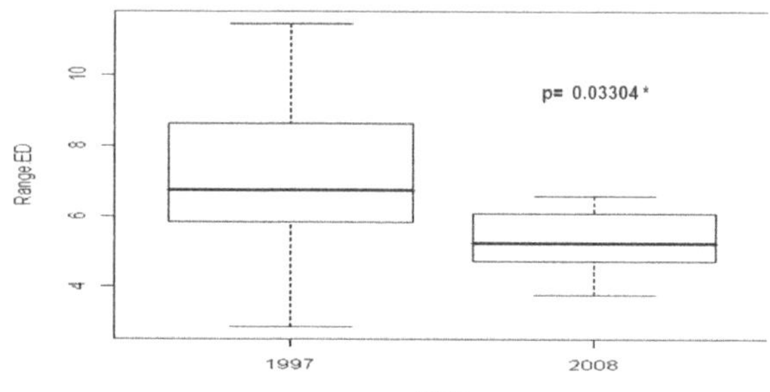

Figure 2. Rangeland Edge Density (ED) in 1997 and 2008. Bold horizontal lines show the median, boxes show the interquartile range, and the whiskers show the maximum and minimum values.

	1997	2008
Rangeland. CA	179678.8	135462.1
Rangeland.ED	7.055	5.266
Range.PLAND	65.703	45.802
Forest.CA	14784.6	13974.11
Forest ED	1.965	1.560
ForestPLAND	14.281	9.216

Table 1. Comparison of landscape metrics in 1997 and 2008 in Zayandehroud river basin

The increase of built-up land and bare land accounted for the most obvious transformation in the study area. This increase was traced to the conversion from rangeland, which experienced the most drastic decrease by 20% in Zayandehroud basin, in a period of eleven years (1997–2008). The results also indicated that water quality were significantly correlated with both the proportions and configuration of Rangeland areas. However, there was no significant correlation between water quality variables and proportion of Forest in Zayandehroud basin. Total edge of range land area had positive effects on water quality, especially on BOD_5 and Ec ($p<0.01$). In particular, concentrations of BOD_5 and Ec were more likely to be high when range land areas in watersheds were fragmented into smaller patches. These results suggest that unregimented large rangelands in watersheds might reduce the concentrations of BOD_5 and Ec in the river. Lee *et.al* also found a similar result about effect of range lands ED on water quality [4]. PLAND and LPI metrics of range land had also positive effect on decreasing nutrient (NO_3, PO_4) of water in this river ($p<0.05$). It was shown that degradation of range land lead in to degradation of water quality which indicated the importance of rangeland conservation [4, 11], Thus, human land uses might degrade water quality not only by transforming natural areas into urban or agricultural areas generating pollutants and nutrients, but also by degrading the quality of remnant range land patches in watersheds with fragmented and isolated range land patches.

4. Discussion and conclusion

Many studies have reported that forest and vegetation cover like rangeland and forests play primary roles in protecting water quality in adjacent aquatic systems [4, 11, and 13]. Results of this study also revealed that the degradation of range land lead in to degradation of water quality which highlights the importance of rangeland conservation in water quality management at landscape scale. However, a significant relationship between Forest areas and water quality was not observed in this study, because the Zayanderoud basin is located in a semi-arid area of Iran, where forests are very limited and are occurred only in small patches with low density. Thus, in semi arid areas like Zayanderoud watershed, range lands conservations play more important role for management of water quality. Results of this study can be used in establishing and implementing effective water management at landscape scale, in this region. In addition, the information on the hydrologic effects of land use can provide guidelines, not only for resource managers in restoring the aquatic ecosystem but also for policy makers in evaluating alternate land management decisions.

5. References

[1] Meghan, B. R. and M. JoAnn, 2008. Long-Term Effects of Changing Land Use Practices on Surface Water Quality in a Coastal River and Lagoonal Estuary ,*Environmental Management*, 44:505–523

[2] Cretaz, A. L., Barten, P. K. , *Land Use Effects on Streamflow and Water Quality in the Northeastern United States*, CRC Press

[3] Caryn, E., 2004. Land Conservation: A Permanent Solution for Drinking Water Source Protection, U.S. Geological Survey, Trust for Publication Land, *Water Protection Series*. 56p.

[4] Lee, S.W., S. Hwang and S.B. Lee, . 2009. Landscape ecological approach to the relationships of land use patterns inwatersheds to water quality characteristics, *Landscape and Urban Planning*,92:80–89

[5] Marker, J. and Y. Trevor, 2007. R. Fayed. Recognition of Biological Signal Mixed Based on Wavelet Analysis. In: Y. Jiang, et al (eds.). *Proc. of UK-China Sports Engineering Workshop*. Liverpool: World Academic Union, pp. 1-8.

[6] Griffith, J. A., 2005, Geographic techniques and recentapplications of remote sensingto landscape-water quality studies", *Water Air Soil Pollut*, 138: 181–197

[7] Salemi, H.R, 2004, An overview of the Hydrology of the Zayandeh Rud Basin, Iran.", *Water and Wastewater (Isfahan).* , 1(50): 2-13

[8] Royappan, M., 2001, "Survey of effect of change of land use to runoff by SWAT model in South Africa", *Kentucky university*, Vol. 482, pp

[9] McGarigal, K., 1995, *Fragstats: spatial pattern analysis program for quantifying landscape structure. General Technical Report PNW-GTR-351. USDA Forest Service, Pacific Northwest Research Station, Portland.*

[10] Olsen, A. R., 2009, *Introduction to R Statistical Software*, wetern ecology division, corvalis

[11] Tong, S.T and Y., Chen, 2002.Modeling the relationship between land use and surface water quality. J. Environ. Manage. 66 (4), 377–393.

[12] Collinge, S.K., 1996. Ecological consequences of habitat fragmentation: implications for landscape architecture and planning. Landscape Urban Plan. 36, 59–77.

[13] White, M.D. and K.A., Greer, 2006. The effects of watershed urbanization on the stream hydrology and riparian vegetation of Los Pe˜ nasquitos Creek, California. Landscape Urban Plan. 74, 125–138.

15

Flood Disaster Management in South Africa: Legislative Framework and Current Challenges

Bongumusa M. Zuma[1,2,*], Catherine D. Luyt[2],
Tatenda Chirenda[2], Roman Tandlich[2]

1 Institute for Environmental Biotechnology, Rhodes University, Grahamstown 6140, South Africa

*corresponding author

2 Environmental Health and Biotechnology Research Group, Division of Pharmaceutical Chemistry, Faculty of Pharmacy, Rhodes University, Grahamstown 6140, South Africa

Abstract

In South Africa, the annual risk of flooding is 83.3 % and the population vulnerability is high due to economic factors and geographical location. Before 1994 the Civil Protection Act No. 67 of 1977 governed disaster management, but its framework was inadequate as demonstrated by 104 deaths in Lainsburg floods of 1981. Thus a major push came towards improvement of institutional capacity and the legislative framework to deal with disaster management after 1994. The 1996 Constitution of South Africa defined the law-making powers and the responsibilities at the national, provincial and local levels of government. The Disaster Management Act No. 57 of 2002 constitutes the institutional capacity at all levels of government. Response to flooding occurring between December 2010 and February 2011 is used to examine the functionality and drawbacks of the current disaster management system. Impacts included damages to drinking water infrastructure, potential for cholera outbreaks and material losses. The response was adequate at the national level, but district municipalities struggled due to skills shortages and lack of disaster management structures. Remedial strategies are proposed using the current novel legislative tools. Research into vulnerabilities and risk must be strengthened.

1. Introduction

South Africa has semi-arid to arid climate and a total land area of 1.2 million square kilometres [1]. It spans between latitudes of 35° and 22° South [2] with a population of 50.5 million inhabitants [3]. Around 38 % of the population is concentrated on 2 % of land area in mainly urban centres [4] and the growth of the urban population places excess pressure on public services [4]; [5]. This potentially decreases disaster resilience of households [6]. The country ranks among the bottom 30 nations from around the world with respect to the population's ability to provide satisfactory food and shelter at the household level and the Health Adjusted Life Expectancy stands currently at 48 years [7]. Vulnerability of the population is further indicated by the minimum living level [8]. Using the dependency ratios, the highest vulnerability is found in provinces of the Eastern Cape, KwaZulu-Natal, the North-West and Limpopo [6]. There have been 77 flood disaster events in South Africa between 1980 and 2010 [9]. A total of 1068 lost their lives in floods with the maximum of 506 in 1987 [9]. Based on these statistics, the risk of a flood occurring in a given year can be calculated at 83.3 %. In the context of climate change, the above mentioned facts pose problems in disaster management. Therefore the current legislative framework, examples of responses to recent disaster and suggestions for improvement in flood disaster management are presented in this paper.

2. National Disaster Management System of South Africa

2.1. Legal Framework

Prior to 1994, disaster management activities were governed by the Civil Protection Act No. 67 of 1977 [10]. However, its framework proved inadequate as demonstrated by 104 deaths in Lainsburg floods of 1981 [11]. This resulted in the legislative and organisational efforts for system improvement and integration in disaster management. The 1996 Constitution of South Africa in Part A Schedule 4 defines that the role of the national, provincial governments and local governments [12]. The first integrated policy on disaster management was the Green Paper on Disaster Management which was published as an extension of the Civil Protection Act [13]. Next was the White Paper on Disaster Management which enshrines the cooperative, proactive and integrated approach to disaster risk management, training and awareness through participation of all relevant stakeholders [14]. Specific goals of the White Paper included creation of the National Disaster Management Centre, improvement of disaster prevention in the poor and disadvantaged areas, creation of an adequate funding system; and information channels to communities [14]. The guiding principles of disaster management were then summarised and responsibilities defined in the Disaster Management Act No. 57 of 2002 [15].

According to the DMA, President of the Republic of South Africa establishes the Intergovernmental Committee on Disaster Management (IGCDM) as stated in chapter 2 section 4 paragraph 1a [15]. According to chapter 1 section 3, the President appointed the Minister of Cooperative Governance and Traditional Affairs to be in charge of the disaster management at the national government level [15]. Other members of the IGCDM include the National Ministers of Water and Environmental Affairs, Health, Defense, Finance, Presidency, Justice and Constitutional Development; Defence; Education; Police; Provincial Ministers in charge of disaster management; representatives of the South African Local Government Association (SALGA). The IGCDM is accountable to the cabinet which it advises on disaster management according to chapter 2 section 4 paragraphs 2 and 3 [15]. At present, a surrogate, namely Joint Disaster Management Committee covers the main responsibilities of the IGCDM [16]; [17].

The National Disaster Management Centre (NDMC) is established in terms of chapter 3 part 1 section 8 of DMA [15]. Its functions include monitoring of disasters, mobilisation of resources and coordination and response to disasters; maintaining a repository of information on disasters, and database of relevant stakeholders (see chapter 3 section 15 paragraph 1 of DMA) [15]. Section 20 in chapter 3 describes that role of the NDMC in prevention and mitigation of disasters [15]. The NDMC is required to publish annual reports on its activities and submit this to the Minister, who passes them onto the parliament within 30 days (according to chapter 3 part 1 section 24 of

DMA) [15]. All organs of state must submit disaster management plans to the NDMC according to the guidelines developed by the centre (see chapter 3 part 1 section 19 of DMA) [15]. Regular reviews and updates of these plans are mandatory and any update must be communicated to the NDMC [15]. The NDMC is preparing the GIS-linked and web-accessible indices to predict risks from floods.

Role of the Provincial government in disaster management by chapter 4 of the act; and it generally mirrors the structure and responsibilities of the national structures inside the particular province [15]. Premier of a given province puts one Provincial Minister in charge of disaster management in that province (see chapter 2 section 4 paragraph 1b; DMA, 2002). At the municipal level of local government, disaster management is governed by chapter 5 of DMA [15]. District municipalities must have a disaster management committee/centre, while local municipalities should appoint a disaster officer [12]. As with the provincial structures, the municipal disaster management structures and their responsibilities are analogical to the national counterparts, but have different jurisdictions. As of March 2011, most of the district and local municipalities from around South Africa have not yet implemented disaster management structures as mandated by the DMA [12]. The National Department of Water Affairs (DWA) is responsible for the management of the water resource over 49 % of the total land mass of South Africa, with special attention paid to the Vaal and Orange River systems [18]. Coordination of the activities during flood disasters takes places in a central command centre at the DWA, i.e. the Flood Room [18]. Management of flooding and overall quality of water resources has been conducted by the governmentally-funded organisation such as Working for Wetlands and Working for Fire [19]. These decrease the risk of disease outbreaks during flooding.

2.2. Involvement of Civil Society and NGOs

National Disaster Management Advisory Forum (NDMAF) is convened and appointed by the Minister in term of chapter 2 section 5 of the Disaster Management Act [15]. Mandatory members of the NDMAF include the Head of the NDMC who serves as Chair, high-ranking officials from all relevant government departments which are part of the IGCDM, municipal officials chosen by SALGA and representatives of provincial departments responsible for disaster management (see chapter 2 section 5 paragraph 1) [15]. Other members can be appointed by the National Minister of Cooperative Governance and Traditional Affairs from the Chamber of Mines, the business community, trade unions and NGOs (see chapter 2 section 5 paragraph 1e) [15]. The NDMAF provides a platform for input from all stakeholders; and serves as a consulting and advisory panel to the Intergovernmental Committee on Disaster Management. This mandatory forum has been in existence since 2007 [20]. The national government published the Disaster Management Guidelines for Municipalities as the Government Gazette Notice as No. 1689 of 2005 [21]. These guidelines allow local government to set up volunteer units to assist with disaster management and govern the rules and scope of possible duties [21]. These pieces of legislation and structures defined in them provide an avenue for the participation of the NGOs in disaster management.

3. Flood disaster response in South Africa

Between December 2010 and February 2011, widespread floods hit South Africa. The total damages were estimated at approximately 1.1 billion USD [22] with 103 fatalities occurring [16]. The particular impacts included the following [18]: flooding of water pumping infrastructure in the provinces of the Free State, North-West and Kwazulu-Natal, waterborne diseases in the North-West Province, washing away of pump motors in the Free State Province, blockage in the pumping of raw water for treatment into the water treatment works in the Free State Province and the possibility of the cholera outbreak in the Limpopo Province. Dams at the main river systems in the country were reported to be 92-115 % of their capacity [18]. To deal with the floods and their impacts, the National Minister of Water and Environmental Affairs released 488 000 USD from the Ministry's Emergency Relief Fund to effectively manage dam levels; raise awareness and to funds to relief operations [22]. To address the humanitarian needs, the National Department of Social Development provided 43.2 million USD from the Social Relief of Distress fund, the Emergency Relief Fund, the Disaster Relief Fund and the National treasury allocation [22].

The Flood Assistance Scheme was activated by the Department of Agriculture, Forestry and Fisheries; and 3.4 million USD was released for damage assessment and agricultural relief to farmers [17]. Departments of Public Works and Defence and Veterans Affairs built 12 low-cost bridges [22]. The IDT Programme has allocated 31.2 million USD for building of mud schools and road repair round the disaster affected areas [22]. The National Government Disease Outbreak Monitoring by the Department of Health was used to monitor drinking and surface water quality, while 18.8 million USD spent on emergency housing by the National Department of Human Settlements [22]. Therefore the national government response was deemed adequate. However, research into understanding the risks of flooding is still limited and must be strengthened [23]; [24]. Focus must be placed on investigating the risks and vulnerabilities during storm surges/ coastal flooding [25] and implications from the acid-mine drainage [23]. Dissemination of the disaster-related information must also be improved at the national level [25].

Major problems exist in disaster management at the local government level [16]. In 2011, 50 % of local municipalities in South Africa lacked the disaster management structures, while 68 % of local and 25 % of district municipalities did not have the disaster management advisory forums [12]. Disaster management roles were often assigned to existing civil defense structure in local municipalities [12]. Thus the fire-and-rescue services and police become overburdened, i.e. diminishing the efficiency of the flood response. At the same time, the stakeholder involvement was also not possible due to the lack of the advisory forums and volunteer units. Constitution of the mandatory disaster management structures must be made a priority by provincial and national governments. Training of local officials by SALGA and other relevant institutions will also have to be strengthened. Where this is not feasible the disaster management advisory forums and volunteer units should be established and made operational as a priority. Local municipalities have been urged to implement a variety of early warning systems for disasters [26].

4. References

[1] South African General Information (SA Geo, 2011). Available at: http://www.southafrica.info/about/geography/geography.htm (website accessed 9th November 2011).

[2] Engelbrecht, F., Landman, W. (2010). Current Climate. In: *South Africa risk and vulnerability atlas;* the South African National Department of Science and Technology Eds., CPD Print: Pretoria, South Africa, pp. 2-5.

[3] Statistics South Africa. (SSA, 2011). Mid-year population estimates 2011. Statistical release P0302, Statistics South Africa, Pretoria, South Africa, pp. 18, 2011.

[4] South African Cities Network, the Department of Provincial and Local Government and The Presidency. (SACN, 2009). National Spatial Trends Overview. Available at: http://www.sacities.net/what/programmes areas/inclusive/spatial/projects/659-spatial (website accessed on 30th April 2012).

[5] Van Huyssteen, E., Oranje, M., Robinson, S., Makoni, E. (2009). South Africa's City Regions: A Call for Contemplation… and Action. *Urban Forum* 20(2): 175-194.

[6] le Roux, A.; van Huyssteen, E. (2010). Socio-Economic Landscape: The South African Socio-economic and Settlement Landscape, In: South Africa risk and vulnerability atlas; the South African National Department of Science and Technology Eds., CPD Print: Pretoria, South Africa, pp. 15-21.

[7] The Legatum Prosperity Index (LPI, 2011). Available at: http://www.prosperity.com/country.aspx?id=ZA (website accessed 9th November 2011).

[8] The Presidency of South Africa (2006). National Spatial Development Perspective. *The Presidency,* Pretoria, South Africa.

[9] Prevention Web. (PRW, 2011). South Africa - Disaster Statistics. Available at: http://www.preventionweb.net/english/countries/statistics/?cid=160 (website accessed 10th November 2011).

[10] Civil Protection Act No. 67 of 1977 (Civil Protection Act, 1977). Enacted by Parliament of the Republic of South Africa, Cape Town, South Africa.

[11] South African Weather Disaster Information System. (SAWDIS, 2009). The Laingsburg Flood Disaster: January 1981. Available at: http://saweatherobserver.blogspot.com/2009/10/laingsburg-flood-disaster-january-1981.html (website accessed on 10th November 2011).

[12] South African Local Government Association (SALGA, 2011). Disaster Risk Management Status Assessment at Municipalities in South Africa, Pretoria, South Africa.

[13] Green Paper on Disaster Management in South Africa (GPDM, 1998). Available at: http://www.info.gov.za/view/DownloadFileAction?id=68922 (website accessed on 10th November 2011).

[14] South African Government Gazette Notice no. 23 of 1999 (WPDM, 1999). White Paper on Disaster Management. Pretoria, South Africa.

[15] Disaster Management Act no. 57 of 2002. Available at: http://www.mangaung.co.za/Legal-Services/Documents/Disaster%20Management%20Act.pdf (website accessed on 10th November 2011).

[16] Portfolio Committee on Cooperative Governance and Traditional Affairs (2011). Flooding Disaster Management: December 2010/ January 2011: Update by Deputy Minister of Cooperative Governance. *Minutes from the committee meeting held on 28th February 2011,* Parliament of South Africa, Cape Town, South Africa.

[17] Portfolio Committee on Cooperative Governance and Traditional Affairs (2011). Flood Disaster Management; Local Government Municipal Systems Amendment Bill: Departmental briefing. *Minutes from the committee meeting held on 25th January 2011,* Parliament of South Africa, Cape Town, South Africa.

[18] Maswuma, L. Z. (2011). Management of Floods / Disaster. *Presentation to the Portfolio Committee: Water & Environment*, Meeting took place on 21st February 2011, Parliament of South Africa, Cape Town, South Africa.

[19] Portfolio Committee on Water and Environmental Affairs (2011). Department of Environmental Affairs 2011 Strategic Plan continued: Meeting minutes. *Meeting held on 12th April 2011*, Parliament of South Africa, Cape Town, South Africa.

[20] National Disaster Management Centre (2007). Annual report 2006-2007. *National Disaster Management Centre*, Pretoria, South Africa.

[21] Department of Local and Provincial Government (DMR, 2005). The Disaster Management Regulations: Disaster Management Act 57 of 2002. *Government Gazette Notice no. 1689 of 2005*, Department of Local and Provincial Government, Pretoria, South Africa.

[22] Africa, E. (2011). Presentation to the Portfolio Committee: Update on the Management of Recent Disaster Incidents. Presentation given the Parliamentary Portfolio Committee on Cooperative Governance and Traditional Affairs. Presented on 3rd March 2011, Parliament of South Africa, Cape Town, South Africa.

[23] Portfolio Committee on Water and Environmental Affairs (2011). Acid Mine Drainage: hearings (day 1). Meeting minutes. *Meeting held on 20th June 2011*, Parliament of South Africa, Cape Town, South Africa.

[24] Portfolio Committee on Water and Environmental Affairs (2011). White Paper on Climate Change: Public hearings. *Meeting held on 14thNovember 2011*, Parliament of South Africa, Cape Town, South Africa.

[25] Portfolio Committee on Water and Environmental Affairs (2011). Climate Change Green Paper 2010: Public Hearings Day 1. Meeting minutes. *Meeting held on 2nd March 2011*, Parliament of South Africa, Cape Town, South Africa.

[26] Portfolio Committee on Water and Environmental Affairs (2011). Climate Change White Paper: Department of Environment Affairs briefing. Meeting minutes. *Meeting held on 17th October 2011*, Parliament of South Africa, Cape Town, South Africa.

16

A Probabilistic Model of Rainfall-Induced Shallow Landslides

Ali Talebi

Faculty of Natural Resources, Yazd University, Yazd, Iran

Abstract

Shallow land sliding is a stochastic process, and understanding what controls the return period is crucial for risk assessment. In this paper, we present the new probabilistic model to describe the long-term evolution of colluvial deposits through a probabilistic soil mass balance at a point. Further building blocks of the model are: an infinite-slope stability analysis; a more realistic description of hollow hydrology (hillslope storage Boussinesq model, HSB); and a statistical model relating intensity, duration, and frequency of extreme precipitation. Long term analysis of shallow landslides by the presented model illustrates that all hollows show a quite different behavior from the stability view point. In hollows with more convergence, landslide occurrence is limited by the supply of deposits (supply limited regime) or rainfall events (event limited regime) while hollows with low convergence degree are unconditionally stable regardless of the soil thickness or rainfall intensity. Overall, our results show that in addition to the effect of slope angle, plan shape (convergence degree) also controls the subsurface flow and this process affects the probability distribution of landslide occurrence in different hollows.

Keywords: probabilistic model, shallow landslides, complex hollows

1. Introduction

Recently D'Odorico and Fagherazzi [1] have presented a probabilistic model of rainfall-triggered shallow landslides in hollows and showed that landslide frequency is linked to the rainfall intensity-duration-frequency characteristics of the region. They developed a stochastic model that computes the temporal evolution of regolith thickness in a hollow and hollow hydrologic response to rainfall based on a steady-state kinematic wave model for subsurface flow. In this research, we will use some elements of this model (stochastic soil mass balance) to simulate the soil production (colluvial deposit) and soil erosion (landslide) in time for hollows with complex shapes. Although our model is similar to that presented by D'Odorico and Fagherazzi [1] in that it is a probabilistic model of rainfall-induced shallow landslides, there is an important difference. Convergent plan shapes or concave profile curvatures cause the kinematic wave model to perform relatively poorly even in steep slopes (Hilberts et al., [2]). Troch et al. [3] observed that hillslope plan shape rather than mean bedrock slope angle determines the validity of the kinematic wave approximation to describe the subsurface flow process along complex hillslopes. Therefore, incorporating a more realistic description of hollow hydrology in the stochastic landslide model is needed, as hollows are generally convergent and hollows with more convergence have more potential for landslide occurrence.

To relax the KW assumptions, in this paper we substitute the linearized steady-state HSB model in the work of *D'Odorico and Fagherazzi* [1] for complex hollows (hollows with different length, slope angle and convergence degree). In fact, using an exponential width function, hollows with different convergence degree is presented and then for each hollow the critical soil depth, the minimum value of landslide-triggering saturated depth and the minimum rainfall intensity needed to trigger a landslide along hollow length are computed. Moreover, the temporal evolution of colluvium thickness is studied through a stochastic soil mass balance. Therefore, by considering the soil production function and hydrology condition in the different hollows, stability of each hollow is analyzed by the infinite slope stability method. Finally, the generalized model helps to investigate the relation between rainfall characteristics (intensity and duration), water table depth and slope stability of colluvial deposits in complex hollows.

2. Model formulation

2.1. Hollow geometry

We consider only hollows with moderate to steep slopes and shallow, permeable soils overlying a straight bedrock where subsurface storm flow is the dominant flow mechanism. Shallow soils are most prone to rain-induced landslides. It is assumed that the plan shape of the hollow can be described using an exponential width function:

$$w(x') = w_0 e^{ax'} \Rightarrow A(x') = \frac{w_0}{a}\left(e^{aL'} - e^{ax'}\right) \tag{1}$$

where w is the hollow width (deposits) along the X' direction, X' is the distance from the outlet of hollow parallel to bedrock), w_0 is the hollow width at the outlet, A is the hollow area, L' is the hollow length and a is a plan shape parameter. Allowing this plan shape parameter to assume either a positive, zero, or negative value, one can define several basic geometric relief forms: $a>0$ for convergent, $a<0$ for divergent and $a=0$ for parallel shapes. As hollows are generally convergent, we will assume a wide range of positive numbers for convergent hollows.

As the purpose of this study is to investigate the effect of hollow geometry and hydrology on landslide probability, we employ the subsurface flow similarity parameter for complex hollows proposed by *Berne et al.* [4]. This dimensionless parameter, the hillslope Péclet number, is defined for subsurface flow as the ratio between the characteristic diffusive time and the characteristic advective time, taken from the middle of the hillslope:

$$Pe = \left(\frac{L'}{2pD'}\right)\tan \mathrm{b} - \left(\frac{aL'}{2}\right) \tag{2}$$

where p is a linearization parameter, D' is the soil depth and β is the bedrock slope angle. As can be seen, Pe is a function of three independent dimensionless groups: $L'/(2pD')$, $\tan\beta$ and ; $aL'/2$; $L'/(2pD')$ represents the ratio of the half length and the average depth of the aquifer (related to the hollow hydrology), and $\tan\beta$ and $aL'/2$ define the hollow geometry.

2.2. Hollow stability Hollow stability

In this study the slope stability model is based on a Mohr-Coulomb failure law applied to an infinite planar slope. The failure condition can be expressed as (e.g. *Montgometry and Dietrich* [5], 1994; *D'Odorico and Fagherazzi*, [1]):

$$g_{sat} D' \sin b = c_t + \left(g_{sat} D' \cos b - g_w h' \cos b j\right) \tan \tag{3}$$

where γ_{sat} and γ_w are the specific weights of saturated soil and water respectively, β is the bedrock slope angle, ϕ is the soil repose angle, c_t is the soil cohesion and h′ is the saturated water depth, with both h′ and D′ (deposit thickness) being measured perpendicularly to the bedrock.

By solving Equation (3) for h′ the minimum value of landslide-triggering saturated depth (h_{cr}) can be obtained as (*D'Odorico and Fagherazzi*, [1]):

$$h_{cr} = \frac{g_{sat}}{g_w} D' \left(1 - \frac{\tan b}{\tan j}\right) + \frac{c_t}{g_w \tan j \cos b} \tag{4}$$

When the soil depth (D′) is equal to h_{cr}, the critical soil depth or immunity depth (D_{cr}) is given as follows

$$D_{cr} = \frac{c_t}{g_w \tan j \cos b + g_{sat} \cos b \left(\tan b - \tan j\right)} \tag{5}$$

In the case of relatively steep slopes ($\beta > \phi$), h_{cr} decreases linearly (i.e. stability decreases) with an increase of soil depth D′ (see Equation (4)). The soil depth D_{max} for which shallow landsliding can occur without saturated throughflow (corresponding to $h_{cr}=0$) is (*Iida*, [6]):

$$D_{max} = \frac{c_t}{g_{sat} \cos b \left(\tan b - \tan j\right)} \tag{6}$$

2.3. Hollow Hydrology

Hillslope hydrological response has traditionally been studied by means of hydraulic groundwater theory (*Troch et al.*, [3]). In many regions, groundwater flow is the main source of streamflow between rainfall events. The basic macroscopic equation describing the movement of water in the soil is known as the three-dimensional Richards' equation.

Troch et al. [3]) reformulated the continuity and Darcy equations in terms of storage along the hillslope, which leads to the hillslope storage Boussinesq (HSB) equation for subsurface flow in hillslopes. Extending Brutsaert's [7]) analysis, they linearized this equation as:

$$\frac{\partial S'}{\partial t} = K \frac{\partial^2 S'}{\partial x'^2} + U \frac{\partial S'}{\partial x'} + Nw \tag{7}$$

with $K = \frac{k_s pD' \cos b}{f}$ and $U = \frac{k_s \sin b}{f} - aK$ where S′ is the subsurface saturated storage, N is the recharge to the ground water table, k_s is the saturated hydraulic conductivity and f is the drainable porosity (note that the value of p is determined iteratively as pD′ should be equal to the average water table height $\int_0^{L'} S'(x')dx'/(Af)$ where A is the hollow drainage area).

According to the definition of the storage S′, the mean groundwater table height (over the hill-slope width) is:

$$\overline{h}'(x') = \frac{S'(x')}{fw(x')} = \frac{Ne^{-ax'}}{af}\left[\frac{e^{aL'}}{U}\left(1 - e^{\frac{U}{K}x'}\right) + \frac{1}{(Ka+U)}\left(e^{\frac{U}{K}x'} - e^{ax'}\right)\right] \tag{8}$$

Again, for parallel hillslopes this reduces to:

Now, we can obtain the maximum groundwater table depth in each hillslope (which is critical for landslide occurrence):

$$\overline{h}'(x'_m) = \frac{N}{fa(aK+U)}\left\{e^{aL'}\left[1 + \frac{U}{aK}\left(1 - e^{-aL'}\right)\right]^{-\frac{aK}{U}} - 1\right\} \tag{9}$$

Equating $\overline{h'}(x'_m)$ and h_{cr}, the critical rainfall intensity for triggering landslides (R_{cr}) can now be calculated as:

$$R_{cr} = \frac{h_{cr} fa\,(aK + U)}{\left\{e^{aL'}\left[1 + \frac{U}{aK}\left(1 - e^{-aL'}\right)\right]^{-\frac{aK}{U}} - 1\right\}} \tag{10}$$

3. Results and discussion

Figure 1 shows long term simulations of deposit thickness evolution in the four hollows (from top to bottom) and illustrates how shallow landsliding occurs when the soil thickness (D′) ranges between D_{cr} and D_{max}. The left figure shows the time series of deposit thickness for the HSB model. As can be seen, this figure1 (left) indicates how, as a function of the hollow geometry from steep slopes (top) to gentle slopes (bottom), the landslide probability is changed as well. For instance in hollow (where $T_r >> T_{im}$), landslides never occur and the system can be termed "unconditionally-stable".

Figure 1 (right) illustrates the probability distribution of colluvium thickness when a landslide occurs as simulated by the HSB model (right column) (D_{slide}). This histogram shows that the different hollows have different distributions of scar depth. As can be seen, the probability of distribution of D_{slide} is concentrated close to the immunity depth (D_{cr}) for the supply-limited case, whereas it is concentrated at significantly larger depths for the event-limited cases.

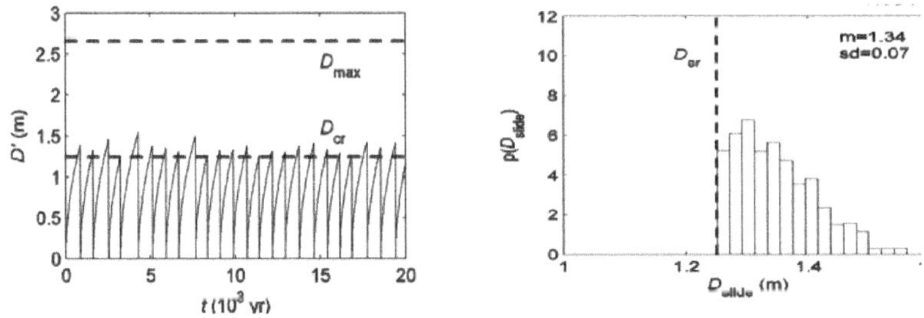

Figure 1. Long term simulation of deposit thickness (left) and Probability distribution of scar depth (colluvium thickness when a landslide occurs) for hollows (right).

Figures 2 (left) and (right) indicate how the probability distributions of the interarrival of the landslide-producing rain events (T_{slide}) and the corresponding rainfall intensities (R_{slide}) vary for the different hollows. These results show that in hollow (which has less convergence and a larger area), T_{slide} is close to T_{im} (supply limited regime)., while in other hollows (which has more convergence and a smaller area), TS_{lide} moves in the direction of T_r (event limited regime).

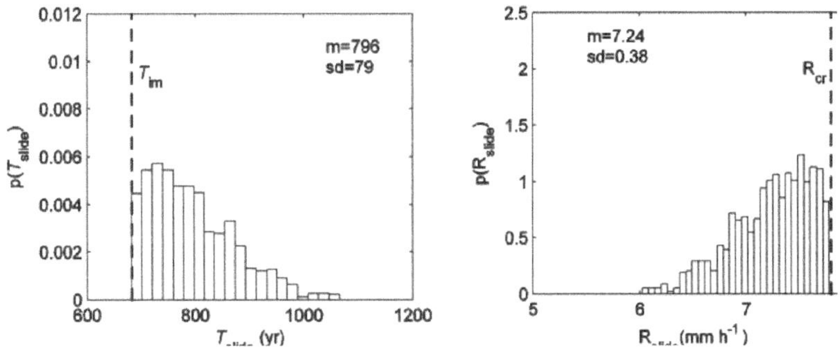

Figure 2. Probability distribution of landslide return period (left) and probability distribution of the landslide triggering rainfall intensity for hollows.

4. Conclusions

The following conclusions can be drawn from our rainfall-induced landslide stability analysis in response to deposit thickness evolution in complex hollows:

(i) Although shallow landslides in hollows are mainly triggered by high rainfall intensities, deposit thickness also plays an important role in stability.

(ii) With other site variables constant, shallow landslides usually occur when the soil depth (deposits thickness) is between D_{cr} and D_{max}.

(iii) Given a deposit thickness, for each hollow there exists a critical rainfall intensity leading to the highest water table and subsequent landslide occurrence.

(iv) In general, when convergence degree of hollows increases, the time period between land slides (T_{slide}) decreases. This means that hollows with more convergence degree are generally more susceptible to landsliding.

(v) In addition to the effect of slope angle, plan shape also controls the subsurface flow and this process affects the probability distribution of landslide occurrence in complex hollows and should be considered in hollow stability analysis.

5. References

[1] D'Odorico, P., and S. Fagherazzi (2003), A probabilistic model of rainfall-triggered shallow landslides in hollows: A long-term analysis, Water Resour. Res., 39: 1262. doi: 10.1029/2002WR001595.

[2] Hilberts, A., E. Van Loon, P. A. Troch and C. Paniconi (2004), The hillslope-storage Boussinesq model for non-constant bedrock slope, J. Hydrol., 291, 160-173.

[3] Troch, P. A., C. Paniconi and E. Van Loon (2003), Hillslope-storage Boussinesq model for subsurface flow and variable source areas along complex hillslopes: 1-Formulation and characteristic response, Water Resour. Res., 39(11):1316, doi: 10.1029/2002WR001728.

[4] Berne, A., R. Uijlenhoet and P. A. Troch (2005), Similarity analysis of subsurface flow response of hillslopes with complex geometry, Water Resour. Res., 41, W09410, doi:10.1029/2004WR003629.

[5] Montgomery, D. R., and W. E. Dietrich (1994), A physically based model for the topographic control on shallow landsliding, Water Resour. Res., 30, 1153-1171.

[6] Iida, T. (1999), A stochastic hydro-geomorphological model for shallow landsliding due to rainstorm. CATENA, 34, 293-313.

[7] Brutsaert, W., (2005), Hydrology: An Introduction, Cambridge University Press.

17

Environmental Benefits of Organic Farming

Farhad Nejadkoorki

Department of Environmental Engineering, Yazd University, Yazd, Iran

Abstract

Predictions of human population and its requirements to generate new farmlands are unavoidable. On the other hand there have been significant concerns over threats of the agriculture expansion over the next 50 years globally. This is due to public concerns on quality of agricultural products and environmental concerns. Organic farming is kind of agricultural that provide the consumers, with fresh, tasty and reliable food while regarding natural life-cycle systems. There are tremendous attentions in organic farming and foods nowadays both in developed and developing countries. In addition to health benefits of organic products for consumers, there are vital environmental benefits for the earth. An organic farming keeps biodiversity and reduce environmental pollutions such air, water. And soil. This paper investigates and highlights these environmental concerns.

Keywords: Organic food, environment, benefits

1. Introduction

Organic farming is kind of agricultural that provide the consumers, with fresh, tasty and reliable food while regarding natural life-cycle systems. In order to reach organic farming a number of practices should be implemented. Unnatural substances such as chemical synthetic pesticide and synthetic fertiliser livestock antibiotics, food additives and processing aids should be limited. The use of genetically modified organisms should be prohibited. Taking advantage of on-site resources, such as livestock manure for fertiliser or feed produced on the farm. Choosing plant and animal species that are resistant to disease and adapted to local conditions. Raising livestock in free-range, open-air systems and providing them with organic feed.

There have been significant concerns over threats of the agriculture expansion over the next 50 years globally. Predictions of human population and its requirements to generate new farmlands are unavoidable. This means changing more land use from farmland and rangeland to agriculture lands. As a consequence biodiversity is expected to decrease. Furthermore using more machinery and chemical pesticides and herbicides world will be facing more environmental pollution (water, air, soil).

Nowadays, organic farming has received increasing attention in agricultural policy and rural development. With growing public concern for food quality and safety, animal welfare and natural resources, the organic farming philosophy and practice become more accepted [1]. Organic market has been welcoming by developed countries [2].

Organic farming as an environmentally friendly version of agriculture is been selected especially by people of developed countries. It provides organic food which is healthier because it does not contain synthetic pesticide traces. The soil structure on organic farms is much better leading to less pollution from nitrate and is healthier for the crop plant, and environmentally organic is better than the other forms and is chemical free [5]. In contrast agriculture with use of pesticides and other chemical materials have been reported to produce foods leading to cancer [5].

2. Biodiversity

Natural ecosystems have been providing a home for plants and animals. The current diversity in species is result of million years of evolution of such systems. However, when we simplify natural ecosystems to anthropogenic ones, home of unwanted plants and animals appear to be limited. A contemporary agriculture system destroys complex ecosystem such as forests and rangelands through clear cutting. Furthermore using chemical based substances to get rid of pests and weeds make the problem doubles.

Organic farming have less impact on hedge bottom vegetation, with hedges on organic farms displaying significantly higher species diversity than those on conventional farms [2].

Evidence from comparative studies under arable regimes indicated a general trend for higher earthworm abundance under organic management. There have been reports that the presence of grass-clover leys within organic rotations is the principal reason for the significantly higher non-pest butterfly, spiders, beetles abundance [2].

3. Air pollution and climate change

A major theme in organic practices is to operate in tight nutrient cycles to minimise losses to the air and water reserves [3]. There is a reduction in air pollution not just from the lower carbon footprint but also from the absence of chemical sprays which get into the atmosphere. There have been tremendous amount of chemicals which are used to direct lands to yield only desired products and not pests and weeds. This is especially can be tracked after agricultural revolution through using planes and tanks of materials.

Agriculture is both cause and victim of climate change. According to the Intergovernmental Panel on Climate Change (IPCC), the annual amount of greenhouse gases emitted by the agricultural sector is estimated about six giga- tonnes CO_2 in 2005. This represents approximately 10-12% of total greenhouse gases. As a consequence an organic farming system is only substitute to produce healthy products without any side effects locally (air pollution) and globally (climate change).

4. Water and soil pollution

Intensive aquaculture may leave substantial amount of nutrients and poisons to water bodies [4]. Water pollution is largely associated with the use and discharge of water in both animal and plant farming. For instance in a fish pond each time water is exchanged, wastewater is discharged to the surrounding surface waters. The wastewater carries a number of pollutants, reflected in the selected indicators. These pollutants ultimately stem from chemicals, fertilizers and feed added to the ponds [4]. Therefore in an organic farming, water pollution is lower, as there is much reduced eutrophication of chemical inputs. Soil structure on organic farms is much better leading to less pollution from nitrate and is healthier for the crop plant, and that environmentally organic is better than the other forms and is chemical free [5].

5. Summary and conclusion

This paper has highlighted the most environmental benefits of organic farming compared to conventional agriculture. It was discussed that the main environmental costs of non-organic farming are narrowing biodiversity and increasing different types of pollution (e.g. water, air, soil). However, if an organic farming is preferred a number of environmental benefits will be met. A holistic assessment can be seen in table1 [6].

Organic agriculture perform :	Much better	Better	The same	Worse
Biodiversity and landscape		X		
• Floral diversity		X		
• Faunal diversity		X		
• Habitat diversity			X	
• Landscape			X	
Soil		X		
• Soil organic matter		X		
• Biological activity	X			
• Soil structure			X	
• Soil erosion		X		
Ground and surface water		X		
Nitrate leaching		X		
Pesticides	X			
Climate and air			X	
• CO		X		
• NO			X	
• CH_4			X	
• NH_3		X		
• Pesticides	X			
Farm input and output		X		
• Nutrient use		X		
• Water use			X	
• Energy use		X		

Table 1. Assessment of organic farming compared to conventional agriculture

The author suggests that more local studies should be taken to compare and quantify the economical and environmental trade off between organic and non-organic farming. A life cycle assessment is an efficient approach to compare these two by addressing a holistic approach.

6. References

[1] Kerselaers, E., et al., Modelling farm-level economic potential for conversion to organic farming. Agricultural Systems, 2007. 94(3): p. 671-682.

[2] Hole, D.G., et al., *Does organic farming benefit biodiversity?* Biological Conservation, 2005. **122**(1): p. 113-130.

[3] Hansen, B., H.F. Alr¸e, and E.S. Kristensen, *Approaches to assess the environmental impact of organic farming with particular regard to Denmark.* Agriculture, Ecosystems & Environment, 2001. **83**(1): p. 11-26.

[4] Anh, P.T., et al., Water pollution by intensive brackish shrimp farming in south-east Vietnam: Causes and options for control. Agricultural Water Management, 2010. **97**(6): p. 872-882.

[5] Trewavas, A., A critical assessment of organic farming-and-food assertions with particular respect to the UK and the potential environmental benefits of no-till agriculture. Crop Protection, 2004. **23**(9): p. 757-781

[6] Stolze M., PIorr A., Haring A. and Dabbert S. (2000) The Environmental Impacts of Organic Farming. University of Hohenheim. Stuttgart, Germany. 6. 127.

18

Application of SWOT Analysis in Strategic Environmental Planning: A Case Study of Isfahan/Iran

Hakimeh Khalifipour, Alireza Soffianaian, Sima Fakheran*

Department of Natural Resources, Isfahan University of Technology, Isfahan, Iran

* Corresponding Author

Abstract

SWOT analysis, which mainly analyzes the strengths, weaknesses, opportunities and threats of target object or place, is a useful method in strategic planning. Strategic planning is an extended tool for regional development and can be defined as a systematic form of preparing for change and for the future of a city. Urban planning is influenced by changes within internal and external operational environments. SWOT is a useful tool for analyzing internal and external factors. In this paper a SWOT analysis is done regarding the urban management approach for Isfahan City in Iran. Based on our findings the Location of city in country, Rich cultural history and civilization, Various historical attractions, The presence of Zayanderood river in Isfahan are the most strengths factors which can make great opportunities for tourist attractions, however the high rate of urban expansion and industrial development, increasing water demands and degrading Zyandehroud water quality, air pollution and heavy traffic, High rates of immigration to the city, Landuse/cover change and natural habitats fragmentation should be considered as weaknesses and threats for strategic environmental planning.

Keywords: Environmental Planning, Strengths, Weaknesses, Opportunities, Threats, Isfahan

1. Introduction

Strategic planning is an extended tool for regional development and can be defined as a systematic form of preparing for change and for the future of a city. Strategic planning takes into account the socio-economic and environmental context. Nowadays, Environmental analysis is a critical part of the strategic management planning process. Environmental Planning is the process of facilitating decision making to carry out development with consideration on the natural environmental, social, political, economic and governance factors and provides a holistic frame work to achieve sustainable outcomes [1]. Environmental planning with strategic approach is necessary as decision support tool and is a way to achieve sustainable development. Urban planning is influenced by changes within internal and external operational environments. SWOT the acronym standing for Strengths, Weaknesses, Opportunities and Threats analysis is a useful tool for analyzing internal and external factors in order to attain a systematic approach and support for a decision situation. SWOT analysis has been widely applied in the fields of land-resource planning, urban strategy planning, tourism planning, etc.

Due to dramatic population growth in addition to industrial and agricultural development in Isfahan Province, the question to be answered is whether, in the future development could be sustainable. To answer this question, our planning schemes should be environmentally responsible toward the major elements of the environment. Assessment of the strengths, weaknesses, opportunities and threats to a city forms a basis for the preparation of a city strategic development plan. In this study, to highlight the Isfahan city constraints, future potentials and challenges a SWOT (strength, weakness, opportunity and threat) analysis has been used.

2. SWOT analysis

A SWOT analysis is a technique commonly used to assist in identifying strategic direction for an organization or practice. SWOT model is a classic strategic analysis tool for strategic management, first proposed by Ken Andrews (Andrews, 1971). The benefits of such an analysis tool is that it can better balance all internal and external aspects of enterprises, ensuring that analysis is more comprehensive.The strengths and weaknesses of a system are determined by internal elements, whereas external forces dictate opportunities and threats. Strengths can be defined as any available resource that can be used to improve its performance. Weaknesses are flaws/shortcomings of any system that may cause to lose a competitive advantage, efficiency or financial resources [3,4].

3. Methods

3.1. Study area

Isfahan city is the capital of Isfahan Province in center of Iran (Figure1). The city of Isfahan, accounted in 1996, for about 32.2 percent of the total population of the province and 43.4 percent of its urban population. Isfahan is also the third most populated city in the country.The total land area is 157,706 square kilometers. The city is located in the lush plain of the Zayandehrood River, at the foothills of the Zagros mountain range. Zayandehrood River, which is the most important river in central of Iran, divides Isfahan city into north and south parts. The Isfahan metropolitan area had a population of 1,791,069 in the 2010, the second most populous metropolitan area in Iran after Tehran [2]. Dramatic population growth in addition to industrial and agricultural development over the past decades, have resulted in the rapidly increasing pollutions and degrading environmental quality in Isfahan. More than 50 percent of major national industries, such as petrochemical and steel factories are located in Isfahan province, out of which about half of those industries are located near Isfahan city.

Fig 1. Location of Isfahan Province in Iran

3.2. Data Preparation and analysis

This research is descriptive – analytical study. For obtaining necessary information, required data have been collected through library-based studies and interviews with experts and prepared a questionnaire.This study is based on internal and external urban conservation factors. Internal factors can be classified as strengths (S) or weaknesses (W), and external factors can be classified as opportunities (O) or threats (T).

Internal and External factors were based on three dimensions of sustainable development (Ecological, socio-economic and cultural factors).

4. Results

The most important Internal and External factors based on three dimensions of sustainable development, which were found for Isfahan City are listed in Table 1.

5 strength factors and 6 opportunity factors as advantages and 6 weakness factors and 7 threat factors as constraints are facing Isfahan city.

	Strength	Weakness
Internal factors	(S_1) Special Location of city in country (S_2) Rich cultural history and civilization (S_3)Various historical attractions, natural and man-made (S_4) The presence of permanent rivers (Zayanderood river) (S_5)Tourist attractions)	(W_1) Concentration of population (W_2) Air pollution from vehicles and industry (W_3)Shortage of water resources and increasing water demands W_4)Successive droughts in recent years (W_5) Inversion temperature (W_6)Heavy traffic of transport system

	Opportunity	Threat
External factors	(O_1)Trend to change the energy consuming Structure replacement of clean energy (solar and wind) (O_2) Potential ability of preventing pollution (O_3)Considered to Decentralization of city (O_4) Increasing importance of tourism in the country (O_5)Improving public transportation system) O6)Expansion of green space)	(T_1) High rates of immigration to the city) (T_2) Concentration of economic activities) (T_3) Inter basin Water transferring projects) (T_4) Air Stability in most days of the year) (T_5) Loss of rare species of animal and plants (T_6) Decreasing precipitation in recent years (T_7) Land use/cover change and natural habitats fragmentation

Table 1. internal and external factors

5. Conclusion

In this paper a SWOT analysis was done regarding the urban management approach using the case of Isfahan City. As described, this research is being done with purpose of environmental management analysis in Isfahan based on internal and external factors and strategic planning tool (SWOT). Based on our findings the high rate of urban expansion and industrial development in Isfahan had a negative impact on most of the heritage and natural environment of the city. Controlling Urban and Population growth and industrial area expansion is a critical issue in Isfahan. Decreasing water demands and Improving Zyandehroud water quality as well as reducing environmental pollutions, improving linkages to city core with subway networks, and increasing public environmental awareness should be considered for strategic environmental planning.

6. Acknowledgments

This project was supported by a grant from Iran National Science Foundation.

7. References

[1] Beer, A. 1977. Environmental Impact Analysis: A Review Article. Town Planning Review, 48(4):p.389

[2] Statistical Center of Iran. http://www.amar.org.ir

[3] Wang, K., and Hong, W, 2011. Competitive advantage analysis and strategy formulation of airport city development—The case of Taiwan. Transport Policy, 18 :p. 276–288.8.

[4] Ying, Y, 2010.SWOT-TOPSIS Integration Method for Strategic Decision. International Conference on E-Business and E-Government.

19

Analysis of Landscape Pattern Changes in Isfahan City During the Last Two Decades

Neda Bihamta Toosi, Sima Fakheran* and Alireza Soffianian

Department of Natural Resources, Isfahan University of Technology, Isfahan, Iran

*Corresponding author

Abstract

Urbanization and urban land-use transition are a global concern and one of the greatest challenges for ensuring human welfare. The landscape patterns resulting from urbanization influence processes at local, regional, and global scales. Quantifying the spatio-temporal pattern of urbanization is important for understanding its ecological impacts and can provide basic information for appropriate decision-making. The main goal of this study was to quantify the changes of landscape patterns in Isfahan city, in Iran, during the last two decades. For monitoring spatial pattern changes, land use and land-cover maps of the area were prepared using supervised maximum likelihood classification of Landsat Thematic Mapper (TM) images taken in 1990 and 2010. Five classes of land use including Water, Agricultural land, Urban area, Undeveloped area, and Bare land were identified. The changes of landscape structure were analyzed using several landscape metrics including: Percentage of landscape (PLAND), Number of patches (NP), Largest Patch Index (LPI) and Contagion (CONTAG), which were derived from spatial analysis software FRAGSTATS. The results indicated that the proportion of undeveloped area decreased from 41.87% to 39.65% and proportion of urban areas was significantly increased from 9.88% to 28.73% during the last two decades, mainly due to reduced agricultural area in Isfahan.

Keywords: Urbanization, Isfahan, Spatio-Temporal Changes, Landscape metrics, FRAGSTATS

1. Introduction

Urbanization, urban expansion and urban land-use transition are a global concern and one of the greatest challenges for ensuring human welfare. Over 50% percent of the world population lives in urbanized areas. Urbanization has profoundly transformed natural landscapes throughout the world, which inevitably has resulted in various effects on the structure, function, and dynamics of ecological systems at a wide range of scales. For example, land transformations associated with urban expansion can significantly affect biodiversity, energy flows, biogeochemical cycles, and climatic conditions at local to regional scales [3]. To improve understanding of urban landscape changes, remote sensed imagery, multiple models, and scenario analysis approaches are widely becoming used. Analyses of land use/cover changes (LUCC) are fundamental for understanding numerous social, economical and environmental problems [5]. To establish this correlation, the first step is to quantify landscape patterns. Landscape metrics are approaches to quantify landscape patterns, mainly, applied to categorical data with spatial interruption [2]. A wide range of metrics are available for the examination of relationships between spatial structure, ecological function, and landscape change, The selection of the metrics depends first of all on the purpose of the study (mostly ecological process) and also on the landscape characteristics [1]. The common usage of the term 'landscape metrics' refers exclusively to indices developed for categorical maps. Landscape metrics are focused on the characterization of the geometric and spatial properties of categorical map patterns represented at a single scale [4]Applying these landscape metrics, this study attempts to quantificationally analyze the landscape pattern changes of Isfahan during the last two decades.

2. Materials and methods

2.1. Study area

The study area is Isfahan city which is located in centre of Iran and covers 340 km^2 (Fig. 1). The city is located in the lush plain of the Zayanderood River, at the foothills of the Zagros mountain range. Zayanderood River divides Isfahan city into north and south parts. Isfahan is one of the most important cities of Iran because of its historical and economic values. Isfahan attracts a large number of tourists each year. The mean annual temperature of Isfahan is approximately 16 $^{\circ C}$. The altitude of the study area is 1580 meters above mean sea level. The Isfahan metropolitan area had a population of 1,791,069 in the 2010, the second most populous metropolitan area in Iran after Tehran. Urban expansion, population growth, in addition to industrial development, have resulted in degrading environmental quality in Isfahan.

2.2. Data and preprocessing

For monitoring spatial pattern changes, land use and land-cover maps of the area were prepared using supervised maximum likelihood classification of Landsat Thematic Mapper (TM) images taken in 1990 and 2010. Five classes of land use/cover including Water, Agricultural land, Urban area, Undeveloped area, and Bare land were identified. We applied spatial pattern analysis software FRAGSTATS 3.3 to calculate landscape metrics of each class type and total landscape. We chose several class-level metrics including: Percentage of landscape (PLAND), Number of patches (NP), and Largest Patch Index (LPI), and also some landscape-level metrics such as Number of patches (NP), Patch Density and Contagion (CONTAG) were calculated to qualify the landscape pattern dynamics of Isfahan city.

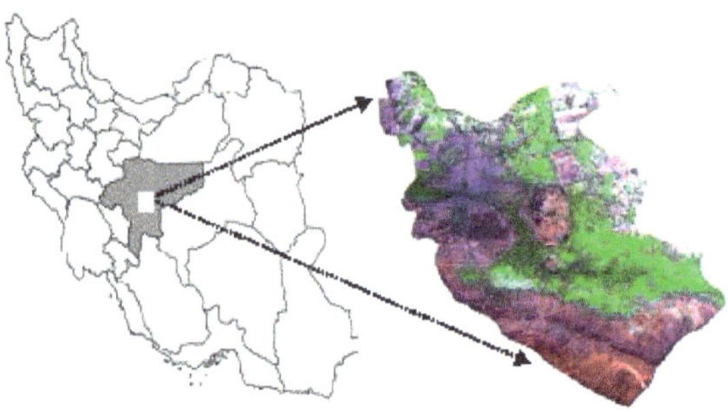

Figure 1. Location of Isfahan Province which is in the center of Iran (left) and Isfahan city extent (right).

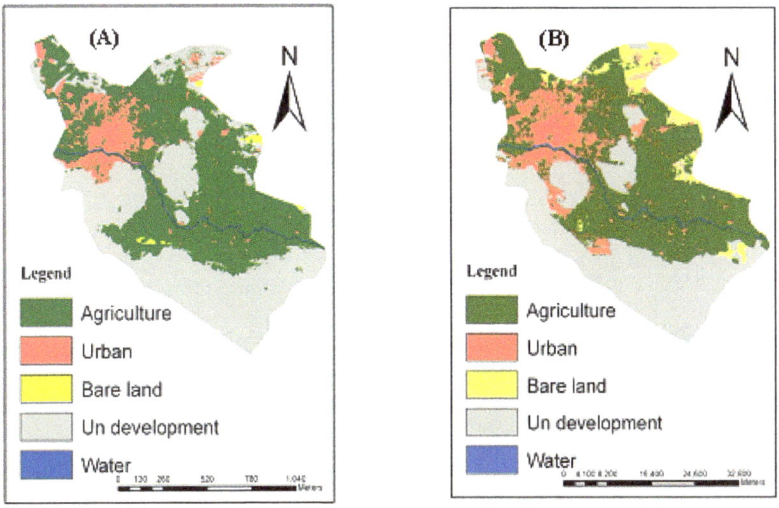

Figure 2. Land cover of Isfahan city: A.1990, B.2010

3. Result

Analysis of PLAND at the class level provides a general representation of landscape composition. The temporal change of PLAND can be used to obtain an overall idea of landscape change in Isfahan city. In 1990, the agricultural area of Isfahan comprised 46.85% of the total area. From 1990 to 2010, the area of agriculture kept declining (Fig.3). The proportion of undeveloped area decreased from 41.87%S to 39.65% in the period of study (Fig.3). Urban growth in the Isfahan maintained a rapid pace during the study period, increasing the percentage of urban area to 28.73% in 2010 (Fig.3).

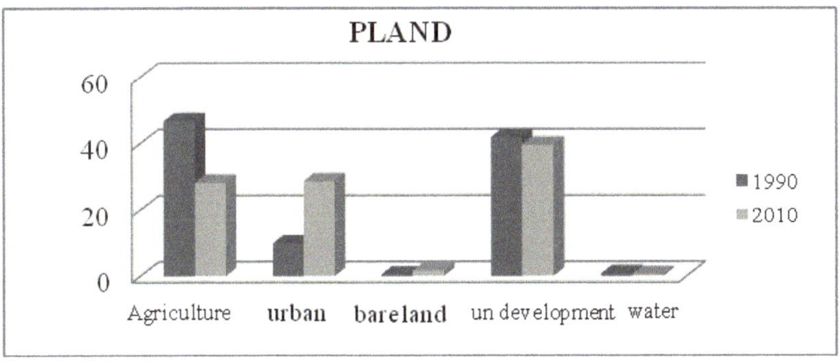

Figure 3. Comparison of Percentage of landscape in Isfahan city in 1990 and 2010.

The number of patches (NP) of all land covers types increased from 1990 to 2010. The most changes in Number of patch are related to agriculture area, with increasing from 1335 to 2979 in this period of time (Fig. 4).

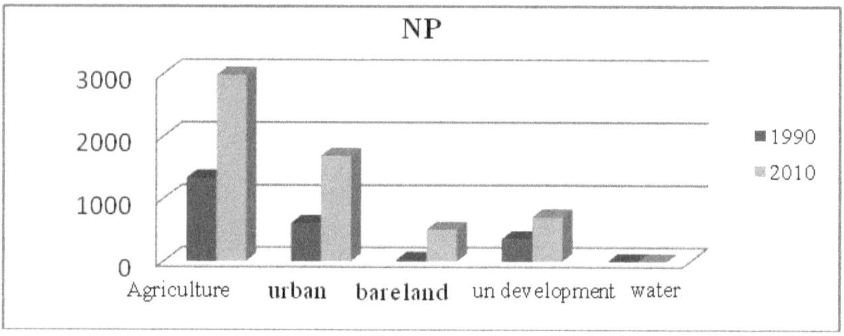

Figure 4. Comparison of Number of patches (NP) in Isfahan city in1990 and 2010 .

Undeveloped area had the highest values of LPI metric (Fig.4). From 1990 to 2010, the largest variation of LPI metric is associated to agriculture area, with decreasing from 26.22 to 7.6 % (Fig.5).

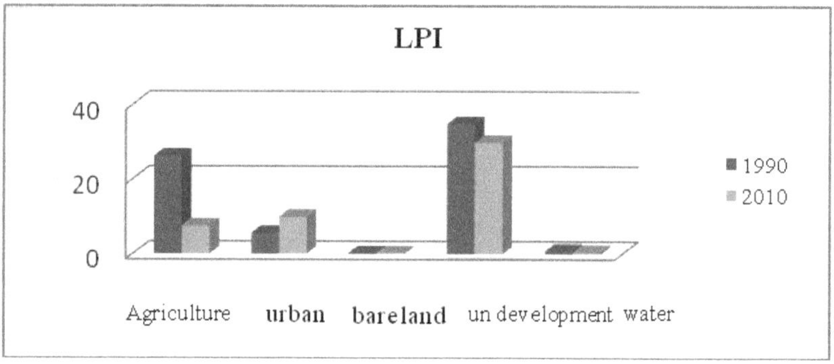

Figure 5. Comparison of Largest Patch Index in 1990 and 2010 in Isfahan city

The synoptic analysis of selected metrics for landscape level indicated the metric values of NP and PD increased from 2333 to 5898 and 1507.53 to 3418.73, respectively in past two decades (table1). The metric values of CONTAGE reduced from 67.63 to 63.22 (table 1).

Year	Spatial metrics		
	NP	PD	CONTAG
1990	2333	1507.53	67.63
2010	5898	3418.7	63.22

Table 1. Synoptic analysis of spatial metric for Isfahan city

4. Conclusion

The effectiveness of information on land covers change for spatial planning and decision making. Satellite images and landscape metrics, can be extremely useful for planners in assessing and monitoring the ecological consequences of landscape patterns. The proportion of agriculture reduced from 46.85 to 28.31, The proportion of urban was significantly increased from 9.88 to 28.73 during the study period, mainly due to reduced agricultural area in Isfahan. The synoptic analysis of selected metrics in landscape level showed the landscape has been more fragmented in Isfahan during the last two decades.

5. Acknowledgments

This project was supported by a grant from Iran National Science Foundation.

6. References

[1] Evelyn, s, Roosaare, J., Oja, T., Mander, U., 2011, *Analyzing the spatial structure of the Estonian landscapes: which landscape metrics are the most suitable for comparing different landscapes?, Estonian Journal of Ecology*,60: p. 70-80.

[2] Jin, s., Deng, Wang, k., Hong, Y., Qi, j., 2009, *Spatial-temporal dynamics and evolution of land use change and landscape Pattern in response to rapid urbanization, Landscape and Urban Planning*, 92: p. 187–198.

[3] Luck, M., and Wu, J., 2002, *A gradient analysis of urban landscape pattern: a case study from the Phoenix metropolitan region- Arizona- USA, Landscape Ecology*, 17: p. 327–339.

[4] McGarigal, K and Marks, B.J., 1995, FRAGSTATS: *spatial pattern analysis program for quantifying landscape structure. General Technical Report PNW-GTR-351 USDA Forest Service, Pasic Northwest Research Station, Portland*. P.961-978.

[5] Rafaela, P., Leone, A., Boccia, L., 2009, *Land cover and land use change in the Italian central Apennines: A comparison of assessment methods, Applied Geography*, 29: p. 35–48.

20

The Role of Trees in Improving the Urban Landscape, (Case Study Vli Asr Street of Tehran city)

Fardad Edalatkhah[1], Mahdi Shafaghati[2,*], Akram Norouzi[2], Somaye Motaghi[2], Maryam Mirbahaei[3]

1 Municipality of Tehran, Tehran, Iran

2 Payame Noor University, Taleghan, Iran

3 Payame Noor University, Varamin, Iran

* Corresponding author

Abstract

City landscaping is the art of visually and structurally integrating the complex of buildings, roads and elements present in them, as well as all those spaces which give shape to urban environments. These include natural spaces inside city precincts which play a decisive role in creating urban landscapes with a decisively different appearance. Among natural elements influencing urban landscapes, trees are of paramount importance owing to their particular structure and spatial function. Like buildings, trees have structures which help create plant-based architectural styles. In lieu with overall changes that a city experiences in the course of its subsequent historical stages, the selection criteria for the varieties of trees to be planted in the city change, too. Meanwhile, almost all cities in the world today host a large proportion of open space with distinctive physical characteristics each. These include streets, squares, and spaces adjacent to car parks. Tree selection criteria for each of these spatial categories vary accordingly as determined by professional city landscaping architects. Based on research findings and the field study carried out, this article tries to assess and highlight the role played by landscaped and natural elements in the greater urban architecture, and draw up certain conclusions as a result.

Keywords: urban landscape, tree species, Vali Asr Street Tehran, Urban environment

1. Introduction

Exactly as the establishment of developed urban areas is considered to be one of the greatest achievements of human civilization, city landscaping is can also be taken as a measure of the degree and nature of a civilization and the collective psyche of a nation. Although the view which regards cities as phenomena independent of the human will is not held as widely to date (3), there is still a deep-seated approach to the subject which has strong following and tries to interpret and analyze the city and her landscapes regardless of humankind as their simultaneous creator and perceiver (7).

2. Concepts

The urban landscape is a concept that lingers in the human memory even after one has evacuated a landscaped spot and results from the interaction between man (the onlooker) and his living environment. (1)

The most decisive feature of such a concept is its being subjective-objective, human-concrete, social-spatial which can be comprehended solely via the human experience and his interaction with his surrounding environements.

There are three major approaches to the qualitative entity of the urban landscape:

a) The view that regards the urban landscape as a quintessential feature of the physical environment of a city, independent of the humankind as the onlooker and the perceiver;

b) The insight which regards the urban landscape an entirely subjective element created by the onlooker, independent of the structure and other characteristics of the physical environment;

c) The approach which considers the urban landscape to be an event emanating from the interaction between palpable physical characteristics of the surrounding environment and the existing paradigms, cultural symbols and intellectual competencies of the observer. In this phenomenological approach, the concept of the urban landscape can be interpreted in various ways. As compared to the previous two approaches, this is the most credible take on the concept of the urban landscape nowadays (7), (3).

3. The case study on Vali Asr Avenue in Tehran

This is a major thoroughfare 18 kilometres long linking the northern and southern tips of the city along an almost straight north-south vertical line. Yet, perhaps the most important reason behind the outstanding role of this road in Tehran is the fact that it acts like a historical bridge connecting the junctures the city went through before and after the road was built. This road is, in effect, a showcase of Tehran's social, cultural, political, religious, etc developments, especially, over the past 100 years, and therefore, her historical backbone.

Figure 1. New picture from vali Asr street in Tehan city,Iran

4. The plane trees of Vali Asr Avenue

These have been lining the road ever since it was built over 80 years ago and now amount to a total of 11,000. The plane trees of the Avenue are long considered to be a unique symbol of Iran's capital city, a national treasure and a cultural heritage which for many residents of the older generation revive endless nostalgic memories of the city's rich and exciting history.

5. The current situation of the trees

Unsettling reports have been published in ever-growing numbers on the immanent extinction of Tehran's historic plane trees. One of the best-researched among reports published very recently blames the following factors for the growing threat that has put the life of these priceless trees on the edge:

a) The demolition of the surface- and ground-water runways of the city;

b) Mismanagement in irrigation;

c) The release of sewage and other pollutants into the running water resources of Tehran;

d) The disposal of a wide range of toxic chemicals such as detergents, industrial chemicals, cement and other construction materials, industrial oils and gasoline in the city's soil and water resources;

e) Physical damage incurred on the root system of the trees as a result of constant diggings, curb lining and concrete injection projects;

f) Changes in top soil levels exposing the trees' root systems to ambient, erosive climatic factors;

g) Physical mistreatment of the trees by humans;

h) High-rise buildings which block solar radiation reaching the canopy of the trees;

i) Highly polluted air all the year round (6), (2).

Figure 2. Old picture from vali Asr street in Tehan city,Iran

6. Spotting optimal locations for landscaping

Landscaped plots must be centrally positioned, that is, they must be located at the heart of a neighborhoods, zone or district of the city. Moreover, such plots should be consistent with the physical structure of the environments that surrounds them. Another crucial criterion is accessibility. City landscapes must be linked with the traffic and transport grid in their four directions to become easily accessible for the greatest number of people and to become safer places for the public as enforcing public security measures in such locations is optimized thus.

7. The role and effects of trees in urban landscapes

City landscaping is the art of visually and structurally integrating the complex of buildings, roads and elements present in them, as well as all those spaces which give shape to urban environments. These include natural spaces inside city precincts which play a decisive role in creating urban landscapes with a decisively different appearance. Among natural elements influencing urban landscapes, trees are of paramount importance owing to their particular structure and spatial function. Like buildings, trees have structures which help create plant-based architectural styles. In lieu with overall changes that a city experiences in the course of its subsequent historical stages, the selection criteria for the varieties of trees to be planted in the city change, too. Meanwhile, almost all cities in the world today host a large proportion of open space with distinctive physical characteristics each. These include streets, squares, and spaces adjacent to car parks. Tree selection criteria for each of these spatial categories vary accordingly as determined by professional city landscaping architects.

8. Recommendations

Any successful management of urban green space requires the serious enforcement of the following measures and/or principles:

1. **Granting a greater role to the public as the direct benefactors:** This can be realized through raising public awareness and strengthening public responsibility among others;
2. **Boosting the ecological productivity of urban green space:** this can be realized through the protection and management of urban green space, the informed selection of the right plant varieties, integrated management systems in planting, growing and training the selected plant varieties, urban woodland management, intelligent plot spotting for the expansion of urban green space, evaluating the feasibility of extending urban green space to abandoned plots and mountainous ecosystems;
3. **Raising the social productivity of urban green space:** this can happen via ensuring safety and security, the provision of basic facilities and amenities, ease of access, consistency and cohesion with the greater physical environment, proactive management, balanced geographical distribution, etc.

9. Conclusion

Both from an architectural-aesthetic point of view and in order to raise the quality of the essential elements in urban livelihoods, paying attention in urban management processes to the image of urban areas overall and urban landscapes in particular is of paramount importance. However, the scientific study of these elements and their contribution to urban development in view of their economic as well as budgetary repercussions is no less significant. In other words, image- and landscape-related projects should not be taken into account barely from an architectural and aesthetic but also from an economic standpoint. As regards important traffic axes, strategic decisions must be made in accordance with the prominent position of spatial planning and landscaping projects in urban areas. In other words, the simple fact that budgetary provisions are available to municipal managers must not automatically lead to the illusion that such resources can be allocated to imagined beautification projects arbitrarily to boost development efforts in urban areas in turn.

10. References

[1] Soltani,B.2010,view to green area of Tehran city, journal of environmental science,environment organization press.

[2] Bijan zad,M,2010,evaluation of green area of Tehran city, jihad daneshgahi ,press, Tehran.

[3] Satari,J,2006,Myth of Tehran city, office of cultural research, environment organization press.

[4] Rasouli, A, 2009, buttonwood, holy tree in Iran, cultural heritage press, Tehran

[5] Riyazi, B, 2009, Development of green space in Tehran, proceeding of green space seminar, Tehran city parks and green space organization, Tehran

[6] Majnoniyan, H, 2010, Debates about parks, green space and promenade, Tehran city parks and green space organization, Tehran

[7] http://www.darioush-shahbazi.com

[8] McPherson, E.G & Nowak,D.G, 2011,The Chicago urban forest climate project.

21

Selective Catalytic Reduction of NO with Ammonia over Nanostructure H-ZSM-5 Supported Transition Metal Oxide Catalysts

A. Niaei*, D. Salari, P. Nakhostin Panahi, S. M. Mousavi

Department of Applied Chemistry & Chem. Eng., Faculty of Chemistry, University of Tabriz, Tabriz, Iran,

*Corresponding author

Abstract

A series of transition metal oxides (Co, Cr, Mn, Fe and Cu) promoted H-ZSM-5 catalysts were prepared by wet impregnation method using dilute soltions of metal nitrate precursors. These were characterized by X-ray diffraction (XRD) and scanning electron microscopy (SEM). XRD and SEM images approved that the formed metal species are in the nanometer size range and well dispersed. The catalytic activity of these materials was evaluated for the selective catalytic reduction (SCR) of NO with NH_3 as reductant in the presence of excess oxygen (5 vol.%). The results revealed that the catalytic activity of Cu-ZSM-5 for NO removal was about 80% at 300 °C, which was the best among various promoted oxides. Then effect of catalyst preparation parameters for Cu-ZSM-5 optimum catalyst was studied.

Keywords: ZSM-5, transition metal oxides, nitric oxide (NO), selective catalytic reduction (SCR)

1. Introduction

Nitrogen oxides (NOx) remain a major source of air pollution. They contribute to photochemical smog, acid rain, ozone depletion, and greenhouse effects. Nearly all NOx (95%) derives from transportation (49%) and power plants (46%) [1]. Due to the increasing threat of NOx to our survival, many approaches have been developed to reduce its emission, among which selective catalytic reduction technique (SCR) is proven to be an effective way compared with other NOx abatement technologies, such as nonselective catalytic reduction technique, storage, and thermal decomposition. The most common reductants for SCR are ammonia, urea, CO, H_2 and hydrocarbons like methane, ethane and propylene but ammonia is still found to be a suitable reductant for NO in the presence of oxygen, in spite of being difficult to handle, because ammonia reacts selectively with NOx to produce N_2 without consuming an excess O_2 [2].

In the past few decades, the backbone of SCR technology is the development of SCR catalysts such as noble metals, supported metal oxides, zeolites and others [3, 4]. The conversion efficiencies of NOx in the catalytic reactions depend mostly on the supports and the nature of active sites. Therefore, the choices of supports and catalysts are especially important [5]. Zeolite-based SCR catalysts have received a great deal of attention because of their high activity for the reduction of

NOx. Some of the extensively studied metal cations exchanged into zeolites to prepare lean NOx catalysts include Co, Pt, Pd, Fe, Ni, and Cu [6].

In the present study we have screened a number of transition metal oxides (Co, Cr, Mn, Fe and Cu) supported on H-ZSM-5 for SCR of NO with NH_3. The objectives of this work are to compare the activity of various metals (Cu, Mn, Fe, Co,...) loaded on H-ZSM-5 and carry out the reaction selectively in a wide temperature range. The prepared catalysts were thoroughly characterized by various physicochemical techniques. Among various catalysts, the Cu-ZSM-5 catalyst showed very promising catalytic activity for SCR of NO with NH_3 by exhibiting high conversions over a wide temperature window and the reaction was highly selective in the presence of oxygen.

2. Experimental

2.1. Catalysts Preparation

H-ZSM-5 obtained from Zeochem Int with $SiO_2/Al_2O_3 = 50$ was used for preparation of the catalysts. The properties of H-ZSM-5 are listed in Table 1. All catalysts were prepared through incipient wetness impregnation method. The aim of this procedure is to deposit metal oxide on the zeolite surface, whereas a proton exchange is not explicitly intended and with this technique an accurate load of metal is adjusted. The precursors of different metals were $Cu(NO_3)_2.3H_2O$, $Co(NO_3)_2$, $Cr(NO_3)_2$, $Fe(NO_3)_2$, and $Mn(NO_3)_2$, and the total metal content was kept at 5 wt%. The prepared catalysts were dried in an oven at 100°C overnight and then calcined at 500°C for 4 h in the air; under these conditions complete decomposition of metal nitrate into metal oxide occurs.

2.2. Catalysts characterization

2.2.1. X-ray diffraction (XRD)

The structure of the catalysts were analysed by powder XRD at room temperature with a D500 Siemens diffract meter using CuK_α radiation (λ= 1.54050 Å). The X-ray tube was operated at 35 kV and 30 mA and the X-ray pattern was scanned with a step size of crystallites sizes of 0.016° (2θ) from 5 to 50° (2θ) and counting time of 1 s per step.

2.2.2. Scanning electron microscopy (SEM)

SEM images for H-ZSM-5 and M-ZSM-5 (M= Co, Cr, Mn, Fe and Cu) were obtained using SEM model EQ-CL 1 instrument to observe particle size and surface homogeneity.

2.3. Catalytic activity measurement

Catalysts activities in NOx reduction were studied at atmospheric pressure in a fixed bed reactor. The reactor consisted of a 0.9 cm i.d. glass tube was located inside in a furnace which was electrically heated. The reactant gas feed, consisting of NO (1000 ppm), NH_3 (1000 ppm), O_2 (5 vol%) and Ar as balanced gas, was mixed in a mixing chamber and introduced to the reactor at a total flow rate of 200 ml/min. In each run, 0.2 g of the catalyst powder was dispread between quartz wool plugs, yielding a gas hourly space velocity (GHSV) of 12000 h^{-1}. Before starting each run, zeolites were pretreated with Ar at 150°C in order to eliminate possible compounds adsorbed on the zeolite surface. After this pretreatment, the reactor was cooled to 100°C and activity tests were performed from 100 to 400°C with a step of 100°C. The concentration of N_2 (as selective product) in the outlet of the reactor was measured by a gas chromatograph (shimadzu) equipped with a thermal conductivity detector (TCD) with Molecular sieve column to separate N_2 and N_2O.

3. Results and Discussion

3.1. Characterization of catalysts

Figure 1 shows the XRD patterns of the series H-ZSM-5 and M-ZSM-5 (M = Cu, Mn, Fe, Co,...). XRD patterns of the prepared samples M-ZSM-5 are similar to that of H-ZSM-5 which suggests that the original structure of H-ZSM-5 is not destroyed during the process of impregnation and calcination. All The characteristic peaks of HZSM-5 were observed in impregnated ZSM-5 samples. XRD peaks remain sharp and intense for M-ZSM-5, but a slight decrease in intensity of main peaks can be observed. This reveals a decrease in crystallinity of catalysts compared to H-ZSM-5. Decreases of peaks intensity imply the entrance of metal species into the channels. Also there is not any other peak in XRD patterns of M-ZSM-5 catalysts compared to XRD pattern of H-ZSM-5 indicating that metal species (i.e. oxide, cations,...) are well dispersed through the zeolites structure.

Figure 2 shows SEM micrographs of H-ZSM-5 and M-ZSM-5 (M = Cu, Mn, Fe, Co,...) catalysts. This figures indicates the formation small size of metal particles, the high dispersion on support and the uniform size distributionin in M-ZSM-5 catalysts that increase activity. Also SEM images approved the nanostructure of catalysts (<100 nm).

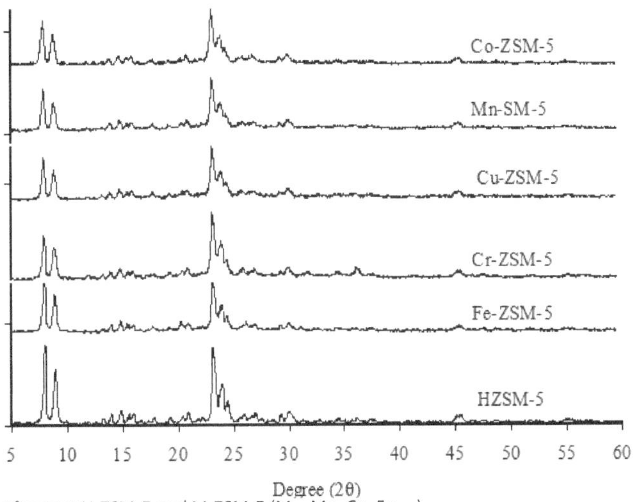

Figure 1. XRD of parent H-ZSM-5 and M-ZSM-5 (M = Mn, Cr , Fe,...)

Figure 2. SEM images of (a): parent H-ZSM-5 and (b): Fe-ZSM-5 (c): Mn-ZSM-5 (d): Cr-ZSM-5 (e)Cu-ZSM-5 (f): Co-ZSM-5

3.2. Effects of different metals for NH_3 – SCR activity

Catalytic activity results for NO conversion to N_2 of NH_3 – SCR reaction over H-ZSM-5 and various transition metal oxides supported on H-ZSM-5 are presented in Figure 3. No obvious NO conversion was observed over pure H-ZSM-5 support up to 300ºC. When the transition metal oxide phases were introduced, the catalytic activities were enhanced significantly over the whole range of temperature investigated. For all catalysts, NO conversion increased with increasing temperature and all catalysts were highly selective. Under identical operating conditions Cu-ZSM-5 showed excellent performance giving 80% NO conversion at 300°C, which was the best among catalysts. High activity for Cu-ZSM-5 also were shown by sultana et al. [7], who reported high selective and activity for Cu/NaZSM-5 and Cu/HZSM-5 in NOx reduction by NH_3.

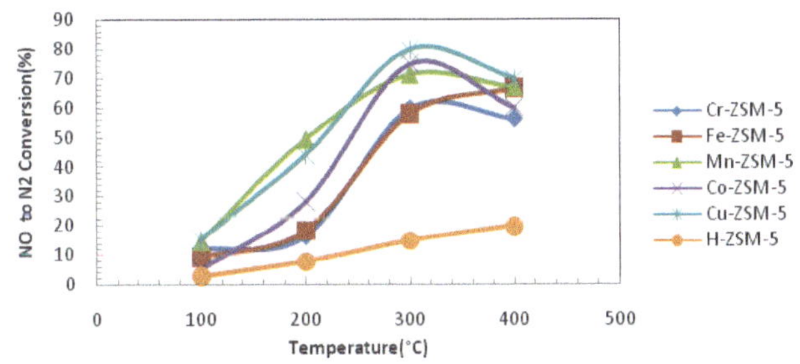

Figure 3. NO to N2 conversion as a function of temperature on M-ZSM-5 (M= Fe, Cr, Co, Mn and Cu) catalysts.

3.3. Catalytic activity of Cu-ZSM-5 catalysts with various Cu-loadings

Cu-ZSM-5 showed more activity than other catalysts thus we focus additional studies on Cu-ZSM-5. To further improve the performance of the Cu-ZSM-5 catalyst, the composition of the catalyst was optimized by varying Cu loading and calcinations temperature. Figure 4 compares the NO conversion to N_2 as a function of temperature over Cu-ZSM-5 catalysts with different loading (1, 3, 5, 7, 9 wt.%). The result indicates that below 5wt.%, increasing Cu content in Cu-ZSM-5 resulted in an increase in conversion but Further increase in Cu content was found to weaken the NO conversion to N_2, therefore Cu loading of 5 wt.% was found to be the optimum loading in this study. The low activity in Cu-ZSM-5 with loading of 1 and 3 wt.% can be attributed to low quantity copper oxide species. A decrease in conversion with an increase in Cu content after that was attributed to excessive metal agglomeration leading to the formation of larger metal particle also further loading of copper causes to block the pores and active sites of zeolite, leading to decrease the catalytic activity of catalyst [8].

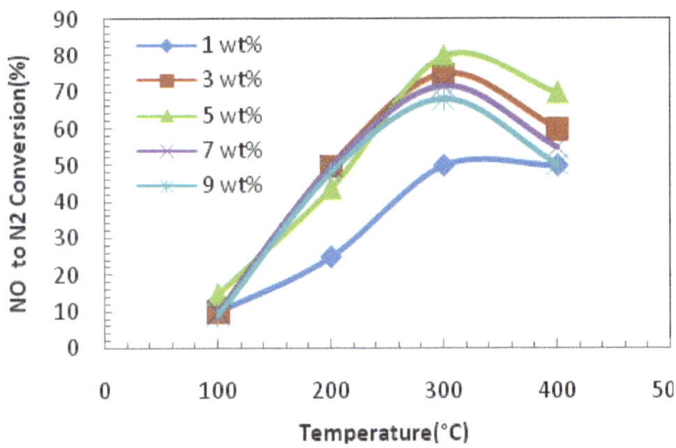

Figure 4. NO to N2 conversion as a function of temperature on Cu-ZSM-5 with different Cu loadings.

3.4. Effect of calcination temperature on the catalytic activity Cu-ZSM-5

Figure 5 shows the NO conversion to N_2 as a function of temperature over 5 wt.% Cu-ZSM-5 at different calcination temperatures. The result indicates that the calcination temperature influenced the SCR activity. From figure 6 we can see that the catalyst calcinated at 550°C had the highest activity. The NO conversion over Cu-ZSM-5 catalysts decreased in the order of Cu-ZSM-5 (550)> Cu-ZSM-5 (500)> Cu-ZSM-5 (600)> Cu-ZSM-5 (650) Cu-ZSM-5 (450)> Cu-ZSM-5 (400). Different calcination temperatures result in different oxidation states of copper, so the calcination temperature affects the activity and selectivity of SCR of NO by NH_3. Further increase in calcination temperature above 600°C caused decreasing NOx conversion. This can be due to the sintering to some extent at high calcinations temperature [9].

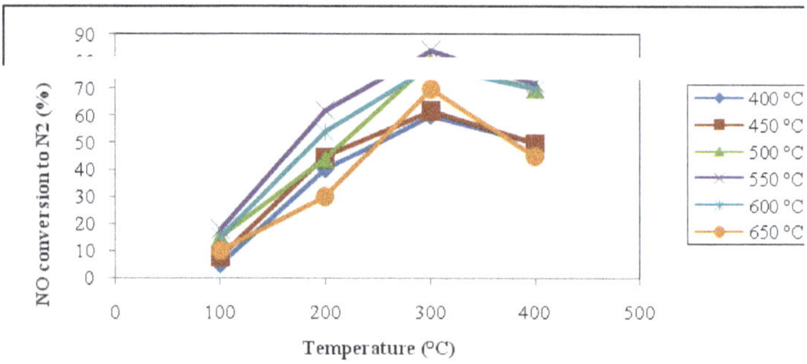

Figure 5. I) to N2 conversion as a function of temperature on Cu-ZSM-5 with different calcination temperatures.

4. Summary and conclusion

This study confirmed higher catalytic activity of M-ZSM-5 (M = Co, Cr, Mn, Fe and Cu) for the selective catalytic reduction (SCR) of NO with NH3 as reductant and revealed that introduction of metal ions onto ZSM-5 caused improvement in catalytic activity of M-ZSM-5 catalysts compared to parent HZSM-5, which confirms catalytic role of metal ions for selective catalytic reduction. Under identical operating conditions Cu-ZSM-5 showed excellent performance giving 80% NO conversion at 300°C, which was the best among catalysts. Also, It was concluded that Cu loading of 5 wt.% be the optimum loading and catalyst calcinated at 550°C had the highest activity in this study.

5. References

[1] Gongshin Qi, Ralph T. Yang, Ramsay Chang. 2004. MnOx-CeO2 mixed oxides prepared by co-precipitation for selective catalytic reduction of NO with NH3 at low temperatures. Applied Catalysis B: Environmental (51) : p. 93–106.

[2] Min Kang a, Eun Duck Park b, Ji Man Kim c, Jae Eui Yie. 2007. Manganese oxide catalysts for NOx reduction with NH_3 at low temperatures. Applied Catalysis A: General (327) : p. 261–269.

[3] javad amanpour .2011. Catalytic reduction of NOx over active carbon coated with Cupper and Vanadium. MSc Thesis, Faculty of chemistry, Tabriz University, Tabriz, Iran.

[4] masood samandari.2011. Optimization of properation condition of some transition metal nanocatalyst based on active carbon in NOx catalytical reduction. MSc Thesis, Faculty of chemistry, Tabriz University, Tabriz, Iran.

[5] Kui-Hao Chuang , Zhen-Shu Liu , Yu-Hao Chang , Chi-Yuan Lu , Ming-Yen Wey. 2010. Study of SBA-15 supported catalysts for toluene and NO removal: the effect of promoters (Co, Ni, Mn, Ce). Reac Kinet Mech Cat (99) : p. 409–420.

[6] M. Mečárová N.A. Miller, N.C. Clark, K.C. Ott, T. Pietraß. 2005. Selective catalytic reduction of NOx with ammonia on gallium-exchanged ferrierites. Applied Catalysis A: General. (282) : P. 267–272.

[7] Asima Sultana, Tetsuya Nanba, Masaaki Haneda, Hideaki Hamad. 2009. SCR of NOx with NH3 over Cu/NaZSM-5 and Cu/HZSM-5 in the presence of decane. Catalysis Communications (10) :p. 1859–1863.

[8] Niaei Aligholi, Salari Dariush, Hosseini Seyed Ali, Khatamian Masumeh, Jodaei Azadeh. 2009. Catalytic Combustion of Ethyl Acetate Over Nanostructure Cobalt Supported ZSM-5 Zeolite Catalysts. Chinese Journal of Chemistry, (27): p.1—10.

[9] Azadeh Jodaei, Aligholi Niaei†, and Darush Salari. 2011. Performance of nanostructure Fe-Ag-ZSM-5 catalysts for the catalytic oxidation of volatile organic compounds: Process optimization using response surface methodology. Korean J. Chem. Eng.

Monitoring Land Use/Cover Changes Using Different Change Detection Techniques (Case Study: Falavarjan Area, Isfahan, Iran)

Maliheh Alsadat Madanian, Alireza Soffianian and Sima Fakheran*

Department of Natural Resources, Isfahan University of Technology, Isfahan, Iran

*Corresponding author

Abstract

Land use/cover change mapping is one of the basic tasks for environmental monitoring and management. In recent years, a variety of change detection techniques have been developed. This research compares three change detection techniques, including image differencing, image rationing, and image regression to study land use/cover changes in Falavarjan /Iran. The data sources used in this study were Landsat Multi-Spectral Scanner (MSS) and AWiFS images taken in September 1972, and September 2008, respectively. First, images were geometrically and radiometrically corrected. The root mean square (RMSe) obtained 0.5 pixels for each images. The Three change detection methods were performed. Then, a supervised maximum likelihood classification was used as a crossclassification to detect "from-to" change which allowed to assess the accuracy of each change detection technique. Based on accuracy assessment, the image differencing method was the most accurate one with an overall accuracy of 85% in detecting land use/cover changes in Falavarjan area. This was followed by the image rationing technique with an accuracy of 84%.

Keywords: Change detection, Image differencing, Image rationing, Post-classification.

1. Introduction

Change detection is the process of identifying differences in the state of an object or phenomenon by observing it at different times [1]. A variety of algorithms have been developed for change detection including, image overlay, image differencing, image regression, image rationing, vegetation index differencing, principal components analysis, spectral/temporal classification, post-classification comparison, change vector analysis, and background subtraction [1,2].

Among the different change detection techniques, image differencing, image rationing, image regression and change vector analysis (CVA) are widespread [3-6]. In theses algorithms, selecting threshold is necessary to determine the changed areas. Petit et al. (2001) found the combination of image differencing and post-classification was better than the only single method in determining "from–to" change in south-eastern Zambia [7]. Berberoglu and Akin (2009) and Prakash and Gupta (1998) compared different change detection methods. They found that each algorithm have its own merits and advantages [8,9]. Angelici et al. (1977) applied the difference of band ratio data and a threshold method to separate change and no change areas [10]. Jensen and Toll

(1982) and Chavez and Mackinnon (1994) found the usefulness of visible red band data in change detection analysis in both vegetated and urban environments [11,12]. Ridd and Liu (1998) applied four change detection algorithms, including image differencing, regression method, tasselled cap transformation, and Chi-square transformation for urban land-use change detection in the Salt Lake Valley area. They indicated that the regression of TM band 3 was the most accurate for detecting changes [13].

This research compares three techniques, including image differencing, image rationing and image regression to evaluate the most accurate one for change detection in the study area.

2. Methodology

2.1. Study area and data

The study area is Falavarjan area in western part of Isfahan city, which covers approximately 17550.6 ha (Fig.1). It is located in 32°29′-32°37′N and 51°20′-51°35′E. Falavarjan city, located in the center of the study area, is on the bank of Zayandehrud River. Zayandehrud River emanates from Zardkuh Mountain and flows in eastern Falavarjan. The climate is hot and dry with an average temperature of about 16.4°C and average annual rainfall of 162 mm/year. The study area includes agricultural fields, Zayandehrud River, bare lands and urban areas.

In the present study, Landsat Multi-Spectral Scanner image (MSS) 4, 1972 and high spatial resolution (56m) Indian remote sensing satellite (IRS-P6) AWiFS sensor data acquired on September, 2008 were used to detect changes over a period of 36 years.

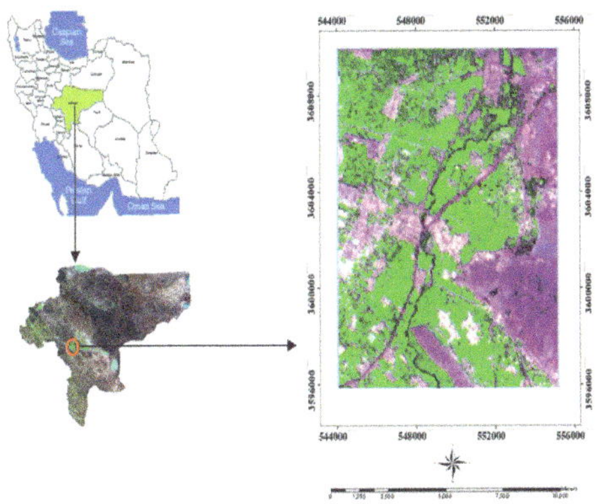

Fig 1. Study area: AWiFS image of Falavarjan area, taken inSeptember 2008 (right) in the west of Isfahan (below left) in central of Iran .

2.2. Image pre-processing

The images were geometrically corrected and geocoded to the Universal Transverse Mercator (UTM) coordinate system using 20 ground control points (GCPs). Resampling was applied using a nearest neighbor method. Root mean square (RMS) error obtained 0.5 pixels for each image.

Radiometric normalization was necessary to reduce differences because of atmospheric or a sensor variation between the two dates. In this paper, the images were radiometrically normalized based on the method developed by Markham and Barker (1986) [14].

2.3. Change detection

In order to detect land cover changes, three common methods, Image differencing, Image rationing, Image regression were applied.

2.3.1. Producing change images with three techniques

Image differencing was applied with each of different bands. The four difference images (Dif1, Dif2, Dif3, Dif4) were created by subtracting the 1972 image from the 2008 image. In this method, digital numbers in the resultant difference image are often considered to be normally distributed where pixels with small change are observed around the mean. Pixels which have been changed largely are distributed in the tails of histogram [1]. For image rationing, the bands were rationed for each image pair on a pixel-by-pixel basis and four change images (Ratio1, Ratio2, Ratio3, Rati4) were produced. The assumption in the image regression technique is that the later image is a linear function of the earlier image. The MSS band 2 was considered as the independent variable and the AWiFS band 3 was taken to be the dependent variable. It was observed a linear relationship between these two images. Then, the predicted image and the base image were subtracted from each other.

2.3.2. Optimal threshold determination

Threshold levels, ranging from 0.1 to 3.0 standard deviations from the mean, were tested on the change images in order to determine the optimal threshold values. Consequently, 1σ was identified as the most accurate one among others as determined from the aerial photographs and ground data. Then, the change images were reclassified into two classes. The value '0' was assigned for 'no change' areas and '1' for change areas.

2.3.3. Classification

Post-classification comparison is an important method in improving the quality of classifications [15-17]. A supervised maximum likelihood classification method was performed for 1972 and 2008 to classify land cover in the study area. Four land use/cover classes including river, bare land, agriculture and urban were observed. This method provides a "from-to" matrix of change information. The change category is divided into five subcategories as shown in Table 1.

Category	From (1972)	To (2008)
0	No Change	
1	Agriculture	Bare Land
2	Agriculture	Urban
3	Bare Land	Urban
4	Bare Land	Agriculture

Table 1. Categories of land use /cover change (1972-2008)

2.4. Accuracy assessment

In order to assess change detection accuracy, an error matrix and a kappa analysis were utilized. The error matrix is the most common method for accuracy assessment [18]. To properly generate the error matrix, Ground data set, air photos and field survey records and RGB composites were used.

3. Results

3.1. Change detection

For accuracy assessment, changed and unchanged pixels were cross-tabulated against the resultant images derived from the different algorithms. Overall accuracies were calculated by dividing the total number of correctly classified pixels to the total number of pixels. Accuracy of change images were estimated at change/no change level. At level change/no change detection the overall accuracies were 85.02% (image differencing), 84.13% (image rationing) and 75.46% (image regression), respectively.

The result from the image differencing and image rationing techniques were very similar (Fig.2). These methods were very effective in separating change from no-change with the visible bands. The MSS band 2 and the AWiFS band 3 had the best accuracies. Therefore, the changed images derived from these bands are more practical than the others for change detection in this study area.

Because the image differencing, image rationing and image regression methods do not provide the detail information about the kinds of land cover change, the outcome of post-classification was crossclassified with the each three techniques to identify "from-to" change and to assess the accuracy of the three change images in detecting the four kinds of change (table 2).

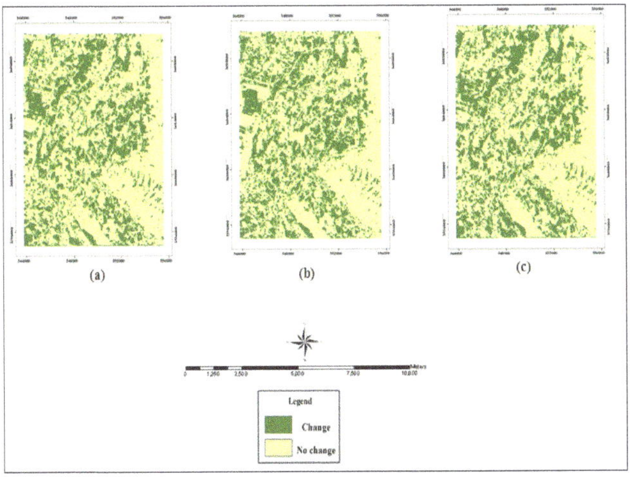

Fig 2. Change images derived from the (a) image differencing, (b) image rationing and (c) image regression.

Based on producer's accuracy, the percent correctly classified for each category, is listed in Table 2. For change type 1, agriculture to bare land, the highest accuracies are from the image differencing and image rationing. For category 2, agriculture to urban, the best result is from the image differencing and image rationing, too. For category 3, bare land to urban, image differencing is the best with an accuracy of 15.44%. For category 4, bare land to agriculture, image differencing at 55.5% is the best, followed by image rationing at 55.23% and image regression at 50.55%. It is observed that the results of image differencing and image regression techniques are very similar. Image regression technique identified all of the categories of change with the least accuracy.

Change detection techniques	Categories of land use/cover change			
	1(%)	2 (%)	3 (%)	4 (%)
Image differencing	93.1	79.81	15.44	55.5
Image rationing	91.81	77.32	8.84	55.3
Image regression	87.93	76.5	14.6	50.50

Table 2. Producer's accuracies of the change images for detecting four kinds of land cover/use change

4. Conclusion

Change detection algorithms have long attracted the attention of the researchers and scientists. In recent years, a variety of approaches have been applied for the monitoring land use/cover change. Each method has some advantages and disadvantages. Many factors such as selection of suitable change detection approach, suitable band and optimal threshold, may affect the success of a classification [19,20].

This research aimed to examine the utility of three techniques, including image differencing, image rationing and image regression in detecting land use /cover changes from 1972 to 2008. Among the different bands, the MSS band 2 and the AWiFS band 3 had the highest accuracies. The optimal threshold was 1 standard deviation from the mean. Results showed that the image differencing and image regression techniques had the highest accuracy in separating change and no change areas. However, these techniques cannot provide a complete matrix of change detection. Therefore, the post-classification method was performed in order to provide details about the nature of changes. In fact, the combination of image differencing, image rationing and image regression with post classification was used. It showed that this technique can provide better change detection results than simple method.

5. Acknowledgments

This project was supported by a grant from Iran National Science Foundation.

6. References

[1] Singh, A., 1989. Digital change detection techniques using remotely sensed data. *International Journal of Remote Sensing*, 10, 989-1003.

[2] Jensen, J. R., 1983. Urban/suburban land use analysis. *Manual Remote Sensing*, 2: 1571-1666.

[3] Woodwell, G.M., R.A. Houghton, T.A. Stone and A.B. Park, 1986. Changes in the area of forests in Rondonia, Amazon Basin, measured by satellite imagery. In: Trabalka, J.R., Reichle, D.E. (Eds.), The Changing Carbon Cycle: A Global Analysis. Springer-Verlag, New York, pp. 242-257.

[4] Coppin, P.R., M.E. Bauer, 1996. Digital change detection in forest ecosystems with remote sensing imagery. *Remote Sensing Reviews*, 13, 207-234.

[5] Fung, T., 1990. An assessment of TM imagery for land-cover change detection. *Transactions on Geoscience and Remote Sensing*, 28, 681-684.

[6] Cohen, W.B., M. Fiorella, 1998. *Comparison of methods for detection of conifer forest change with Thematic Mapper imagery*. In: Lunetta, R.S., Elvidge, C.D. (Eds.), Remote Sensing Change Detection: Environmental Monitoring Methods and Applications. Ann Arbor Press, Michigan, 89-102.

[7] Petit, C. C., and E. F. Lambin, 2001. Integration of multi-source remote sensing data for land cover change detection. *International Journal of Geographical Information Science*, 15, 785-803.

[8] Berberoglu, S., A. Akin, 2009. Assessing different remote sensing techniques to detect land use/ cove r changes in the eastern Mediterranean. *International Journal of Applied Earth Observation and Geoinformation*, 11, pp. 46-53

[9] Prakash, A., R.P. Gupta, 1998. Land-use mapping and change detection in a coal mining area- a case study in the Jharia Coalfield, India. *International Journal of Remote Sensing*, 19, 391-410.

[10] Angelici, G., N. Brynt and S. Friendman, 1977. Techniques for land use change detection using Landsat imagery. *Proceedings of the 43rd Annual Meeting of the American Society of Photogrammetry and Joint Symposium on Land Data Systems*, Falls Church, VA, USA (Bethesda, MD: American Society of Photogrammetry), 217-228.

[11] Chavez, P.S.JR, and D.J. Mackinnon, 1994. Automatic detection of vegetation changes in the southwestern United States using remotely sensed images. *Photogrammetric Engineering and Remote Sensing*, 60, 571-583.

[12] Jensen, J.R., D. L. Toll, 1982. Detecting residential land use development at the urban fringe. *Photogrammetric Engineering and Remote Sensing*, 48, 629-643.

[13] Ridd, M. K., J. Liu, 1998. A comparison of four algorithms for change detection in an urban environment. *Remote Sensing of Environment*, 63, 95-100.

[14] Markham, B.L., J.L. Barker, 1986, Landsat MSS and TM post-calibration dynamic ranges, exoatmospheric reflectance and at-satellite temperatures. EOSAT Landsat Technical Notes, 1, 3-8.

[15] Stefanov, W.L., M.S. Ramsey and P.R. Christensen, 2001. Monitoring urban land cover change: an expert system approach to land cover classification of semiarid to arid urban centers. *Remote Sensing of Environment*, 77, 173-185.

[16] Harris, P.M., S.J. Ventura, 1995. The integration of geographic data with remotely sensed imagery to improve classification in an urban area. *Photogrammetric Engineering and Remote Sensing*, 61, 993-998.

[17] Murai, H., S. Omatu, 1997. Remote sensing image analysis using a neural network and knowledge-based processing. *International Journal of Remote Sensing*, 18, 811-828.

23

Demand for Gasoline in United Arab Emirates

Meena Al-Mansoori, Aydin Basarir, and Sherin Sherif

Agribusiness and Consumer Sciences Department, United Arab Emirates University, UAE

Abstract

Despite being ranked30th world high in terms of the HDI, the demand for gasoline in the UAE is increasing every year, aided by a very high per-capita income that ranks UAE as high as 6th in the world. The number of autos operating in the country is almost approaching the number of population causing carbon dioxide emissions to be alarming. The demand for gasoline in UAE was analyzed by utilizing log-log and ARIMA models for the 1995-2012 period and forecasted up to the year 2020. In the log-log model, quantity demanded of gasoline was regressed on its own price, population, and per-capita income, number of vehicles in UAE, HDI, and a lagged dependent variable. Results showed that population, real income per capita, and number of vehicles had positive impacts on gasoline consumption. On the other hand, real gasoline prices, human development index (HDI), and lagged dependent variable had negative effects on the demand for gasoline. Forecasting analysis showed that gasoline demand for the years following 2012 wills not significantly change. However, because of high standard deviation for each additional year, the estimated confidence interval of gasoline demand amount grew wide.

Keywords: Gasoline Consumption, Log-Log Model, ARIMA Model, Human Development Index (HDI), UAE

1. Introduction

United Arab Emirates (UAE) is made up of seven emirates; namely, Abu Dhabi, Dubai, Sharjah, Ajman, Umm Al-Quain, Ras Al-Khaimah, and Al-Fujairah. UAE is a member of the Gulf Cooperation Council (GCC). Its GDP per capita was estimated at $39,623 in 2011. The population of the country is nearly 8.1 million and its human development index (HDI) is a very high 0.846 (30th world rank) [1]. The country is considered one of the world's major oil producers ranking 6th worldwide in terms of proven oil reserves and 7th in terms of natural gas reserves. Revenues from oil is a major funding source for the UAE economy[2].

The country, however, suffers from high carbon dioxide rates in its air because of excessive use of petroleum products, mainly gasoline, and natural gas[3]. The main objective of this study is to analyze the demand for gasoline in UAE and to identify the factors that affect its consumption. In addition, the paper aimed at obtaining a clue on the future demand for this important product through forecasting. Based on the results obtained, suggestions are lined up.

2. Literature Review

Price elasticity of gasoline demand has been one of the most analyzed topics in energy studies. Havranek, Irsova, & Janda[4]conducted a quantitative survey to determine the elasticity reported for various time periods. By using a mixed-effects multilevel meta-regression it is found that while the long run elasticity of gasoline demand reached–0.31; the short-run elasticity reached only -0.09. By using household-level data, Kayser[5] realized a close relationship between household's car-portfolio and gasoline demand decisions. Since there was only short-run price and income elasticities in gasoline demand, gasoline tax was not likely to result in large decreases in gasoline consumption.

Lise and Van Montfort [6] used a cointegration analysis to unfold the linkage between energy consumption and GDP for Turkey with annual data over the period 1970–2003. Results showed energy saving would not harm economic growth in the country. Nugroho et al.,[7]developed a model of household energy consumption based on in-home and out-home activities.It is found that household vehicle fuel consumption was mostly dominated by the use of cars in Jakarta. Model showed positive influence of vehicle fuel consumption on in-home energy consumption. Sene[8] estimated the aggregate demand for gasoline in Senegal from 1970 to 2008. Results indicated to having short run elasticity smaller than long-run elasticity and that gasoline demand was inelastic with respect to both price and income for both the short and long runs. Shittu et al.[9] examined a cross-section survey of 90 Nigerian households in estimating a system of energy demand equations and elasticities. Income, household ownership of electrical/electronic appliances and automobiles, and household head's age showed significant influence on the relative shares of some of the seven energy commodities in household budgets. Demand for petrol, diesel, and domestic gas were income elastic. Wadud et al. [10]used more flexible semiparametric techniques in studying USA gasoline demand. It is shown that price responses vary with demographic variables such as income, multiple vehicles holding, and presence of multiple wage earners in rural or urban residential locations. Households' responses to a price change decreased with higher income. Dilaver[11] estimated an industrial electricity demand function for Turkey by applying the structural time series technique to annual data over the period 1960 to 2008. Results suggested that output and real electricity prices had an important role to play in driving Turkish industrial electricity demand. The output and price elasticities were estimated to be with low values of 0.15 and –0.16, respectively. In addition to the above, there have been a lot of energy demand studies in the literature. However, the researchers were not aware of any studies that have been conducted in the UAE on the subject matter. In addition, the researchers were unaware of any gasoline-demand study which used HDI in demand analysis. Forecasting has rarely been used in gasoline demand studies. The study in hand took into consideration some of the shortcomings of the published research to cover the topic in the UAE from a socioeconomic perspective.

3. Data and Methods

As a result of the unavailability of UAE monthly or quarterly time series data, annual data was used for this study. Further, appropriate data could not be found before 1995. Consequently, the series was taken after 1995 until first 5 months of 2012. Some of the published data for 2011 and 2012 were estimates of the sources.

The data was collected from different sources such as UAE National Statistical Bureau (UAENSB) [12], Gulf Corporation Council (GCC)[13],Food and Agricultural Organization (FAO)[14],U.S. Energy Information Administration (EIA)[3], International Energy Agency (IEA)[15], and United

Nations Development Program (UNDP)[1]. In order to get real value of per capita income and gasoline prices, the variables have been deflated by the consumer price index (2000=100).

Since the emphasis was on gasoline demand elasticity, the log-log model seemed to be the most appropriate one to use. Following Judge, et al.,[16] the model could be written as follows;

$$logY_t = \beta_0 + \beta_K logX_{tK} + \varepsilon_t \quad (1)$$

Where Y_t the annual amount of gasoline demanded is, X_{tK} stands for the independent variables such as per capita real income, population of the country, the real price of gasoline per gallon, Human Development Index, and lagged values of demand for gasoline. The last term of the equation, $\varepsilon_t\varepsilon_t$, is error term which assumed to have "0" expected value and σ^2 variance. The calculated elasticity of gasoline demand (β_K) in such a model is constant. That is why the equation can be named as constant elasticity model, which is very convenient for economists.

In addition, a forecasting of the demand for gasoline for the next 8 years until 2020 was made. The reason for choosing just 8 years was because of having relatively short time series data and the need to obtain more meaningful estimation. The forecasting of gasoline demand was done by conducting a simple time series model in which the demand for gasoline was regressed on time as follows;

$$Y_t = \beta_0 + \beta_1 t + \varepsilon_t \quad (2)$$

Where tt is time which takes value of "0" for the base year and increases by "1" for each additional year up to the end of time period.

In addition to the simple time series model given in equation 2, an autoregressive integrated moving-average model (ARIMA) was used to forecast the demand for gasoline in UAE. The ARIMA model includes autoregressive (AR), moving average (MA) and nonstationary integrated (I) time series models. That is why the model is more efficient than the AR, MA, and I; and provides more reliable results[17].FollowingJudge, et al., [16]the ARIMA model can be written as follows;

$$Y_t = \delta + \theta_1 y_{t-1} + \cdots + \theta_p y_{t-p} + e_t + \alpha_1 e_{t-1} + \cdots + \alpha_q e_{t-q} \quad (3)$$

Where pp order is for AR, qq is order for MA part of equation 3.

By using the Box-Jenkins methodology with equation 3, forecasted future values of gasoline demand could be obtained. The Box-Jenkins methodology is more useful for short term forecasting[17]. Since we relatively have short term predictions, the methodology could be said to be appropriate.

The data used in this study coveredthe 1995-2012 periods. The variables and their coding in regression analysis could be briefly explained as follows:

Demand for Gasoline (=LNGASDMD): is the total annual quantity demanded of gasoline in UAE.

Income (=LNINC) :is per capita real income.

Population (=LNPOP): is the annual total population of the country.

Price (=LNPRICE): is real price of gasoline per gallon.

HDI (=LNHDI): is human development index of UAE.

CARS (=LNCARS): is the total number of passenger cars in the country for a given year.

LNGASD-1: is the lagged value of quantity demanded of gasoline.

4. Results

In this paper, mainly two different analyses were conducted. The first one was the utilization of log-log model for estimating UAE demand for gasoline by using time series data covering the years from1995 to 2012. The second analysis was forecasting the amount of gasoline that will be demanded for the next 8 years by using simple time series and ARIMA models.

Variables	Estimated Coefficients	Standard Error	T-statistic	Significance Level
Constant	-46.046	11.485	-4.009	0.0001
LNINC	2.202	0.587	3.751	0.0002
LNPOP	1.901	0.643	2.956	0.0031
LNPRICE	-0.242	0.108	-2.239	0.0251
LNHDI	-5.761	2.124	-2.712	0.0067
LNCARS	0.949	0.239	3.968	0.0001
LNGASD-1	-0.107	0.957	-1.118	(0.2636) Insignificant
RHO	-0.418	0.220	-1.900	0.0575

Table 1. Estimation Results of UAE Gasoline Demand

The regression results of the log-log model for gasoline demand model are given in Table 1. Results showed that the estimated coefficients (which are at the same time elasticities) are mostly significant on their effects on the quantity demanded of gasoline as anticipated. As per capita income, population of the country and the numbers of cars increase the quantity demanded of gasoline increases as well. On the other hand, as real price per gallon of gasoline, HDI, and the gasoline demand for the previous year increase the quantity demanded of gasoline decreases as expected, as well.

Variables	Estimated Coefficients	Standard Error	T-statistic	Significance Level
Constant	8036.690	804.145	9.99	< 0.0001
Time	1745.540	80.752	21.62	< 0.0001

Table 2. Forecasting of UAE Gasoline Demand for the Next 8 Years Using Simple Time Series Model (Thousand Barrels per Year)

R^2=0.97, F Ratio = 467.26

The second step was to forecast the demand for gasoline for the next 8 years. The results of simple time series and ARIMA models are given in Tables 2 and 3. The results of the simple time series model showed that the quantity demanded of gasoline increases by 1745.54 thousand barrels for each additional year, as indicated in Table 2.

Variable	Observed Demand for Gasoline	Forecasted Demand for Gasoline	Standard Error	95% Confidence Interval for Forecasted Demand
1995-1999	11325.86			
2000 -2004	19913.06			
2005-2009	29851.59			
2010-20012	35425.09			
2013		36639.2177	3042.5505	30675.9283 - 42602.5070
2014		36639.6372	4263.3769	28283.5721 - 44995.7024
2015		36639.6295	5205.9406	26436.1735 - 46843.0855
2016		36639.6297	6002.2611	24875.4141 - 48403.8452
2017		36639.6297	6704.6597	23498.7382 - 49780.5212
2018		36639.6297	7340.1488	22253.2024 - 51026.0570
2019		36639.6297	7924.8412	21107.2264 - 52172.0329
2020		36639.6297	8469.2639	20040.1775 - 53239.0818

Table 3. Forecasting of UAE Gasoline Demand for Next 8 Years Using ARIMA (Thousand Barrels per Year)

The results of ARIMA model are given in Table 3. The first 3 rows in column 2 are the average gasoline demand for a five-year period. The following row in the same column is the average for 3 years. The forecasted values for gasoline demand are given starting from the year 2013. The forecasted values arevery close to each other and for the majority of years are almost the same. Because of high standard deviation, confidence intervals are getting larger values for each additional year of forecasting.

5. Summary and Conclusion

In this study, the demand for gasoline in the UAE, an oil producing country, has been analyzed. Having positive effects of income, population growth, and number of vehicles and negative effects of the prices on gasoline demand was expected. On the other hand, the negative effect of HDI on the demand for gasoline consumption was considerably important. HDI is an indicator of the country's development. So as the country develops, the awareness of negative impacts of gasoline on the environment is realized. This result is in favor of sustainable development concept. According to the international energy agency, CO_2emissions from the consumption of petroleum were 75.7208 Million Metric Tons in 2009. The country is one of the leading economies which suffer a lot from CO_2 emission. Developing policies to increase environmental awareness and decrease demand for gasoline in the UAE will help enhancing sustainable development.

6. References

[1] UNDP. Human Development Index [Online]. Available: http://hdrstats.undp.org/en/countries/profiles/ARE.html

[2] SCAD, "Statistical Yearbook for Abu Dhabi www.scad.ae," 2012.

[3] EIA. International Energy Statistics [Online]. Available: http://www.eia.gov/cfapps/ipdbproject/IEDIndex3.cfm?tid=5&pid=5&aid=2

[4] T. Havranek, Z. Irsova, and K. Janda, "Demand for gasoline is more price-inelastic than commonly thought," *Energy Economics*, 2011.

[5] H. A. Kayser, "Gasoline demand and car choice: estimating gasoline demand using household information," *Energy Economics*, vol. 22, pp. 331-348, 2000.

[6] W. Lise and K. Van Montfort, "Energy consumption and GDP in Turkey: Is there a co-integration relationship?," *Energy Economics*, vol. 29, pp. 1166-1178, 2007.

[7] S. B. Nugroho, A. Fujiwara, J. Zhang, K. Kanemoto, S. S. Moersidik, and S. Abbas, "Development of a Household Energy Consumption Model for Megacities in Asia."

[8] S. Ousmane Sene, "Estimating the demand for gasoline in developing countries: Senegal," *Energy economics*, vol. 34, pp. 189-194, 2012.

[9] A. M. Shittu, A. O. Idowu, A. O. Otunaiya, and A. K. Ismail, "Demand for energy among households in Ijebu Division, Ogun State, Nigeria," *Agrekon*, vol. 43, pp. 38-51, 2004.

[10] M. A. Wadud, "An analysis of meat demand in Bangladesh using the almost ideal demand system," *Empirical Econ. Lett*, vol. 5, pp. 29-35, 2006.

[11] Z. Dilaver and L. C. Hunt, "Industrial electricity demand for Turkey: a structural time series analysis," *Energy Economics*, vol. 33, pp. 426-436, 2011.

[12] UAENSB. UAE National Statistical Bureau [Online]. Available: http://www.uaestatistics.gov.ae/EnglishHome/tabid/96/Default.aspx

[13] GCC. Digital Library [Online]. Available: http://sites.gcc-sg.org/DLibrary/index-eng.php

[14] FAO. Food and Agriculture Organization [Online].

[15] IEA. Statistics by country/region [Online]. Available: http://www.iea.org/stats/index.asp

[16] G. G. Judge, W. E. Griffiths, R. C. Hill, and T. C. Lee, "Theory and practice of econometrics," 1985.

[17] R. C. Hill, W. E. Griffiths, and G. G. Judge, *Undergraduate econometrics* vol. 2: Wiley, 2001.

24

Ecotourism Planning of Murat Mountain (Usak, Kutahya/Turkey)

Abdurrahman Dinc[1,*], Nurhan Kocan[2]

1 Department of Tourism Guide, Tourism Faculty, Konya University, Konya

2 Department of Landscape Architecture, Aggriculture Faculty, Ege University, Izmir

*Corresponding author

Abstract

In today's life, people are tightening urban environments and search of recreation in natural habitat that turning to mountainous areas and natural resources for tourism activities. Mountain areas with a wide variety of use options for different users these areas of the state makes a focal point for tourism. In this study were investigated original natural and cultural values of Murat Dagi and the immediate surroundings where geographical located in Usak and Kutahya cities. For the ecotourism of the existing and improved features of the area were included of high potential for tourism activities. The environmentally sensitive tourism development and tourism potential of the area and immediate surroundings to the suggestions were made ensure the sustainable development of rural areas.

Keywords: Ecotourism, alternative tourism, mountain tourism, sustainable development, Murat Dagi (Usak, Kutahya/Turkey).

1. Input

Difficult with each passing day ecological conditions in urban areas directs the movement of people and ecotourism. These movements, especially in tourism as the activities carried out in regions rich in natural and cultural values. These visits are not for people just to rest and have fun at the same time developing person's thoughts, providing socialization, enhance awareness of nature, is in the form of visits, including cultural activities. Thus, for natural area tourism activities and resources of significant interest in recent years.

Protection of natural and cultural values, tourism, preserve, evaluate, and play a leading role in the promotion. However, uncontrolled tourism, bringing with it a steadily growing interest in the fields of the original nature, mountainous areas of natural and cultural values is corrupt. Long-term attractiveness and diversity of the mountains, resources, stable, based on prudent and sustainable management. Therefore, the management of mountainous areas, requests for tourists, local people's needs and the need to ensure a sustainable balance between conservation of natural resources [1]. Alternative to natural and cultural resource values of Murat Mountain is a significant potential for tourism.

However, to avoid damage to the structure of ecological tourism, is an absolute must operate within the plan. Murat Mountain surroundings with the aim of this study demonstrate the potential of ecotourism, tourism planning and emphasize the necessity of sustainable development.

1.1. Ecotourism and Mountain Tourism

Removed from the influence of contemporary living and urban ecosystems in time living with nature, human beings found in nature, but it belongs at every opportunity created by nature, if only temporarily, returning to break [2]. Mountains, water, energy, agriculture, forestry and tourism activities are important resources in terms of biodiversity has a key role to. Mountain tourism in mountainous environment, having the cares of beaches and mountain resorts, including recreation and tourism movement. Mountain tourism capital of the mountain and nature itself. The mountains, rocky, forested, mountain ecosystems and water of various qualities, including areas for different activities, allowing each season. Recreation and tourism activities in the purpose of vacation, as well as migration of health tourism in these areas, nature studies, nature and winter sports training is also widely used for such purposes. Usually the mountains from a height of 800 meters above sea level to 2000 meters altitude ranging from the healthy generation and protection of human health to climate areas are attractive places. Heights of 1650-3000 meters in the mountain, winter sports, and sports fields daily heights of 1600-2000 meters, 2000-3000 meters, the planned residential areas, and thus the heights of the mountain is high up in each of the different alternatives are awarded for a tourism activity [3].

Today, tourism and increase the variety of tourism activities and tourism, countries have made steps to extend the season to be benefited from the mountain and winter tourism facilities have engaged effectively. In 2002, the United Nations by the "World Year of Mountains" were announced. Mountaineering, consisting of only the top of a mountain or a rock face climbing but hiking, mountain sports, camping, wildlife viewing, such as a wide range, covering the various activities. Mountain tourism is an activity that can be places where people live close to the environment such as the activities with accommodation in remote areas is carried out in [1]. Positive effects on human health, clean air and microclimate in mountainous areas is a natural demand for the environment increases. Will be possible with ecotourism planning in order to meet this demand.

1.2. Tourism Planning and Sustainable Development

Increasing environmental awareness of the changing trends in world tourism and the development of natural areas, tourism activities necessitate a controlled manner. Tourism planning a rational relationship between requirements and organizes the resources for such purposes. Tourism planning purposes this. relationship between rational resources requirements Ensuring the sustainable use of natural and cultural resources to the evaluation of the physical planning decisions in the field of ecological planning strategies.

The potential that exists to protect nature and take advantage of the maximum level, improving the offer to benefit future generations, do not conflict with each other, basing it on a plan it is possible to determine the usage options [4]. Ecology, sustainable use of natural and cultural resource values are increasing, the need for basic land use planning. This obligation, which continued to increase pressures on natural and cultural environment (land speculation, population growth, environmental degradation, inadequate legal and clearances, improper resource management, etc.). Has become a necessity due to [5]. Important criteria for use of natural and cultural areas, and determination of the planning and promotion, tourism potential and the development of these regions provides an opportunity for economic development of rural areas [6]. One of the most important source of tourism values of the local culture. Crafts and foods, such as peculiar to the region by increasing the demand for local products as a means to protect the local culture, take on an important task [7].

2. Material Method

Murat Mountain and vicinity workspace and material creates. Area of natural and cultural elements carry with attractive features is the potential for tourism and recreation activities. In this study, pre-made materials to help research, domestic and foreign sources and internet data used to establish the relevant literature. To investigate the structure of the natural and cultural areas of field observations made at different times, area photos and on-site evaluations have been taken. Topographic map of the area and the data obtained from field studies revealed the potential for ecotourism Murat Mountain. And sustainable development in light of the evaluations under the proposals have been made for the use of ecotourism.

3. Findings

3.1. Murat Mountain Location and limits

Murat Mountain and the surrounding area has a history older. The old name of Murat Dağı Ddymus. MÖ.1200 'lü years later the Phrygians in Macedonia has experienced with the Byzantine Empire Period. Murat Mountain Germiyanoğulları Turks, and later the Ottoman Empire upon has taken place in the territory. Murat Dagi in the name of the hero of the Murat Gazi \ martyred 1313 from falling. National Park is within the eastern extensions of the mountain, the Commander in Chief. Murat has been fighting in the War of Independence in 1922 foothills [8].

Geographical Features: Murat Mountain, Aegean Region, Department of Domestic Gediz Basin is located in western Anatolia. Kütahya and Usak Usak province in the southern parts of Muratdagi forming the boundary between the northern sections remain within the boundaries of the city of. The study area-29057′38″ 29046′50″ 38049′54″-38055′02″ east longitude and is situated between north latitudes Kütahya [9]. In General, a Northwest-Southeast direction, extending East of the Başkomutanlık National Park, to the West of Mount Murat Gediz Valley and to the South of the mountain, and the Valley of Banaz Çayı Şaphane Elmadag Mountain is located in the North with the Skillet (Figure 1).

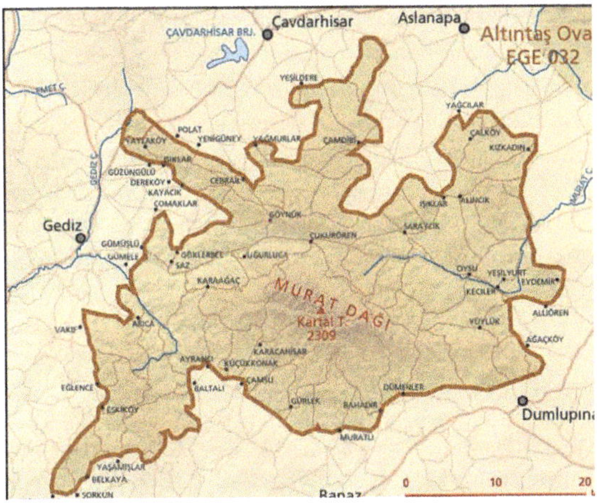

Figure 1. Murat Mountain location (taken from www.dogadernegi.org)

Geomorphological Features: Murat Mountain geomorphological high mountainous areas, valleys and low plains can be considered. Average of Murat Mountain is located on a plateau of the Inner Western Anatolia between 1000-1250 m. Creates the highest point of the mountain, 2309 m and Kartaltepe. Mountains; Kırkpınar T (2218 m), Öküzkayası T (2213 m), Tinaz T (2097 m), Çatmalmezar T (1990 m), Kazıkbatmaz T (1857 m) the high peaks are also available. The average height of 1586 m and geomorphological aspects are of a high mountainous terrain [10]. Faults is limited to the northern edge of the upper part of Muratdagi flattened, and this has led to separation of the plains and plateau areas in the form of steep slopes. Field BAF (Western Anatolian Fault) line is located within the earthquake zone is present. Circus in the higher parts of the mountain lakes (Lake location, Lamb Lake) were also observed.

Climate Features: Muratdagi continental climate, with some features of the transition climate of Mediterranean and Black Sea climate type. Due to its geographical position and topographical characteristics of the study area in summer and winter seasons remain under the influence of different pressure centers. Winters are cold, depending on the rainfall seen in the frontal activity, cool and wet summers than in the Mediterranean climate, the climate characteristics of the transition is observed [8]. Which is closest to the mountain meteorological stations (Usak, Kutahya and Gediz) analyzed data from an annual average temperature values were determined to be between 10.5 to 12.3 0C. Soaring heights of Mount Murat is estimated that this value falls to C 3-4. Field is exposed to winds, especially in the northern sector. During the summer winds called etesian effect. Ranged between 534-571 mm of annual rainfall of the area in the northern slopes are much higher than the amount of rainfall [11].

Hydrography: Murat Mountain is located within three major river catchment. Gediz Basin in western and northwestern parts of the field (Murat River) eastern and northeastern parts of the Sakarya River Basin (Badger Creek) in the southern parts of Great Menderes Basin (Great Menderes River) enters into. There are five lakes around Mount Murat [8].

Vegetation: The vegetation is different from the northern and southern slopes. Northern slopes of the Black Sea or the European-humid and semi-humid forests of Siberia, biogeography, dry forests of the southern slopes of the Iran-Turan seen biogeography. 22 000 34 000 hectares and hectares of forested Muratdagi high forest, 12 000 hectares of forest is damaged or coppice form. The study area in terms of elevation-climate zones 3 major red pine (900-1150), Larch (1150-1550), Scotch Pine (1600-1750) and two intermediate Red Pine+Larch, and Larch+yellow pine(1550-1600m) divided into generation. Type of climate transition Murat Mountain ensured that type of climate is very rich in endemic species. Taksondan 114 were grown in 890 endemic for Turkey. 3 One of these species (*Alyssum davisiasum, Verbascum coropifolium, Sedum hispanicum*) Murat Mountain endemic for [12].

3.2. Potential for Tourism and Ecotourism Events Murat Dagi

Murat Mountain, with the decision of the Council of Ministers No. 11 608 dated 18.03.1987 "Tourism Center" has been announced. Ottoman history dating back to 1450 meter high mountain is located in the hot springs. The total flow of spring water 45.26 l / s, the temperature is 29-50 0C sulfate, calcium, magnesium and contains bromide. Kocahamam, Hacethane, Sarıkız, five baths are available with the name of Governor and Scabies Bath. The center of the spa 18 bungalows, 154 rooms, 13 apartments, 80 wooden houses 616 beds, 2 baths, 2 swimming pool, guest house and conference center with 15 beds serving 200-bed camp. According to the records more than 15000 visitors per year of thermal tourism center, and visitors coming for tourist purposes' type have been the residence of Murat Mountain is an area rich in permanent and seasonal water sources. Murat Mountain is the source of three major rivers. An important source of the rivers for water-based tourism activities. Value as a major tourist flows through the northeast part of the Badger Creek Kesiksöğüt Pınarı'ndan source of the water, forms a part of the source of the Gediz River flows towards the northwest. Menderes River in the south of the mountain, the source of the River of Tepelene [8].

They are created by splitting the beds of rivers waterfalls deep in the field (Gürlü Dere) important elements of the landscape. Near the summit of the mountain formed part of a circus element in lakes is a major tourist attraction, especially Lake Lamb. Initially, caused by thermal tourism Murat Mountain and then in a large pine forests because of the rich vegetation and clean air has become the center of mountain tourism plays an important role. Gökoluk Murat Mountain tourism center, Municipal, Dokuzçam, Cherry and the ability to stay on the plateau is called Kıcıroğlu. Tourism around the yellow flowers close to the center, İkizce, Arif Tosun, there are plateaus Söbealan and other so-called Karaağıl [10].

Murat Mountain in winter other climatic charm. Plateau and Ridge Söbealan Karakötek length, width, ideal for winter tourism in terms of examination. Changes in climatic conditions within short distances seen Murat Mountain. This area has made a rich floristic cover. Areas in the Mediterranean, Black Sea and is a land climate, plant species. Space enthusiasts have flocked especially in spring and summer herbs. Tepelene Baklan Pine hill south of the so-called black pine (Pinus nigra) are memorial trees. Memorial tree 11 meters long, 3.05 meters in diameter, a large tree is 500 years old.

As an alternative to other types of tourism on Mount Murat is made; hiking, nature observation, plant / animal observation, mountain biking, rock climbing, paragliding, moto cross, horseback riding, fishing, camping, picnicking, photo safari, nature photography, documentary shooting, scenic viewing, religious tourism (Murat Dede Mausoleum) such activities (Figure 2, 3).

Figure 2. Nomad life and nature walks in Murat Mountain

Figure 3. Nature photography and a picnic in Murat Mountain

Murat Mountain is also important for wildlife. Area suitable for hunting tourism nature. Spa 2 km far away from the speakers are producing farm lara. Murat Mountain animals are protected species of wildlife such as otter and badger includes monitoring and research opportunities. Murat Mountain rocks formed in different geological periods, a tourist attraction and has the geomorphologic structure of visual beauty. Geomorphology and geology is the structure of space-science education and as a practice field for the potential. Murat Mountain and around the village settlements on the plains there are under 1300 meters. Those of authentic village houses in the villages and settlements in the forest farm tourism, farm tourism and provide opportunities for activities such as examination of the traditional ways of life. At these locations, the tarhana soup unique to the region, Gediz kebab, halvah, bending an important food for the tourism sector are foods such as pancakes. In addition, local crafts and gift market of products of wood, embroidery products.

4. Conclusions and Recommendations

In case of physical improvements and provide for tourism and conservation policies in the field of Murat Mountain will be a center for ecotourism, sustainable development, will provide the people of the region.

Murat Mountain is an area up to now has succeeded in preserving the habitat of. However, the current situation in tourism development plan achieved. Tourist area of hardware and infrastructure, the necessary accommodation for organized tours, such as lack of food and beverage facilities include the physical deficiencies. Daily or weekend visitors picnic and camping environment pollution is waste. This issue is not enough supervision. It is not sufficient to control later in this topic. In addition, the lack of awareness of the protection of the field with the introduction of such problems. Field trips and nature walks suitable for tourism, local product outlets, cafes, restaurants, tea garden, tea house, living and recreation areas, kayakevleri, camping and picnic areas will be the charm and the space to portray the socio-cultural life.

Tourism management plan for tourism infrastructure, and the local people's support is required. The scope of the project; village units and sales of products grown locally established guidelines. Local people, lived through the region's natural and cultural values from all walks of life without losing the economic income would be the protector of these values. Documentary films and festivals should be done visually in the form of promotion. Space that makes it an attractive spa services. In order to use resources more efficiently the thermal rearrangement of the existing accommodation facilities in and near the modern way to bring sports facilities will be positive. Field is located near the villages, gardens and farm tourism for tourists from the area of a portion of the improvement can be edited. Walking on the route of tourists and villagers sell their interest in local products such as fruits and vegetables are an expression of this need. All of these natural and cultural resource values in the Inner Western Anatolia, Murat Mountain primarily a source of important alternative for the country.

5. References

[1] Somuncu, M. 2004. Dağcılık ve Dağ Turizmindeki İkilem: Ekonomik Yarar ve Ekolojik Bedel. Coğrafi Bilimler Dergisi, Sayı:2, Cilt:1, S.1-21.

[2] Özkan, M.B. 2002. Kırsal Rekreasyon Alan Planlaması. Ege Üniversitesi Ders Notları, S.52, İzmir.

[3] Ülker, İ. 1992. Dağ Turizmi Planlama Yöntemleri, Yüksek Dağlarımız, Kayak Merkezleri. ISBN: 975-7478-19-9, Ankara.

[4] Başal, M. 1998. Doğalgaz-Yapracık Tesisleri Alan Kullanım Planlaması. Ankara Üniversitesi, Ziraat Fakültesi Yayın No:1104, Bilimsel İnceleme ve Araştırmalar:602, Ankara.

[5] Yeşil, M. 2010. Tozanlı Havzası Tokat-Almus İlçesi Ekolojik Temelli Kırsal Peyzaj Planlaması. Atatürk Üniversitesi, Fen Bilimleri Enstitüsü, Doktora Tezi, S.193, Erzurum.

[6] Mullens, J.B., Bristow, R.S. 2003. Overcomig The Nation's Best Landscaped Sewer Recreator's Perceptions of the Connecticut River. Journal of The American Water Resources Association, P.7-15.

[7] Tisdel, C. 1996. Ecotourism, Economics and the Environment: Observations from China. Journal of Travel Research, Volume:34, Issue:4, P.11-19.

[8] Aytaç, A.S. 2003. Murat Dağı'nın Milli Park Planlaması ve Çevre Eğitimi Açısından Değerlendirilmesi. Dokuz Eylül Üniversitesi, Eğitim Bilimleri Enstitüsü, Yüksek Lisans Tezi, S.160, İzmir.

[9] Altun, L., Yılmaz, E., Günlü, A., Ercanlı, İ., Usta, A., Yılmaz, M., Bakkaloğlu, M. 2007. Murat Dağı (Uşak) Yöresinde Yayılış Gösteren Ağaç Türlerinin (Kızılçam, Karaçam ve Sarıçam) Verimliliğini Etkileyen Kimi Ekolojik Etmenlerin Araştırılması. Kastamonu Üniversitesi, Orman Fak. Dergisi, Cilt:7, No:1, S.71-92, Kastamonu.

[10] Özav, L. 1995. Turizm Açısından Murat Dağı'nın Önemi. Atatürk Üniversitesi, Türkiyat Araştırmaları Enstitüsü, Sayı:4, S.115-141, Erzurum.

[11] Devlet Meteoroloji İşleri Genel Müdürlüğü. 2010. Uşak, Kütahya, Gediz Rasat Verileri. http://www.meteor.gov.tr.

[12] Çırpıcı, A. 1989. Murat Dağı'nın Florası. Doğa Dergisi, Sayı:2, Cilt:13, TÜBİTAK, Ankara.

25

Sustainability Impact Assessment of Watershed Programs

Korous Khoshbakht[1], Hadi Veisi[1,*], and Mohammad Ebrahim Rezai[2]

1 Environmental Sciences Research Institute, Shahid Beheshti University, G.C., Evin, Tehran-Iran,

2 Agronomy and Plant Breeding Department, University of Mohaghegh Ardabili, Iran

*Corresponding author

Abstract

This paper assessed the impacts of watershed programs (WPs) on agro-ecosystemsin Hamedan Province, Iran. This study's methodology was based on criteria for the comparison of agro-ecosystems with and without WPs and nine indicators was selected to assess the sustainability impacts. The results revealed that WPs have contributed in raising the stability, productivity and relative acceptability of the agro-eco-system by improving the indicators for ecological sustainability,generating a better benefit-cost ratio and promoting the quality of life and access to public services for farmers in agro-ecosystems with a project. However, the involvement of stakeholders in program implementation and monitoring has reduced. At the end, we recommend demand-driven WPs rather than supply driven ones to be increase the social acceptability of WPs.

Keywords: Assessment, Agricultural Sustainability, Watershed Programs, Iran

1. Introduction

Concern about widespread soil degradation and scarce, poorly managed water resources has led to the spread of watershed management investment throughout Iran (Golrang et al., 2006). Despite the growing importance of watershed projects as an approach to rural development and natural resource management, to date there has been relatively little research on their impacts. So, despite the high political priority of the sustainable management of the country's land and water resources in Iran and elsewhere, the contribution of these projects to equity, the quality of life, and perceived wellbeing of rural community seems more uncertain (Ahmadvand and Karami, 2009). Therefore, research is needed to ensure that new projects supply agro-ecosystem sustainability.In order to obtain spatially and temporally acceptable indicators here, in view of biophysical and socio-economic conditions in the study area in the previous studies (Farshad and Zinck, 2001; Ali-Beigi and Baboli, 2008; Hayati and Karami, 2007; Ireavani and Darban-astaneh, 2004; Shahi et al., 2009; Hashemi et al., 2010; Hosseini et al., 2007; Allahyari, 2010; Ahmadvand et al., 2009; and Veisi et al., 2012) 9 indicators representing ecological, economic and social impacts of watershed projects were selected.

2. Methods

A causal comparative method was used to understand better the sustainability impacts of the WPs at Gonbad-chai watershed. Two villages – Gonbad-chai and Tahon-abad – were meant to be the beneficiaries of the WPs and so they were included in the study as villages of agro-ecosystem with the watershed program. The villages of Bagche, Sabz-abad and Gorgoz also were selected as the villages in agro-ecosystem without WPs that located in the same watershed basin as the watershed project villages; have similar cultural and social characteristics; and be relatively close to the villages with WPs. Face-to-face interviews were used to collect data with the aid of a questionnaire containing open and closed questions. The sample size for the household survey was determined by using the formula given by Cochran (1977). Altogether, 69 households were surveyed from an agro-ecosystem with WPs and 67 from an agro-ecosystem without WPs, representing above one-third of the households of each area. The social impact items were categorized according to three social criteria. They include: quality of life (7 items); access to public services (6 items); social capital and social structure (6 items). The economic impact item was marginal cost-benefit ratio. The ecological impact items were categorized into five ecological criteria as follows: crop management (10 items), biodiversity (5 items), soil health (9 items), hydrological processes (2 items) and energy (5 items) (Table 2). Weighting the data was undertaken using conjoint analysis such as Sydorovych and Wossink (2008) and the data were standardized; the weight of each indicator is shown in Table 1. A 'T' test was employed to test the differences between agro-ecosystems with and without WPs.

3. Results

3.1. Impacts of the WPs across the social criteria

Quality of life: The analysis of the impact of WPs on rural quality of life indicated that there were no significant differences in quality life between agro-ecosystems with ($X= 43.07$) and without WPs ($X= 40.81$, $T = 1.59$; $P> 0.05$). It is concluded that the respondents in villages with WPs believed their quality of life has improved. This is somewhat congruent with the findings of Ahmadvand and Karami (2009), regarding the impact of sustainable land and water resources management on the quality of life of the rural community. However this result is opposite to the causal conclusions of Kerr and Chung (2001), which expressed that landless people felt harmed by the projects, but to create jobs for these people in the study area is different from the results.

Access to public services: Equality of access to public and support services can ensure social stability and encourage farmers to improve production while conserving resources (Dang, 2001). The findings suggested that there was a significant difference between agro-ecosystems with ($X= 47.87$) and without WPs ($X = 38.53$) with regard to access to public services ($T=4.79$; $P<0.001$). Increase in access to public services in agro-ecosystem with WPs is relation to the distance between this villages and urban areas. The status indicator has not been reported in other studies.

Social capital: Comparison of social capital showed significant difference ($T=2.65$; $P<0.01$) between agro-ecosystems with ($X= 72.69$) and without WPs ($X = 78.57$). The respondents in agro-ecosystem with WPs perceived that their social capital had declined. These results are consistent

Indicators	Weight indicators	Watershed projects				P	T
		Without		With			
		SD*	Average	SD*	Average		
Quality of life	.069	40.81	8.43	43.07	8.09	-1.597	.113
Access to public services	.084	38.53	11.78	47.87	10.98	-4.786	.000
Social capital and social structure	.070	78.57	10.77	72.69	14.71	2.654	.009
Cost-benefit ratio	.015	49.15	13.09	65.34	13.67	-7.050	.000
Crop management	.019	43.90	5.46	46.06	5.24	-2.351	.020
Biodiversity	.045	46.86	20.76	61.74	17.74	-4.486	.000
Soil health	.108	54.63	6.85	58.76	5.72	-3.821	.000
Hydrological processes	.091	44.13	6.63	54.27	14.89	-5.104	.000
Energy	.064	55.90	9.95	69.21	11.17	-7.333	.000

Table 1. Comparing the sustainability criteria and the weight of each indicator

with the findings of Ahmadvand and Karami (2009), and in contrast with the results of Brouwer and Van Ek (2004) which who concluded that the new policy of water management in The Netherlands improved the social capital of farmers.

3.2. Impacts of WPs across the economic criteria

Cost-benefit ratio: Comparison of the cost-benefit ratio showed significant difference ($T=7.05$; $P<0.001$) between agro-ecosystems with ($X= 65.34$) and without WPs ($X=49.15$). In other words, the marginal cost–benefit ratio was raised in agro-ecosystem with WPs. The patterns of these findings are congruent with recent development efforts in rural Iran, such as those described by Ahmadvand and Karami (2009), Golrag et al. (2006), Parizanganeh et al. (2008) and Sadeghi et al. (2003).

3.3. Impacts of WPs across the ecological criteria

Crop management: Comparison of crop management showed strategies significant difference ($T= 2.35$; $P< 0.05$) between agro-ecosystems with ($X=46.06$) and without WPs ($X=43.90$). It seems that improvements in farm management practices are the effect of an increase in economic power, better access to agricultural experts and an increase in knowledge and education. The status indicator has not been reported in other studies, but some components of the indicator, such as farm size, agree with the results of the study of Parizanganeh et al. (2008).

Biodiversity: The above facts clearly indicate that the WPs altered the land use system in favour of hourticultural crops, mostly fruits. The findings susggested that there was significant difference in agro-ecosystem with ($X=61.74$) and without WPs ($X= 41.86$) with regard to access to public services(T=2.35; P< 0.05). This confirmed with the findings of Palanisami and Kumar (2009) and Singh and Prakash (2010).

Soil health: The analysis of the effect of WPs on soil health indicated that there were significant differences in soil health between agro-ecosystem with (X=58.76) and without WPs (X=54.63, T=3.82; P<0.001).This was in agreement with the findings of Palanisami and Suresh Kumar (2009), who noted the positive impact of a watershed management project to protect, fertility and organic matter in the soil.

Hydrological processes: Better-performing projects have been based on promoting communities' traditional water harvesting and conservation practices (Sharma, 2003: 76). The findings suggested that there was a significant difference in agro-ecosystems with (X = 54.27) and without WPs (X = 44.13) with regard to hydrological processes (T=5.10; P<0.001). Also, improvement in hydrological conditions also conforms to the results of studies Ahmadvand and Karami (2009) and Palanisami and Suresh Kumar (2009).

Energy: Comparison of energy showed a significant difference (T=7.33; P<0.001) between agro-ecosystems with (X=69.21) and without WPs (X=55.90). Improved energy indicators were seen, thus increasing yield, reducing energy used for pumping water and use of surface water for irrigation. The status indicator has not been reported in other studies.

In order to achieve an adequate integration and synthesis of the results, on AMOEBA diagram was used (Lopez-Ridaura et al., 2002). This diagram shows, in qualitative terms, to what extent the objective has been met for each indicator and it enables a simple, yet comprehensive, graphical comparison to be made of the advantages and limitations of WPs under evaluation (Fig. 1). Comparing the results of agro-ecosystems with and without project revealed that project had negative impacts on some social criteria, including development of social capital and social structure. Also, the project had a positive impact on indicators of access to public services, rural and agricultural economic conditions, crop management procedures and conservation of community resources, such as the conservation of biodiversity and water resources.

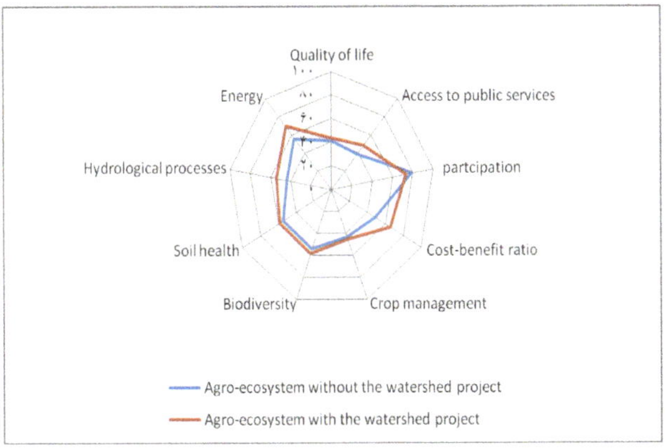

Fig 1. Integration of sustainability indicators for two agro-ecosystems using an AMOEBA diagram

4. Conclusions

The analysis showed that the impacts and benefits of the watershed program were significantly greater in those agro-ecosystems with WPs as compared to agro-ecosystems without WPs. Based on the findings of this study, it can be said that WPs have contributed to raising the stability of the agro-eco-system by improving the ecological sustainability indicators such as biodiversity, soil health, hydrological process, energy use and crop management. This is in accordance with the conclusions obtained by Joshi et al. (2008) who cited that a watershed program provides multiple benefits in terms of conserving soil and water resources. This finding also verifies the conclusions of Altieri (2009) who stated that the farmer can increase the stability of the agro-ecosystem by adopting key ecological management practices. In terms of economic efficiency, WPs generated a better benefit-cost ratio. Following Joshi et al. (2008) and Sreedevi et al. (2006), it can be concluded that the watershed program is a vehicle of development to raise farm productivity, so that investing in a watershed program will increase good net present value and internal rate of returns, while maintaining a good environment and soil –to- water balance in the watershed. Concerning social indicators, in agreement with the conclusions of Ahmadvand and Karmai (2009), the findings indicated that the projects had negative impacts on farmers' participation and it had positive impacts on quality of life and access to public services for farmers in agro-ecosystems with a project. Thus, projects initiated in the study area to improve the conditions of rural communities in dry regions appear to be socio-economically acceptable but socio-culturally undesirable. According to these findings, although increased income from agriculture alongside better access to public services enabled farmers to improve their standard of living; their level of participation declined. These findings are contrary to expectations that the project would have positive social impacts on rural communities. We interpret this anomaly with regard to the fact that this generation of WPs in the study area was supply-driven, so that government officials identified locations and decided on various activities for the implementation of WPs (Sreedevi et al., 2006). In contrast, Joshi et al. (2008) asserted that important conditions of people's participation are related to (1) demand-driven watershed projects rather than supply driven ones, (2) involvement of all stakeholders in program implementation and monitoring and (3) decentralization of the decision-making process.

5. References

[1] M. Golrag , J.Ghoddosi, M.Mashayekhi. Socio-economic and environmental impact assessment of watershed management plans (Case study: plan, organize Znjanrvd). *Geog. Res.* 2006, 21 (4): 93-110.

[2] M.Ahmadvand, E. Karami. A social impact assessment of the floodwater spreading project on the Gareh-Bygone plain in Iran: A causal comparative approach. *Env. Impact Asses. Rev.* 2009, 29: 126–136.

[3] A.Farshad, A.J. Zinck. Assessing agricultural sustainability using the six-pillar model: Iran as case study, in: Gliessman, S.R. (Ed), *Agroecosystem sustainability: developing practical strategies*. CRC Press LLC, Boca Raton, London, New York and Washington, D.C. 2001, pp. 137-151.

[4] A. Ali-Beigi, M. Baboli. Evaluation of the Agricultural Sustainability of irrigated wheat farmers of Sarpol-Zahab: application of DSR. Model. *Iran J Agric Res*. 2008, 39(1): 111-122.

[5] D.Hayati, A. Karami. Knowledge structures that influence the sustainability of farming and cropping systems (case study in Fars Province). *Agri. Sci. Nat. Resour.* 2007, 3 (2): 21-33.

[6] H.Ireavani, A. Darban-astaneh. Measurement, analysis exploitation of the sustainability of farming systems (case study: wheat production, Tehran province). *J. Agri. Sci.* 2004, 35 (1): 39-52.

[7] H.Shahi, H. Irvani, K.H. Kalantari. Assess the sustainability of conservation farming cooperatives of wheat producers covered by the Fars province, *Eco. Res. Agri. Dev.* 2009, 2-40(2): 135-143.

[8] F.S.Hashemi, K. Khoshbakht, A. Mahdavi Damghani, H. Veisi, H. Liaghati. A survey of agrobiodiversity in Gachsaran county and influence of climatic factors. *J. Agroecology*. 2010, 2(1): 1-12.

[9] M.Hosseini, K.H. Kalantari, M.K. Naderi. Stability of ecological systems in sub-owner Saleh-Abad, Hamedan. *Agri. Sci.*, 2007, 2-38 (1), 91-98.

[10] M.S. Allahyari. Social sustainability assessment of fishery cooperatives in Guilan Province, Iran, *Fish. Aqu. Sci.* 2010: 5, 216-222.

[11] M.Ahmadvand, E. Karami, G. Zamani, F. Vanclay. Evaluating the use of Social Impact Assessment in the context of agricultural development projects in Iran. *Env. Impact Asses Rev.* 2009, 29, 399–407.

[12] H. Veisi, K. Khoshbakht, H. Sabahi. A Participatory assessment of sustainability of the agro-ecosystem of Absard, Iran. *Int. Jo. Agri. Sus.* 2012, 10 (4): 1–17.

[13] Cochran, W.G. *Sampling Techniques*. Wiley, New York, 1977.

[14] O. Sydorovych, A. wossink. Anpplication of conjoint analysis agricultural sustainability assessment. *12th EAAE congress (People, Food and Environments: Global Trends and European Strategies)*. Gent (Beglgium). August, 2008, 26-29.

[15] J. Kerr, K. Chung. *Evaluating watershed management projects*. International Food Policy Research Institute, Washington, D.C. 2001.

[16] N.V. Dang. *Farm household development in the midlands of Vietnam: a case study of Doan Hung District, PHU THO province*. Master's thesis. Asian Institute of Technology, Bangkok, Thailand, 2001.

[17] R. Brouwer, R. Ek.Van. Integrated ecological, economic and social impact assessment of alternative flood control policies in the Netherlands. *Eco. Econ.* 2004, 50, 1-21.

[18] A.H. Parizanganeh, H.J.M. Yazdani. Socio-economic and environmental impacts of the watershed management projects: a case study of the Zanjan River restoration project. *Proceedings of TAAL -2007: The 12th World. Lake Conference,* Jaipur, India. 2008, 1214-1219.

[19] S.H.R.Sadeghi, F. Sharifi, E. Forootan, M.Rezaee, Quantitative performance evaluation of watershed management measures (Case Study: Keshar Sub-Watershed*). Pajouhesh & Sazandegi.* 2003, 65, 96-102.

[20] K. Palanisami, , D. Suresh Kumar. Impacts of Watershed Development Programmes: Experiences and Evidences from Tamil Nadu. *Agri. Econ. Res. Rev.* 2009, 22 (Conference Number), 387-396.

[21] S. Sharma. Rethinking watershed development in India: Strategy for the twenty-first century, in: Achouri, M., Tennyson, L., Upadhyay, K., White, R., (Eds.), Preparing for the next generation of

watershed management programmes and projects. *Proc. the Asian Regional Workshop, Kathmandu, Nepal*, 2003, pp. 69-77.

[22] J. Lopez, J.Scott. *Social structure*. Open University Press, Buckingham, 2000.

[23] J P.Koshi, A.K. Jha, P.Wani Suhas, T.K. Sreedevi, F.A. Shaheen, Impact of Watershed Program and Conditions for Success: A Meta-Analysis Approach. *Global Theme on agro-ecosystems Report no. 46*, international Crops Research Institute for the Semi-Arid Tropics, Patancheru, Andhra Pradesh, India, 2008, pp. 502-324.

[24] M.A.Altieri, P. Koohafkan, *Enduring Farms: Climate Change*, Smallholders and Traditional Farming Communities. Third World Network, Malaysia, 2008.

[25] T.K.Sreedevi, S.P.Wani, R.Sudi, Patel, M.S., T.Jayesh, S.N.Singh, T. Shah. On-site and Off-site Impact of Watershed Development: A Case Study of Rajasamadhiyala, Gujarat, India. *Global Theme on Agro-ecosystems Report no. 20*, International Crops Research Institute for the Semi-Arid Tropics, Patancheru, Andhra Pradesh, India, 2006, p. 44.

Efficacy of Different Plant Extracts Against Diamondback Moth, *Plutella xylostella* (L.) on Cauliflower

Jan M Mari

Department of Plant Protection Sindh Agriculture University, Tandojam, Pakistan

Corresponding author

Abstract

The results of studies on the efficacy of different bio-pesticides against 2nd and 3rd instar *Plutella xylostella* larvae on cauliflower under laboratory conditions were carried out in the Department of Entomology. After 24 h the neem extract was found to be the most effective treatment with maximum (14.67%) mortality followed by tobacco extract, datura, akk and control treatments it was 13.33,10, 11.33 and 0.67%, respectively. It is evident from the data that mortality of 2nd instar larvae after 48 hours was observed increasing with the rate of 23.33, 16.67, 12.67, 14 and 0.67% for neem, tobacco, datura, akk extract and control treatment, respectively. It can be also observed from the data that on 72 hours increasing trend was remained continue with maximum 32, 25.33, 16.66, 20.66 and 4% for neem, tobacco, datura, akk extract and control treatment, respectively. The reductions in mortality in 2nd instar larvae, among plant extracts were observed after 96 hours and it was 21.33, 18, 13.33, 16.66 and 4% when treated with neem, tobacco, datura, akk extract and control, respectively. It was concluded from the data that mortality of L3 after 48 hours was observed increasing with the rate of 20.66, 14, 11.33, 12 and 0.66% for neem, tobacco, datura, akk extract and control treatment, respectively and after 72 hours increasing trend was remained continue with maximum 26, 19.33, 14, 17.33 and 3.33% for neem, tobacco, datura, akk extract and control treatment, respectively. The reductions in mortality in L3, among plant extracts were observed after 96 h and it was 18, 16, 12, 15.33 and 4% when treated with neem, tobacco, datura, akk extract and control, respectively.

Keywords: efficacy, plant extracts, diamondback moth and cauliflower

1. Introduction

The diamondback moth (DBM), *Plutella xylostella* (Linnaeus), is an important and multicultural pest that feeds exclusively on crucifers [1, 2]. It was [1] reported that in the tropics and subtropics region *P. xylostella* has become a common and a major obstacle in economic growth of cruciferous vegetables. Crucifers' grown-up in extensively hot and humid areas, where *P. xylostella* continues to cause severe loss and often causes a complete loss of the crop [3, 4 and 5]. It is the greatest threat to crucifer production in many parts of the world, sometime causing more than 90% crop loss [6]. Pesticides have been the primary means to control *P. xylostella* for more than 40 years [7].

Due to indiscriminate use of pesticides it has developed a resistance, now it has been very difficult to control his increasing population, particularly in Southeast Asia and the Far East [8, 9]. It is also time intensifying concerns about the long-term environmental impact of vegetable production, focusing particularly on the heavy use of pesticides with local health hazards, pesticide residues for consumers, the build-up of resistance and contamination of the environment. These problems have increased the interest in alternative control methods, such as integrated pest management (IPM), crop rotation and biological control [10]. New technologies, such as plant extract pesticides, are starting to overcome the problem of resistance. Keeping the view the work was done in laboratory and studies were carded out specifically to determine the effect of different plant extracts on survival of 2nd and 3rd instar larvae of *P. xylostella* on cauliflower

2. Material and Methods

The present studies on the efficacy of different plant extracts against diamondback moth, *Plutella xylostella* (L.) on cauliflower under laboratory conditions were carried out in the Department of Entomology.

2.1. Culture of the host

P. xylostella adults were obtained from the laboratory of Agriculture Research Institute Tandojam to observe the efficacy of bio-pesticide on 2nd and 3rd instar larvae of *P. xylostella*. The Chinese cabbage leaves were boiled in a pan. The stock was used to soak parafilm which was then allowed to dry. This was placed inside a container ready for oviposition by adult Plutella. The adults were removed after 24 hours and the eggs given time to develop at 25 °C and 70% relative humidity. Upon hatching the first instar were transferred in screen cages (42cm X43cm X55cm) in laboratory room at a temperature of 25 (±5) ^{0}C. Their larvae were fed with Chinese cabbage leaves after emerging 2nd instar they were shifted to leaf discs in a 5 cm Petri dishes. Moist filter paper was laid underneath to the disc to delay desiccation. Four hundred µl of water applied to the filter paper to moisten it. Five larvae were placed into each Petri dish acting as one sample unit or replicate. Same procedure were used for 3rd instar

2.2. Plant materials

Extracts were prepared from 4 plants leaves of akk (*Calotropis procera*), datura or jimson weed (*Datura stamonium*), neem (*Azadirachta indica* A. Juss) and tobacco (*N. tabacum*) were put in the local grinder (manually used). After that the extracts were set in the muslin cloth and squeezed them. The 100% of extracts preserved in well-cleaned bottles. Stock of distilled water was also obtained from local market for preparation of the solutions of suspension 2% solutions of extract were separately mixed with 98% distilled water of akk, datura, neem and tobacco respectively steeped in water.

3. Leaf disc bioassay

Bioassays were conducted with 2nd and 3rd instar larvae of *P. xylostella* on Chinese cabbage leaf discs. Test solutions were prepared in 98 ml distilled water with 2 ml pure extract as an additional surfactant. Each leaf disc (4.8 cm dia.) was immersed in a test solution for 10 s. For control, leaf disc were dipped in surfactant solution without any test solution. The leaf discs were placed in individual Petri dishes (5 cm dia.) containing a moistened filter paper and Five larvae (L2) were placed in each dish 6 replicates were made per treatment including the control. The mortality was assessed after 24, 48, 72 and 96 h. same procedure was used for L3. For all residual analysis data were corrected for mortality, using Abbott formula [11]

4. Results

4.1. Effect of different plant extracts on mortality of 2nd instar larvae

The regression analysis depicted that efficacy of extracts was less in initial hours against 2nd instar larvae and there was positive correlation between extracts efficacy with time intervals, it increased with increasing time and reached their highest on 72 hours. The regression model further indicated that the mortality reached their peak just few hours before 72 h on 67.58 h for neem followed by tobbaco, dhatora and akk it was 69.46, 69.68 and 68.61. R-squire of regression was about 0.75 for neem followed by tobbaco, dhatora and akk it was 0.49, 0.44 and 0.43, respectively. It revealed that 75, 49, 44 and 44% variation in mortality was counted by time interval.

The neem extract was found to be the most effective treatment with maximum (14.67%) mortality followed by tobacco extract, datura, akk and control treatments 13.33,10, 11.33 and 0.67%, respectively after 24 h. It is evident from the data that mortality of 2nd instar larvae after 48 hours was observed increasing with the rate of 23.33, 16.67, 12.67, 14 and 0.67% for neem, tobacco, datura, akk extract and control treatment, respectively. It was also observed from the data that on 72 hours increasing trend was remained continue with maximum 32, 25.33, 16.66, 20.66 and 4% for neem, tobacco, datura, akk extract and control treatment, respectively. The reductions in mortality in 2nd instar larvae, among plant extracts were observed after 96 hours and it was 21.33, 18, 13.33, 16.66 and 4% when treated with neem, tobacco, datura, akk extract and control, respectively.

4.2. Effect of plant extracts on mortality of 3rd instar larvae

The regression analysis also depicted that efficacy of extracts was less in initial hours against 3rd instar larvae and reached their peak on 72 hours. The regression model further indicated that the mortality reached their peak just few hours before 72 h on 66.08 h for neem followed by tobbaco, dhatora and akk it was 73.68, 78.62 and 95.37. R-squire of regression was about 0.66 for neem followed by tobbaco, dhatora and akk it was 0.54, 0.19 and 0.52, respectively and it revealed that 66, 54, 19 and 52% variation in mortality was counted by time interval, respectively.

Extracts derived from various plant leaves affected the survival of 3rd instar during 24 hours. The neem extract was found to be the most effective treatment with maximum mortality (14.66%) followed by tobacco extract, datura, akk and control treatments 11.33,10, 11.66 and 0%, respectively (fig- 2).

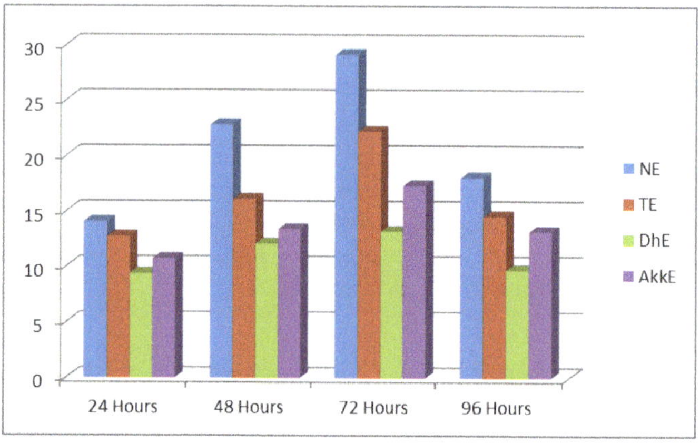

Fig 1. Efficacy of different plant extracts against 2nd instar larvae on different time intervals

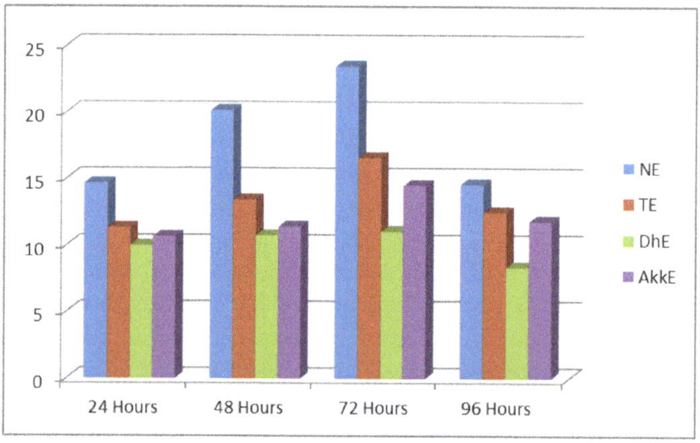

Fig 2. Efficacy of different plant extracts against 3rd instar larvae on different time intervals

5. Discussion

Different plant extracts caused different percentage of reductions of the target pest on different time intervals. The present study showed that neem extracts found to be more effective control of 2nd and 3rd instar larvae of the of *P. xylostella* when compared with other plant extracts.

The results of our study, which suggest neem extract as a better pest control it is in accordance with previous studies, where neem extract marginally increased the mortality of *P. xylostella* larvae [12]. It is also reported by many researchers that neem extracts limiting different pest species in the world [13, 14]. It was [14] observed that neem extracts can be successfully used as an excellent substitute to synthetic insecticides.

In our studies the tobacco extracts were also found to be the most active against these larval instars, affecting their survival. The results supported by the findings of the earlier workers [12] they reported that the use of tobacco extract is effective for control of larval instar of diamondback moth.

During the study it was observed that there is effect of akk extract on the survival of 2nd and 3rd instar larvae of diamondback moth. It is also in agreement with [15] who reported that leaf extract of *C. procera* decreased mosquito larvae population may be due to the different compounds present in the extract possessing different bioactivities.

The study also indicate that dhatora extracts effect the activity of larval instars during the experiment conducted in the laboratory It is clearly proved by the researchers that dhatora plant extracts are less expensive and highly effective for the control of insect pests [16, 17]. The extract of *D. alba* could be used as an effective botanical insecticide to be included in the Integrated Pest Management Programme for *P. americana* and other insect pests as well.

6. References

[1] Talekar, N. S, and A. M. Shelton, 1993. Biology, ecology, and management of the diamondback moth. Annu. *Rev. Entomol.* 38, 275-301.

[2] Idris, A. B, and E. Grafius, 1996. Effects of wild and cultivated host plants on oviposition, survival, and development of diamondback moth (Lepidoptera: Plutellidae) and its parasitoid *Diadegma insulare* (Hymenoptera: Ichneumonidae). *Environ. Entomol.* 25: 825–833.

[3] Sastrosiswojo, S, and S. M. Sastrodihardjo, 1986. Status of biological control of diamondback moth by introduction of parasitoid *Diadegma eucerophaga* in Indonesia. *In* : Talekar, N. S. & Griggs, T. D. (ecls.), Diamondback moth management: Proceedings of the first international workshop. pp. 185-194. Asian Vegetable Research and development Center, Shanhua, Taiwan.

[4] Talekar, N. S., J. C. Yang, Liu, M. Y, and P. C, Ong, 1990. Use of parasitoids to control the diamondback moth, *Plutella xylosteUa*. In : Mochida, O., Kiritani, K., & Bay-Peterson, J. (eds.), The use of natural enemies to control agricultural pests., pp. 106-114. FFTC Book Series No. 40, Food and Fertilizer Technology Center for the Asian and Pacific Region, Taipei, Taiwan.

[5] Kibata, G.N., 1996. The diamondback moth: A problem pest of Brassica crops in Kenya. In: Sivapragasam, A., Kole, W.H., Hassan, A.K., Lim, G.S. (Eds.), The Management of diamondback moth and other crucifer pests. Proceedings of the third international workshop, Kuala Lumpur, Malaysia. pp. 47–53.

[6] Iqbal, M., R.H.J. Verkerk, M.J. Furlong, P.C. Ong, A.R. Syed, and D.J. Wright, 1996. Evidence for resistance to Bacillus thuringiensis (Bt) subsp. Kurstaki HD-1,Bt subsp. Aizawai and Abamectin in field populations of Plutella xylostella from Malaysia. Pestic. Sci. 48, 89–97.

[7] Syed, A.R., 1992. Insecticide resistance in diamondback moth in Malaysia. In: Talekar,N.S. (Ed.),Manageme nt of DBM and other Cruciferous Pests. Diamondback moth and Other Crucifer Pests: Proceedings of the Second International Workshop,Tainan, Taiwan,AVRDC. Shanhua,Taiwan. pp. 437–442.

[8] Talekar, **N. S,** and T. D. Griggs, 1986. Diamondback moth management: Proceedings of the first international workshop, Asian Vegetable Research and Development Center, Shanhua, Taiwan. 471 p.

[9] Tang, Z. H., K. Y. Gong, Z. P. You, 1988. Present status and countermeasures of insecticide resistance in agricultural pests in China. Pestic. Sci. 23, 189-198.

[10] Ooi, P.A.C and G.S. Lim, 1983. Introduction of exotic parasitoids to control the diamondback moth in Malaysia. J. Plant Prot. Trop. 6,103–111.

[11] Abbott, W., 1925. A method of computing the effectiveness of insecticides. Journal of Economic Entomol. 18 (2): 265-267

[12] Waiganjom, M.M., S.S. Sithanantham, L.M. Gitonga, and J.M. Mueke, 2008. Preventive control alternatives to routine foliar spray against Thrips tabaci in onions. Journal of Agricultural Science and Techno. 10 (1): 26-42.

[13] Nadia, Z. D, and G. H. Schmidt, 1992. Efficacy of Neem-Azal and Margosan against the bean aphid, Aphis fabae Scop. J. Pest Sci. 65 (4): 75-79.

[14] Nadia, Z. D., A. A. Barakat, E. F. Abdalla, H. E. El-Metwally, and A. M. E. Abd El Salam, 1995. Evaluation of two neem seed kernel extracts against Liriomyza trifolii (Burg.) (Dipt. Agromyzidae). J. Pest Sci. 68 (2): 39-41.

[15] Singh, S., 2003. Effect of aqueous extract of neem seed kernel and azadirachtin on fecundity, fertility and post embryonic development of melon fruit fly, Bactrocera cucurbitae and B. dorsalis (Diptera: Tephritidae). J. Appl. Entomol. 27 (9/10): 540-547.

[16] Sripongpun, G., 2008. Contact toxicity of the crude extract of Chinese star anise fruits to house fly larvae and their development. Songklanakarin, J. Sci. Technol. 30: 667-672

[17] Cavalcanti, ESB, SM. deMorais, A.L.M. Ashley, P.S.E. William, 2004. Larvicidal activity of essential oils from Brazilian plants against Aedes aegypti L. Memorias do Instituto Oswaldo Cruz, 99: 541-544.

[18] Farkhanda, M., N. Ghazala, S. Sadia, A. M.Saadiya, 2011. Effect of ethanolic plant extracts on three storage grain pests of economic importance. *Pak. J. Bot*. 43(6): 2941-2946

Toxic Effect of Fertilizers on Inferior Plants Resed as Biological Models

Fadila Khaldi[1, 2], Houria Berrebbah[2] and Med.Réda. Djebar[3]

1 Department of Biology,University of Mohammed Cherif Messaadia, Souk-Ahras, Algeria

2 Laboratory of Cell Toxicology, Department of Biology, Faculty of Sciences, Algeria

3 University of Badji Mokhtar, Annaba, Algeria

Abstract

The inopportune throws out of diverse substances in the atmosphere, constitutes without any doubt the obvious of environmental pollution by man. Among these substances, we are interested in the NPK (nitrate –phosphate-potassique).Nitrate fertilizers widely used in farming in our region - Annaba located in the eastern part of Algeria – and manufactured in the same region. In fact, the excessive fertilization, the intensive spreading of animal faeces and the industrial pollution are the accumulation sources of nitrate in vegetables, drilling and the underground waters.

The treatment by NPK affect the respiratory metabolism of mosses as well as the measure of the consumption of the oxygen shous the abviousness contrasted with a dampening of respiration but also of the photosynthesis.The perturbation of the respiration and photosynthesis of mosses can explain the degradation of the plant material and the disappearance of certain species from our ecosystem.

The effect of NPK indicate also the perturbation of enzymes antioxidants fonctions : GSH and GST.

Keywords: NPK, mosses, Cytotoxicity tests, respiratory and photosynthetic metabolism, Biomarkers; Antioxidant enzymes ,GSH,GST.

1. Introduction

Bryophytes are particularly suitable organisms for the study of metal and organic pollutants. They owe this to their anatomical efficiency (high ratio surface / volume or surface area / mass, no waxy cuticle, of conducting vessels and real root system, easy to identify the annual growth) and physiological (photosynthetic activity continues at year round). They are therefore subject to the impact of pollutants in both dry depositions. Bioaccumulation of pollutants in plant species is an indicator of exposure. Indicators of effects of these pollutants can also be measured; they may be more defined, especially in the form of various biochemicals or physiological parameters (biomarkers) [1].

2. Methods

2.1. Sampling procedure of the lower plants

The samples of Mousses (species *Leucodon sciuroides)* were taken in the area of Séraïdi, located at 850m above the sea (Annaba,Algeria). Our choice was made on this area because it is a zone considered as not polluted.

2.2. Tests of cytotoxicity for the moss

NPK fertilizer was tested with four concentrations:10 ,20 ,30 and 40 mM. The solutions prepared with the various concentrations of NPKs are used for the imbibition of the samples of mousses. Approximately 1g of thallus was soaked in 100 ml of solution during 3 days [2].

2.3. Determination of Glutathione (GSH) and activity Glutathione S-transferase (GST)

The glutathione was assayed by the method of [3], based on measuring the absorbance of the 2-nitro-5 mercapturic resulting from the reduction of the acid 5-5 'thiol-bis-2-nitrobenzoic acid (DTNB) by the thiol groups (-SH) glutathione. The glutathione S-transferase activity is performed by the method of [4]. It is based on the conjugation reaction between GST and a substrate, CDNB (1-chloro 2, 4 dinitrobenzene).The GSH and GST biomarkers are expressed in µm/mg of protein. The protein level was measured according the method of [5].

2.4. Proportioning of chlorophyls

The extraction of chlorophyls at summer was carried out according to the method of [6]. The formula related to solvent, enables us to calculate the values of chlorophylls [7].

2.5. Polarographic study

The apparatus used is an electrode with oxygen (HANSATECH) which allows the measurement of the production of the oxygen uptake during a reaction. Its sensitivity makes it possible to detect concentrations of about 10µM [8].

2.6. Statistical study

The statistical analysis was performed by Student t test used to compare between two samples (control and treated).This test is performed using the analysis software statistical processing of data: Minitab version 16.1.0. , n = 5 [9].

3. Results

After 3 days of treatment, we found that glutathione-S-transferase tends to increase a dose-dependent manner. This increase was highest in the treaties with 40 mM concentration where the rate is: (0.103 (± 0.003)) µmole / min / mg of protein)(Fig.1).According to Fig.2, we find that glutathione levels decreased dose-dependent manner. Thus at 40 mM concentration, the GSH level is low (16.09 (± 0.49)) µmole / mg of protein) compared to the control of which is: (31.27 (± 0.21)) µmole / mg of protein).

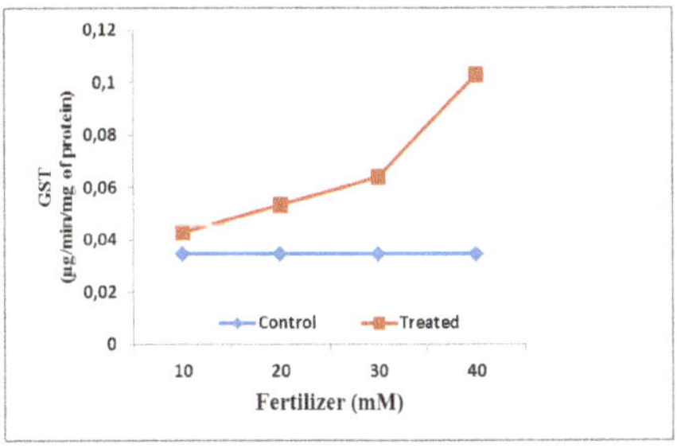

Figure 1. Evolution of GST activity according to the fertilizer concentrations ($P \leq 0,001$).

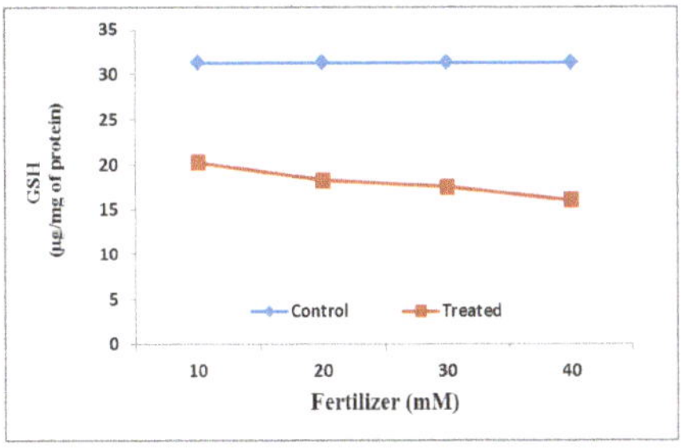

Figure 2. Evolution of GSH based on fertilizer concentrations ($P \leq 0,001$).

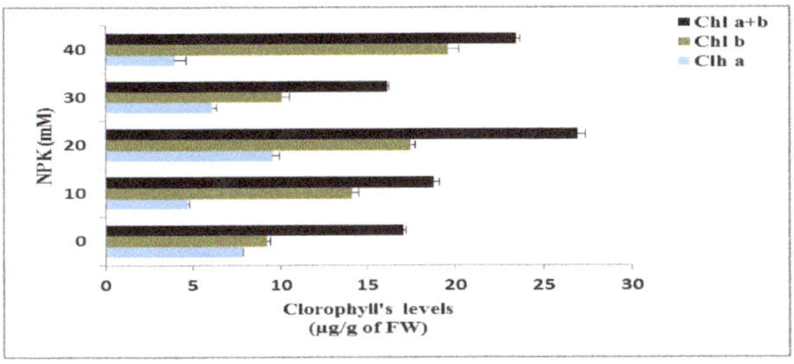

Figure 3. Changes in chlorophyll (a, b, a + b) in Leucodon sciuroides treated by different concentrations of NPK.

Fig. (3), highlights the changes in rates of chlorophyll *a, b* and (*a* + *b*) as a function of increasing concentrations NPK.Statistical analysis revealed a significant difference between control and treated with the concentration (30 mM) for (Chl *b*) ($P \leq 0.05$), while very highly significant differences for all treated and all concentrations (10,20, 30.40 mM) (Chlorophyll *a, b* and *a* + *b*) ($P \leq 0.001$) compared with controls always.

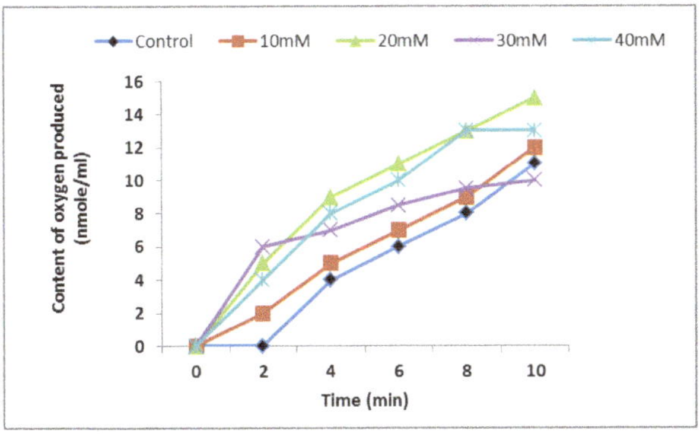

Figure 4. Effects of NPK on photosynthetic metabolism of Mousses (Leucodon sciuroides) ($P \leq 0,001$).

This figure (4), illustrates the effect of different concentrations on the photosynthetic metabolism NPK foams where there has been a release of O_2 in the medium for control samples and for samples treated with the four selected concentrations. 0n a marked increase in the amount of O_2 produced in the middle as the fourth minute of recording for control samples. The minimum of this amount was recorded at the time (10min) for the sample treated with 30 mM concentration, which reached 10 nmol O_2 / ml (Lower than the control: 11 nmol O_2 / ml). While the sample treated with 20 mM concentration shows the highest amount produced until the time (10 min) or 15 nmol O_2/ml.On samples treated with other concentrations (10 and 40), have produced higher amounts of oxygen to those of the control resulting in stimulation of photosynthesis.

Time (min) Mousses	NPK treatments (mM)				
	0	10	20	30	40
0	340.00	340.00	340.00	340.00	340.00
2	340.00	332.00	328.00	324.00	320.00
4	355.00	324.00	316.00	308.00	300.00
6	330.00	316.00	304.00	292.00	280.00
8	325.00	308.00	292.00	276.00	260.00
10	320.00	300.00	280.00	260.00	240.00

Table 1. Oxygen consumption (nmole/ml) in moss response to the NPK treatment (mM) ($P \leq 0{,}001$).

The observation of the table (1) shows that the foams have a witness who starts breathing 340nmole O_2 and reached 320 nmol O_2 after 10min, the oxidation rate is an average of 2 nmol O_2 / min.Ce treatment causes an acceleration observed from the 2nd minute especially in samples treated with 40 mM concentration where the rate of oxidation is 10 nmol of O_2 / min). Indeed, this speed is about 4, 6 and 8 nmol of O_2 / min, respectively, in samples treated with 10, 20 and 30 mM of NPK.

4. Discussion

We propose in this work to proceed with the demonstration of the effect of NPK on foams, where we found a decrease in dose-dependent manner in the presence of GSH NPK. This condition can be explained by the direct connection of glutathione to the atoms of xenobiotic (NPK) as glutathione has a carboxylic acid group, an amine group, a group sulfhydryl (-SH) and two bypass likely peptide to be involved in reactions with other atoms. Its functional group-SH would then play an important role in binding to the xenobiotic [10].Our results agree with those of [11] and [12] in which the GSH level is decreased with increasing tolerance the accumulation of the pollutant for low concentrations .Our results show a significant increase of GST, in mosses in the presence of NPK; this increase is a response to oxidative stress caused by the presence of xenobiotics in the plant cell. The biotransformation enzymes are among the first to respond to the presence of a pollutant in a living organism [13]. This increase indicates a high rate of conjugation of atoms

NPK with glutathione. Our hypothesis is that induction of GST enzyme system can be explained by the entry of Xenobiotics (NPK) in plant cells (foam) and induction of detoxification system.

The other aspect of our work was to measure the mean levels of chlorophylls *a, b* and *(a + b)*, parameters that can tell us a possible state of stress due to the presence of a pollutant in mosses. In general, chlorophyll appears to be affected by the xenobiotic (NPK). This perturbation in the mean levels of chlorophyll *a, b* and *(a + b)*, in these plants, explains the attenuation of photosynthetic activity. Our results agree with our previous work [2], which have demonstrated a disruptive effect of nitrate of ammonium on the biosynthesis of chlorophylls. Our results are quite revealing, and the NPK causes a stimulation of photosynthesis in mosses, is excessive production of oxygen in the culture medium with a clear stimulation of respiratory metabolism.

Air pollution exposes plants to various forms of nitrogen that can be highly toxic (nitrogen dioxide, ammonia and ammonium). Among the reactions to the toxic effects of these compounds include: defoliation, training of larger cells thin-walled, yellowing, the lesions on some organelles of the plant and the reduction of drought resistance [14].The most important direct effect on vegetation results from the interaction of these various forms of nitrogen with other pollutants and impaired balance with other nutrients.

5. Summary and conclusion

We can conclude that the NPK disrupts the photosynthetic metabolism and respiratory mousses. Our results are in perfect agreement with our previous work [2] and [15] which have demonstrated a stimulation of photosynthesis and respiration in these plants. But in higher plants one of the mechanisms of defense against air pollution is indeed a decrease in respiration and photosynthesis.

6. References

[1] Bargali R, 1998a. Mosses as passive and active monitors of trace elements.In Bargali R. Trace elements in terrestrial plants. An ecophysiological approach to biomonitoring and biorecovery. Springer, Berlin,207-236.

[2] Khaldi F, 2003. Toxicité du nitrate d'ammonium NH_4NO_3 sur trois modèles biologiques : les paramécies, les mousses et les lichens. Effet sur leur métabolisme respiratoire. Mémoire de Magister en Biochimie Appliquée. Université Badji Mokhtar, Annaba : 86pages.

[3] Weckberker, G., and Cory, G, 1988. Ribonucléotide reductase activity abd growth of glutathione depleted mouse leukemial 1210 cells in vitro. *Cacer letters* 40: 257-264.

[4] Habig, W.H., M.J. Pabst and W.B. Jakoby, 1974.Gluthation-S-transferases: the first enzymatic step inmercurapturic acid formation. Journal of Biologica lChemistry, 249: 7130-7139.

[5] Bradford M , 1976 . A rapid and sensitive method for the quantification of microgram quantities of protein utilizing the principle of protein-dye binding. Anal.Biochem.72:248-254.

[6] Holden M,1975 .Chlorophylls in chemistry and biochemistry of plant pigment.2-nd Edition Academie press, New York : pp.133.

[7] Arnon DI , 1949. Copper enzymes in isolated chloroplasts polyphenoloxidase in Beta vulgaris. Plant physiology 24 :1-15.

[8] Djebar, M.R, 1989. Compte rendu de l'académie des sciences. Paris III.

[9] Dagnelli P.(1999).Statistiques théoriques et appliquées. Tome 2 : références statistiques à une et à deux dimensions. *Bruxelles. Univ De BOECK et LARCIER,* 659 p.

[10] Anderson M E, 1985. Determination of glutathione and glutathione disulfide in biological samples. Method. Enzymol.113: 548-554.

[11] Zhu YL, Pilon-Smits E, Tarun AS, Jouanin L, Terry N , 1999. Overex pression of glutathione synthetase in indian mustard enhances cadmium accumulation and tolerance. Plant Physiology 119: 73-80.

[12] Freeman JL, Persans MW, Nieman K, Albrecht C, Peer W, Pickering I, Salt DE ,2004. Increased glutathione biosynthesis plays a role in nickel tolerance in Thlaspi nickel hyperaccumulators. Plant Cell 16: 2176-2191.

[13] Pesch R. et Schröder W, 2006. Integrative exposure assessment through classification and regression trees on bioaccumulation of metals, related sampling site characteristics and regions, *Ecological informatics,* 1, p. 55-65.

[14] Cape J.N, Vander Eerden L.J, Sheppard L.J, Leith I.P and Sutton M.A, 2009. Evidence for changing the critical level for ammonia.Environmental Pollution 157:1033-1037.

[15] Bensoltane S, Khaldi F, Djebar H and Djebar M.R , 2005 .Toxicity of the ammonium nitrate NH_4NO_3 on the respiratory metabolism of three biological models : *parmeciums,* mosses and lichens. Comm.Appl.Biol.Sci,Ghent University, 70/4 :1043-1051.

28

A GIS Based Digital Land Resources Framework for Optimal Soil Management in Barda and Awaje Basin, Syria

Said Sawy, Ali Abdel-Hameed and Kais A. Sultan*

Faculty of Agriculture, Cairo University, Egypt

*Corresponding author

Abstract

Barada and Awaje Basin is located in the southwest part of Syria and includes the capital of the country, the city of Damascus and its suburbs. It covers an area of approximately 8596 km^2 and constitutes a hydrologically closed basin. It has a far higher population density than any of the Syrian regions. The estimated population living within the borders of this region is more than 4 millions.

Population growth in the study area causes competition on land resources between different sectors and pressure on limited water resources. The national development plans aim to conserve arable lands, improve its productivity, and sustain the land and water resources.

Therefore, providing accurate and integrated information about land resources is a must, especially with the accelerated progress of information technology. Such information would be the base for planning, decision making and research needs. Various available information and database systems were employed (e.g. ArcGIS, ERDAS IMAGINE and ENVI). Data of previous soil survey activities were the bases for the created GIS digital database. Soils survey maps (SCALE 1:100,000), were prepared as GIS ready maps. Urban settlements were updated using recent ETM+ and SPOT satellite images. An intensive field investigation was performed in the study region, with the purpose of representing the soil units and collecting ground control points and soil samples for laboratory analyses. The created digital land resources database was used to figure the distribution of soil units and to evaluate and map land suitability on the bases of FAO, 1985 [1].

It was found that the Aridsols soil order characterize most of the alluvial fan soils of Barada, while Inceptisols were found in the western plains and intermountain areas. The soil orders Entisols were found dominating the western mountain areas. The results showed that 28.6 % of the areas are classified as highly suitable for irrigated agricultural production, corresponding with Typic Haplocambids and Typic Haploxerepts soil sub-great group. The moderately suitable soils, exhibit 14.7 % of the areas. The marginally suitable soils represent 14.6 % of the areas. The not suitable soils dominate the soils of eastern lacustrine and desert deposits, representing 21.2 % of the areas. It could be pointed out that achieving such detailed digital land resources database for local administrations is a great step towards the implementation of sustainable development and management programs. It is characterized by its comprehensiveness, geographical accuracy and updatability. In the information technology, such data can be handled, enhanced and exchanged by different users and authorities. The most striking findings noticed was the urban encroachment on the account of most fertile soils; hence shrinkage in areas of high agricultural capabilities. On the other hand, urbanization doesn't extend largely to low capable land.

Keywords: GIS, Remote Sensing, Soil Survey, Land Evaluation.

1. Introduction

Population increasing in the study area (*Barada and Awaje Basin)* caused competition on land resources between different sectors and pressure on limited water resources. The national development plans aim to conserve arable lands and to improve its productivity, Moreover to sustain the land and water resources. However, realizing these purposes requires availability of accurate documented data for integrated natural resources. Techniques of remote sensing and Geographical Information Systems (GIS) provide suitable means for inventory monitoring and documentation of natural resources, as they are characterized by satisfactory resolution and multi-spectrality. Also, distribution of the natural resources and detection of their changes are accessible by multi-temporal space data nature. The objectives of the former are to build database of available natural resources data and combine them into suitable format and make them ready for use by the land use planning recommendation component.

2. Material and methods

2.1. Study area

Barada and Awaje Basin is located in the southwest part of Syria (a country in southwest Asia bordered by Turkey, Iraq, Jordan, Lebanon and the Mediterranean Sea) and includes the capital of the country, the city of Damascus and its suburbs (Fig.1). It covers an area of approximately 8596 km^2 and constitutes a hydrologically closed basin (there is no excess water flowing out of the basin). It has a far higher population density than any of the Syrian regions. The estimated population living within the borders of this region is more than 4 millions. This basin is a rewarding subject for investigation, as it is a very heterogeneous landscape containing many different types of climates, topography, soil, vegetation and land uses. The drainage system of this basin is represented mainly by two rivers (*Barada River* and *Awaje River*) and a few valleys that have dry river beds filled with water only during the rainy season. There are also two dry lakes (*Al-Outaibe* and *Al-Haijaneh*) which are usually salty and drainless and only during the rainy season are covered by a thin layer of water. The topography is characterized by low mountain systems outlining the flat central Damascus depression with its lowest elevation point 600 above the sea level. Maximum height of the mountain is 2814 m (Fig.2).

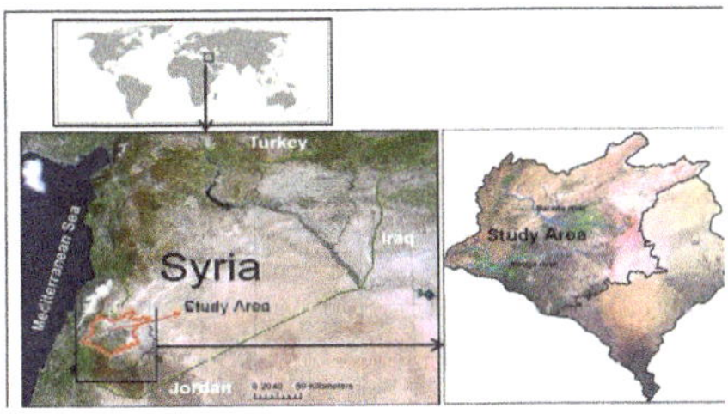

Fig 1. Shows location of the study area.

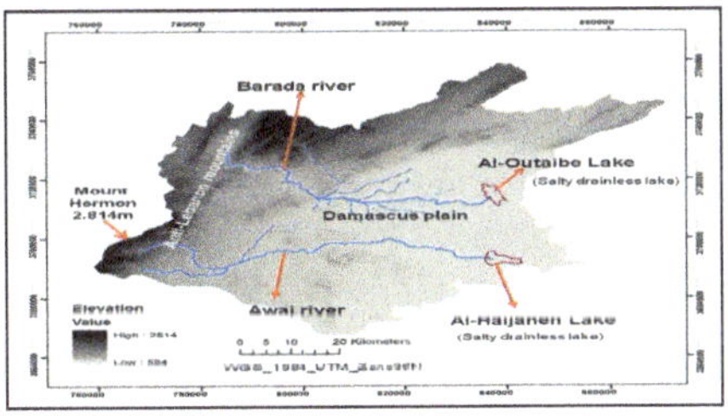

Fig 2. Shows a Digital Elevation Model of the study area along with vector data (river channels and lakes) draped over it.

The major part of study area lies in the transition zone between the arid climate on the plains to the moderate humid climate at the heights above 2000 m. The year is subdivided into seasons:

- The dry summer season (from April to October) with high average monthly air temperatures from 25-27 C° on the plains to 19 C° and less at the heights above 2000 m; with a low precipitation rate and dry winds.

- The humid cold season (from November to March) with the average monthly air temperatures from 10-15 C° on the plains to zero and less at the heights above 1500-2000 m; with periodic precipitation (up to 80-90 % of the annual rate) and cold north-eastern winds.

The soils of the drainage basins of the Barada and Awaje rivers relate to the soil-bioclimatic formation of cinnamon and low-alkaline, grey brown soils of a dry subtropic climate. Diversity of the relief, climatic, hydrogeological and other environmental conditions account for a mixed character of the soil mantle of the area.

2.2. Materials used

Soil survey to a scale of (1:100,000) was performed by Leningrad State Institute for Design of Water Resources Development Projects (LENGIPROVODKHOZ) as part of the project of Water Resources Use in Barada and Awaje Basins For Irrigation of Crops (USSR, 1986). Soil survey covers 3408 km^2 of the study area, a total of 556 soil observations pits and profiles were done. The soil maps exist as analogue format, the collected maps consist of two map sheets (sheet for soil mapping units and sheet for soil observation points) at scale of 1:100,000. Soil survey data were interpreted and classified according to Russian soil classification USSR. The soil maps of Barada and Awaje basin are the main materials collected and converted to the digital format. Satellite ETM+ image of 2006, a scene (Path174 / Row37), covering the whole basin, was collected and processed to be included in the GIS land resources database and be used in thematic mapping processes. The recent SPOT images of 2009 were obtained from General Organization of Remote Sensing (GORS) in Syria, and processed for updating different thematic maps and detecting changes in urban area. The Shuttle Radar Topographic Mission (SRTM) images of 30 pixel size resolution, in addition to driven Digital Elevation Module (DEM) for the study area and its surrounding were consulted to represent the area landscape [2] and [3].

2.3. Methodologies

2.3.1. Coding the Soil Units

In order to input the soil maps in the geographic information database, it was necessary to codify the mapping units. The original map units classified according to Russian soil classification USSR [4]. Soil data was reclassified according to American Soil taxonomy USDA [5]. Coding the soil units ought to be indicative to the soil Taxonomy, properties, as well as the landscape.

2.3.2. Satellite images processing

Pre-processing activities were performed, to reduce some undesired variations/noises and to enhance other desired features. It commonly comprises a series of sequential operations, including radiometric correction or normalization, image registration, geometric correction, masking and image enhancement (e.g., for clouds, water, irrelevant features). Geometric rectification of the imagery was applied using ERDAS IMAGIN. This becomes especially important when scene to scene comparisons of individual pixels in applications such as change detection are being sought [6]. Image enhancement techniques (e.g. contrast stretching, Gray-level threshold, Level slicing, and spatial stretch) were tested to improve the visual interpretability of the used images by increasing the apparent distinction between the features.

2.3.3. Building GIS database

The digitizing specification of maps was defined according to the available themes. The different digital maps were corrected from different errors and edge-matched after the geo-referencing processes. Edge matching, as a spatial adjustment process that aligns features in adjacent map sheets, was performed according to Tomlin, 1990. This process was applied on the soil maps (scale 1: 100,000). The descriptive thematic data related to all layers were attached as additional attribute tables.

2.3.4. Spatial adjustments

It was noticed, after edge matching, that there is a kind of deviation (constant in many places in its direction and magnitude) between the produced maps and the well registered topographic ones as well as the satellite images of the study area. It was possible to attribute the deviation to two reasons; lack of coordinate system in some original map sheets and the rubber-sheeting accompanied the edge-matching task [7]. In order to overcome this problem another spatial adjustment (transformation) has been performed. Well registered topographic maps and accurately georeferenced satellite images have been used to perform the transformation process. The transformation tools of ArcGIS systems were found to be very effective in performing the spatial adjustment of the soil maps [8] and [9].

2.3.5. Compilation of laboratory analysis results

A number of 12 soil profiles, representing the soil units of the studied regions were morphologically described according to FAO [10]. A number of 56 disturbed soil samples were collected for laboratory analysis, following the laboratory methods manual [11]. The results of these analyses have been compiled in the database and then incorporated into the attribute tables of the digital GIS soil maps [12].

3. Results and discussions

3.1. Soil Resources

Soil Taxonomy has been used to classify the soils of the study area up to family and phases of families (Soil Survey Staff, 1999 and 2010). Soil Taxonomy is a hierarchical system of soil classification that identifies six levels. At the highest level, twelve orders are recognized worldwide, but only three have been recorded within the study area: Aridisols, Entisols, and Inceptisols. At the subsequent levels of classification eight suborders, fourteen great groups, and thirty subgroups have been recorded. The identified great groups and their coverage are: Haplocambids (comprising 34.7 % of the study area); Haplocalcids (2.9 %); Haplogypsids (2.7 %); Petrogypsids (2.19 %); Aquisalids (1.1 %); Haplosalids (1.85 %); Calcigypsids (0.77 %); Torriorthents (11.78 %); Torrifluvents (3.7 %); and Haploxerepts (11.88 %). Bedrock occupies about 18 % of the study area.

Aridisols cover almost all of the central and eastern part of the basin where the annual precipitation drops below 250mm. In addition, Aridisols are characterized by an aridic (hot and dry) soil moisture regime, and they have light colour as there is not enough vegetation to add organic matter to the soil profile. Furthermore, they often accumulate calcium carbonate, gypsum, and other materials that are readily leached from soils in more humid environments [13].

Entisols are soils that have little or no indication of development of pedogenic horizons [5]. This soil order includes recently developed soils, which do not have the requirements of the other soil orders. Entisols cover the western north mountain in the study area.

Inceptisols are soils of semiarid to sub-humid environments that generally show only moderate degrees of soil weathering and development. Even though they are better developed than Entisols, they are still young soils and resemble very closely the parent material [14].

3.2. Urban areas

The urban area, according to soil survey that performed in 1986, was occupying about 2.24 % of the study area (77.18 km^2). Whereas, urban area, according to visual interpretation of Spot image satellite taken in October 2009, was occupying about 4.17 % of the study area (143.46 km^2). For example, the area of Damascus city in 1986 was 58.03 km^2 and the area of Damascus in 2009 was 124.49 km$^{2.}$ it is important to indicate that the extension of urban area especially Damascus city occupy the soils of alluvial fan that have high potential for agriculture. The urbanization rate in the study area reaches to 100 % through 23 years (from 1986 to 2009).

3.3. Land Suitability Classification of the Barada and Awaje soils for Irrigated Agriculture

Rating criteria (land use requirements for irrigated agriculture) were developed based on an international and regional review of the arid and semi-arid conditions prevailing in the neighboring Countries. These include JAZPP project (2001) in Jordan [15]. Evaluation analysis preformed according to FAO (1985). These requirements were matched with land attributes that were derived from soil survey data using average and mode method which utilizes the soil map units and the observation points, (by averaging land characteristics within soil mapping units). Simple limitation method was used in suitability analysis. Suitability analysis was performed by using Arcview 3.2, Query comment (select by attribute). About 28 % of study area was classified highly suitable for irrigated agriculture (S1), 15 % of study area was classified moderately suitable (S2), 15 % of the study area classified as marginally suitable (S3), and 21 % of the study area classified not suitable for irrigated agriculture. Most of soils Typic Haplocambids and Typic Haploxerepts classified as highly suitable for agriculture production. Soil depth, Rocks outcrops, stones in the surface horizon, salinity are the most important limiting factors that lowering suitability classification of 20.9 %, 14.5 %, 8.1 % and 10.3 % of the study area respectively. Figure.3 shows distribution of suitability classes within the study area.

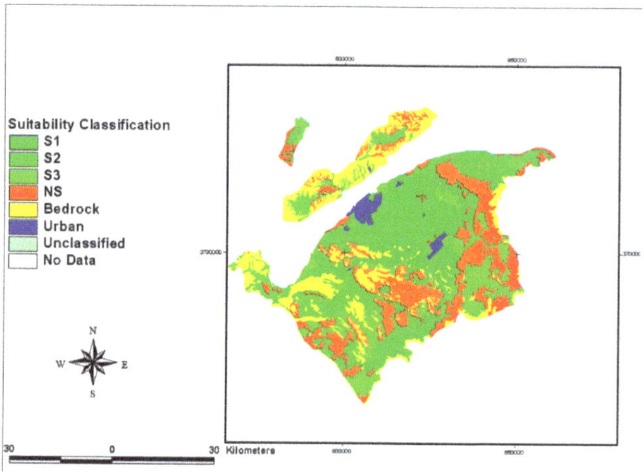

Fig 3. Land suitability classification for irrigated agriculture.

4. Conclusions

It could be concluded that the creation of land resources database is rather important in documenting the environmental themes. Such documentation leads to data harmonization and maximization of its value. It also allows an easy data processing, and updating. The land resources databases are rather useful in elaborating site selection for sustainable development projects, in addition to decision support and early warning. Remote sensing, with its multi-concept approach, provides up-to-date information on different themes. Multi-dates images allow detecting the changes occurring in the different environmental conditions. Also, the multi spectral satellite images reflect the environmental elements characterized by a variety of spectral signature. Moreover, GIS and its integrated functional nature with remote sensing, facilitate the creation and developing land resources databases.

5. Acknowledgements

We thank the Syrian Ministry of Irrigation for providing reconnaissance soil map (scale: 1:100.000). The comments of the editor and the referees are highly appreciated.

6. References

[1] FAO (1985). "Land evaluation for irrigated agriculture" soils bulletin 55, FAO, Rome.

[2] Dobos, E. Norman, B. Bruee, W. Luca, M. Chris, J. and Erika, M., 2002. "The Use of DEM and Satellite Images for Regional Scale Soil Database"17th World Congress of Soil Science (WCSS), 14-21 August 2002, Bangkok, Thailand.

[3] NCSA (2005). "Extracting topographic features from shuttle radar topography mission (SRTM) Images" Technical Report, alg. 05-002, July 18, 2005, National Center for Supercomputing Applications (NCSA), 605 East Springfield Avenue, Champaign, IL 61820.

[4] Shishov, L.L., Sokolov, I.A., 1990. In: Genetic classification of soils in the USSR. Soil classification. USSR State Comm. For Environ. Protection, Moscow, pp. 77–93.

[5] USDA (2006). "Keys to Soil Taxonomy" United State Department of Agriculture, Natural Resources Conservation Service (NRCS) tenth edition, 2006.

[6] ERDAS, 1999 Practical excercises, University of Leicester, UK, 1999.

[7] Boullie, F. (1978). Structuring cartographic data and spatial processes with the hypergraph-based data structure. In: First International Advanced study symposium on Topological Data Structures for Geographic Information Systems (ed. G. Dutton) Vol. 5. Laboratory for Computer Graphics and Spatial Analysis, Harvard.

[8] Burrough, P.A. (1986). Principals of geographic Information Systems for land resources assessment. Clarendon, Oxford.

[9] Tomlin, C. D. (1990). Geographic Information Systems and Cartographic Modeling. Prentice Hall, Englewood Cliffs, N.J.

[10] FAO (2006) FAO. "Guidelines for soil description" Fourth edition, FAO, Rome, ISBN 92-5-1055211.

[11] Soil Survey Staff. 2003. Keys to Soil Taxonomy. 9th ed. USDA Natural Resources Conservation Service, Washington, DC.

[12] Nguyen Quec Dinh (2001). Cave database development, spatial analysis and 3D visualization with GIS – Case study in Son La (Vietnam).VUB, Master dissertation (unpubl.).

[13] Edwards-Jones, G. (2001): Final Report on Agricultural Policy and the environment in Syria: An examination of impacts and suggestions for policy reform.

[14] Harpstead, M.I. and Sauer, T.J. and Bennett, W.F. (2001): Soil science simplified. Wiley-Blackwell.

[15] Hatten, C.J. and Taimeh, A.Y. 2001. Improvement of agricultural productivity in arid and semi-arid zones of Jordan. In A.Y. Taimeh and B.I. Hattar (eds). A Cooperative Project between Ministry of Agriculture and the European Union. Volume 1, Main Report. University of Jordan Press, Amman, Jordan.

29

Rare Plant Species of the Protected Area of Kalmand-Bahadoran, Yazd Province, Iran

Ali Akbar Karimian

Faculty of Natural Resources, Yazd University, Yazd, Iran

Abstract

In this study, we collected and determined rare plant species from protected areas of kalmand-Bahadoran located in 30-105 km SE of Yazd City in the Yazd Province. Analyses of the flora showed that, there is 148 vascular plant species in this area. Threatened species of this region analyzed, according to the IUCN criteria. On the basis of this study, four categories of rare species so called Endangered, vulnerable, lower risk and data deficient are determined and the list of these species has been presented. Result showed that, there is34 threatened plant species in this protected area. Finally, floristic composition, and species richness of this area is discussed.

Keywords: Flora, Biologic form, rare species, Kalmand-Bahadoran.

1. Introduction

Nowadays the procedure of investigation, recognition, and maintenance of herbal species, especially useful and rare ones has gained vital importance in the world. It constructs a foundation for sustainable development and presents principle and logical utilization of nature and natural resources and is defined as a basis for protecting and maintaining herbal species and genetic treasure. Therefore, aiming at recognition and introduction of rare and useful species of plants and animals all over the world and adopting necessary approaches to prevent from their extinction, International Union for Conservation of Nature and Natural Resources (IUCN) has been established.

Among studies conducted in Iran by Iranian and foreign botanists about collection and identification of plants we may refer to Iran's flora[12], Iranica flora [14], Orientalis flora [3], Iran's herbs [9], Iran's flora [2], Iran'Asastragalus bisulcatus(astragalus spp) [8], study of Iran's desert flora and herbs [7]and etc.; however, although some floras are argent for research and educational purposes, there not onlydo existed documented floras published regarding various parts of Iran but also there are less studies about rare species according to patterns and criteria of IUCN organization.

To achieve this goal, one protected zone of Yazd province i.e.Kalmand-Bahadoran protected zone is selected. The reason for selecting this zone stem from this fact those plants of this area is to some extent being preserved from livestock foraging and human destruction and therefore study of this area flora is applicable and beneficial. Up to now flora of some protected zones like Turan and Kavir[14, 15], Arasbanan[1], are being investigated. In Yazd province there are some investigations conducted on province total-floras [10] and various regions including Kalmand-

Bahadoran and KuheBafgh protected zones [6]as well as other locations. However, there is not any documented report about Yazd's protected zones plants.

2. Study area

Kalmand-Bahadoran protected zone with approximately 255 thousands hectare situated 30 to 105 kilometers far in southeast of Yazd alongside Yazd-Kerman road at 31° and 20′ of north latitude and 54° and 30′ of east longitude. The zone mean attitude is 1616 above sea level and it is claimed as Kalmand-Bahadoran protected zone in 1994. The highest point is Medvar Mountain with altitude of approximately 3290 meters and the lowest part is positioned in Mahdiabad plain with an altitude of 1400 meters. Based on meteorological statistics annual precipitation average is 100 millimeters. The highest amount of rainfall occurs in January–February (22/21 millimeter). As it is observed in ambro-thermic diagram of the region drought period is started from the second half of March (beginning of Farvardin) and continues till second part of November (beginning of Azar). Maximum temperature average is 29/44 and coldest average is 5/22°C; moreover annual humidity average is 30%.

3. Methods of investigating rare and endangered species

Researchers use various criteria for identification, investigation, and classification of rare species like limited geographical propagation and low population; in 1985, Grime, beside above criteria, considered hard bio-environmental conditions and heavy environmental changes as important factors in determining rare species. Rabinowitz(1981) identified some rare species based on geographical range of propagation, habitat features, and population size. Fielder and Ahouse(1992) described rare classes according to species, spatial dispersion and their chronological resistance. In order to determine and classify rare species, some researchers use various routinesliketaxonomy, chronology, endemism, the quality of settlement and natural proliferation, the manner of plants utilization by human beings, livestock, wildlife, and finally illness and diseases, lack of bio-reactions that causes population reduction and therefore species extinction.

In this study, some criteria such as limited geographical propagation, human utilization of plants, livestock, wildlife, population amount, biologic form, how to settle and natural reproduction, have been used in determining rare species classes of investigated zone. Among 8 classes ofrare plants, based on IUCN classification principles, we have identified 4 classes as follows:

1. Endangered species facing extinction (En)
2. Vulnerable species (Vu)
3. Lower Risk species (LR)
4. Data Deficient species (DD)

4. Results

Preliminary Results achieved from KalmandBahadoranprotected zone show that there are 148 herbal species in this zone. Among them rare species were extracted and their biologic form featureswere determined as in table 1.

With respect to the fact that there is not complete information available on the regions' flora in the past years and yet no exact investigation is being carrying out regarding rareness of species in Iran, about 34herbalspecies were identified as endangered class in this region that mostly classified as lower risk category. Table 1 shows rare and vulnerablespecies of the above mentioned region.

Scientific Name	Rare class	Biologic form
Apiaceae		
PrangoscheilanthifoliaBoiss.	LR	He.
Asteraceae		
Centaureagaubae (Bornm.) wagenitz	LR	He.
CentaureaispahanicaBoiss.	LR	He.
Cirsiumspactabilis DC.	LR	He.
Cousiniapiptocephala Bunge.	LR	He.
EchinopsceratophorusBoiss.	LR	He.
JurineabungeiBoiss.	DD	He.
Jurinea radians Boiss. Subsp radians	DD	He.
Boraginaceae		
OnosmastenosiphonBoiss.	LR	He.
Brassicaceae		
Alyssum bracteatumBoiss. , Buhse	LR	Th.
IsatisrugulosaBge.exBoiss.	LR	Th.
SamerariaelegansBoiss.	LR	Th.
Sterigmostemumlongistylum(Boiss).Bornm.	LR	Th.
Caryophyllaceae		
AcanthophyllumchloroleucumRech.f ., Aell	DD	Ch.
Lamiaceae		
Nepetasaccarata Bunge.	LR	Th.
NepetaSatureioidesBoiss.	LR	Th.
ZatariamultifloraBoiss.	LR	Ch.
Liliaceae		
Allium chloroneurumBoiss.	LR	Cr.
Papilionaceae		
Astragalus (Choronopuse) jesdianusBoiss. , Buhse	LR	He.
Astragalus (choronopus) vanilla Boiss.	LR	He.

Scientific Name	Rare class	Biologic form
AstragalusbakaliensisBge.	LR	Th.
AstragalusbiovulatusBge.	LR	Th.
AstragaluscampylanthusBoiss.	LR	He.
AstragaluseriostomusBornm.	Vu	Ch.
AstragalusglaucacanthusFisch.	LR	Nph.
AstragalusmicrophysaBoiss.	LR	Ch.
GlauciumcalycinumBoiss.	LR	He.
Plumbaginaceae		
AcantholimonscorpiusBoiss.	LR	Ch.
Polygonaceae		
CalligonumbungeiBoiss	LR	Nph.
Primulaceae		
DionysiajanthinaBornm. , Winkler	Vu	He.
Resedaceae		
Reseda macrobotrysBoiss.	LR	He.
Tamaricaceae		
Reaumuriaoxiana (Ledch.) Boiss.	LR	Ch.
TamarixroseaBge.	Vu	Ph.

Tab 1. List of Rare plant species of Kalmand Bahadoran Yazd Province

Rare categoriesinclude:*Endangered species facing extinction (En),vulnerable species (Vu), and Lower Risk species (LR), Data Deficient species (DD)*
Biologic form include: *Throphytes(Th),Criptophytes(Cr),Hemicryptophytes(He), Phanerophytes(Ph.), Nanophanerophytes (Nph)*

5. Summary and Conclusions

The investigated region is situated in the heart of Iran's central plateau and is assumed as IranoT ouranian'svegetativeregion. Totally, regarding plants, this regionis less enriched than the whole country. The existence of limited number of tree and shrub species even in sparse and sporadic-form shows that this area is poor concerningvariety and amount of woody species. Based on information achieved from this research most of the herbal species included at firstperennialshrub species that may tolerate drying conditions and regarding low precipitation they can continue their survival and finely reproduce in rainy years and secondlyannual species that are drought escape and whendrought is prevailed they biologically turn to dormant.Herbal speciesof the region are mainly belonging to *Astraceae*, *Paplionaceae*,and*Brassicaceae*families. Herbal species were to some extent immune from destruction in recent years since they are protected by Environmental Protection Agency and because livestock manager were removed from the region; however, recent droughts caused their population and frequency to be declined meaningfully.According to Zohri view point important Iran's locations regarding floristic enrichment, percentage of exclusive and rare species, are Alborz and Zagrosmountain ranges and some single mountains like Karkas, Shirkooh in Yazd, and south Kerman mountains. The investigated region is relatively near Shirkooh and its endangered species is not so much as indicated in table 1. The endangered species are mostly among lower Risk class and their biologic form character is described as hemicryptophyte47%, Therophyte23.5%, Chameophyte12%, and rarely Phanerophyteand NanoPhanerophyte.

6. References

[1] Assadi, M., 1988. Plants of Arasbaran Protected area, N.W. of Iran (Part II). Jour. Bot.4 (1): 1.59.

[2] Assadi, M., 1989-1996. Flora of Iran, volume, 1-14, Ministry of forest and rangelands publication, Iran.

[3] Fielder, P.L. and Ahouse, J.J. 1992, Hierarshiesof cause: toward an understanding of rarity in vascular plant species. In: Fieldr, P.L. and Jain, S.K. (Eds.). Conservation Biology, the Theory, and Practice of Nature Conservation, Preservation and Management. Chapman and Hall, New-York.

[4] Grime, P.P., 1985. The C-S-R model of primary plant strategies- Origins, implications and tests. In Gottlieb, L.D. and Jain, S.K. (Eds.). Plant Evolutionary Biology. Chapman and Hall, London.

[5] IUCN. 1981 How to use the IUCN Red Data Book Categories. Threatened Plants Committee Secretariat. IUCN, Kew.

[6] Karimian, A.A. 2001. Investigation of plant species in KalmandBahadoran Protected Area, Ministry of Environmental protection Publication, Iran, Pp. 44.

[7] Leonard, J., 1981 – 1992. Contribution an' I' etude de la Flore et de la Vegetation des deserts de I' Iran. Fasc. 1-10. Jard, Botanique National de Belgique.

[8] Masumi. A.A. 1986-1995. Astragaluses of Iran, Volume 1-3, Ministry of forest and rangelands publication, Iran.

[9] Mobayen, S. 1975-1994. Vegetation's of Iran, (Vascular Plants) volume 1-4, Tehran University Publication, Iran.

[10] Mozafarian, V., 2001. Flora of Yazd, Yazd publication, Iran, Pp.473.

[11] Parsa, Ahmad, 1948-60. Flore de I' Iran. Vol. 1-8, Ministered de I 'Education. Teheran.

[12] Parsa, Ahmad, 1978-80. Flora of Iran, Vol. 1-2, Ministry of Culture and Higher Education of Islamic Republic of Iran., Tehran.

[13] Rabinowitz, D., 1981. Seven forms of rarity, In: Synge, H. (Ed.). The Biological aspects of rare plant conservation. John Wiley and Sons, New-York, pp.: 205-217.

[14] Rechinger, K.H. (Ed.), 1963-1993. Flora Iranica. Vol. 1-172. AkademischeDruck-U Verlagsanstalt, Graz, Austria.

[15] Rechinger, K. H., 1977. Plants of the Touran protected Area (Iran). Iran. Jour. Bot. 1(2): 155-180.

Influence of Lactose and Sucrose on Growth and Acetaldehyde Production by Three Strains of *Streptococcus thermophilus*

Rabha Bennama [1], Victor Ladero [2], Miguel A Alvarez[2], María Fernández [2], Ahmed Bensoltane [3]

[1] Department of Biotechnology, University of Mostaganem, Mostaganem, Algeria

[2] Instituto de Productos Lácteos de Asturias (IPLA-CSIC), Villaviciosa, Spain

[3] Department of Biology, University of Oran, Algeria

Abstract

This investigation describes three strains of *Streptococcus thermophilus* on the basis of production of acetaldehyde as aroma compounds. The strains under study (BN1, BN2 and BN3) were isolated from Algerian raw milk and were identified according to microbiological, biochemical, and genetic criteria. The growth of the strains and the determination of acetaldehyde were performed on M17 medium added to 0.5 and 3% (w/v) of lactose and sucrose. It was observed that the produced biomass (log cfu/ml), reached high values in the presence of 3% (w/v) of lactose and sucrose compared to that posted with 0.5% (w/v). The strains appeared to produce acetaldehyde. This production was more powerful in the case of concentration 3% (w/v) of lactose and sucrose. Strain BN1 produced approximately 205±75μmol and 218±90μmol of acetaldehyde respectively in the presence of 3% (w/v) of lactose and sucrose. This ratio was significantly higher ($P<0.01$) compared to that quantified with strains BN2 and BN3.

Keywords: *Streptococcus thermophilus*, growth, lactose, sucrose, acetaldehyde production.

1. Introduction

Fermented dairy products have become commonly consumed food in many countries around the world. These products were industrially developed using lactic acid bacteria, which were at the origin of an individual transformation process that affected the texture, flavour, quality, and the conservation of fermented dairy products [1]. *Streptococcus thermophilus* is one of the species that plays a great, interesting role for its contribution to the rapid transformation of lactose milk in lactate, the secretion of exopolysaccharides, synthesis of vitamins like folic acid, and production of some flavour compounds such as acetaldehyde [2]. Acetaldehyde is the major component responsible for the typical flavour of yogurt and a number of cheeses [3-4]. It is produced by the two yoghurt bacteria: *S. thermophilus* and *Lactobacillus bulgaricus*, but *S. thermophilus* species is deemed to be a good acetaldehyde producer [3]. The exact mechanism for the production of acetaldehyde from *S. thermophilus* is not well understood. In general, it is formed directly from the pyruvate decarboxylation through the action of the pyruvate decarboxylase or indirectly from the acetyl – CoA, through the pyruvate dehydrogenase and aldehyde dehydrogenase [5-6]. Moreover, acetaldehyde can be produced through the serine hydroxyl-methyl transferase

(SHMT), which catabolizes threonine into acetaldehyde and glycine [3]. SHMT is not the only shunt involved in the formation of acetaldehyde in yogurt, but also in the formation of glycine, serine, and significant amounts of folic acid [7].

Concerning this compound, the focus was on improving among other things of biosynthesis conditions, the composition of the culture media. In this context, this work is interested to the screening of three strains of *S. thermophilus* by studying the effect of carbon source (lactose and sucrose) on the growth and production of this aromatic compound.

2. Materiel and Methods

2.1. Strains and Culture Conditions

Three strains of *S. thermophilus* BN1, BN2, and BN3 isolated from Algerian raw milk were used in this study. Strains were identified by phenotypic and biochemical criteria and confirmed by molecular methods as described by Bennama et al. [8]. They were routinely grown on M17 medium [9] containing 0.5% (w/v) of lactose (LM17) and incubated anaerobically at 42°C.

2.2. Fermentation and Growth Parameters

Experimental cultures for growth were established using M17 medium containing as sole carbon sources 0.5 and 3% (w/v) of lactose or sucrose (LM17 or SM17) (Biochemika); adjusted to pH 7.0. Fermentation was initiated by inoculating the media with 1% (1×10^8 cfu/ml) overnight cultures of the studied strains. After 8h of fermentation at 42°C, viable bacterial counts were performed by serial dilution in peptone-saline water [[(1 $g.l^{-1}$) and NaCl (8.5 $g.l^{-1}$)]. Selected dilutions were then plated onto LM17 agar. Plates were incubated at 42°C for 48h; growth is expressed as log cfu/ml and analyzed by comparison to the initial rate of inoculation. Acidification that developed in the cultures was measured with a pH meter (Hanna Instruments, pH210 microprocesor pH meter). All experiments were triplicated.

2.3. Acetaldehyde Estimation

Estimation of acetaldehyde was carried out under the same conditions outlined above with LM17 or SM17. In order to avoid evaporation, it should be noted that cultures destined for these experiments were prepared in centrifuge tubes hermetically sealed. However, acetaldehyde was determined after 8h of fermentation at 42°C. It was measured by spectrophotometer (Jenway J7305) using an assay kit (R-Biopharm: enzymatic bioanalysis, Germany), based on the reduction of the NAD to NADH in the presence of aldehyde dehydrogenase. All assays were repeated three times.

2.4. Data Analysis

In all experiments mentioned above, the values of results are the mean ±standard error. Statistical analysis was done with student's test.

3. Results and Discussion

3.1. Growth of Strains in the Presence of Lactose and Sucrose

Bacteria obtain energy by multiple ways. Most of this energy is used in the biosynthesis of many metabolites or anabolism. Performance of different metabolic reactions depends essentially on the nature of the carbon source and its concentration. Thus in this work, the importance of the carbon source was checked with strains BN1, BN2, and BN3 of *S. thermophilus* by studying their growth in the presence of lactose and sucrose. These have been incorporated to M17 medium at final concentration of 0.5 and 3% (w/v). It is well known that *S. thermophilus* exhibits a highly affinity to grow on lactose; sucrose can also be used nevertheless with lower efficiency than lactose [10-11]. This affinity towards both carbon sources is related to the presence of a large variety of important genes and enzymatic equipment of sugar metabolism and central carbon pathways [11-12].

The results related to the increase in biomass (log cfu/ml) and pH variations after 8h of growth on LM17 and SM17 media are illustrated in figures 1 and 2. From these results, it appears that strains displayed a high level of growth with lactose and sucrose used at 3% (w/v). Comparatively to the initial rate of inoculation, the strains showed an average increase in the biomass of 3.20±0.09 and 3.10±0.14 log cfu/ml respectively in the presence of 3% (w/v) of lactose and sucrose. A significant difference ($P<0.05$) in the biomass was only observed between the sucrose used at 0.5 and 3% (w/v).

During the growth, a significant decrease in the pH of the cultures was observed. The pH values decreased, in average, from 7.0 to 4.5 in the presence of 3% (w/v) of both carbon sources (figure 2). The pH decreasing over time raises the existence of a specific metabolic activity in the different fermentative pathways used by *S. thermophilus*, which acidifies the medium by the production of lactate or other acids [11]. Furthermore, it is important to point that the level of growth of the strains was not influenced by the decrease in pH, even after 24h of incubation (data not shown).

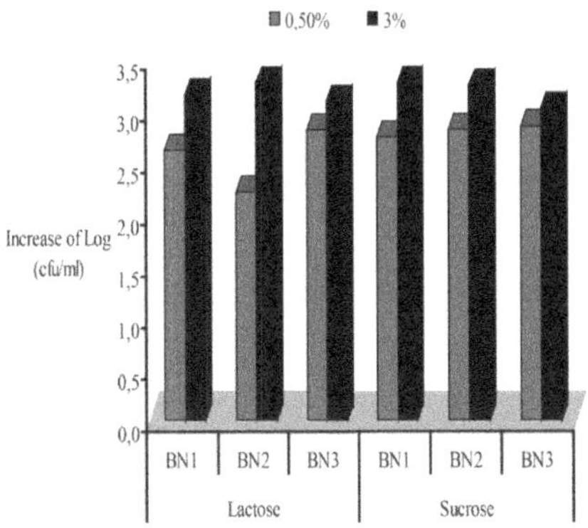

Fig 1. Mean values of increase in biomass (log cfu/ml) of the strains BN1, BN2 and BN3 after 8h of fermentation at 42°C in LM17 and SM17 media.

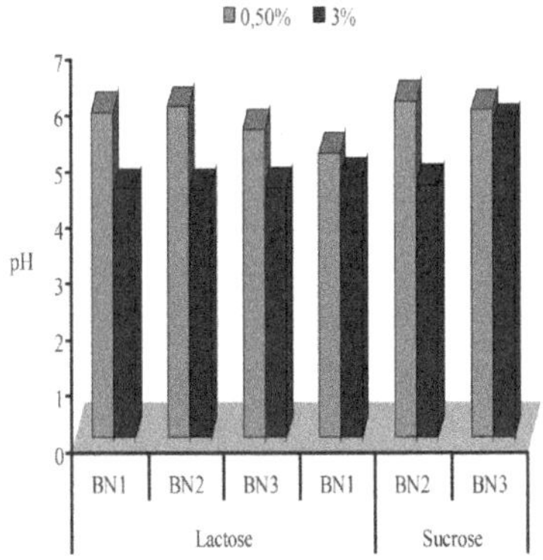

Fig 2. Mean values of pH measured after 8h of fermentation at 42°C in LM17 and SM17 media inoculated with strains BN1, BN2 and BN3.

3.2. Acetaldehyde Production

The ability to produce acetaldehyde from a carbon source was determined in strains BN1, BN2, and BN3 in the presence of 0.5 and 3% (w/v) lactose and sucrose. The acetaldehyde amounts were quantified after 8h of incubation at 42°C (figures 3, 4). The results indicated clearly that the three strains produced acetaldehyde, but the production appeared closely related to the concentration of carbon source and strain-dependent. The amounts of acetaldehyde formed by the strains were proportional to the concentration of carbon source. In the presence of 3% (w/v) of either lactose or sucrose, the strains formed a high amount of acetaldehyde compared to the 0.5% (figures 3, 4), with the exception of the strain BN2, which formed about 83.50±4.90 and 85.20±4.00 μ mol respectively with 0.5% and 3% (w/v) of sucrose (P> 0.05). However, the most potent acetaldehyde producer was BN1 strain. It was capable of producing up to 205±75μmol and 218±90μmol of acetaldehyde with 3% (w/v) of lactose and sucrose respectively. This ratio was significantly higher (P <0.01) than the one produced by the BN2 and BN3 strains, which produced 70.50±3.70, 85.20±4.00, 54±12 and 80±7μmol, respectively. These results indicate that there is variability in the amounts of acetaldehyde formed by the strains of *S. thermophilus* under study. This variability has been widely reported by many authors [3-13]. Moreover, Ayhan et al. [13] noticed a large variability in the amounts of acetaldehyde produced by 30 strains of *S. thermophilus*. Chaves et al. [3] signaled that the production of acetaldehyde appeared to be strain specific and variable. The same authors reported that the high levels of acetaldehyde were obtained when L-threonine was added to the culture medium. This observation was also reported by Bennama et al. [4].

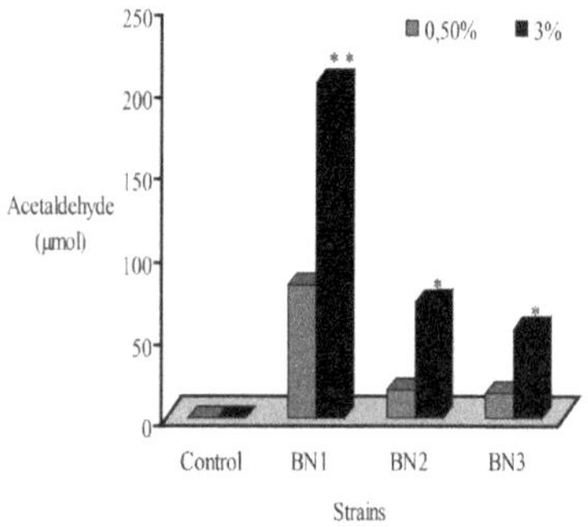

Fig 3. Acetaldehyde amounts (μmol) formed by the strains BN1, BN2 and BN3 after 8h of fermentation at 42°C in LM17 medium.

**: Highly significant difference (P<0.01) compared to mean values of acetaldehyde obtained with BN2 and BN3 strains.

*: No significant difference.

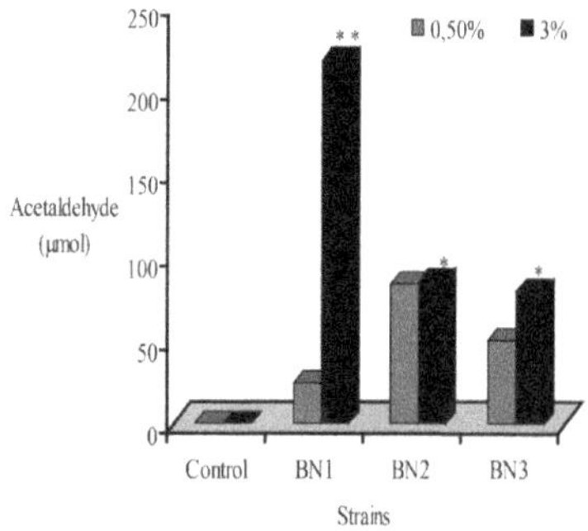

Fig 4. Acetaldehyde amounts (μmol) formed by the strains BN1, BN2 and BN3 after 8h of fermentation at 42°C in SM17 medium.

**: Highly significant difference (P<0.01) compared to mean values of acetaldehyde obtained with BN2 and BN3 strains.

*: No significant difference.

According to Ott et al. [5], the production of acetaldehyde in milk by lactic acid bacteria seems to be strain-dependent too. These authors also showed that glucose appeared as the main precursor of acetaldehyde in milk fermented with *S. thermophilus*. They confirmed that the major production of acetaldehyde was related to the glycolitic pathway. These observations explain the origin of acetaldehyde amounts formed by strains under study especially by BN1 strain. Furthermore, Oïzer and Atasoy [14] reported that lactose hydrolysis induced by the β-galactosidase caused a significant increase in the level of acetaldehyde in yoghurt samples prepared using viscous starter cultures. The results of this study show that production of acetaldehyde by BN1 strain seemed efficient with lactose and sucrose, whereas for BN3 strain, the production was more important with 3% (w/v) of sucrose.

4. Conclusion

In this study, it was found that strains synthesized significant amounts of acetaldehyde in the presence of 3% (w/v) lactose and sucrose. However, with 0.5% (w/v) small quantities were formed. The results reveal that acetaldehyde production was strain-specific and influenced by the concentration of carbon source added to the medium. Following these promising findings, it appears that depending on the strains of *S. thermophilus*, a determined concentration of carbon source constitutes one of the optimal conditions for the synthesis of acetaldehyde in this thermophilic species. This property, mainly for sucrose, could be useful in dairy technology to enhance natural production of this aroma compound by *S. thermophilus*.

5. References

[1] Tamime, A. Y., 2002. Fermented milks: a historical food with modern applications-a review. European Journal of Clinical Nutrition, 56 (4): p. S2-S15.

[2] Delorme, C., 2008. Safety assessment of dairy microorganisms: *Streptococcus thermophilus*, International Journal of Food Microbiology, 126: p. 274- 277.

[3] Chaves, A.C.S.D., M. Fernandez, A.L.S. Lerayer, I. Mierau, M. Kleerebezem, and J. Hugenholtz, 2002. Metabolic Engineering of Acetaldehyde production by *Streptococcus thermophilus*. Applied and Environmental Microbiology, 68 (11): p. 5656-5662.

[4] Bennama, R., Rechidi-Sidhoum, N. and A. Bensoltane, 2011. Effect of threonine on growth and acetaldehyde production by *Streptococcus thermophilus*. World Applied Sciences Journal, 15 (2): p. 160-163.

[5] Ott, A., J. E. Germond, and A. Chaintreau, 2000. Origin of acetaldehyde during milk fermentation using ^{13}C-labeled precursors. Journal of Agriculture and Food Chemistry, 48: p. 1512-1517.

[6] Bongers, S. R., M.H.N. Hoefnagel, and M. Kleerebezem, 2005. High-Level acetaldehyde Production in *Lactococcus lactis* by metabolic engineering. Applied and Environmental Microbiology, 71 (2): p. 1109-1113.

[7] Ott, A., L.B. Fay, and A. Chaintreau, 1997. Determination and origin of the aroma impact compounds of yogurt flavour. Journal of Agricultural and Food Chemistry, 45: p. 850-858.

[8] Bennama, R., M. Fernández, V. Ladero, M. A. Alvarez, N. Rechidi-Sidhoum, and A. Bensoltane, 2012. Isolation of an exopolysaccharide-producing *Streptococcus thermophilus* from Algerian raw cow milk. European Food Research Technology, 234: p. 119-125.

[9] Terzaghi, B. E., and W. E. Sandine, 1975. Improved medium for lactic streptococci and their bacteriophages. Applied and Environmental Microbiology, 29: p. 807-813.

[10] Van Den Bogaard, P.T.C., P. Hols, O.P. Kuipers, M. Kleerebezem, and M.W. De Vos, 2002: Sugar utilization and conservation of gal-lac gene cluster in *Streptococcus thermophilus*. Systematic and Applied Microbiology, 27: p. 10-17.

[11] Hols, P., F. Hancy, L. Fontaine, B. Grossiord, D. Prozzi, N. Leblond-Bourget, B. Decaris, A. Bolotin, C. Delorme, S.D. Ehrlich, E. Guédon, V. Monnet, P. Renault, and M. Kleerebezem, 2005. New insights in the molecular biology and physiology of *Streptococcus thermophilus* revealed by comparative genomics. FEMS Microbiology Reviews, 29: p. 435-463.

[12] Iyer, R., S.K. Tomar, T. Uma Maheswari, and R. Singh, 2010. *Streptococcus thermophilus* strains: Multifunctional lactic acid bacteria. International Dairy Journal, 20: p.133-141.

[13] Ayhan, K., F.Durlu-Ozkaya, and N. Tunail, 2005. Commercial important characteristics of Turkish origin domestic strains of *Streptococcus thermophilus* and *Lactobacillus delbrueckii subsp.bulgaricus.* International Journal of Dairy Technology, 58 (3): p. 150-156.

[14] Ozer, B., and F. Atasoy, 2002. Effect of addition of amino acids treatment with ß-galactosidase and use heat shocked on the acetaldehyde level in yoghurt. International Journal of Dairy Technology, 55: p. 1-6.

The Alkaline Phosphatase Levels in the Seminal Plasma and Sperms of Sub-Fertile Patients and Normospermic Men

Faris N. A. Alhady Alibawi [1], Sahib Y. Al-Morshidy [2], Ali G. Alhuweizi [1]

[1] College of Science, University of Babylon, Babylon, Iraq

[2] University of Kufa, Najaf, Iraq

Abstract

This study examined 110 semen specimens collected from sub-fertile and Normospermic men after a period of abstinence from 3-5 days. The samples were collected in Fertility Center Laboratories/ Al- Saader Hospital/ Najaf province in the period from November 2009 to May 2010.

This study aimed to comparison between the concentrations of the enzyme alkaline phosphatase (ALP) in seminal plasma of different in sub-fertility groups include: Asthenospermia (AS), Oligo-asthenospermia (OAS), Azoospermia and Normospermia (NS) to study the effect of ALP levels in the quality of sperm in sub-fertile patients.

The results revealed that a significant increase ($p < 0.01$) in the concentration of ALP in the seminal plasma compared to the levels of this enzyme in sperm per each group of this study: NS, AS and OAS. The comparison study of the seminal plasma ALP enzyme level indifferent groups of this study, revealed a significant decrease in the levels of ALP enzyme in the seminal plasma of Oligoasthenospermic($p < 0.05$) and Azoospermic ($p < 0.01$) patients compared to seminal plasma of Normospermic men, while this decrease was insignificant in the seminal plasma of Asthenospermic patients . The results showed non significant decrease ($p > 0.05$) in ALP enzyme activity of the asthenospermic patient`s sperm, while a significant decrease ($p < 0.05$) of this enzyme in Oligoasthenospermic sperms compared to the levels of this enzyme in Normospermic sperm.

It was concluded that the level of ALP in the seminal plasma and sperms correlated with the concentration of the sperms.

1. Introduction

The alkaline phosphatase secret in seminal fluid by Epididymis of rabbet and dogs (1;2), while it was secreted by prostate and testis in human (3). All Isoenzymes of this enzyme is encoded by three genes in the body, which includes the placental alkaline phosphatase , intestinal and Osteohepatic, this differences of the isoenzymes refer to the addition of gluco groups to the peptide chains to form glucoproteinsand thus these isoenzymes differ in their different characteristics such as behavior which is expressed during the Electrophoresis (4). According to the differences of isoenzymes of alkaline phosphatase can diagnosis of seminal fluid quality (5). The alkaline phosphatase is intended to be more effective in the Leyding cell , transitional and Fibroblast in seminiferous tubules (6), but on the other hand not see any effectiveness of this enzyme in Sertoli cell.

The alkaline phosphatise from the enzymes that cause loss of the phosphorus Group, which is effective in several tissues, including bone, liver, kidney, bowel, lung, and placenta in addition to the reproductive system. The amount of most alkaline phosphatise in the bulls excreted from the seminal vesicles in addition to the testis and epididymis, which constitutes a small percentage of concentration in semen (7).Another study indicated the existence of significant difference in the concentration of the alkaline phosphatise in dogs that underwent a process of Vasectomization for those that did not take place in this process (8). Also observed significant decrease in the levels of one isoenzyme of alkaline phosphatase secreted from germ cells, which called Placental-like alkaline phosphatase, in men who have undergone the process of Vasectomy So that it is possible to use the measuring of this enzyme concentration in the semen of these patients, to see the success of vasectomy (3;5). The strains of rabbits that have high fertility, have higher levels of seminal phosphatase enzymes compared to those that have low fertility (9;10).

The seminal alkaline phosphatase in boar inhibited by Theophylline, Caffeine and Pentoxifylline which lead to the improvement of sperm parameters (11) but the same research added that this result may be different when studied on human sperm.

This study aimed to study the correlation between the level of Alkaline Phosphatase and male infertility

2. Materials and Methods

Semen specimens

Semen specimens were collected from Normospermic (NS), Asthenospermic (AS), Oligoasthenospermic (OAS) and Azoospermic patients by masturbation after three days of sexual abstinence. The specimens were allowed to liquefy at 37C°, and then seminal fluid analysis was performed to determine the sperm parameters include: sperm concentration, sperm motility percent, grade activity and abnormal sperm morphology percent.

Preparation of Seminal Fluid specimens for estimation of Alkaline Phosphatase

One ml of each specimen was centrifuged (3000 rpm for10 minutes) after the Seminal Fluid Analysis was performed to obtain the seminal plasma as supernatant and pellet. The supernatant (Seminal Plasma) was transferred to the other tube and then used for estimation the level of ALP enzyme. The pellet was washing by using one ml of Normal Saline and mixed well , then re-centrifuged again (3000 rpm for 5 minutes), then the supernatant was removed completely. The ALP enzyme connected with the sperm plasma membrane, so that the sperm should be crushed by wood stick, then 0.5 ml of distilled water was added to the specimens contain the sperms for disruption the cell membrane due to the different of osmotic pressure . The seminal plasma and sperm suspension was froze untile the test.

Alkaline phosphatase concentration:

The ALP enzyme level in the seminal plasma and sperms was estimated by colorimetric method by using a kit manufactured by Biolabo: France Company.

Statistical analysis:

Analysis of data was performed by usingStatistical Package for (SPSS) (Version). Results are expressed as mean ± S.E. Statistical differences were determined by Least Significance Differences (LSD) test for multiple comparisons between different groups.

3. Results

The results showed significant decrease in sperm concentration in (OAS) group ($p<0.01$), and ($p<0.05$) in (AS) group compared to (NS). The sperm motility percent and grade activity in (AS) and (OAS) groups were significant decrease ($p<0.01$) compared to (NS), also there was a significant increase ($p<0.01$) in abnormal sperm morphology percent in all infertile groups compared to normospermia.

The comparison between the ALP concentration in seminal plasma and sperms showed significant increase ($p<0.01$) of ALP level in the seminal plasma compared to sperm in (AS), (OAS) and (NS) specimens. Also the results revealed a significant decrease ($p<0.01$) in seminal ALP concentration in (OAS) and azoospermia compared to (NS), while this decreasment is non significant ($p>0.05$) in (AS). Also there was a significant decrease ($p<0.01$) of sperm ALP level in (OAS) compared to (NS) Table-1-.

Seminal fluid parameters Mean ± SE	Normospermia Mean ± SE	Asthenospermia Mean ± SE	OligoAsthenospermia Mean ± SE	Azoospermia Mean ± SE
Sperm concentration (million/ml)	74.285 ± 4.257	57.826 ± 5.258 *	7.700 ± 1.955 **	---
Sperm motility percent	65.00 ± 2.182	28.913 ± 2.981**	25.227 ± 2.662**	---
Grade activity	3.714 ± 0.106	1.913 ± 0.168**	1.500 ± 0.177**	---
Abnormal sperm morphology percent	59.761 ± 3.423	76.956 ± 2.82**	87.391 ± 2.492 **	---
ALP Level in seminal plasma (IU/L)	273.6 ± 29.9	218.5± 26.8	179.2± 13.0 * *	156.7± 15.0 *
ALP Level in sperms (IU/L)	68.6 ± 6.2	54.4 ±3.1	45.1 ±26.8 * *	------

* Significant difference ($p<0.05$)
** Significant difference ($p<0.01$)

Table 1. Seminal Fluid Parameters and Alkaline Phosphatase (ALP) Level in Infertile groups and Normospermia.

4. Discussion

The results of the current study showed, the parameters of semen and sperm for patients with infertility significant decrease ($p < 0.05$) in the sperm concentration for (AS) and (OAS) patients compared to (NS) men, this result is agree with other studies (12; 13). The concentration of sperm in the (NS) is higher than the other infertility groups due to several factors, including: a defect in the action of hormones or bacterial infections that affect the male reproductive system, or varicocele.

Also the results revealed significant increase in abnormal sperm morphology percent in (AS) and (OAS) patients compared to (NS), this result agree with the study of (14), that the normospermic men have 75.67% normal sperm morphology as compared to other groups of infertility.

The results showed significant increase ($p < 0.01$) in the concentration of the alkaline phosphates (ALP) enzyme in the seminal plasma of AS and OAS patients compared to the concentration of ALP enzyme in sperm. The place and the amount of secretion of the ALP enzyme different according to the type of organism varies from one species to another (1). The studies conducted with human have indicated that the secretion of ALP enzyme from the prostate and testis (3) , while the another study revealed that this enzyme in rabbits and dogs is secreted by the epididymis (2). There is another study conducted on one of thecattle breeds showed an inverse relationship between the ALP concentration and sperm concentration (15), so this variation may be due to the presence more than one source of the ALP enzyme secretion (16;17).

The comparison study of the concentration of seminal plasma ALP between the different infertility groups and normospermia showed non significant decrease of ALP in (AS) and significant decrease in (OAS) and Azoospermia. The significant differences of ALP concentration refer to sperm number only not for the sperm motility percent or grade activity, this result agree with the other study (3), which are revealed to the existence of a positive relationship between sperm counts and concentration of this enzyme in the semen samples. Another study found that the concentration of spermatozoa increase with the reduction of ALP, while the sperm concentration increase with the increasing of ALP , and the same study showed that the lowest activity of ALP enzyme in azoospermic patients compared to another groups of infertility (16). The levels of ALP enzyme increased in the first split ejaculate compared to the second split is because the largest amount of the first split of ejaculate secreted from prostate (18).

The ALP enzyme in the sperm showed a significant decrease in sperm of AS and (OAS) patients compared to (NS) , this significant decrease in (OAS) may be refer to the decline sperm count in this group of patients. Low and Saltiel; 1988 (19) revealed, that the ALP enzyme linked to the sperm cell membrane by Phosphatidylinositolglycan located on the outer surface of the sperm. The site of ALP enzyme is in the plasma membrane in addition to cytoplasmic droplet and acrosome body of the sperm (20; 21). The ALP enzyme act through hydrolysis of phosphate ester of the nucleotide , sugars and ATP and has a potential role in removing phosphorus from Adenosine Monophosphate (AMP), also works to prevent the addition carbohydrates groups to glycoproteins in the surface of sperm (22). The another study indicated that the ALP enzyme present in the chloride channels in the sperm (23), and that have a role in the acrosome reaction (24), and thus there is a positive linear relationship between the level of this enzyme and sperm concentration in the semen samples of NS, AS and OAS.

5. References

[1] Frenette, G.; Dube, J.Y. and Tremblay, R.R. (1986) . Origin of alkaline phosphatase of canine seminal plasma. Archives of Andrology.;*16*:235-241.

[2] Muller, B. (1983) Genital tract proteins in the male rabbit: Alkaline phosphatase enzyme action and site of synthesis. Andrologia.*15*: 676-681.

[3] Lewis-Jones, D. I., Johnson, P. M. Desmond, A. D. and McLaughlin, P. J. (1992) Germ cell alkaline phosphatase in human seminal plasma following vasectomy. Br. J. Urol. *69*: 418-420.

[4] Seargeant, L. E. and Stinson, R. A. (1979) Evidence that three structural genes code for human alkaline phosphatase. Nature *281*: 152-154.

[5] Kutzler, M. A.; Solter P.F.; Hoffman, W. E. and Volkmann, D. H.(2003) Characterization and localization of alkaline phosphatase in canine seminal plasma and gonadal tissues. Theriogenology, *60*:299-306.

[6] Gunawardana, VK. (1990). Ultrastructural localization of alkaline phosphatase in intertubular tissue of the testis in domestic fowl. Tissue cell.; *22*:113-121.

[7] Zakrzewska, H.; Udala, J. and Blaszczyk, B. (2002). In vitro influence of sodium fluoride on ram semen quality and enzyme activities. Fluoride. *35*: 153-160.

[8] Stornelli, A.; Arauz, M.; Baschard, H. and R-L de la sota.(2003) Unilateral and bilateral vasectomy in the dog. Blackwell Synergy. Reproduction in domestic animals. *38*(1):1.

[9] Elkomy, A.E.; El-SeibyM.E. and K.I. Kamel,(2008)Comparative study of semen quality and free aminoacid content in seminal plasma between high andlow motile sperm rabbit bucks. Egypt. Poult. Sci.,*28*(2): 633-649.

[10] El-Seiby, M.E.; Elkomy A.E. and K.I. Kamel,(2008)Evaluation of spermatological parameters and free amino acids composition of Black Baladi, New Zeland White andV-Line rabbits buck under winter egyptian condition. Egypt. Poult. Sci., *28*(2): 617-631.

[11] Glogowski, J.; Danforth, D.F. and Ciereszko, A.(2002) Inhibition of alkaline phosphatase activity of boar semen by petoxifylline,caffeine and theophylline. J. Andrology, *23*(6).

[12] Chia, S-H.; Lim, S-T.A.; Tay, S-k. and Lim, S-T.(2000).Factors associated with male infertility: a case- control study of 218 infertile and 240 fertile men. BJ. *101* (1): 55-61.

[13] Okonofua, F.; Menakaya, U.; Onemu, S.O.; Omo-Aghoja, L.O. and Bergstorm, S.(2005). A case-control study of risk factors for male infertility in Nigerria Asian J. Andrology, *7*(4): 351-361.

[14] Pal, P.C.; RajalakShmi, M.; Manocha, M.; Sharma, R.S.; Mittal, S., and Rao, O.N.(2006). Semen quality and sperm functional parameters in fertile India men. Andrologia, *38*:20-25.

[15] Bell, D.J. and Lake, P.E.(1962).Comparison of phosphomonoesteraseactivities in the seminal plasmas of domestic cock, tom turkey, boar, bull, buck, rabbit and man. J ReprodFertil , *3*: 262-268.

[16] Singer,R.;Barnet, M.;Allalouf, D.;Schwartzman, S.; Sagiv, M.; Landau, B.; Segenreich,E.; Servadio,C.(1980) Some properties of acid and alkaline phosphatase in seminal fluid and isolated sperm. Archives of andrology, *5*:195-199.

[17] McLaughlin, P. J.; Lewis-Jones, I.; Hutchinson, G. E. and Johnson, P.M. (1986) Placental-Type alkaline phosphatase in human seminal plasma from fertile and infertile men. Fertil. Steril. *46*:934-937.

[18] Lewin, L.M.; Golan, R.; Soffer, Y.; Kaufman, S.; Yulzary, Y. and Zaidman, J.(1993) Alkaline phosphatase in human semen: An investigation using enzyme inhibitors and gel electrophoresis. Eur. J. Clin. Chem. Clin. Biochem.*31*: 811-814.

[19] Low, M.G. and Saltiel, A.R. (1988). Structural and functional roles of glycosylphosphatidylionsitol in membranes. Science, *239*:268-275.

[20] Soucek, DA.; Vary, JC.(1984) Some properties of acid and alkaline phosphatase from boar sperm plasma membranes. Biol. Reprod.*31*:687-693.

[21] Parks, JE.; Arion, JW. and Foote, RH.(1987) Lipids of plasma membrane and outer acrosomal membrane from bovin spermatozoa. Bio Reprod; *37*: 1249-1258.

[22] Yuan, YY.; Shi, QX. and Srivastava, PN.(1995) Inhibition of rabbit sperm acrosomal enzymes by grossypol. MolReprodDev; *40*:228-232.

[23] Becq, F.; Fanjul, M.; Merten, M.; Figarella, C.; Hollande E.; Gola, M.(1993) Possible regulation of CFTR-chloride channels by membrane-bound phosphatase in pancreatic duct cells. FEBS Lett.*327*: 337-342.

[24] Meizel S.(1997) Amino acid neurotransmitter receptor/chloride channels of mammalian sperm and the acrosome reaction. Biol. Reprod.56: 569-574.

Honey Bee Venom Modulates Hyperglycemia in Response to Hyperandrogenism in Polycystic Ovarian Syndrome-Induced Wistar Rats

M Nabiuni[1], S Nasri[2], F Poyanmanesh[2], L Karimzadeh[1], Z Nazari[1]

1 Department of Cell and Molecular Biology, Faculty of Biological Sciences, Kharazmi University, Tehran, Iran.

2 Department of Biology, Faculty of Agriculture and Basic Science, Payamnoor University, Tehran, Iran.

Abstract

Polycystic Ovarian Syndrome (PCOS) is inflammatory disease characterized by hyperandrogenemia, hyperthecosis and chronic anovulation. Honey bee venom (HBV) contains a variety of biologically active components having various pharmaceutical properties. This study was designed to detect the possibility of HBV application as an anti-inflammatory therapeutic agent. To induce PCOS, 2mg/100gr B.W Estradiol Valerate (EV) was subcutaneously injected to induce PCOS in mature Wistar rats then ovaries and serum from three groups of EV-induced PCOS, HBV-treatment and normal intact animals were collected for histological comparison and blood sugar test. As a result, a significant increase in ovarian weight was observed in experimental group rather than controls. Furthermore, in HBV-treated group a significant decrease was observed in ovary weight comparing with experimental group ($P<0.01$). The results obtained from Chemo Luminesance Immuno Assay (CLIA) declared that testosterone and Estradiol levels in experimental group significantly increased ($P<0.001$). These hormones were decreased in animals treated with HBV. Blood sugar level showed reduction in HBV-treated rats. Thickness of theca layer, number and diameter of cysts significantly decrease in HBV group comparing to PCOS group. Moreover, corpus luteum, as a sign of ovulation, was observed in HBV-treated ovaries. In conclusion our results suggest that beneficial effect of HBV against PCOS may be mediated by the inhibitory effect of HBV on TNF-α level.

Keywords: Polycystic ovarian syndrome, honey bee venom, blood sugar, theca layer

1. Introduction

Polycystic Ovarian Syndrome (PCOS) is an inflammatory disease characterized by hyper androgenemia, hyperthecosis, hyperglycemia and chronic anovulation(1-6). Honey bee venom (HBV) contains a variety of biologically active components like peptides (Melittin and Apamin), enzymes and biologically active amines (histamine, epinephrine. It has shown that HBV has analgesic, anti-cancer and anti-inflammatory activity. Melittin, the major active ingredient of BV, has been reported to induce apoptosis and to possess anti-proliferation effects (7-9). TNF-α is a key inflammatory stimulus which plays a main role in regulating normal activity of ovary in follicular growth and luteal stages. It's over expression in adipose tissue leads to obesity and insulin-resistance in humans and rodents. This factor by stimulating mitotic activity in undifferentiated theca cells and increasing steroidogenic cells causes PCOS (10-13). Whereas the anti-inflammatory and anticancer effects of HBV were proved, we examined sugar levels in rats with PCOS before and after treatment via HBV. We also measured plasma testosterone and theca layer thickness, the commonly used index of PCOS (1). We hypothesized that HBV can modulate hyperglycemia in bee venom-treated rat with PCOS in response to androgenemia, compared with age-matched controls.

2. Materials and Methods

Experiments were performed on Female Wistar rats (170±20g). Before and during the experiment they were housed in special cages with a standard space and under controlled cycle of light and darkness (lights on from 06:00 to 20:00), humidity 55±15% and temperature range of 20-24°c and free access to water and commercial food (Behparvar Com., Iran). Induction of PCOS was administered using 1mg/100gr B.W intramuscular injection of EV. After verifying the induction of PCOS, experimental group was divided into two groups: PCOS group and PCOS+BV group. PCOS+BV received 0.5 mg/kg BV sc for 14 days, continuously (7). PCOS group in this period of time received physiological saline solution. At around 09:00 am, trunk blood was collected and the serums were separated using 4000 RPM centrifuge for 10 min. The ovaries were separated from the twisted oviduct tubes and were placed in bouin fixative for histological analysis; fixed samples were kept in alcohol solutions of 20 to 100% for a period of 45 min for dehydration and afterwards in alcohol/xylen (50:50) and xylen (3 times) for clearing and blocked in paraffin. The samples were sliced in 7 micron thickness using a microtome and the sections were placed on slides previously coated with gelatin and then stained with hematoxylin-eosin for histological observation. Serological analysis was performed to measure sugar level and hormone alterations. Testosterone and estradiol detected by Chemo Luminesance Immuno Assay (CLIA). In order to detect suger, glucose kit (GOD_PAP 90014) was used. The one-way ANOVA and INSTAT software were used to determine the statistical significance of differences between the values for the experimental and control groups. Data are expressed as means ± standard errors (S.E.M) and the results are taken from at least three independent experiments performed in triplicate. P-values of 0.05 or less were considered statistically significant.

3. Results

The ovaries were also precisely weighted and a significant increase was observed in experimental group rather than controls. Furthermore, in bee venom-treated group a significant decrease was observed in ovary weight comparing with experimental group ($P<0.01$). The results obtained from CLIA declared that testosterone and Estradiol levels in experimental group significantly increased ($P<0.001$). These hormones were decreased in animals treated with bee venom, and comparing with animals in control group they were regulated. Reduction observed in testosterone and Estradiol levels was significant with $P<0.05$. These data including raise in androgens showed that induce of syndrome was absolutely successful. What more is bee venom managed to reduce Estradiol and testosterone.(Table1) sugar level fundamentally adjusts its production, which in this study, PCOS induction by means of Estradiol volerate led to a significant raise in sugar level (P<0.01), and its reduction in rats treated with honey bee venom was outstanding (P<0.05). (Diagram1).

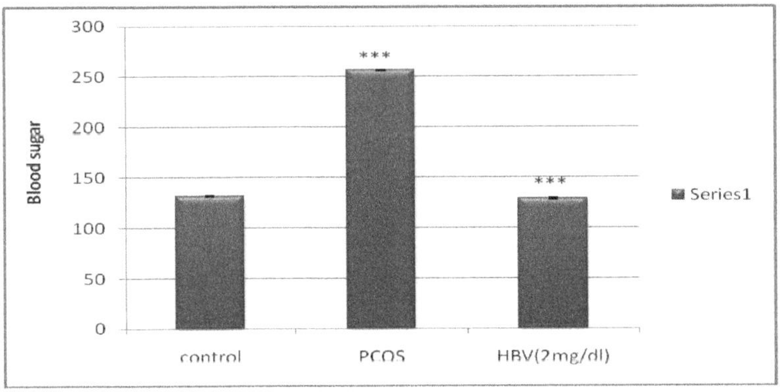

Diagram 1. Different sugar level in control and polycystic rats (n=10). In polycystic group, a significant increase was observed in sugar level. ***P< 0.001, PCOS vs. control, HBV vs. PCOS group.

	Cont	SD	SEM	PCOS	SD	SEM	PCOS+BV	SD	SEM	con vs pcos	con vs pc+bv	pcos vs pc+bv
Estradiol (pg/ml)	14.5	1.958	0.6191	54.778	8.599	2.866	21.889	4.485	1.495	*** P<0.001	* P<0.05	*** P<0.001
Testosteron (ng/dl)	62.5	9.925	3.138	345.9	112.99	35.73	115.9	39.145	12.379	*** P<0.001	ns P>0.05	*** P<0.001
Ovarian weight (mg)	13	3.232	1.022	20.5	4.577	1.447	15	4.714	1.491	** P<0.01	ns P>0.05	* P<0.05
Theca layer -late antral follicles (micrometer)	99.8	11.144	3.524	157.2	44.183	13.972	110.7	23.281	7.362	*** P<0.001	ns P>0.05	** P<0.01

Table 1. Bee venom treatment effects in polycystic ovarian syndrome (PCOS). Baseline parameters of polycystic ovarian syndrome (PCOS) rats (n=10) and control (n=10) and bee venom –treated rats (n=10).

In order to determine follicular development, follicles were classified based on morphology and diameter into 6 groups consisting of: primordial, primary and preantral (<600μm), antral (600-1000μm), cystic follicles, and corpus luteum. Decrease in the number of primary follicles, antral follicles and corpus luteums was significant with P<0.001, in primordial follicles with P<0.01, and in preantral follicles with P<0.05. In PCOS ovaries, some large cystic follicles with thick theca layer were observed. In this group, no corpus luteum as a sign of ovulation was seen.

In rats treated with honey bee venom, the number of primordial and preantral follicles and corpus luteums increased, whereas, the number of cysts and thickness of theca layer in antral follicles decreased, which these changes were significant in comparison with sham group. Also, some corpus luteums were observed in this group which was considered as a sign of relative improvement in PCOS ovaries. (Diagram 2 and 3).

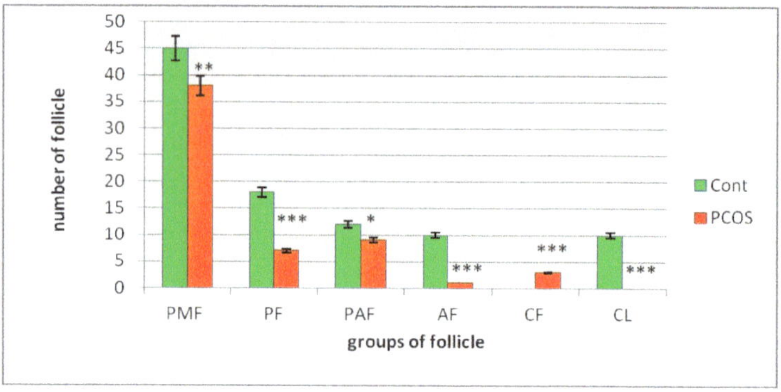

Diagram 2. Different follicular groups in control and polycystic ovaries (n=10). In ovaries of polycystic group, a significant increase and decrease was observed in number of cysts and number of corpus luteums respectively. ***P< 0.001, **P< 0.01, *P< 0.05. (PMF, Pre Mordial Follicle; PF, Primary Follicle; PAF, Preantral Follicle, AF, Antral Follicle; CF, Cystic Follicle; CL, Corpus Luteum.)

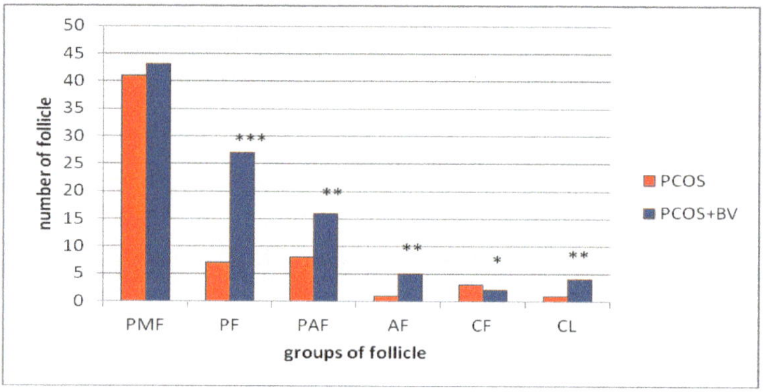

Diagram 3. Different follicular types in ovaries of polycystic and bee-venom treated polycystic group. In polycystic ovaries treated with honey bee venom (n=10), a significant increase was observed in all follicular clusters (except primordial follicles) rather than polycystic group. The primordial follicles were not significantly increased. Moreover, a significant decrease was seen in the number of ovarian cysts. ***P<0.001, **P< 0.01. (PMF, Pre Mordial Follicle; PF, Primary Follicle; PAF, Preantral Follicle, AF, Antral Follicle; CF, Cystic Follicle; CL, Corpus Luteum.)

4. Summary and conclusion

Insulin sensitivity and hyperglycemia is directly related to androgen levels (14). These findings suggest that hyperandrogenism play a role in the development of insulin resistance and hyperandrogenism in PCOS. It is considered metformin as a treatment for PCOS which has been shown to inhibit the NF-κB activation and also reduce sugar level in PCOS woman (15). Then we can also name honey bee venom as a similar factor decreasing sugar level. Histological changes observed in ovary after bee venom treatment, can also be considered as a confirmation for this syndrome progress. Our results confirm that Bee venom caused a decrease in follicular theca layer in PCOS rats, which is actually because of increased lipolysis and decreased hypertrophy of this layer. Due to this decrease, the androgens and steroids produced by this layer also decrease and consequently the total levels of serumic estrogen and androgens reduce by honey bee venom. In this regard we have demonstrated that bee venom injection produces a significant anti-hyperglycemia effect in PCOS Wistar rats.

5. References

[1] Repaci, A., Gambineri, A., Pasquali, R., 2011. The role of low-grade inflammation in the polycystic ovary syndrome. Molecular and Cellular Endocrinology 335, 30–41.

[2] Sathyapalan, T., Atkin, L., 2010. Mediators of Inflammation in Polycystic Ovary Syndrome in Relation to Adiposity. Mediators of Inflammation. Article ID 758656, 5 pages.

[3] Panidis D, Kita M, Katsikis I, Karkanaki A, Karayannis V, Rousso D., 2006. Mechanisms of infertility in polycystic ovary syndrome. Aristotle University Medical Journal. 33(2), 67-77.

[4] Kelly, C., Lyall, H., Petrie, J., Gould, G., Connell, J., Sattar, N., 2001. Low Grade Chronic Inflammation in Women with Polycystic Ovarian Syndrome. J Clin Endocrinol Metab. 86(6), 2453–2455.

[5] Benson, S., Janssen, O., Hahn, S., Tan, S., Dietz, T., Mann, K., Pleger, K., Schedlowski, M., Arck, P., Elsenbruch, S., 2008. Obesity, depression, and chronic low-grade inflammation in women with polycystic ovary syndrome. Brain, Behavior, and Immunity. 22, 177–184.

[6] Baravalle, C., Salvetti, N., Mira, G., Pezzone, N., Ortega, H., 2006. Microscopic characterization of follicular structure in letrosole-induced poly cystic ovarian syndrome in the rat. Archives of medical research. 37, 830-839.

[7] Luo, H., Zuo, X., Li, T., Zhang, J., 2006. Effect of bee venom on adjuvant induced arthritis in rats. Zhong Nan Da Xue Xue Bao Yi Xue Ban. 31(6), 948–951.

[8] Park, H., Son, D., Lee, C., Choi, M., Lee, U., Song, H., Lee, J., Hong, J., 2007. Melittin inhibits inflammatory target gene expression and mediator generation via interaction with IκB kinase. Biochemical Pharmacology. 73(2), 237-247.

[9] Son, D., Lee, J., Lee, Y., Song, H., Lee, C., Hong, J., 2007. Therapeutic application of anti-arthritis, pain-releasing, and anti-cancer effects of bee venom and its constituent compounds. Pharmacology & Therapeutics. 115, 246–270.

[10] Spaczynski, R., Arici, A., Duleba, A., 1999. Tumor Necrosis Factor-a Stimulates Proliferation of Rat Ovarian Theca-Interstitial Cells. Biology of Reproduction. 61, 993–998.

[11] Norata, G., Tibolla, G., Seccomandi, P., Poletti, A., Catapano, A., 2006. Dihydrotestosterone Decreases Tumor Necrosis Factor- a and Lipopolysaccharide-Induced Inflammatory Response in Human Endothelial Cells. J Clin Endocrinol Metab. 91(2), 546–554.

[12] McGrath, K., McRobb, L., Heather, A., 2008. Androgen therapy and atherosclerotic cardiovascular disease. Vascular Health and Risk Management. 4(1), 11–21.

[13] Demissie, M., Lazic, M., Foecking, E., Aird, F., Dunaif, A., Levine, J., 2008. Transient prenatal androgen exposure produces metabolic syndrome in adult female rats. Am J Physiol Endocrinol Metab. 295, 262–268.

[14] Gonzalez. F., Rote, N., Minium, J., Kirwan, J., 2006. Increased Activation of Nuclear Factor _B Triggers Inflammation and Insulin Resistance in Polycystic Ovary Syndrome. J Clin Endocrinol Metab. 91(4), 1508–1512.

[15] Diamanti-Kandarakis, E., Paterakis, T., Alexandraki, K., Piperi, C., Aessopos, A., Katsikis, I., Katsilambros, N., Kreatsas, G., Panidis, D., 2006. Indices of low-grade chronic inflammation in polycystic ovary syndrome and the beneficial effect of metformin. Human Reproduction. 21(6), 1426–1431.

Effects of Heavy Metals on the Snails *Helix aspersa* Bioindicators of the Environment Pollution for Human Health

Nedjoud Grara[1], Amira Atailia[2], Mounir Boucenna[2],
Fadila Khaldi[1], Houria Berrebbah[2], Mohamed Reda Djebar[2]

1 Laboratory of Sciences and Technology of water and the environment, University of Mohammed cherifMessaadia, Souk-Ahras, Algeria

2 Laboratory of Cellular Toxicology. University of Annaba, Algeria

Abstract

In this study we were interested in the evaluation of toxicity sub-chronicle of the metal dust collected on the level of the iron and steel complex of EL-Hadjar**(Eastern Algeria) on** accumulating organizations bio and bio indicator of pollution *Helix aspersa*. The first results on the metabolic level show that metal dust causes a significant increase in proteins with a significant reduction in the Carbohydrates and lipids on the level of the two studied bodies (digestive Gland and the kidney). With regard to the bio markers we highlighted a reduction in the acetyl cholinesterase (AChE) activity at the level of the head. In addition, the exposure of *Helix aspersa*to metal dust induced a lipidic peroxidation with release of malondialdehyde (MDA) to the level of the studied bodies.

Keywords: *Helix aspersa*, dust metal, biomarkers, pollution, MDA, AChE, bioaccumulation, digestive gland, kidney.

1. Introduction

The transfer of pollutants in the trophic networks is not limited to the organic compounds. The increase in the concentrations in elements traces metal (ETM) in the grounds – mainly due to the human activities[1].The central model of this study is the snail *Helix aspersa* for its capacities to accumulate the ETM with significant concentrations in its fabrics[2].The objective of this work is to study the effects of the stress oxidizing induced by metal dust of the iron and steel complex of EL-Hadjar (Annaba) on an accumulating metal bio, the gastropod terrestrial *Helix aspersa*.

2. Materiel and Methods

2.1. Biological Material

The biological material used is a terrestrial gastéropode: the snail Helix*aspersa*collected area of Guelma**(Eastern Algeria)**. The snails (of average Weight of 8,5 ± 0,15g) are high under the following optimal environmental conditions[2].

2.2. Chemical material (The Metallic Releases)

The iron and steel complex of EL-Hadjar (Annaba) is at 15 km of the town of Annaba on the trunk road N44 (**(Eastern Algeria)** Metal dust in the study was collected manually .One analyzes chemical by atomic absorption was used to determine the composition of this dust. This analysis determined the presence of 07 heavy metals (Cu, Zn, Pb, Cr, Ni, Mn, Fe) [3].

2.3. Mode of treatment

Treatment of the animals at summer carried out by addition of the increasing concentrations of metal dust in the food. We retained 4 concentrations and a pilot medium (100, 500, 1000, 1500 food μg/g).[2].After (28days) of treatment, the snails are dissected and digestive Gland (DG),the kidney (K) and Headare taken.

2.4. Measured parameters

The proteins are quantified to the method of [4], the proportioning of the carbohydrates is carried by the method of [5] and lipid level's is given according to the method of [6] .The MDA is proportioned according to the method of [7] , the proportioning of the (AChE) is carried out according to the method of[8].

2.5. Statistical analysis of the results

The results are compared by the nonparametric test of Kruskal-Wallis, This test is carried out using software of analysis of the data: Minitab (Version 14.0) [9].

3. Results

3.1. Effect of the metal rejections on the total proteins

Figure (1) illustrates the variations of the total protein on the level of the digestive gland and the kidney in the presence of metal dust. We note that at the treaties, the rate of total proteins tends to increase in manner proportions dependent in the digestive gland and kidney.

3.2. Effect of the metal rejections on the total of Carbohydrates

Figure (2) illustrates the variations of the total of Carbohydrates on the level of the digestive gland and the kidney in the presence of metal dust. It is noticed that the rate of the carbohydrates at the level of digestive gland and kidney to decrease it in snails treated with the concentrations compared to the control.

3.3. Effect of the metal rejections on the Total lipids

Figure (3) illustrates the variations of the total lipid levels on the level of the digestive gland and the kidney in the presence of metal dust. We note that in the presence of xenobiotic the lipid level in the digestive gland and kidney decreases in a significant way for the treaties by concentrations compared to the control.

3.4. Effect of the metal rejections on the malondialdehyde (MDA)

Figure (4) illustrates the variations of the rate of MDA on the level of digestive gland and of the kidney in the presence of metal dust, we note that in the presence of xenobiotic, the rate of MDA tends to increase in a manner proportions – dependent and very highly significant between the rate on MDA for the treaties by the various concentrations on the level of the two bodies and this always compared to the control.

3.5. Effect of the metal rejections on the activity Acetylcholine Esterase (AChE)

Figure (5) illustrates the variations of the rate of AChE on the level of the head of snails; we note that in the presence of xenobiotic, the acetylcholine esterase rate tends to decrease in a manner proportions – dependent. The statistical analysis reveals a difference very highly significant compared to the control for the treaties by the various concentrations tested.

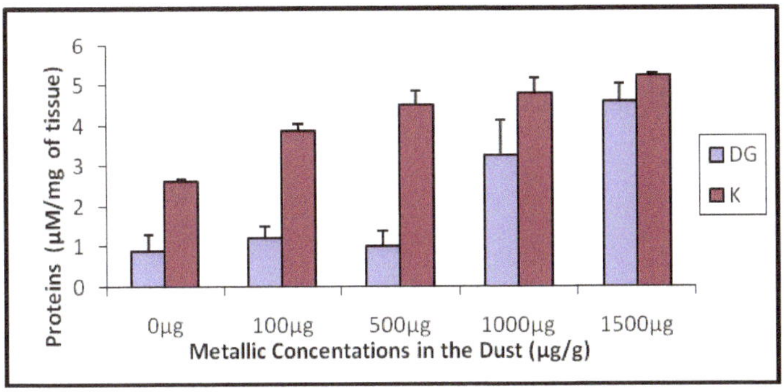

Fig 1. Evolution of the total proteins rate according to the increasing concentrations in metal dust.

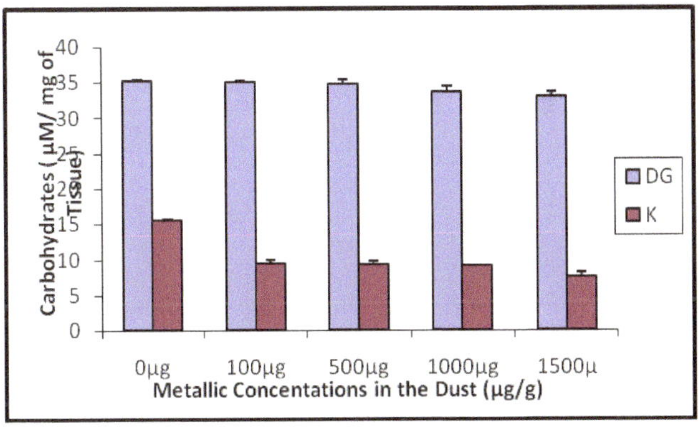

Fig 2. Evolution of the Carbohydrates rate according to the increasing concentrations in metal dust.

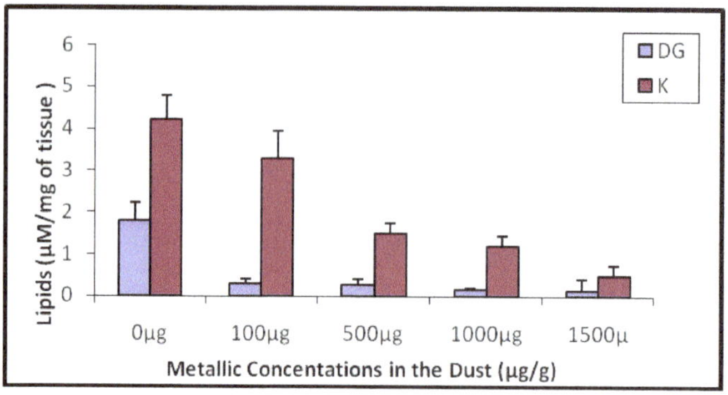

Fig 3. Evolution of the lipids rate according to the increasing concentrations in metal dust.

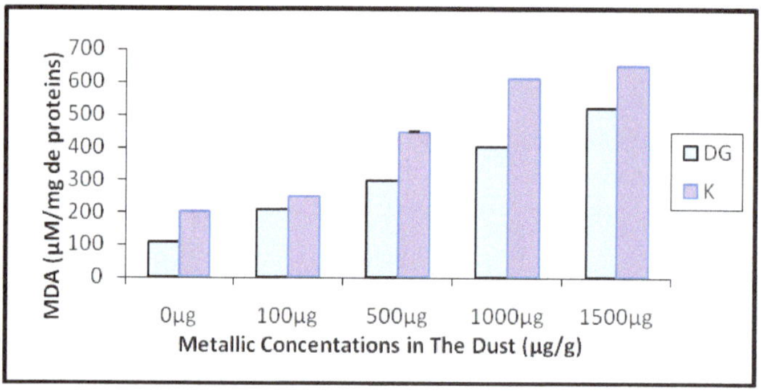

Fig 4. Evolution of the MDA according to the increasing concentrations in metal dust.

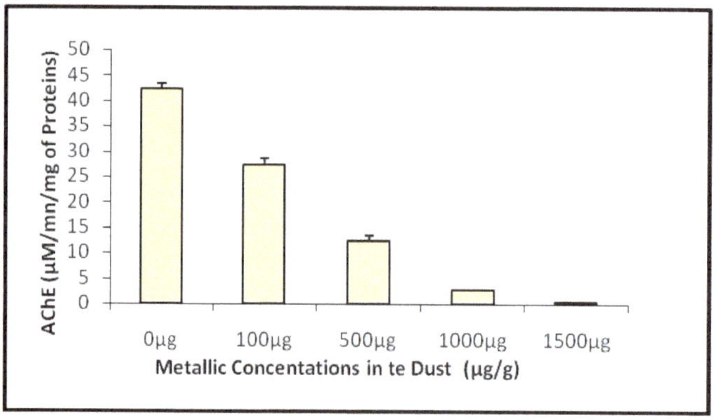

Fig 5. Evolution of AChE activity according to the increasing concentrations in metal dust.

4. Discussion

In our work, we highlighted that the rate of proteins in the two bodies increases in a manner proportions –dependent in the presence of metal dust and of cadmium, these results go in the same direction as those of Masaya and *al.*,(2002)which highlighted a significant increase in the total protein rate under the effect of a chemical stress at different biological models[10].

The results concerning the evolution of the lipid level in the two bodies highlight a significant re duction in the lipids in treated snails and this in manner proportions – dependent by the various concentrations , Aurousseau(2002) suggest the free oxygenated radicals are toxic via the degradation of the lipids of which the β-oxidation [11].concerning the evolution of the Carbohydrates , we noted that this rate decreases in a manner proportions –dependent in the presence of metal dust in the two bodies chosen, this reduction would be due to the oxidation of the carbohydrates in the presence of the metal ions leading to the release of aldehyds and hydrogen peroxide, under the condition of stress, the reserves of carbohydrates are exhausted to satisfy the energy demands, these results are in conformity with those of EL-Wakil and Radwan (1991), which suggested that exposed to Endosulfane, the methyl parathion, of the quinalphos and to Nuvan (pesticides) would be the consequence of the direct use of glycogen for the generation of energy[12].

In the present study, toxicity in metal dust is at the origin of an increase in the rate of MDA which is the principal active aldehyde of theperoxidation of acid membranes.

The MDA is also under product of the biosynthesis of the prostaglandin [13] . Our results are in agreement with those of Viarengoand *al.*,(1990) [14] which studied the toxic effects of heavy metals on the peroxidation of the lipids at *Mytillusgalloprovinciallis*. Our results concerning the evolution of the rate of AChE during the various experiments, highlighted a reduction proportions dependent on the activity of this enzyme dice the weakest metal concentrations. Our results are in agreement with the study of Antonio and *al.*, (2003)[15] The studied snails are edible species for the man and can cause significant metal concentrations within the organization of the human[1].The metals transferred at the humandy the consummation of gastropoda can be at the origin of oxidative stress which represents one of the factors potentiating the genesis ofplurifactorielles diseases such as the cardiovascular diseases, the diabetes, rheumatisms, asthma, cancer,the neurodegenerative diseases and disease of Alzheimer[16].

5. Conclusion

Our experiments show that the snails answer the criteria of the bioindicator to take part in the bio monitoring of the environment.De, more, these are edible species for the man, therefore it is appropriate attentive on the origin of snails to be collected in nature because, being likely to contaminate the human.

6. References

[1] ScheiflerR.Evaluation de la biodisponibilité et des transferts de polluants métalliques et organiques dans les réseaux trophiques "sol-plante-invertébrés". Thèse de doctorat, Université de Franche-Comté, Besançon, France, 2002,196 p.

[2] Coeurdassier M., Saint-Denis M., Gomot-de Vaufleury A., Ribera D., BadotP.M.The garden snail *(Helix aspersa)* as bioindicator of organophosphorus exposure: effects of dimethoate on survival, growth and acetylcholinesterases activity. *EnvironmentalToxicology and Chemistry*2001; 20, 1951-1957.

[3] Kleche M. 2002. Etude de la pollution des poussières aciéries du complexe sidérurgique d'EL Hadjar sur la germination et le métabolisme respiratoire des végétaux. Mémoired'ingénioratUniversité d'Annaba.35 p.

[4] Bradford, M.M., 1976. A rapid and sensitive method for the quantitation of microgram quantities of protein utilizing the principle of protein dye binding, Analytical Biochemistry, 72, 248-254.

[5] Duchateau, G., Florkin. , 1959. For treahalosemie of insects and *its* signification.Arch. Insect. Physiol. Biochem, 67, 306-314.

[6] Goldsworthy, G.J., Mordue, W., Guthkelch, J., 1972. Studies on insect adipokinetic hormones. Gen. Comp. Endocrinol, 18 (3), 545.

[7] Draper H.H., Hadley M., Malondialdehyde determination as index of lipid peroxidation. *Meth. Enzymol*1990; 186, 241-431.

[8] Ellman G.L., Courtney K.D., Andres V. Featherstone R.M.A new and rapid colorimetricdetermination of acetylcholinesterase activity. *Biochem. Pharmacol* 1961; 7, 88-95.

[9] Dagnelie P. Statistiques théoriques et appliquées. Tome2 : références statistiques à une et à deus dimensions. Bruxelles.Univ De boeck et Larcier;1999 ; 659p.

[10] Masaya, M., Yoshinobu, H., Ai, Y., Maki, K., Yasuo, O., 2002. Determination of cellular levels of nonproteinthiols in phytoplankton and their correlation with susceptibility to mercury .Journal of Phycology , 38(5) ,983.

[11] Aurousseau, B., 2002. Les radicaux libres dans l'organisme des animaux : Conséquence sur la reproduction, la physiologie et la qualité de leurs produits. INRA Prod. Anim , 15(1), 67-82.

[12] EL–wakil, H.B., Radwan, M.A., 1991. Biochemical studies on the terrestrial snail *Eobaniavermicula* (Müller) treated with some pesticides, J.Environ .sci .Health, B26, 479-489.

[13] CoeurdassierM.Utilisation de mollusques gastéropodes pulmonés terrestres (*Helixaspersa*) et aquatiques (*Lymniastagnalis* et *Lymniapalustris*) comme indicateurs de pollution par les éléments métalliques et les xénobiotiques .Thèse de doctorat, université de franche comté 2001, France.281p.

[14] Viarengo ,A., Canesi ,L., Pertica ,M., Poli ,M., Moore ,M.N., Orunesu,M.,1990. Heavy metal effects on lipid peroxidation in the tissues of *Mytilusgalloprovincialis.* lam.comp.biochem.physiol, C79, 37- 42 .

[15] Antonio M.T., Corredor L., Leret M.L. Study of the activity of several brain enzymes like markers of the neurotoxicity induced by perinatal exposure to lead and/or cadmium, Toxicology Letters 2003; 143, 331-340.

[16] NzengueY.Comparaison des mécanismes de toxicité redox du Cadmium, du cuivre et du zinc: place des métallothionines et de P53. Thèse de doctorat, Université JOSEPH FOURIER – GRENOBLE 2008, France .299 p.

34

Characterization of *Quercus* Species Distributed in Jordan Using Molecular Markers

Mohammad S. Jawarneh[1], Mohammad H. Brake[2], Riad Muhidat[1], Hussein M. Migdadi[3], Jamil N. Lahham[1], Ahmad Ali El-Oqlah[1]

1 Department of Biological Sciences, Faculty of Science, Yarmouk University

2 Science Department, Faculty of Science, Jerash University

3 Plant Production Department, Collage of Food and Agricultural Sciences, King Saud University, Saudi Arabia

Abstract

Genetic diversity among 25 natural populations of three different species of *Quercus* in Jordan at molecular levels using Random Amplified Polymorphic DNA (RAPD) primers was assessed. Significant molecular variations among and within 25 *Quercus* populations were estimated. Twenty-seven polymorphic markers and 5917 scored bands were generated using six RAPD primers. Based on RAPD data, the populations were grouped together in the same cluster according to species regardless to local of collections. This study has emphasized the ability of the molecular markers in the determining the genetic diversity among and within the populations of *Quercus* and the resulted high genetic variability could be utilized in implications of improving conservation, restoration, and reforestation strategies of *Quercus* in Jordan.

Keywords: Quercus spp, genetic diversity, RAPD markers, conservation, restoration

1. Introduction

Quercus L. (Oak) is one of the exceptionally important woody genera worldwide. It is a large genus in the family Fagaceae with about 600 species growing in a wide range of habitats and distributed in temperate and subtropical regions of the northern hemisphere (Yilmaz et al. 2011). Members of the genus grow as shrubs and trees and form prominent deciduous forests or evergreen woodlands with a range of distribution extending from cold latitudes to tropical Asia and the Americas (Manos et al. 1999).

In Jordan, the genus constitutes an important component of the forest ecosystems in the Mediterranean topographic zone. Three species of the genus are known to occur naturally in this region with a general range of distribution extending from Aum Qais in the North to Tafilah in the South. These species are *Quercus calliprinos L., Quercus ithaburensis, and Quercus infectoria* (Kasapligil 1956, Long 1957, Zohary 1961; 1962; 1973). *Q. calliprinos* is the most widespread and evergreen species; distributed throughout the Mediterranean region from Ajloun in the North through Salt and Fuhais to Tafilah and Shobak in the South. *Q. infectoria* is the least abundant and deciduous species; restricted largely to the Northern parts of the country. *Q. ithaburensis* is intermediate in terms of distribution and abundance; it forms deciduous forests in the Northern and middle parts of the country, mainly around cities of Irbid, Jarash, Salt, and Fuheis. Information on levels of genetic diversity within and among populations of Quercus species in Jordan per se is lacking. Knowledge of the genetic variation of this important genus provides a robust framework for follow-up systematic studies and facilitates its use in genetic conservation and rehabilitation. In addition, this information will help understand the dynamics of the population genetics of *Quercus*, its evolutionary trends, and its responses to changes in the environment.

In this study, the classical technique of random amplified polymorphic DNA (RAPD) is employed to investigate for the first time levels of genetic diversity of natural populations of the genus *Quercus* in Jordan. In Palestine the *Quercus boissieri* Reut. an associated species within the *Quercus calliprinos*–Pistacia palestine association of the Mediterranean sclerophyllous broad-leaf forests. The use of random amplified polymorphic DNA (RAPD) markers,within- and among-populations genetic diversity of *Quercus boissieri* in Palestine , as influenced by geo-climatic parameters (Schiller, et al., 2006).

2. Materials and methods

Samples of the three different Jordanian *Quercus* species were collected from the field during October, November and December of 2008. Samples were randomly selected from different populations distributed over various geographical regions in Jordan. Samples were identified cautiously and taxa names were confirmed following Zohary, (1962). Fresh leaf samples were stored at -20 for DNA extraction.

Leaves collected from each individual tree were manually ground in liquid nitrogen with a mortar and pestle, to a fine powder and DNA was extracted according to the protocol of Genomic DNA Purification Kit form Fermentas and then the DNA samples were stored at -20ºC until use.

Six DNA samples were tested using 60 primers from Operon kits A,B and D. Primers that amplified consistently reproducible polymorphisms were selected and used to analyze all of the 25 *Quercus* populations. Only six RAPD primers were used to amplify the 25 populations.

RAPD reactions were performed in total volume of 15 μl according to standared protocol (Sambrook *et al.*,1989). The amplification products were loaded using 1.8% agarose gel electrophoresis at 100 volts for 2 hrs using horizontal gel electrophoresis apparatus. The amplified products were visualized and documented by gel documentation system. 100 bp ladders were used as a DNA marker to estimate the molecular weights of the amplified products.

Data generated from RAPD analysis were analyzed using Jaccard similarity coefficients (Jaccard, 1908). These similarity coefficients were used to construct dendrograms using the unweighted pair group method with arithmetic average (UPGMA) employing SAHN (sequential, agglomerative, hierarchical, and nested clustering) using the NTSYSpc (ver.2.10) program, (Rohlf, 2005).

3. Results

A total of 60 RAPD primers were evaluated for their ability to amplify polymorphic regions from six randomly selected populations. Of the 60 primers, 6 amplified consistently reproducible polymorphisms, and so these were used to analyze all of the 25 *Quercus* populations. The features of the primers across the tested populations are summarized in Table 1. The 6 primer generated a total of 27 polymorphic markers (alleles). In total, 5917 data points (bands) could be scored with an average of 986.2 bands per primer pair across the genotypes, thereby confirming the high multiplex ratio expected for the RAPDs. The ability of different primer to generate RAPD markers varied from 4 to 5 markers with an average of 4.5 markers per primer pair across all genotypes. On a per-population basis, the number of markers generated by the primer pairs ranged from 32.4 for OPD 17 to 47.9 for OPA 20 with an average of 39.4 markers per primer. The percentage of polymorphic among the primers generated was100% polymorphic markers.

Polymorphic bands ranged in size from 250 to 790 bp. The size out of this range was not considered in the analysis). The densely stained markers were considered in scoring. The total bands for each primer ranged from 810 for primer OPD-17 to 1197 bands for OPA-20 using 389 plants representing 25 *Quercus* populations.

Primer	Total markers[a]	Average bands[b]	Polymorphic markers[c]	Polymorphic markers %	Size range (Bp)	Total no. of bands[d]
OPA 12	4	43.4	4	100	280-700	1086
OPA 17	5	42.0	5	100	250-750	1050
OPA 19	4	35.5	4	100	300-790	887
OPA 20	5	47.9	5	100	250-680	1197
OPB 5	4	35.5	4	100	400-780	887
OPD 17	5	32.4	5	100	380-750	810
Total	27		27			5917
Average	4.5	39.4	4.5	100		986.2

Table 1. The features of RAPD primers selected in *Quercus* genetic diversity

a Total number of differently sized RAPD markers amplified across all 25 populations, b Average number of RAPD bands scored per population, c Total number of RAPD markers found to be polymorphic across the 25populations, d Total number of RAPD bands (data points) scored for all populations

Based on the Jaccard coefficients index (Jaccard, 1974), a genetic similarity matrix was constructed using the RAPD data to assess the genetic relatedness among the 25 *Quercus* populations. The within population means were used to construct the similarity matrix. The mean similarity indices ranged from 0.24 between population 4 and population 11 to 0.84 within the population number 8 and 0.48 for over all populations.

The results showed based on RAPD product data that populations from the different locality represent species tend to grouped together in the same cluster figure (1). The species of the 25 *Quercus* populations clustered into two main clusters; the first cluster consists of the populations belong to *Quercus ithaburensis* and the second cluster consists of the populations belong to the *Quercus infectoria* and *Quescus Calliprinos* species.

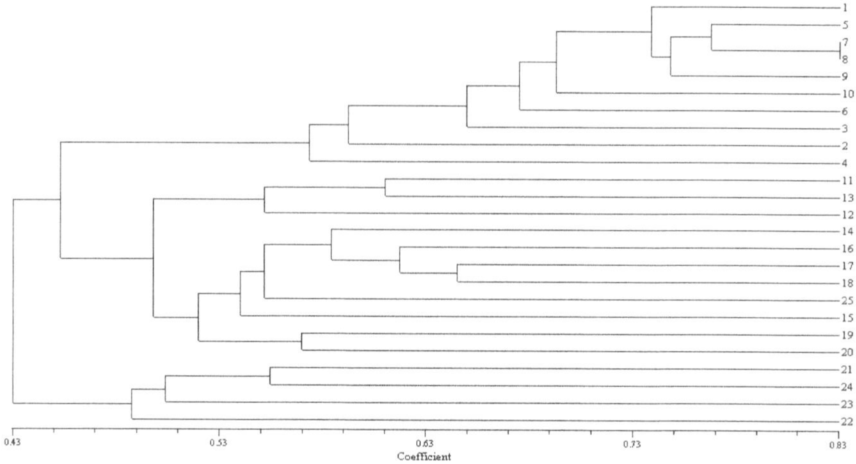

Figure 1. Dendrogram of 25 *Quercus* L. (1=Ber Adbagat, 2= Al-Rashadia, 3= Heisha, 4= Dana,5= Achtifina, 6=Ebein, 7= Enbeh,8= Anjara, 9= Rahaba,10= Fuheis,11= Gelad, 12= Kufour Houda, 13= Zobia, 14= Bargesh 15= Ebein, 16= Ashah, 17= Umm Qaiss, 18= Aosra, 19= Jeneen Safa, 20= Aqraba, 21= Kufour Kifya, 22= Makhraba, 23= Alouk, 24= Jobbah, 25= Gelad) generated by UPGMA cluster analysis of the genetic similarity values.

4. Discussion

The 25 populations analyzed in this study represented *Quercus* species from a wide range of geographical areas in Jordan. In this work we followed the nomenclature used in the previous workers (Kasapligil 1956, Long 1957; Zohary 1962, 1973). Our results showed that Jordan has at least three *Quercus* species and each has its morphological characters.

The current study uses the RAPD-PCR based protocol to assess genetic variability of the *Quercus* species in Jordan. Genetic diversity determines the adaptive potential of a species and is an essential component of the stability of ecosystems. Analysis of within- and among-population genetic diversity is a fundamental step in the development of strategies for conservation of genetic resources and, consequently, of their adaptability. With its oak forests that comprise *Quercus ithaburensis*, *Quercus boissieri*, and *Quercus calliprinos*, is in a geographically peripheral position to the main area of distribution of these species in the Mediterranean basin (Awishi, 1967). According to (Safriel et al. 1994), unlike core populations, peripheral ones may be tolerant to environmental extremes and changes because of their higher genetic variability, which has resulted from fluctuating selection. It is also likely that peripheral populations evolve resistance to extreme conditions; therefore, they should be treated as a biogenetic resource, to be used for rehabilitation and restoration of damaged ecosystems. Owing to their long life cycle, forest trees are among the species that cannot migrate or adapt quickly enough to cope with the rapid changes imposed on the environment by human activity, and this could create ecological and forest management problems. Thus, attention should be given to in situ and ex situ conservation of the varieties of

Quercus ithaburensis genetic material represented by the three main assemblages of its distribution in this region.

In this study it was possible to show that the amplification products from six random primer RAPD assay were sufficient to discriminate among and within population of *Quercus* species for each location. Also, the assay was useful in discriminating among plants of the same location. The ability to distinguish between closely related individuals was simply a function of the observed number of RAPD bands.The results of RAPD markers were compared in a genetic diversity of *Quercus* species the differences in the level of polymorphism detected by the markers and evaluating the potential of these markers in assessing the genetic variation in 25 population of *Quercus* to three species *Quercus ithaburensis, Quercus infectora,* and *Quercus calliprinos.* that matching result come into view the morphological result. The classes of molecular markers adopted in this study deserve additional discussion. The key of the success of multilocus PCR-based markers has to found in their high multiplex ratio. In fact, owing to their own genetic nature, RAPD assays detect simultaneously many loci randomly distributed in the genome. Moreover, compared to SSRs, these marker systems allow a more precise estimate of marker allele frequencies at single loci and faster estimate of population polymorphisms over several loci.

From the similarity matrix the highest values of similarity between populations was found between the populations Aqraba, Makhraba, Ashah and Umm Qiass these population are cluster together in the hierarchical cluster constructed on the base of the genetic similarity values. These population belong to the same species *Quercus ithaburensis* and the population are found in the same region and located at the same elevation. The results obtain confirm once again the great versatility, reliability and precision of the techniques based on molecular markers, which can be used to aid the classical evaluation of the differentiation between population based on the observation of morphological characteristics. Our molecular results also in agreement of those Cottrell et al. (2003) who used six microsatellite markers found high expected heterozygosity values in Quercus robur and Quercus petraea populations. ranging from 0.87 to 0.92 and from 0.76 to 0.82 in Q. crispula populations (Ohsawa et al., 2007d). And with those of SCHILLER et al 2003 who found that Quercus aegilops L. ssp. ithaburensis populations were aggregated according to main geographic regions.

In conclusion, the variations among *Quercus* species studied at molecular levels indicated that there is a high variation among these populations and the RAPD technique was useful for studying genetic variability of *Quercus*. The wide geographical distribution of *Quercus* populations across different environments means that this species has good genetic resources to fill the gap between northern natural distribution sites with the southern natural distribution site. *In-situ* as well as *ex-situe* conservation, restoration, and reforestation should be done in the nearest populations within the same geographic region.

5. Refrences

[1] Awishi M (1967) Taxonomic–geographic studies in the Middle- East oaks. Dissertation, Hebrew University of Jerusalem

[2] Aykut Y, Uslu E, Babac M (2011) Cytogenetic studies on Quercus L. (Fagaceae) species belonging toIlex and Cerris section in Turkey. CARYOLOGIA 64: 297-301

[3] in Personality 7:358-67

[4] Jaccard, J (1974) Predicting social behavior from personality traits. Journal of Research

[5] Kasapligil B (1956) Report to the Government of the Hashemite kingdom of Jordan on ecological survey of the vegetation in relation to forestay and grazing. FAO Report No. 549

[6] Long D (1957) The biochmatology and vegetation of eastern Jordan. FAO Report 57/2/1109

[7] Manos P, Doyle J, Nixon K (1999) Phylogeny, biogeography, and processes of molecular differentiation in Quercus subgenus Quercus (Fagaceae). Molecular Phylogenetics and Evolution 12:333–349

[8] Rohlf J (2005) NTSYS-pc: numerical taxonomy and multivariate analysis

[9] Safriel U, Volis S, Kark S (1994) Core and peripheral populations and global climate change. Israel Journal of Plant Sciences 42:331-345

[10] Sambrook J, Fritsch E, Maniats T (1989) Molecular cloning: A laboratory manual. Cold Spring Harbor Laboratory Press, NY

[11] Schiller G, Shklar G, Ungar D, Omari A, Zyadin F, Korol L (2006) Genetic diversity assessment by random amplified polymorphic DNA of oaks: 3. Quercus calliprinos Webb. in Israel and Jordan. Israel Journal of Plant Sciences 54:137–148

[12] system, version 2.2. Exeter Software: Setauket, NY

[13] Zohary M (1961) On the oak species of the Middle East. Bull. Res. Council Israel Sect. D. Botany 4:161-186

[14] Zohary M (1962) Plant Life of Palestine, Israel and Jordan. The Ronald Press Company, New York

[15] Zohary M (1973) Geobotanical Foundations of the Middle East. Volumes 1 & 2. Gustav Fischer Verlag, Stuttgart & Swets & Zeitlinger, Amsterdam

Honey Bee Venom Will Differentiate Mesenchymal Stem Cells in to the Osteocyte

Mohammad Nabiuni[1,*], Elham Azimi[2],
Abdolhosein Shiravi[2], Zahra Nazari[1]

1 Department of Cell and Molecular Biology, Faculty of Biological Sciences, Kharazmi University, Tehran, Iran

2 Department of Biology, Faculty of Science, Islamic Azad University, Damghan, Iran

*Corresponding author

Abstract

Umbilical cord (UC) is an important source of multipotential mesenchymal stem cells (MSCs). It has observed that bee venom (BV) is effective in survival and differentiation of the cells. We hypotheses that BV can cause differentiation of MSCs to osteocytes. The cells obtained from mouse UC tissues were digested and suspended in DMEM medium. After 24 hours, to induce osteogenic differentiation, cells were cultured for 14 and 21days respectively in DMEM medium contain different concentrations of BV (1,2,3,4,5,6 μgr/ml). Following the treatment, calcium's level in the cells was determined by Alizarin red staining. Cytotoxic effects of BV on MSCs were tested by MTT assay which are shown that BV inhibits MSCs growth. Furthermore, by Alizarin red test, we found that BV increases calcium level in MSCs on dose and time dependent manner. In conclusion we suggest that the MSCs from UC have differentiation potential to osteocyte under BV treatment and it may be useful in cell therapy.

Keywords: mesenchymal stem cells, Bee venom, osteogenic differentiation.

1. Introduction

Umbilical Cord (UC) is a rich source of multipotential mesenchymal stem cells (MSCs). MSC's are a type of multipotent adult stem cell that was originally described as early as the 1960's in animal experiments [1]. Many studies have demonstrated that MSCs have an enormous therapeutic potential for cell therapy [2]. These cells are also considered to have regenerating potential for certain degenerative conditions. MSCs can be differentiate into bone, adipose, cartilage, muscle, and endothelium if these cells are cultured under specific conditions MSC's is obtainable from placental blood, bone marrow, fatty tissue and amniotic fluid. It is also easily obtainable from the tissue of the cord itself [3,4]. In this study MSC's present in the UC were selected. UC is a low cost source of MSCs with multiple differentiation capacities and so it is a much better source than other sources, such as bone marrow or fatty tissue [5,6]. Bee venom (BV) has been used for the treatment of chronic inflammatory diseases such as rheumatoid arthritis and relief of pain in oriental medicine. More recently, studies indicate that BV has potential role in cancer therapy [7,8]. BV contains a variety of biologically active components like melittin and phospholipase A2

(PLA2) [9]. Previous observations have shown that bee venom or its components are effective in proliferation, survival and differentiation of the cells [10]. In this view the aim of this study was the examination of differentiating potential of BV on MSCs. Our hypothesis is that BV could cause differentiation of MSCs to osteocytes.

2. Materials and methods

UC from 1-2 pregnant mice were obtained from animal lab unit in Kharazmi university which were killed by ether and the harvested pieces of tissue were washed several times in sterile phosphate-buffered saline (PBS) and then mechanically minced and enzymatically digested with 0.25% trypsin-EDTA (Gibco-Invitrogen) for approximately 10 min at 37 °C. After centrifugation (1500rpm for 10 min), cells were collected and cultured in Dulbecco's modified Eagle's medium (Gibco-Invitrogen) supplemented by 10% fetal bovine serum (FBS), 100 U/ml penicillin-streptomycin (Gibco-Invitrogen). Cell cultures were maintained at 37 °C with a water-saturated atmosphere and 5% CO_2. Medium was replaced one to two times every week. Three to 5 days after initiating incubation, to calculate the proper doses of BV that cause differentiation of MSCs and also have low cytotoxicity, the cells were cultured with different concentrations of BV (1–12 μgr/ml) for 24h. Cell viability then measured by MTT assay. In continue, to induce osteogenic differentiation, the cells were treated by different concentration of BV (4,5,6 μgr/ml) respectively for 14 and 21 days. After these times, osteogenic differentiation was analyzed by alizarin red staining.

3. Results

Based on MTT assay, results the IC_{50} values of BV for MSCs were 7.5μgr/ml after 24h (fig.1). Also the results showed that BV induces cell death in MSCs in high doses while, at low doses it could inhibit cell growth and induce differentiation. Maximum BV-induced cytotoxicity was evident after 24 h exposure to 15μgr/ml concentration. Our results from alizarin red staining illustrated that undifferentiated MSCs (control group) showed almost no specific staining. Osteocyte-like cells could be seen after 14 days of osteogenic induction (Fig.2). But after 21 days of cultivation the overall color intensity of the differentiated MSCs was markedly enhanced (Fig.3).

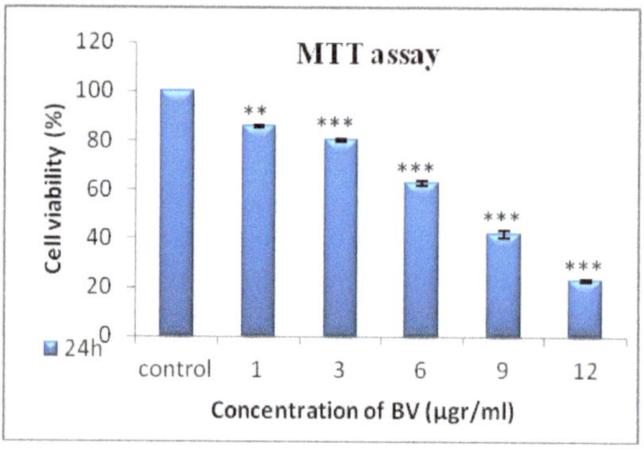

Figure 1. Effect of BV on cell viability. MSCs were treated with different concentration of BV for 24h.The cell viability was then determined by MTT assay, as described under Section 2. Data shown are the mean of three independent experiments ± SEM and are statistically significant in comparison to the control by one-way ANOVA.

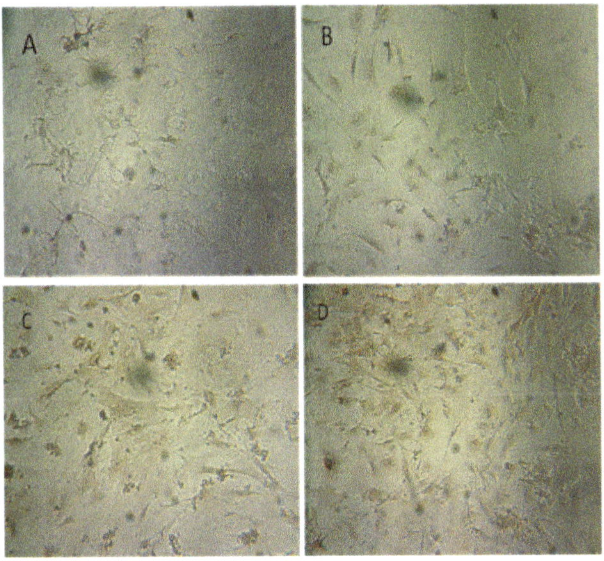

Figure 2. Light microscope micrograph of the alizarin red staining of MSCs following exposure to different concentrations of BV for 14 days. A: control group, B: 4μgr/ml BV treated cells, C: 5μgr/ml BV treated cells. D: 6μgr/ml BV treated cells (200x magnification)

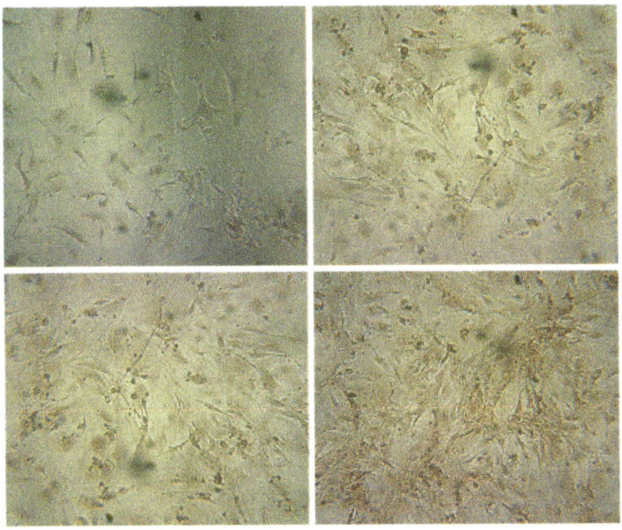

Figure 3. Light microscope micrograph of the alizarin red staining of MSCs following exposure to different concentrations of BV for 21 days. A: control group, B: 4μgr/ml BV treated cells, C: 5μgr/ml BV treated cells. D: 6μgr/ml BV treated cells (200x magnification)

4. Summary and conclusion

UC cells have many advantages because of the immaturity of newborn cells compared with other sources of stem cells. The UC appeared to be a source of fetal cells that could be easily used as multipotent stem cells [3]. Umblical cord derived mesenchymal stem cells can replicate stably in culture, possess the capability to differentiate into a wide variety of tissues after culture [5]. In this study we reported the osteogenic differentiation of MSCs, as characterised by alizarin red staining. By use of key markers of osteogenic differentiation (calcium deposition), we demonstrated that presence of BV can differentiate to osteocyte cells in the absence of osteogenic media. Osteogenic differentiation was greater when cultured-MSCs were treated with BV for 21 days rather than 14 days treatment. In conclusion, we suggest that BV can cause differentiation in the MSCs from UC and it may be useful in cell therapy.

5. Acknowledgements

This project was done in laboratory of Cell and Developmental Biology at Kaharazmi University and the authors wish to thanks Prof. Shahrbanoo Oryan for her support.

6. References

[1] Barry F, Murphy JM. Mesenchymal stem cells: clinical application and biological characterization. Int. J. Biochem. Cell. Biol. 2004; 36:568-84.

[2] Baksh D, L. Song, R. S. Tuan. Adult mesenchymal stem cells: characterization, differentiation, and application in cell and gene therapy J. Cell. Mol. Med. 2004; 8 (3): 301-316.

[3] Romanov Y.A., Svintsitskaya V.A., Smirnov V.N. Searching for alternative sources of postnatal human mesenchymal stem cells: candidate MSC-like cells from umbilical cord. Stem Cells 2003; 21(1): 105-110.

[4] Deans RJ, Moseley AB. Mesenchymal stem cells: biology and potential clinic uses. Exp Hematol. 2000 Aug; 28(8): 875-84.

[5] Wang HS, Hung SC, Peng ST, Huang CC, Wei HM, Guo YJ, Fu YS, Lai MC, Chen CC Mesenchymal stem cells in the Wharton's jelly of the human umbilical cord. Stem Cells 2004; 22(7): 1330-7.

[6] Miao Z, Jin J, Chen L, Zhu J, Huang W, Zhao J, et al.Isolation of mesenchymal stem cells from human placenta: Comparison with human bone marrow mesenchymal stem cell. Cell Bio. Inter. 30 (2006) 681-687.

[7] Jang M.H, Shin M.C, Lim S, Han S.M, Park H.J, Shin I., et al. (2003). Bee venom induces apoptosis and inhibits expression of cyclooxygenase-2 mRNA in human lung cancer cell line NCI-H1299. J. Pharmacal. Sci 91: 95-104.

[8] Suh S.J, Kim K.S, Kim M.J, Chang Y.C, Lee S.D, Kim M.S., et al. (2006). Effects of bee venom on protease activities and free radical damages in synovial fluid from type II ollagen-induced rheumatoid arthritis rats. Toxico. in Vitro 20 (8): 1465–1471.

[9] Son D.J, Lee J.W, Lee Y.H, Song S.H, Lee C.K, Hong J.T. (2007). Therapeutic application of anti arthritis, pain-releasing and anti cancer effects of bee venom and its constituent compounds. Pharmaco. & Therap. 115: 361-763.

[10] Nabiuni M, Hoveizi E, Parivar K, Azarnia M, Khodadadi S. Neural Differentiation and developmental Effects of Honey Bee Venom on PC12 Cells; A Comparison to Nerve Growth Factor (NGF). Journal of Adv. Lab. Res. in Bio. ISSN 0976-7614.

Total Phenolic Content, Antioxidant, Antimicobial and Anticancer Activities of Lespedeza Bicolor Turcz (Papilionaceae)

Samiullah[1, 2, *], Asghari Bano[1], Sisay Girmay[2] and Ghee Tan[2]

1 Department of Plant Sciences, Quaid-i-Azam University Islamabad

2 College of Pharmacy, University of Hawaii at Hilo, USA

*Corresponding author

Abstrat

Anticancer activity against Human lung carcinoma (LU-1) and Human prostrate carcinoma (LnCap) along with antimicrobial and antioxidant activity on DPPH ((1,1)-diphenyl-2-picrylhydrazyl) and Hydrogen peroxide radicals scavenging activity and the contents of total phenolic and flavonoids were assessed in methanol extract of *Lespedeza bicolor*. The highest content of total phenolic content was detected in the arial part of *Lespedeza bicolor* (0.5-1.7 mg gallic acid equiv./g), while the highest content of total flavonoids was found in the aerial part of *Lespedeza bicolor* (0.102-0.148 mg/g D/W). *Lespedeza bicolor* arial parts and root extract showed IC_{50} value of 12.5μg/ml and 50μg/ml against human lung carcinoma (LU-1) whereas, ≤ 12.5 μg/ml and 12μg/ml were calculated against Human prostrate carcinoma (LnCap) cell line. MIC value of 20-35 μg ml^{-1} has been observed gainst *Aspergillus fumigates, Aspergillus niger, Fusarium solani* and *Mucor sp* in comparision with 1-2.5μg/ml of Terbinafine used as a standard fungicide. MIC value of 20 μg/ml and 35 μg ml^{-1} of *Lespedeza bicolor* arial parts and root extract against bacterial pathogen *Klebsiella pneumonia* and 20-50 μg ml^{-1} against *Enterococcus* has been measured. DPPH radical scavenging activity of *Lespedeza bicolor* with IC_{50} values of ≤ 50 μg/ml and ≤ 200 μg ml^{-1} was observed whereas, hydrogen peroxide scavenging activity with IC_{50} values of ≤ 25 μg/ml for arial parts and ≤ 50 μg ml^{-1} for the root extract of *Lespedeza bicolor* has been shown with galllic acid (R2= 0.819) and ascorbic acid (R2= 0.728). These data suggested that the methanolic extract of *Lespedeza bicolor* could be potential candidates for natural antioxidants and anticancer.

Keywords: Lespedeza bicolor, anticancer activity, antioxidant, Antimicrobial

1. Introduction

Lespedeza bicolor Turcz (Papilionaceae) commonly called; bush clover has been collected from natural high saline and arid habitat of District Mardan, Pakistan (34° 05′ to 34° 32′ north latitudes and 71″ 48′ to 72° 25′ east longitudes. According to [1] Six pterocarpans isolated from the root bark of *Lespedeza bicolor* has exposed significant levels of bacterial neuraminidase inhibitory activity with IC_{50} = 0.09-3.25μM. *Lespedeza bicolor* constituents including flavonoids, alkaloids, terpenes, organic acids, and stigmasterols have been screened for anti-inflammation, reducing blood sugar, antioxidation anti-radiation, anticancer, and anti-tumor by [2]. The work of [3] showed that the total amount of hydrolyzed amino acid was 148.95 mg/100g, free amino acids were 106.39 mg/100g and that of γ-aminoisobutyric acid was recorded 12.57 mg/100g in *Lespedeza bicolor* stem extract. The contents of neutral lipids, glycolipids, and phospholipids in *Lespedeza bicolor* seed detected by [4] were 71.75%, 23.26% and 4.99% respectively. 12 flavonoids including Quercetin, kaempferol, trifolin, isoquercetin, homoorientin, and orientin has been isolated from *Lespedeza* bicolor by [5]. N, N-dimethyltryptamine isolated from *Lespedeza bicolor* **var. japonica** has uterus contracting action in 1-2 × 10^{-6} dilution. The [6] work revealed that the Leaves, shoots and inflorescences of *Lespedeza bicolor* have been used in the treatment of acute and chronic nephritis, azothemia and dieresis.

2. Materials and Methods

Extraction

Fresh aril parts and root of *Lespedeza bicolor* (300g) were collected, rinsed with distilled water and air dried for 12 days. The leaves were ground into powder, then soaked in 80% methanol and incubated for two weeks at room temperature (25 °C). The mixture was filtered twice, using whatman-41 filter paper. The extracts were dried by removing the methanol using a rotary film evaporator.

Preliminary phytochemical screening

Phytochemical screening of the *Lespedeza bicolor* was performed to detect the presence of different classes of constituents, such as alkaloids, flavonoids, saponins, steroids, terpenes, Coumarins, Anthraquinone, phlobatannins, Cardiac glycosides and tannins [7]. Total phenolic contents of *Lespedeza bicolor* were determined by the Folin-Ciocalteu colorimetric method [8]. Tannin content was determined by using [9] method. Total Flavonoids content was determined according to the standard protocol [10]. The absorbance was measured immediately at 510 nm spectrophotometer. Alkaloid content was determined by [11] method using 10% acetic acid followed by concentrated ammonium hydroxid. Saponin contents were calculated as percentage of the dried fraction using [12] method.

Antibacterial and antifungal assays

Antibacterial activity of *Lespedeza bicolor* crude extracts was determined by the agar well diffusion method [13]. The agar tube dilution method was used for determination of antifungal activity of methanolic extracts of *Lespedeza bicolor*.

Antioxidant potential of Lespedeza bicolor

The antioxidant activity of *Lespedeza bicolor* crude extract was assessed in DPPH radical scavenging system using gallic acid and ascorbic acid as a positive control, and the decrease in absorbance was determined at 517 nm [14]. The ability of the extracts to scavenge hydrogen peroxide was determined according to the method of [15].

Anticancer activities

The cytotoxic potential of the total methanolic extract of *Lespedeza bicolor* was determined in the human lung carcinoma (LU-1) and human prostrate carcinoma (LnCaP) cell line at the highest concentration of 20 μg/mL with sulforhodamine B (SRB) method [16].

3. Results and discussion

Several groups of polyphenols (anthocyanins, tannins, flavanones, isoflavones, resveratrol and ellagic acid) are currently used in nutraceuticals industries and functional foods [17]. MIC of *L.bicolor* crude extract against *E. coli* and *B. subtilus,* was found 0.5 mg/ml [18].

Several flavonoids and tannins isolated from medicinal plants have been discovered for their significant role in antibacterial, antifungal and anti-inflammatory activities. It is, therefore, possible that the present activities observed with this extract in the study may be attributable to its total phenolic, total flavonoids and tannins contents.

In table 1 total phenolic content (TPC) was shown in the range of 1.23-1.70 mg/g of the *lespedeza bicolor* extract using a standard curve of gallic acid (R2= 0.783). The total flavonoids are in the range from 0.102-0.148 mg/g D/W shown in Table 1.

Metabolites	*Lespedeza* Arial	*Lespedeza* Root
Tannins (mg/g. D/W)	0.193±0.014	0.064±0.326
Total Flavonoids (mg/g. D/W)	0.148±0.003	0.102±0.001
Alkaloids (mg/g. D/W)	1.8±0.150	1.4±0.255
Saponins (mg/g. D/W)	2.0±0.215	2.2±0.137
Total phenolic content (mg/g. D/W)	1.669±0.06	1.23±0.121

Data are expressed as mean±SEM (n = 3) of three independent experiments

All data expressed as (mg/g. Dry Weight)

Table 1. Tannins, total Flavonoids content (TFC), Alkaloid and Saponins content of methanolic extract of Lespedeza bicolor arial parts and root

The methanolic extracts of *Lespedeza bicolor* arial parts and root were significantly active against the fungal pathogens studied. The arial parts of *Lespedeza bicolor* showed the broadest spectrum of activity against *Aspergillus fumigates, Aspergillus niger, Fusarium solani* and *Mucor sp* with MIC value of 20-35 μg ml^{-1} than the root extract shown in table 2. MIC value of 20 and 35 μg ml^{-1} of *Lespedeza bicolor* arial parts and root extract against bacterial pathogen *Klebsiella pneumonia* and 20-50 μg ml^{-1} against *Enterococcus* has been shown in table 2. Penicillin and Chloramphenicol with MIC value of 1.5-2.5 μg ml^{-1} has been used as a positive control against *Klebsiella* and *Enterococcus* specie.

To better understand the antioxidant potential of *Lespedeza bicolor* extracts of root and arial parts were evaluated for radical scavenging activity against DPPH. Fig.1 illustrated a significant decrease in the concentration of DPPH due to scavenging activity of the extract. DPPH radical scavenging activity of *Lespedeza bicolor* arial parts and root extract with IC_{50} values of ≤ 50 and ≤ 200 μg ml^{-1} respectively with galllic acid (R2= 0.871) and ascorbic acid (R2= 0.780) was shown in table. 3 whereas, hydrogen peroxide scavenging activity with IC_{50} values of ≤ 25 for arial parts and ≤ 50 μg ml^{-1} for the root extract of *Lespedeza bicolor* has been shown in the same table with galllic acid (R2= 0.819) and ascorbic acid (R2= 0.728).

Cytotoxicity results against LU-1 and LnCaP cell lines are summarized in table 3. *Lespedeza bicolor* arial parts and root extract showed IC_{50} value of 12.5 and 50 μg/ml against LU-1 whereas, ≤ 12.5 and 12 μg/ml were calculated against LnCaP cell line. Interestingly, *Lespedeza bicolor* possessed the highest inhibition potential against human lung carcinoma (LU-1) and human prostrate carcinoma (LnCaP) cell lines indicating its ultimate potential for biopharmaceutical uses.

Micro-organisms	Tested materials (MIC µg ml⁻¹) ±SEM		
Fungai	***Lespedeza* Arial**	***Lespedeza* Root**	***Terbinafine (µg/ml)**
Aspergillus fumigatus	20±0.381	40±0.241	1.5±0.075
Aspergillus niger	≤ 35±0.305	≤ 45±0.254	1.5±0.075
Aspergillus flavus	≥ 70±0.672	60±0.167	≥ 2.0±0.124
Fusarium solani	≤ 30±0.380	≤ 40± 0.244	2.0±0.122
Mucor Sp	25±0.355	≤ 40±0.239	≤ 2.0±0.191
Klebsiella pneumonia	20±0.401	≤ 35±0.168	-
Enterococcus	≥ 20±0.372	≥ 50±0.388	-

Data are expressed as mean±SEM (n = 3) of three independent experiments
*Terbinafine 1-2.5µg/ml is used as a standred fungicide, *Penicillin and Chloramphenicol 1-3.5µg/ml is used as a standred antibiotics.

Table 2. Antifungal activities (expressed in MIC) of methanolic extracts of Lespedeza bicolor arial parts and root

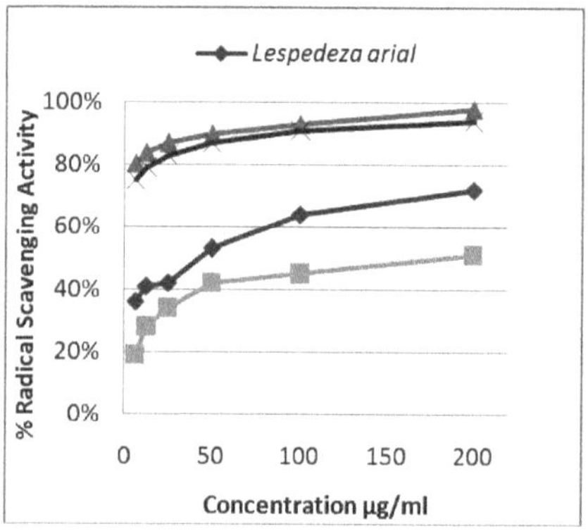

Fig 1. Analysis of DPPH Radical Scavenging activity of methanolic extract of arial parts and root of *Lespedeza bicolor*

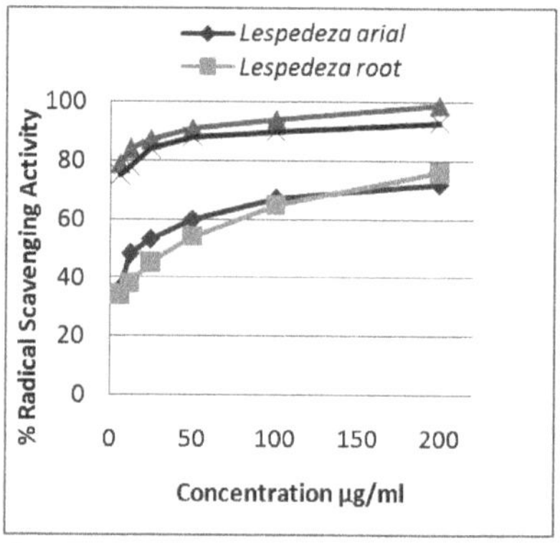

Fig 2. Analysis of Hydrogen Peroxide Radical Scavenging activity of methanolic extracts of arial parts and root of *Lespedeza bicolor bicolor*

Micro-organisms	Tested materials IC_{50} values (µg ml⁻¹) ±SEM	
Anticancer assays	***Lespedeza* Arial**	***Lespedeza* Root**
*[1]LU-1	12.5±0.168	≤50±0.199
*[2] LnCaP	≤ 12.5±0.144	12.5±0.154
Antioxidant assays		
DPPH Radical Scavenging Activity	≤ 50±1.431	≤ 200±0.210
Hydrogen peroxide-scavenging activity	≤ 25± 0.099	≤ 50±0.184

Data are expressed as mean±SEM (n = 3) of three independent experiments
*1Human lung carcinoma
*2Human prostrate carcinoma
Colchicine with IC50 values 0.02±0.002 is used as Standard anticancer drug

Table 3. Cytotoxicity against (LU-1) and (LnCaP) and antioxidant activities of methanolic extracts of *Lespedeza bicolor* arial parts and root expressed as IC_{50} (µg ml⁻¹)

4. Conclusion

In conclusion, the high antimicrobial, antioxidant and cytotoxic potential of *Lespedeza bicolor* highlight the need of further investigations to isolate the active principle and their subsequent evaluation. The results also suggest the presence of biologically active principles which may be worth further investigation and elucidation. Further studies are in fact currently under way to isolate and characterize the active principle(s) of the crude extract.

5. Acknowledgement

The authors would like to thank the Higher Education Commission of Pakistan for providing the necessary financial support for the study.

6. References

[1] Woo Hyun Sim; Kim Dae Wook; Curtis-Long Marcus J; Lee Byong Won; Lee Ji Hye; Kim Jun Young; Kang Jae Eun; Park Ki Hun. 2011. Potent inhibition of bacterial neuraminidase activity by pterocarpans isolated from the roots of Lespedeza bicolor. *Bio-Orgainc and Medicinal Chemistry Letter*. 21 (20): 6100-6103.

[2] Zhang, Fan; Qi, Xiaohua; Zou, Mingqiang; Xie, Ruili; Li, Jinfeng; Zhang, Honggui. 2008. Research progress of extraction and analysis of flavonoids and chemical constituents in Lespedeza bicolor Turcz. *Shizen Guovi Guoyao*. 19(12): 2884-2885.

[3] Chen BH, Li SH, Lin JX, Zhuang HR. J Fujian Normal University 2003;19:86.

[4] Kim, Hyang Ran; Koh, Moo Seok; Yang, Hee Cheon. 1987. Studies on the lipid composition of bush clover (Lespedeza bicolor) seed. *Hanguk Yongyang siklyong Hakhoechi*. 16(3): 75-84.

[5] Glyzin, V. I.; Ban'kovskii, A. I.; Zhurba, O. V.; Sheichenko, V. I. 1970. Flavonoids of Lespedeza bicolor. *Khimiya prirodnykh soedinenii*. 6(4): 473-474.

[6] Trumpe TE, Gulyaev VG. Med Help 1994;2:47.

[7] Wagner H, Bladt S, Zgainski EM. Plant drug analysis. Berlin: Springer; 1984.

[8] Lee YR, Woo KS, Kim KJ, Son JR, Jeong HS (2007). Antioxidant activities of ethanol extracts from germinated specialty rough Rice. Food Sci. Biotechnol. 16(5): 765 – 770.

[9] Van-Burden T.P, Robinson WC (1981). Formation of complexes between protein and tannin acid. J. Agric Food Chen. 1: 77-82.

[10] Sakanaka, S., Y. Tachibana and Y. Okada, 2005. Preparation and antioxidant properties of extracts of Japanese persimmon leaf tea (kakinoha-cha). Food Chem., 89: 569-575.

[11] A. Sofowara. "Medicinal plants and Traditional Medicine in Africa", Spectrum Books Ltd, Ibadan, Nigeria. 1993: p. 289.

[12] Carron, E.A, J.M. Maran, L. Montero, A. Fernandozalgo and A.Dominguez. 1987. Antimicrobial properties of some extracts obtained from some Mediterranean plants of medicinal value. *Plantes Medicinales et Phytotherapie*, 21: 195-202.

[13] Washington, J.A. and V.L. Sutter. 1980. "Dilution susceptibility test: agar and macro-broth dilution procedures, In: E.H. Lennette, A. Balows, WJ. Hausler Jr., and J.P. Truant (Ed,).

[14] Ruch, R.J., Cheng, S.J., Klaunig, J.F., 1989. Prevention of cytotoxicity and inhibition of intracellular communication by antioxidant catechins isolated from Chinese green tea. Carcinogenesis 10, 1003–1008.

[15] Chung YC, Chang CT, Chao WW, Lin CF, Chou ST (2002). Antioxidative activity and safety of the 50% ethanolic extract from red bean fermented by Bacillus subtilis IMR-NK1. J. Agric. Food Chem., 50: 2454-2458.

[16] J. B. Harborne. "Phytochemical methods, London. Chapman and Hall". Ltd., pp. 49-188, 1973.

[17] Oktay, M., Gülçin, ˙I., Küfrevioˇglu, Ö.˙I., 2003. Determination of in vitro antioxidant activity of fennel (*Foeniculum vulgare*) seed extracts. Lebensmittel-Wissenchaft und Technologie, 36, 263–271.

[18] Samiullah, asghari bano, rabia naz and humaira yasmin. 2011, *in vitro* inhibition potential of *lespedeza bicolor* turcz against selected bacterial and fungal strains. journal of medicinal plants research vol. 5(16), pp. 3708-3714.

Rattus rattus Parasites of El-Kala National Park (Algeria)

Farida Becir[1,2,*], Idir Bitam[3], Hadjira Hannachi, M'Barek Chetoui[4] and Zihad Bouslama[2]

1 SNV Institute, Department of Biology, El-Tarf University, Algeria.

2 Laboratory of aquatic and terrestrial ecosystems (Ecostaq), Badji-Mokhtar University, Algeria.

3 Vectorial systems ecology service, Pasteur Institute of Algeria.

4 Laboratory of Animal Ecology, Faculty of Sciences, El-Manar University, Tunis.

*Corresponding author

Abstract

This study was conducted at the National Park of El Kala (Algeria) which has a mosaic of habitats, and located in north-eastern Algeria. This rich landscape induces another one; it is the ecological biodiversity of the park. We are interested in a particular species of mammal that is the Black Rat (*Rattus rattus*) and specifically in identifying different species of parasites that her little body homes and vehicles. The Black Rat's skin is very noisy. We identified mites (*Dermanysus bacoti*), ticks (*Ixodes ricinus*), fleas (*Nosopsyllus fasciatus* and *Xenopsylla chiopis*), lice (*Poluplax sp*) and sandflies (*Flebotomus sp.).*

Keywords: Black Rat, ectoparasites, tick, flea, Algeria.

1. Introduction

The Muridae family is the most diversified at globe level, including more than 700 species and 120 kinds amongst Rattus which accounts 50 species [8]. South – eastern Asian originally species has become cosmopolitan through times. In fact, from the Far East, the black Rat has conquered all the continents following Man everywhere [6]. At the National Park El Kala (North – east of Algeria) we identified *Rattus rattus* (relying on morphological, craniological and caryological approaches). The rodent *Rattus rattus* (Linnaeus, 1958) is a devastating species, reported as resistant to plague *Yersinia pestis*, and also a vector of several pathogens.

The black rat (*Rattus rattus*) which is a small unit of the biological system is a very rich and diverse synopsis. Indeed, this micro-ecosystem supports the installation and development of a mosaic parasite, which exists, co-evolves, multiplies and spreads even in the ecosystem. The parasite is bad for the host; it can only harm his life, and whatever its form is horizontal or vertical parasitism.

The black rat is observed at the park in urban, sub-urban nature reserve, the forest ... etc. We used rat traps that were placed in different environments to capture it. After capture we identified the species according to body size of individuals recovered alive (performed by the method of Chappelier), Then, we collected all suspected parasites kinds of the species.

2. Material and Methods:

Forty three black Rats (twenty five males and eighteen females) captured during the year 2011 in different biotopes of El Kala National Park, whose geographical coordinates are 36° 43′ N. to 36° 57′ N. and of 7° 43′ E. to 8° 37′ E. The park extends on a 78 400 hectars surface [5].

Rattraps were put randomly in different biotopes (cork forests, alder forest, scrubland…etc.) where daily visits where planned.

As soon as captured, samples were cleaned of parasites (direct collecting method). Collected parasites were conserved in alcohol 70 % and identified at Pasteur Institute of Algeria (vectorial systems ecology service) according to identification keys.

3. Results :

Identified parasites are either insects (fleas, lices or phlebotomes) or acarides (ticks or mites), with the majority of fleas (*Xenopsylla cheopis*).

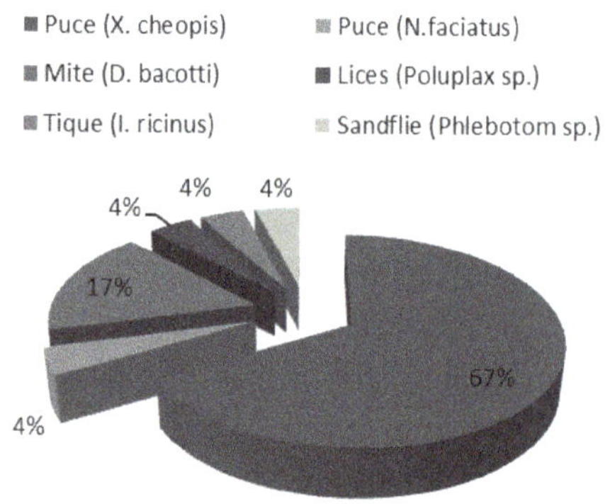

Fig 1. majority of ectoparasites.

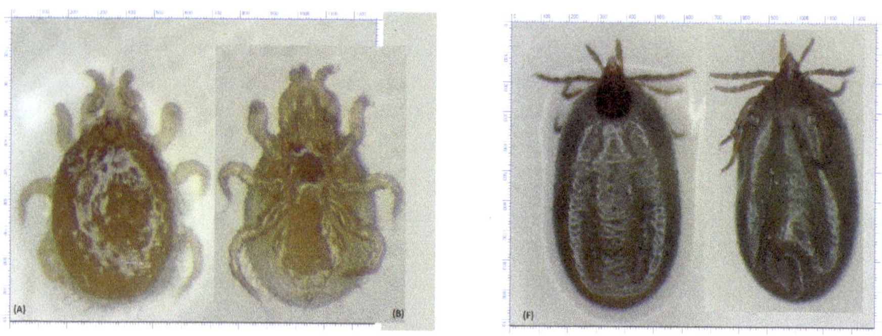

Fig 2. Aacarids : A et B Mite: Dermanyssus bacoti (A : dorsal side, B ventral side) ; F Tick : Ixodes ricinus (adult female)

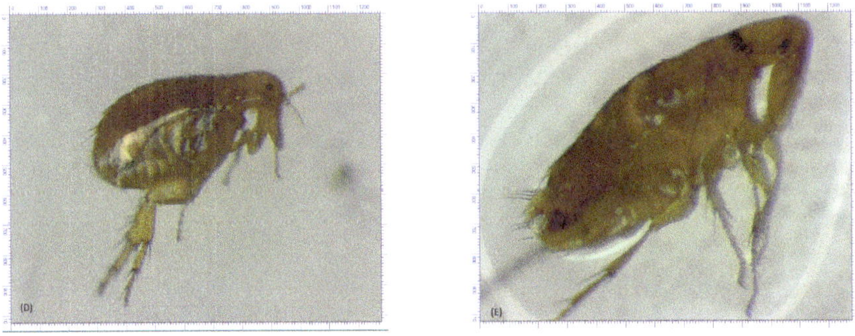

Fig 3. Flea : D : Xenopsylla cheopis (female); E : Nosopsylla sp. (male)

Fig 4. C : Lice: Flebotomus sp.

Fig 5. G: lice (Poluplax sp)

4. Discussion

These parasites are hematophagous arthropods that live on several domestic and wild animals. These parasites are themselves carriers of very harmful dseases [7] . Ticks are considered as vectors of pathogens (Bacterian, viral, parasitarian) the most important all over the world after mosquitos. They can transmit several pathogen organisms like *Borrelia, Rickettsia, Bartonella, Coxiella, Ehrlichia, et Anaplasma.* Fleas and louses can also transmit some pathogens like *Bartonella, Rickettsia* and espcially *Yersinia* for fleas.

In Algeria, a small number published works were realized on the role of pathogen, the biodiversity and the dynamic of ticks, fleas, none on louses of different mammals. The study done by Bitam *et al.* on [1] and [2], allowed the detection of *Rickettsia* by PCR, on fleas, on differents ticks speaces (*Rhipicephalus sanguineus, Hyalomma marginatum, Rhipicephalus turanicus*) and on fleas (*Ctenocephalides canis, Xenopsylla cheopis)* from different mammals. In Bitam *et al.* [3] and [4] allowed the detection by PCR *Yersinia,* on fleas *Xenopsylla cheopis* captured on anthrophile mammals, and *Bartonella* on puces *Xenopsylla cheopis, Archeopsylla erinacei* and *Leptopsylla segnis* collected on insectivores as well as rodents.

5. References

[1] F. Audoin-Rouzeau : Le rat noir (Rattus rattus) et la peste dans l'occident antique et médieval. Manuscrit n° 1963/PLS11.Journée IP en hommage à Paul-Louis SIMOND.

[2] Bitam, P. Parola, K.D. De La Cruz Matsumoto, B. Baziz, J-M. Rolain, M. Belkaid and D.Raoult. First Molecular detection of *Rickettia felis* in fleas from Algeria. *Am. J. Trop. Med. Hyg.*, 74(4), 532-535P. 2006a.

[3] I. Bitam, B. Baziz, P. Parola, J.M. Rolain, K. Matsumoto, K. De La Cruz, M. Belkaid and D. Raoult. First Molecular Detection of de *Rickettsia conorii, Rickettsia aeschlimanii,* and *Rickettsia massiliae* in Ticks from Algeria. *Ann. N.Y. Acad. Sci.*, 1078, 368-372P., 2006b.

[4] I. Bitam, B. Baziz, JM. Rolain, M Belkaid and D. Raoult. Zoonotic Focus of Plague in Algeria. *Emerging Infectious Diseases*, Vol. 12., 2006c.

[5] I. Bitam, K. Dittimar, D. Raoult, P. Parola, and MF. Witing. Fleas and flea-borne diseases. *International journal of infectious diseases.* 1-10P., 2010.

[6] S. Benyacoub. Ecologie de l'avifaune forestière nicheuse de la région d'El-Kala (Nord-Est algérien). Thèse Univ. De Bourgogne. Dijon. 287 p. 1993.

[7] J. Bernard. Les mammifères de Tunisie et de régions voisines. Bulletins de la Faculté d'Agronomie N° 24-25. Univ. Tunis. 1969.

[8] A. Chikweto, M.I. Bhaiyat, C.N.L. Macpherson, C. DeAllie, R.D. Pinckney, C. Richards and R.N. Sharma: Existence of Angiostrongylus cantonensis in rats (Rattus norvegicus) in Grenada, West Indies. Parasitol. (2009), doi:10.1016/j.vetpar.2009.02.020.

[9] D.E Wilson And D.M. Reeder: Mammal species of de word, ataxonomic and geographic reference, 2nd edition, Washington et London: *Smithsonian Institution Press* 1993, 1206p; 1993.

Homocystein and Trace Elements Levels in Patient with Ischemic Heart Disease and some Associated Diseases

Saad Merza Al-Araji, Ala H. Abbase, Zainab F. Hassan

Department of Medical Physiology, College Of Medicine, University Of Babylon, Iraq

Abstract

This paper includes the determination homocysteine level and trace elements magnesium (Mg), zinc (Zn)and iron (Fe) in ppm (part per million); lead (Pb), cadmium (Cd), selenium (Se), chromium (Cr) and germanium (Ge) in ppb (part per billion)) in random serum of patients with pure ischemia, ischemia with hypertension and ischemia with diabetes. Homocysteine level was significantly increased ($P< 0.01$) in pure ischemic patients, ischemia with hypertension and ischemia with diabetes in comparison with control group. A comparison had also been done between male & female groups in patients and control groups and no significant changes ($P >0.05$) were observed. The result of this study showed that concentration of the trace elements (Pb& Cd) were significantly increased ($P< 0.01$) in patients groups in comparison with control group and the concentration of (Mg ,Zn, Se, Cr and Ge) were significantly decreased ($P< 0.01$) among patients groups in comparison with control group.

Keywords: Homocysteine, High performance liquid chromrtography, Trace elements ,Atomic absorption spectrography. Ischemic heart disease .

1. Introduction

Homocysteine (Hcy) is an intermediate of methionine metabolism [1]. Elevated blood homocysteine concentration was an independent risk factor for cardiovascular disease [1, 2]. High-normal serum homocysteine concentrations are associated with an increased prevalence of carotid artery wall thickening [3] . The significance of the contribution of homocysteine to the variation of carotid intima-media thickness suggests a role for homocysteine as an independent risk factor for early carotid artery atherosclerosis in the asymptomatic subjects. Different studies indicated that the elevated level of total homocysteine (tHcy) had increased the risk of cardiovascular diseases and stroke [4]. Homocysteine is elevated in the case of inborn errors of methionine metabolism and excessive amount of homocysteine and its derivatives are found in blood and tissues of cardiovascular patients [5]. Moderate hyperhomocysteinemia up to (30 μmol/l) is a major independent risk factor of a number of diseases characteristic of old ages , primarily occlusive vascular disease [6]. Non metallic elements (hydrogen, oxygen, carbon and nitrogen) make 99% of all the elements in human body, while major elements, which are calcium, magnesium, phosphorus, sulfur and chlorine, make 0.9% of the total. However, essential trace elements provide approximately less than 0.1% of human body [7]. Trace metals are metals in extremely small quantities, almost at the molecular level, that reside in or are present in animal and plant cells and tissue. Trace metals are a group of metals that include both heavy and transitional elements present in micrograms quantities in the blood. There are divided into two groups, essential for health and have no known biological function.

In human body, there are some metals which naturally exist and are essential to human health. These essential metals at trace levels play vital role when present in human body and may cause some diseases when present beyond specific concentrations [8, 9].

2. Designing

MATERIALS AND METHODES

Patients: The study was conducted during the period from Augest, 2008 to May , 2009 in Merjan hospital in Hilla city , Babylon province, Iraq. A total 57 patients were included from urban and rural area(35 males and 22 females) (Table 3-1). Twenty three patients had pure ischemic heart disease (17 males and 6 females), 18 had ischemic heart disease with hypertension (10 males and 6 females),18 had ischemia heart disease with diabetes (8 males and 10 females). The age of the patients varying between 40-85 years old (mean± SD 57.82±10.25).

Control Group: The control group consisted of 40 healthy person who were chosen as healthy, non smokers , didn't have any history of chronic disease and didn't take any treatment for chronic diseases

3. Results

The patients and control were divided according to ages into two groups group 1 in which the age range between 40-60 year and age group two in which the age is > 60 year as shows in table 1.

	Patients	Age	
		40- 60 year	**60 year<**
Ischemic patients	Pure ischemia	11	12
	Ischemia with hypertension	14	12
	Ischemia with diabetes	8	10
	Control	124	16

Table 1. Patients and control number according to age group

The sex distribution of total ischemic patients, was clearly obvious . The highest percentage in male was 61% while female was 39% (figure 1).

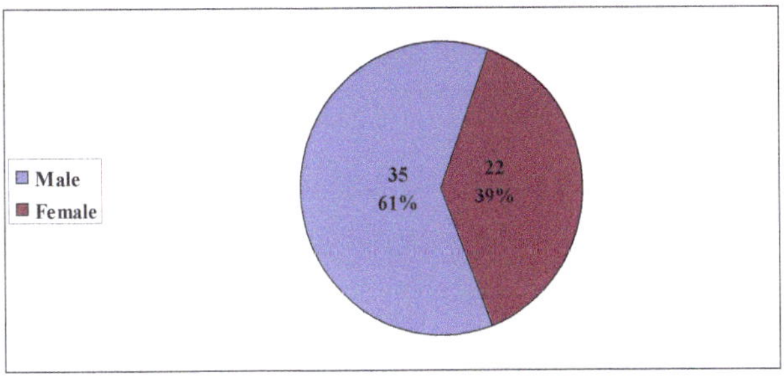

Figure 1. Sex distribution of total ischemic patients.

Table 2 show the sex distribution for all patients in this study. Number of female patients in ischemia with diabetes was more than male patients while in ischemia and ischemia with hypertension male patients were more than female.

	Groups	Number			
		Male	%	Female	%
Patients	Pure ischemic patients	17	31	6	14
	Ischemic patients with hypertension	10	18	6	14
	Ischemic patients with diabetes	8	15	10	24
	Control	20	36	20	48
	Total	55		42	

Table 2. Sex distribution in the study groups

Comparison between ischemic male or female and similar sex healthy controls results indicate an increase in level of homocysteine among patients groups. This elevation in homocysteine concentration value was highly significant ($P < 0.01$). When similar comparisons were done between ischemia with hypertension and ischemia with diabetes and control group a similar behavior were observed ($P < 0.01$).Table 3 shows the mean values of serum level of homocysteine in male and female compared with the mean values in control group. No significant changes was observed when comparison was done between two age groups as shown in table 4.

Groups	Male Mean ± SD	Female Mean ± SD
Control	9.58 ±0.93	9.11±0.65
Ischemic patients	24.11**±5.45	25.4**±2.5
Ischemia with hypertension	20.84**±3.38	23.9**±4.85
Ischemia with diabetic	22.16**±3.59	21.26**±1.44±

** P < 0.01

Table 3. The mean serum level of homocysteine in male and female patients groups and control group

Age group	Pure ischemia	Ischemia with hypertension	Ischemia with diabetes
40-60	23.29±5.3	19.83 ± 1.81	24 ± 4.62
> 60	25.32 ±4.93	22.88 ±4.87	21.02 ±1.23

Table 4. Comparison of levels of homocysteine (μmol/l) between two age groups (age group 1 (40-60) and age group 2 (>60))

There is a highly significant increase in homocysteine level in ischemic patients compared with control values ($P < 0.01$).Also there is significant increase in homocysteine level in patients with ischemia with hypertension & ischemia with diabetes compared with control patients as shown in table 5.

Groups	Mean μg/ml±SD
Control	9.35±0.82
Ischemic patients	24.31 **±5.09
Ischemia with hypertension	22.46**±4.61
Ischemia with diabetic	21.84**±2.96

** P < 0.01

Table 5. The mean serum level of homocysteine(μmol/l) in different patients groups and control groups

There is no significant changes in serum homocysteine level between pure ischemic patients , ischemia with hypertension and ischemia with diabetes ($P > 0.05$).

Measurements of trace elements in serum of two age groups (40-60) and > 60 years shows no significant changes in their levels ($p > 0.05$).Table 6 shows comparison of trace metal concentration in control group between two age groups

	Trace elements	Concentration (mean ± SD)	
		Age groups of 40- 60 year	Age groups of >60 years
μg/ ml	Mg	19.6±1.47	19.59±1.84
	Zn	1.04 ±0.37	1.03±0.13
	Fe	1.06±0.18	1.05±0.13
ng /ml	Pb	0.09±0.15	0.12±0.19
	Cd	0.00	0.00
	Se	102.65±18.03	103.75±10.05
	Cr	55.22±6.79	55.98±6.17
	Ge	42.43±4.41	41.54±4.18

Table 6. Comparison of trace metal concentration in control group between two age groups

Measurements of trace elements in two age groups (40-60) and > 60 years in all patients groups shows no significant changes in their levels ($p > 0.05$) as shown in table 7 .

Age groups	Pure ischemia			Ischemia with hypertension			Ischemia with diabetes		
	Mg	Zn	Fe	Mg	Zn	Fe	Mg	Zn	Fe
40-60	14.43 ±2.33	0.73 ±0.17	1.5 ±1.77	14.25 ±1.08	0.73 ±0.11	1.01 ±0.13	17.9 ±3.73	0.95 ±0.23	0.94 ±0.19
>60	14.32 ±3.03	0.6 ±0.12	1.15 ±0.76	14.5 ±1.36	0.68 ±0.2	1.41 ±1.23	13.14 ±1.95	0.68 ±0.24	0.96 ±0.19
All ages	14.38 ±2.46	0.66 ±0.16	1.33 ±1.36	14.43 ± 1.27	0.96 ±0.18	1.3 ±1.06	15.26 ±3.66	0.8 ±0.26	0.95 ±0.18

Table 7. Comparison of levels of (Mg , Zn and Fe) between two age groups { age group 1 (40-60 year) and age group 2 (>60 year)}.

Trace elements level:

There is a significant decrease in the level of magnesium, zinc ,selenium , chromium & germanium in all patients groups in comparism with control group ($P < 0.01$) & significant increase level of both lead & cadmium in all patients group than that of control group ($P < 0.01$) while there is no significant changes in the level of iron ($p < 0.5$) as shown in table 8.

	Trace elements	Control	Pure ischemia	Ischemia with hypertension	Ischemia with diabetes
g/mlμ	Mg	19.59±1.58	14.38±2.64**	14.44±1.27**	15.25±3.66**
	Zn	1.16±0.22	0.66±0.15**	0.68±0.18**	0.8±0.26**
	Fe	1.06±0.18	1.33±1.36	1.30±1.06	0.95±0.19
ng/ ml	Pd	0.11±0.17	6.39**±2.89	5.63**±1.73	3.58**±1.06
	Cd	0.00±0.00	5.07**±2.69	4.78**±1.83	4.89**±1.75
	Se	103.64±11.95	53.37**±14.85	61.53**±11.17	61.81**±7.19
	Cr	55.21±6.93	32.42**±6.95	32.05**±11.77	31.02**±3.21
	Ge	42.37± 4.76	18.51**±6.14	16.9**±5.19	14.61**± 6

Table 8. Trace elements level in control & patients groups. ** ($P < 0.01$)

4. Discssion

The relationship between increased homocysteine and heart disease is well established in the medical community. Unlike the other three predictors of heart disease which are cholesterol, triglycerides and C Reactive Protein, homocysteine levels are influenced by what the person does not eat rather than what he does eat. This is due to the fact that homocysteine is a sulphydryl-containing amino acid derived from demethylation of methionine. Nutritional deficiencies in the vitamin cofactors (folate, vitamin B12, and vitamin B6) required for homocysteine metabolism may promote hyperhomocysteinaemia [10]. This means that increased homocysteine levels are associated with increased risk of cardiovascular disease and then tHcy measurement will become another useful marker of vascular risk, multivitamin therapy will be another therapeutic option for people at risk of atherothrombotic vascular disease, and fortification of food with folic acid will rise high on the political and public health agenda. Homocysteine level in all patients with heart disease in this study (pure ischemia, ischemia with hypertension and ischemia with diabetes) were found to be elevated significantly($P < 0.01$) compared with control groups. Only about two-thirds of all episodes of symptomatic atherothrombotic vascular disease in developed countries can be attributed to established genetic and environmental vascular risk factors [11]. An additional causal vascular risk factor may be raised plasma levels of homocysteine (hyperhomocysteinaemia).

Mild hyperhomocysteinaemia occurs in approximately 6% of the general population [12,13]. Patients with mild hyperhomocysteinaemia are typically asymptomatic until the third or fourth decade of life when premature coronary artery disease develops, as well as recurrent arterial and venous thrombosis.The elevation of serum level of homocysteine (hyperhomocysteinemia) in this study could be considered as a risk factor for cardiovascular disease. Research has shown that increased homocysteine level is associated with both the hyperinsulinemia seen with insulin resistance and increased urinary albumin excretion. It is also associated with low serum levels of vitamin B12 [14]. However, supplementation with vitamin B12 has resulted in reduction of homocysteine levels, but as failed to show subsequent reductions in incidence of cardiovascular disease [15]. Trace elements are micro-nutrients, present in blood and tissues. They are essential for enzymatic activities and metabolic processes and even for vital functions [16 , 17] . Some trace metals have antiviral activity, while others may alter the genome of the viruses enhancing their virulence [18]. Several trace elements are of great importance in a number of biological processes, mostly through the following terms:

1- Their action as activators or inhibitors of enzymatic reactions,

2- Competing with other elements and proteins for binding sites,

3- Influencing the permeability of cell membranes.

It is therefore reasonable to assume that these minerals would also exert an action, either directly, or indirectly, on the cardiac cell, on the blood vessel walls, on the blood-pressure-regulating centers, or on other systems related to cardiovascular function. Astudy showed that the mean values of serum magnesium were lower in patients with acute myocardial infarction than in non cardiac patients. The authors suggested that a lowering or elevation of serum trace elements could be useful in the diagnosis of recent infarction and could possibly have other implications [19]. Table 6 shows that serum magnesium level was decreased in all patients in comparison with control & this is in agreement with most studies [20]. The disturbances of magnesium metabolism may have profound effect on the contractile state of vascular smooth muscle and as a result on blood pressure. Accordingly, magnesium is particularly important for assessment of hypertension. [20] Reported that magnesium deficiency has a role in the pathogeneses of hypertension.

The results of this study show that there was a reduction in iron serum level of ischemia with diabetes patient and an elevation in patients with ischemia and ischemia with hypertension compared with control group. However, these changes were not significant (P>0.05). This was explained due to the fact that ischemic heart diseases, diabetes and hypertension as diseases occurs with variant risk factors mostly not affect the levels of iron if these diseases occur from risk factors other than those in the previous points. The results reveal a significant increase p<0.01 in serum lead of ischemic heart disease, hypertension and diabetes patients where when compared with the corresponding control values. Till (1997), believed that there was no known biological requirement for germanium (Ge) which was regarded as not an essential element [21, 22]? However, recent study observed that different symptoms in the deficiency of germanium such as cardiovascular disease, atherosclerosis, higher risk for several cancers, osteoporosis, and arthritis weakened immune system, decreased oxygen [23].

5. References

[1] Allon ,N.; Friedman G.; Bostom, Jacob Selhub; Andrew, S. Levey; and Irwin, H. Rosenberg .(2001) The Kidney and Homocysteine Metabolism , J Am Soc Nephrol 12: 2181–2189.

[2] Ford, E. S., Smith, S. J., Stroup, D. F., Steinberg, K. K., Mueller, P. W. & Thacker, S. B. (2002) Homocysteine and cardiovascular disease: a systematic review of the evidence with special emphasis on case-control studies and nested case-control studies. Int. J. Epidemiol. 31:59-70.

[3] Willinek, Winfried, A.; Ludwig, Malte; Lennarz, Martina; Höller, Tobias; Stumpe, Klaus O.(2000). High-normal serum homocysteine concentrations are associated with an increased risk of early atherosclerotic carotid artery wall lesions in healthy subjects Journal of Hypertension: , 18 (4) : 425-430.

[4] Bostom, AG.; Silbershatz, H.& Rosenberg,IH. (1999). Nonfasting plasma total homocysteine levels and ll-cause and cardiovascular disease mortality in elderly Framingham men and women. *Arch Intern Med* . **159**:1077–80

[5] Mudd, SH.;Skovby, F.; Levy ,HL.; Pettigrew, K.D.;Wilcken, B.; Pyeritz, RE.; Andria, G.; Boers ,G.H.; Bromberg, IL.& Cerone, R. (1985) The natural history of homocystinuria due to cystathionine beta-synthase deficiency. Am J Hum Genet 37: 1-31.

[6] Boushey ,CJ.; Beresford ,SA.; Omenn ,G.S. and Motulsky ,A.G. (1995) A quantitative assessment of plasma homocysteine as a risk factor for vascular disease. Probable benefits of increasing folic acid intakes. *J Am Med* Assoc 274: 1049-1057.

[7] Conor R. (2007), Food Science & Technology and Nutrition & Dietetics,Company/Brand: Blackwell Publishing

[8] WHO, World Health Organization, Cadmium, Geneva, 1992

[9] Khurshid, S. J.and Qureshi, I. H.(1984). The role of inorganic elements in the human body, The Nucleus, 21: 3-23.

[10] Auer, J.;Berent , R. and Eber , B.(2001) . Homocysteine and Risk of Cardiovascular Disease. J Clin Basic Cardiol . 4: 261-264.

[11] Stampfer ,M.J.; Malinow, M.R.;Willett ,W.C.; Newcomer, L.M.; Upson ,B.;Stephen, N.;Adler, Dianne, B. Gasbarra.(1999). A Pocket Manual of Differential Diagnosis, Lippincott Williams & Wilkins, ISBN: 0-78171-943-7.

[12] Whisnant JP.(1997). Modeling of risk factors for ischemic stroke: the Willis lecture. Stroke , 28: 1839–43.

[13] Ueland PM& Refsum H.(1989). Plasma homocysteine, a risk factor for vascular disease: plasma levels in health, disease, and drug therapy. J Lab Clin Med; 114: 473–501.

[14] Meigs, James B.(2001). Fasting Plasma Homocysteine Levels in the Insulin Resistance Syndrome: The Framingham Offspring Study. Diabetes Care 24:1403–1410.

[15] Lonn, E.; Yusuf, S.; Arnold, M.J.; Sheridan, P.; Pogue, J.; Micks, M.; McQueen, M.J.; Probstfield ,J.; Fodorm G.; Held ,C.and Genest, J. Jr;(2006).Heart Outcomes Prevention Evaluation (HOPE) 2 Investigators. Homocysteine lowering with folic acid and B vitamins in vascular disease. N Engl J Med. , 13;354(15):1567-77.

[16] Aggett PJ. (1980). Trace elements in human nutrition. J Trop Pediatr .26:43-5.

[17] Goyer AR.(1997). Toxic and essential metal interactions. Ann Rev Nutr .17:37-50.

[18] Chaturvedi, U. C. ; Richa, S. and R. K. Upreti.(2004). Viral infections and trace elements: A complex interaction, CURRENT SCIENCE, 87(11) (1536-1554).

[19] Salem ,N. Bakos , Hamid, K. Ahmed and Tariq A. K. Nasser .(1988). Serum Copper, Magnesium, Zinc, Calcium, and Potassium Changes Following Acute Myocardial Infarction, Angiology, Vol. 39, No. 5, 413-416 .

[20] Durlach,J.;Durlach,V.;Rayssiguier,Bara,Y.M.&GuietBara,A.(1992).Magnesi-um Res.,5 (2),147.

[21] Tao, S.H .and Bolger, P.M.(1997). Hazard assessment of germanium supplements, Regul Toxicol Pharmacol ,25(3):211-9.

[22] Schauss, AG.(1991). Nephrotoxicity and neurotoxicity in humans from organogermanium compounds and germanium dioxide. Biol Trace Elem Res, 29(3):267-80 .

[23] Food and Drug Administration (FDA), US,Germanium, August 2009.

Comparison of the MAKLER & HINRICHS (1993) Technique Versus Application of Hepes Lysis Solvent in Determining the Activities of Plasmodium Lactate Dehydrogenase (pLDH) in Plasmodium berghei-Infected Erythrocytes

Shafariatul A.I*, Hasidah M. Sidek, Salmijah Surif

Department of Biomedical Science, Faculty of Allied
Health Sciences, Universiti Kebangsaan, Kuala Lumpur, Malaysia

Abstract

Plasmodium lactate dehydrogenase (pLDH) is a clinically valuable diagnostic indicator for malaria disease. In this preliminary report, there are some considerable interest in measurements of pLDH assay. All these years, pLDH was evaluated according to method which was established by Makler and Hinrichs in 1993. This method has became very popular among malaria researches . In this study , we tried to describe an alternative way or method for measuring pLDH assay. The method was used Hepes lysis to disrupt the parasitized erythrocyte membrane and thus released the pLDH in the solution. Measurement was done based on the intensity of chromatographic changes in color through ELISA reader at 650nm. The Hepes lysis technique was recently introduced in few years ago. Both Makler (1993) and Hepes lysis techniques have demonstrated their ability and efficiency for assessing the pLDh assay . Based on the results, Makler and Hinrich (1993) technique have shown that pLDH assay activities were higher than Hepes lysis technique significantly (2 way ANOVA). The results were also supported by the microscopic view on both techniques .The experiment were conducted at 10% parasitemia and 30% parasitemia. Further investigation are needed in order to create and have more robust lactate sensing format and simple device with fast and precise respond.

Keywords: Plasmodium lactate dehydrogenase , Makler and Hinrich (1993), Hepes lysis , pLDH assay

1. Introduction

Plasmodium lactate dehydrogenase (pLDH) is an essential metabolic enzyme that responsible for energy production in the parasite and parasite development. It is present in all malarial parasites of man and other animal host. pLDH catalyzes dehydrogenation of lactate and generates pyruvate by using NAD^+ as a cofactor. For more efficient and reliable test for pLDH activity, it can easily be differentiated from host lactate dehydrogenase (LDH) with 3-acetylpyridine adenine dinucleotide (APAD), an analogue of nicotinamide adenine dinucleotide (NAD) (Makler et. al 1993). pLDH can utilize APAD at 200 fold more rapidly and effectively than host LDH isoforms.

The pLDH converts $APAD^{+}$ to $APADH/H^{+}$ and in turn reduces colorless Nitro Blue Tetrazolium (NBT) to blue formazan that regenerates APAD for another cycle of reaction with pLDH. It is apparent that this colormetric assay is very versatile, having the ability to follow drug resistance, monitoring drug therapy and diagnose malaria.

The early method of assessing activity of pLDH was introduced by Makler et.al 1993. This technique has been widely used by many researchers with some modifications. The assay has been proven able to detect the presence of *Plasmodium falciparum* from *in vitro* cultures at parasitemia levels of 0.02% (Makler and Hinrichs, 1993). In recent years, there were some researchers working on pLDH by using Hepes lysis to hemolyzed red blood cells. The technique by While Makler et.al 1993 hemolyzed the red blood cells by four freeze-thaw cycles during which the samples were frozen at -20°C and thawed in a 37°C water bath. These two methods have different steps or procedures in order to obtained the pLDH and determined the formation of APADH at 650nm using a multiwavelength Elisa plate reader. Details of both techniques as explained in material and method section. The aim of this study is to compare the efficiency, precision, sensitivity and ease to evaluate pLDH activities between the two techniques. It is clear that both techniques hold a prominent position as the available sources of method to detect the presence of pLDH activities. It's also a reliable nonmicroscopic screening assay for any plasmodium species at a different level of sensitivity. Previous studies have showed a correlation between levels of parasitemia and the activity of parasite LDH (IMR, 2000; Vander Jagt et.al, 1981& Xu, X.L. et.al, 2007).

In this report, the *Plasmodium berghei* (*P. berghei*)– rodent model has been developed and used in this study to allow the determination of pLDH activities from both techniques. The significance of using malaria rodent instead of cultured human malaria is because it is easy and safe to handle and manipulate any stage of the life cycle in the laboratory (C.J.Janse, 1995). Hence, rodent malaria are still valuable models in several crucial areas of malaria research. The life cycle of *P.berghei* has similarities with *Plasmodium vivax* (*P.vivax*) which provide the first justification for the use of rodent in malaria research and definitely for in this study.

2. Material and method

Malarial Parasites

The *Plasmodium berghei* (NK-65) was obtained from swiss mice that reared in Universiti Kebangsaan Malaysia animal house. The *P. berghei* was used to test for the pLDH activities. Parasites are maintained thorough weekly blood passage in mice. The mice were treated and the experimental works were approval by Universiti Kebangsaan Malaysia Animal Ethics Committee (FSKB/BIOMED/2009/ZAKIAH/20-AUGUST/275-AUGUST-2009-DECEMBER-2009).

Detection of Parasitemia

Parasitemia was determined by light microscopy (1000 x magnification) using Giemsa-stained thin smears. The percentage of parasitemia in mice was monitored until it reached up to required percentage parasitemia which are 10% and 30 %. Then the blood will be collected by orbital sinus or heart puncture. The collected blood was pooled in EDTA tube. Then it will readily to be used for further steps.

Plasmodium Lactate Dehydrogenase Bioassay

The two techniques which are Makler et.al 1993 with some modifications and Hepes Lysis were conducted to determine the activities of pLDH respectively.

According to Makler et. al 1993, the blood samples which consisted of plasma or hemolyzed red blood cells would have gone through freeze and thaw cycle for at least 4 times. The blood samples were frozen at -20°C and thawed in 37°C water bath. Then all aliquots were transferred into 96- well, flat bottomed microtiter plate containing Malstat reagent (20mg of Sodium L-lactate, 5.5mg of basic Tris Buffer (Tris –C) and 3.7 mg of APAD dissolved in 1 ml of distilled water) and NBT-PES mixture (1.6mg of Nitroblue Tetrazolium (NBT) and 0.1 mg of Phenazine ethosulphate (PES) dissolved in 1 ml of distilled water). Absorbance was measured at 650nm using an ELISA plate reader.

In this study, the temperature for freezing was done at -40°C for red blood cells to be completely lysed. The above procedures are directly assayed without incubation process. For incubation process, the samples were incubated in a candle jar for 48h at 37°C, and were subsequently cooled at -40°C to lyse the red blood cells. For the positive control, parasitized red blood cells were used whereas non parasitized red blood cells were prepared for the negative control.

For the technique that used Hepes Lysis, the blood samples were centrifuged at 10 000xg to obtain supernatant samples containing pLDH. The supernatants were then transferred into 96-well microtiter plate containing Malstat reagent and followed by addition of NBT-PES mixture. Then measurements of pLDH were assayed at 650nm using Elisa plate reader. As with the Makler technique, some samples were incubated and some were directly assayed.

3. Results

The min activities of pLDH for both techniques were analysed with 2 ways ANOVA test. Study had shown that activities of pLDH were significantly difference according to percentage level of parasitemia and technique used.

The pLDH activities by Makler technique (1.16) were higher than for the Hepes Lysis technique (0.800) (Fig 1). Both are significantly different with $F (1, 20) = 27.042$, ($p<0.05$). Infected cells with parasitemia at 10% were shown significantly different from normal cells or non parasitized cells (0% parasitemia) by both techniques . Based on Hepes lysis technique, the mean activity of pLDH in non parasitized cells (1.130 ± 0.031) were significantly lower than parasitized cells at 10% parasitemia(1.930 ± 0.027). The Makler technique gave lower mean activity of pLDH in non parasitized cells (1.560 ± 0.040) compared to parasitized cells at 10% parasitemia (2.72 ± 0.040). In comparison, the mean activity of pLDH in non parasitized cells (1.130 ± 0.031) and parasitized cells at 10% parasitemia (1.930 ± 0.027) were significantly lower by Hepes lysis technique than non parasitized (1.560 ± 0.040) and parasitized cells (10% parasitemia)(2.721 ± 0.040) by Makler technique respectively . Briefly, the results from the above study demonstrated that Makler technique was more efficient and reliable than Hepes lysis technique.

The above results were supported by the microscopic view on both techniques. . By doing freeze and thaw cycling, the red blood cells were almost completely lysed, whereas, Hepes Lysis still showed some red blood cells that did not lyse. The hemolysed parasitized cells will release the pLDH in homogenates or supernatant. The more the cells lysed the bigger chance of parasitized cells to be lysed as well.

In this study, densities of infected cells do play a major role to achieve high activity of pLDH assay. As we have seen, the higher parasitemia , the more infected cells containing pLDH to be released. The 30% parasitemia indicated that 30 out of 100 red blood cells were infected with plasmodium .Likewise, 10% parasitemia indicated that 10 of 100 red blood cells were infected.

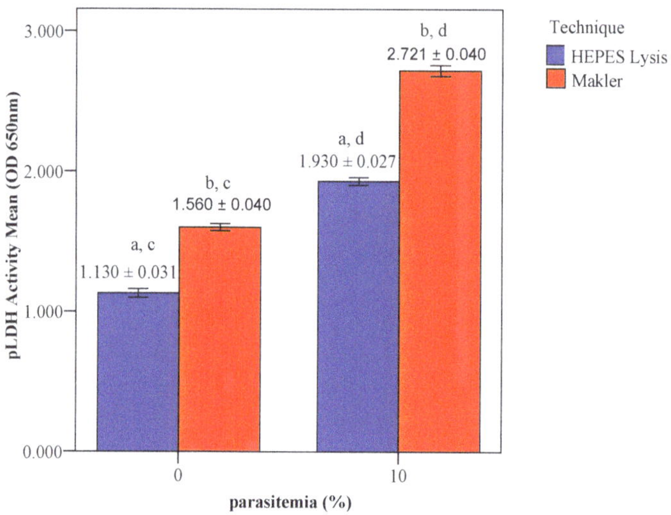

Fig 1. Determination Mean Activities of pLDH Between Makler (1993) Technique And Hepes Lysis

a = Min activity of pLDH in normal cells (0% parasitemia) which are significantly different compared to infected cells (10% parasitemia) by Hepes lysis technique.
b = Min activity of pLDH in normal cells (0% parasitemia) which are significantly different compared to infected cells (10% parasitemia) by Makler technique.
c = Min activity of pLDH in normal cells (0% parasitemia) by Hepes lysis technique which are significantly lower compared to min activity of pLDH in normal cells (0% parasitemia) by Makler technique .
d= Min activity of pLDH in infected cells (10% parasitemia) by Hepes lysis technique which are significantly lower compared to min activity of pLDH in infected cells (10% parasitemia) by Makler technique .

4. Discussion

Initially pLDH assay was developed to detect the presence of Plasmodium in the erythrocytes of malaria patient. The assay uses APAD (3-acetyl pyridine adenine dinucleotide) which is an analog to NAD (β- Nicotinamide adenine dinucleotide) with higher oxidation potential than NAD. It can substitute for NAD as a hydrogen accepting cofactor in many dehydrogenase reactions ; e.g : lactate dehydrogenase from *Toxoplasma* , *Clonorchis* , *Plasmodium* as well as mammalian dehydrogenase. It can also act as a proton acceptor in various transhydrogenation reactions with

NADH or NADPH. In *Plasmodium*, the lactate dehydrogenase works efficiently 200 times higher than human lactate dehydrogenase with APAD.

The assay development was introduced and established by Makler and Hinrichs in 1993. For many years, the assay was used in diagnosing malaria infection. In recent years, it has become popular as a validated target for antimalarial agent development in drug screening. In the meantime, there is an alternative method of pLDH assay by using Hepes Lysis solvent. It was used by some researchers but the performance is still under investigation. Basically, the principal for both methods were lysed the membrane erythrocyte. By doing so, pLDH would be released and allow to be measured by Elisa reader based on colorimetric changes. From these studies both methods have shown significant results. The negative control which was normal cells or non parasitized cells showed lower activities. However, the findings showed that Makler and Hinrichs (1993) technique exhibited higher activities than Hepes Lysis.

According to Makler and Hinrichs technique (1993) , thaw and freeze cycle process will disrupted the structure of erythrocyte membrane . Disruption will lead to cells hemolysate . For parasitized erythrocyte , hemolysate process assisted in releasing the pLDH. The more the parasitized cells lysed , the more the pLDH would released . The lysate erythrocyte turned out to be small debris and scattered in it's surrounding.

By using Hepes lysis solvent , erythrocyte was lysed chemically. Hepes lysis was alkaline solution with pH 8. For normal cells the pH ranging between 7.34 to 7.45. Zeidler and Kim (1977) , reported that alkaline condition will disrupt the stability of the cell membrane protein and eventually lysed. Most of the erythrocytes were not lysed. The morphology of treated erythrocyte were seen similar as in control cells or untreated cells. Probably the Hepes lysis need to be more alkaline to work efficiently.

Based on the overall results, erythrocytes that were treated with Makler and Hinrichs technique (1993) showed significantly higher activities than erythrocytes treated with Hepes Lysis. The results were supported by microscopic view .The activities of pLDH will increased in relation to the higher numbers of parasites. Comparison of pLDH activities were conducted by Makler and Hinrichs (1993) technique at 30% and 10% parasitemia respectively .The parasitized erythrocytes at 30% parasitemia presented significantly higher pLDH activity than 10% parasitemia and 0% parasitemia respectively.

Regarding the morphology of erythrocyte infected-*Plasmodium berghei* , the erythrocytes as the host may had more than one parasites reside in it. The size of the infected erythrocytes is bigger than normal cells. The features were similar to *Plasmodium vivax* .The colour were light purplish red. The Plasmodium could be found in various stages, mostly, were ring and trofozoite form.

On the other hand , the pLDH assay showed higher activities in cells without incubation process compared to cells treated by incubation process. This might be due to the contamination of the cells which occured during the incubation period.

5. Summary and conclusion

In this study, Makler and Hinrichs (1993) technique showed to be more efficient than Hepes lysis solvent to quantitate the activities of *Plasmodium Lactate dehydrogenase* .

6. Acknowledgement

I would like to express my greatest gratitude to Faculty of Science and Technology, UKM and supportive staffs from the Department of Biomedical science, Faculty of Allied Health Sciences UKM.

7. References

[1] Institute For Medical Research.2000. IMR Annual Report 2000. Kuala Lumpur

[2] Janse,C.J and walters,A.P. 2006. The *Plasmodium berghei* research model of malaria.

[3] Makler,M.T., and Hinrichs, A.D.,1993. Measurement of the Lactate Dehydrogenase activity of *Plasmodium falciparum* as an assessment of parasitemia. *American Journal of Tropical Medicine and Hygiene* **48(2)** : 205-210

[4] Makler,M.T., Ries, J.M., Williams, J.A., Bancroft, J.E., Piper, R.C., Gibbins, B.L. and Hinrichs, A.D.,1993. Parasite Lactate Dehydrogenase as an assay for *Plasmodium falciparum* drug sensitivity. *American Journal of Tropical Medicine and Hygiene* **48(6)** : 739-741

[5] Romero, M.R., Garay, F., Baruzzi, A.M. 2008. Design and optimizing of a lactate amperometric biosensor based on lactate oxidase cross-linked with polymeric matrixes. *Sensor and Actuators B.* **131** : 590-595

[6] Vander Jagt, D.L., Hunsaker, L.A., Heidrich, J.E. 1981. Partial purification and characterization of lactate dehydrogenase from *Plasmodium falciparum*. *Mol Biochem Parasitol* **4** : 255-264

[7] Xu, X.L., Yang, R.Y., yang, X.Q., Chen, P.Q. and Zeng, Q.P. 2007. A quantitative assay of recombinant malaria Lactate Dehydrogenase as a platform for screening inhibitors from crude herbal extracts. *Chinese Journal of Biotechnology* **23 (4)** : 593-597

[8] Zeidler, R and kim, H.D . 1977. Preferential hemolysis of postnatal calf red cells induced by internal alkalinization. *J Gen Physiol* **70 (3)** : 385-401

Contribution to the Study of the Impact of Phosphate Fertilizer on Biochemical Parameters of *Triticum durum*

Sabrina Bouchelaghem*, Djebar M. R

Laboratory of Toxicology and cellular bioenergetics, University of Annaba, Algeria

*Corresponding author

Abstract

In this study we are interested in assessing the impact of different regimes of NPK and its effect on wheat Triticum durum. The first results show that the presence of fertilizer causes a decrease in the percentage and speed of germination and an inhibition of growth of wheat. On the metabolic level, the NPK caused a significant increase in mean levels of proline and soluble sugars, and inhibition of protein synthesis.

Keywords: NPK, Proline, Protein, Soluble sugar, *Triticum durum*.

1. Introduction

Adequate fertilization is a prerequisite for modern agriculture in order to meet high yields and optimum quality of crops. The vast majority of mineral elements essential to plant development, is only necessary in minute amounts, which may be provided by most soils without supplements. However, some key elements that are phosphorus (P), potassium (K), sulfur (S) and nitrogen (N), are frequently in short supply in the soil to allow optimal growth of crops. Despite its abundance, atmospheric nitrogen is unusable for most plants. The only plants able to use nitrogen from the air are those that develop a symbiotic relationship with microorganisms in nodules on plant roots. Unlike phosphorus and potassium, plants need large amounts of nitrogen, nitrogen representing 3-4% of their dry matter. Moreover the efficiency of nitrogen fertilization on crop yields contributed to the development of modern agriculture more intensive nitrogen. In addition to heavy loads caused by this item of expenditure, the massive inflow of nitrogen fertilizers started in the middle of last century affects the natural nitrogen cycle, which is not without consequences for the environment and health.

The objective of this study is to evaluate the effects of NPK fertilizer nitrogen on physiological parameters, biochemical and metabolic ones on wheat *Triticum durum*.

2. Materials and methods

2.1. Biological material

The experimental material used in our work is wheat (Triticum durum) Geta come from hard JTGC (demonstration farm and seed production of Guelma).

2.2. Determination of soluble sugars:

The total soluble sugars were determined by the method of Burnett and Schield [2],

2.3. Determination of proline.

Proline is determined by the method of Monnoveux and Nemmar, [3].

3. Results

3.1. Effects of NPK on the physiological parameters of germination and growth

3.1.1. Effects on the germination percentages.

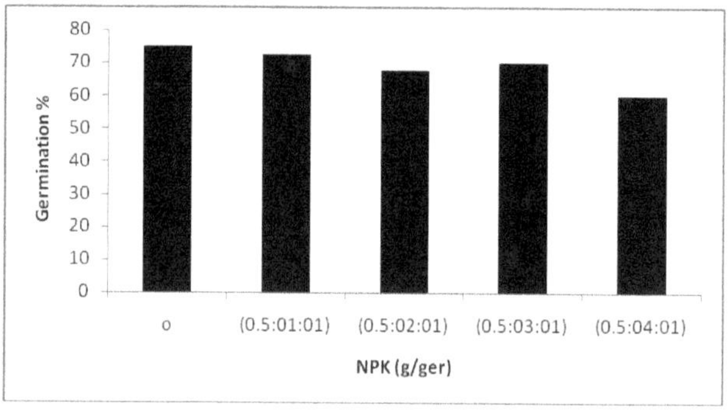

Figure 1. Effects of different regimes of NPK on the average rate of seed germination of wheat germination time = 96h. p = 0.031.

3.2. NPK effects on biochemical parameters of germination.

3.2.1. Effect of NPK on the average content of sugar

The results of treatment effects by the various regimes of NPK on average contents of total sugars in roots and leaves of wheat seeds are shown in Figures 2.

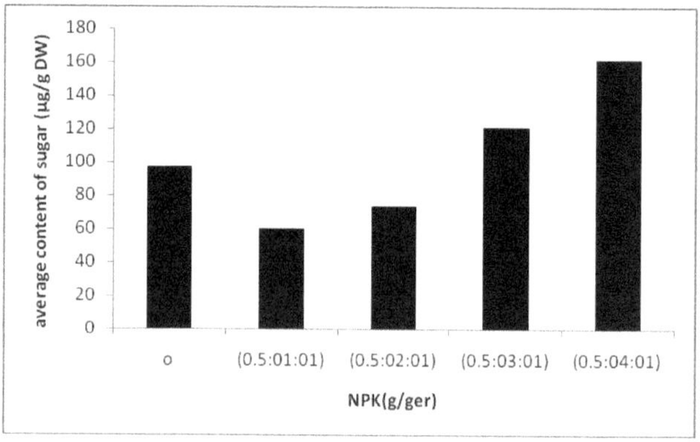

Figure 2

3.2.2. Effect of NPK on the average content of proline:

The results of the effects of treatment with NPK on the mean levels of proline are shown in Figure 3.

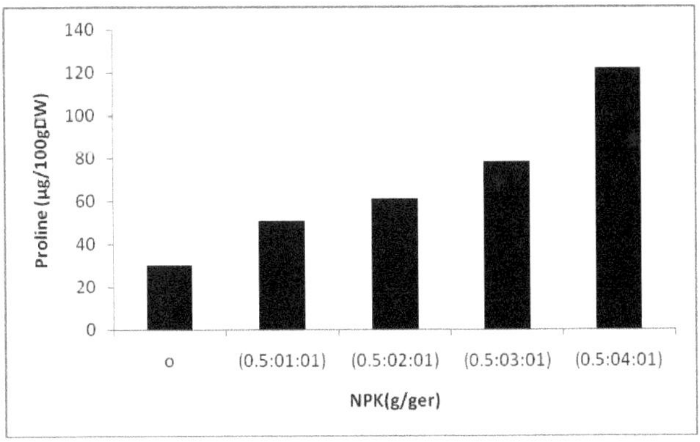

Figure 3. Effect of NPK on the average rate of proline roots from wheat seeds.

4. Discussion and Conclusion

In this section we determined the effects of NPK on germination percentage, there is a slight decrease in germination percentage especially in the plan (0.5:0.4:0.1) (20%). Our result is in perfect agreement with the work of **[4]** (Savoie and Smith, 1997) which showed that germination was reduced with increasing nitrogen.

This disturbance observed in wheat seeds is due to the effects of substances contained in the manure on the outbreak of various enzyme activities originally starting the germination process. The work of other authors **[5, 6]** as support the findings in our work.

In pea *Pisum sativum,* the nature reserve is mainly protein, up to 40% **[7]**. Firstly, it has been shown in *Medicago truncatula* (another legume) that a massive flow of nitrogen is released into the seed during germination. The high concentrations of nitrogen (as ammonium free) are toxic to cells that need to be cleared. The nitrogen is assimilated to form amino acids: this is detoxification.

Indeed, in a second step, it is also proposed that metabolism of amino acids is very active during germination. They would not only protein synthesis, but also provide nutrients for the growth of new tissue and the synthesis of sugars in the seed.

The determination of proline stems and roots exposed to NPK shows a strong increase of proline, this could explain as stress response. This ability of plants to the synthesis and accumulation of proline is not specific only to wheat, it is also for many glycophytes, in pea **[8]**, the barley Hordeum vulgare L. **[9]**, beans **[10]** and Nicotiana tabacum **[11]**.

The results we obtained for the mean levels of sugars showed increasing rates for wheat. This could be due to osmotic stress in response to treatment with NPK. **[12]**. Many studies confirm the accumulation of high levels of soluble sugars in different types of plants under stress conditions: water **[13, 14]**. ; saline **[15, 16]**, osmotic **[17]** and metal **[18]**.

5. References

[1] Kaur J. C. Duffu . 1989. The effect of naf on cereal seed germination and seeding growth. Plant Cell and environnement. 12:154-161.

[2] Schield R. M. Burnett. 1966. Determination of protein bound carbohydrate in serum by a modified anthorons. Method Anal., Chem. 32 : 885-886.

[3] Monneveux P.H. M. Nemmar. 1986. Contribution à l'étude de la résistance à la sécheresse chez le blé tendre (Triticum aestivum) et chez le blé (Triticum durum Desf) : étude de l'accumulation de la proline au cours du cycle de développement. Agronomie 6 :583-590.

[4] Savoie D. A. Smith. 1997. Effet de l'azote sur la production de semences du ray-grass vivace Barrage. Ministère de l'Agriculture, l'Aquaculture et des Pêches.

[5] Clijsters, H. A. Cuypers, A.and J. Vangronsveld. 1999. Physiological responses to heavy metals in higher plants; defence against oxidative stress, Z. Naturforsch. 54 c: 730-734.

[6] Van Assche, F. H. Clijsters. 1990. Effects of metals on enzymes activity in plants, Plant cell. Environ. 13: 195-206.

[7] Lafon J.P. C. Tharaud-Prayer and G. Levy. 1988. Biologie des plantes cultivées. Organisation, physiologie de la nutrition. Paris : Technique et Documentation Lavoisier.1 : 103-124.

[8] Bar-Nun N. and A. Poljakoff-Mayber, 1979. Intervarietal differences in the amino acid composition of pea roots as related to their response to salinity. Ann. Bot. 44: 309 – 314.

[9] Buhl M.B. C.R Stewart. 1983. Effect of NaCl on proline synthesis and utilisation in excised barley leaves. Plant Physiol., 72 : 664 – 667.

[10] Belkhoudja M. 1996. Action de la salinité sur les teneurs en proline des organes Adultes de trois lignées de fève (Vicia faba L.) au cours de leur développement. Acta bot. Gallica, vol. 143(1) : 21-28.

[11] Rhode D. S. Handa. 1989. Amino acid metabolism in relation to osmotic adjustment in plant cells. In Environmental Stress in Plants: Biochemical Mechanism, NATO ASI Series, Vol. G 19 (JH Cherry. Ed. Springer, Berlin, 41-62.

[12] Stiborova M. M. Doubravova. A. Brezinova and F A. 1988. "Effect of heavy metal ions on growth and biochemical characteristics of photosynthesis of barley (Hordeum vulgare L.)." Photosynthetica 20(4): 418-425.

[13] Mefti M.A. Abdelguerfi et A. Chebouti. 1998. Etude de la tolérance à la sécheresse chez quelques populations de Medicago truncatula (L.) Gaertn. Science. 5: 173-176.

[14] Kameli A. et D.M. Losel. 1995. Contribution of carbohydrates and other solutes to osmotic adjustement in wheat leaves under water stress. J. Plant Physiol. 145: 363-366.

[15] Zid E. C. Grignon. 1991. Tests de sélection précoce et résistance des plantes aux stress. Cas des stress salin et hydrique, L'amélioration des plantes pour l'adaptation aux milieux aride.Ed. AUPELF-UREF. John Libbey Eurotext. Paris. pp. 91-108.

[16] Ben Khaled L.A. Morte Gõmez, M.Ouarraqel and A. Oihabi. 2003. Réponses physiologiques et biochimiques du trèfle (Trifolium alexandrinum L.) à la double association Mycorhizes-Rhizobium sous une contrainte saline. INRA, EDP Sciences. Agronomie. 23: 571–580.

[17] Abdelkrim F., R. Djebbar. et F. Aid. 2005. Effet d'un stress osmotique sur la germination et le début de croisssance de deux variétés de colza : Brassica napus L. Eurol et Goeland. 1er Colloque Euroméditerranéen de Biologie Végétale et Environnement, Annaba 28-30 novembre 2005.

[18] Massantini F.R. Fauili. G. Magnani and N. Oggiano. 1990. Soils culture, biotechnology for high quality vegetables, 4 (2): 27-39.

Frost Occurrence Risk Management for Pistachio Industry in Rafsanjan

Alireza Hosseini[1], Mohammad-Saber Fallah Nezhad[1],
Yahya Zare Mehrjardi[1], Reza Hosseini[2]

1 Department of industrial engineering, Yazd University, Iran

2 Division of Biostatistics, University of Southern California, USA

Abstract

This work develops a statistical model to assess the frost risk in Rafsanjan, one of the largest Pistachioproduction regions in the world. These models can be used to estimate the probability that: a frosthappens in a given time-period in the year; a frost happens after 10 warm days in the growing season etc.These probability estimates then can be used for: (1) assessing the agroclimate risk of investing in thisindustry; (2) pricing of weather derivatives. Autoregressive models with different seasonal componentsand lags are compared using AIC, BIC, AICc and cross validation criterions. The optimal model is AR(1)with 12 terms from Fourier series. The long-term trends are also accounted for and estimated from data.The optimal models are then used to simulate future weather from which the probabilities of appropriatehazard events for pistachio yield are estimated.

Keywords: Pistachio; Frost; Weather derivative; Minimum temperature; Time-varying autoregressivecoefficients.

1. Introduction

The greater Rafsanjan area in north of Kerman Province in Iran is a region with the largest pistachioproduction in the world and most of the region's economy relies on pistachio production. In therecent years the most important risk factor for pistachio producers and industry (e.g. farmers,distributers) has been frosts that have destroyed a large proportion of the yield. Therefore methodsthat can estimate the probability of such events is useful. In particular such methods can: (1) assess the agroclimate risk of investing in this industry; (2) pricing of weather derivatives. In factweather derivatives, which may be created as part of a risk management program, can be writtenin terms of the attainment or non-attainment of specific target-values stipulated in the contract.Temperature-related trades account for 80% of the transactions among all weather derivatives [1]. Most of the work in this area has focused on HDD/CDD (heating degreedays/cooling degree days) (e.g. [2,3]). In this paperwe focus on the occurrence of frosts an issue recently considered in [4], foragricultural crops in Canada.

The models developed in this paper can be applied to estimate: the probability that a given period is frost-free; the probability that a given day is the start of a long frost-free period;the distribution of the length of the frost-free period and so on. The same model can be used tocompute the probability that a given day of the year is the beginning of the growing season (the first day that the mean temperature is higher than 5 degrees for 5 consequent days) as well as thelength of the growing season which are important for agricultural applications. For example in thisstudy we estimate the probability of a useful event: *"the minimum temperature goes below zero atleast one day in the period March 27th-April 20th"*. This is an important event because it coincideswith the general flowering time of pistachio trees. Throughout this paper, temperature is measuredin degrees Celsius. Let us denote the minimum temperature series by {Y (t)}, t=0,1,2,..., wheret denotes time. We let F to be the investor's defined frost which we take it to be zero in this work.Then we can define the binary frost process:

$$Y_F(t) = \begin{cases} 1 & Y(t) \leq F \text{ (deg C)} \\ F & Y(t) > F \text{ (deg C)}. \end{cases}$$

In order to study frosts we can use these approaches among others: (a) Fit the continuous-valued Markov model to the Y (t) chain; (b) Fit a binary Markov model to the YF (t) chain. Hosseinisuggests using binary Markov models to avoid assumptions regarding the distribution of temperature and gain robustness for modeling frosts in Alberta, Canada[4]. They show time-varyinghigh-order Markov models with complex seasonal structure are needed and therefore their computations become challenging. Here we investigate Method (a) in fitting such chains and calculatingthe probabilities of frost events. The advantages of Method (a) are: (1) Thefitting can be donewith standard packages such as R with less computational problems; (2) only this method canestimate the probability of complicated events. One such complicated event is: *"the temperaturein March-April is above 5 (deg C) for at least 3 consecutive days and is below zero after"*. A comparison of the two methods in terms of estimation when they are both applicable is left to futureresearch.

2. Data and statistical models

The data in this study are daily minimum temperature values collected at Rafsanjan weather station from 1992 to 2010. At the moment we do not have access to more data from other stationsin the area but we hope to acquire those data for future studies to offer more local predictions. Inorder to model frost occurrences, we introduce statistical models for minimum daily temperaturein Rafsanjan. Several features of the temperature process should be considered in modeling: (1) seasonal trends over time; (2) long-term trends (possibly a result of global warming or volcanicevents etc) (3) dependence over time. Let {Y (t)}, t = 0,1,2,..., T denote the daily minimumtemperature process in centigrade, where t denotes the day starting from March 1st 1992 toDecember 28th 2010. Here we consider autoregressive models with a seasonal component andvarious lags:

$$Y(t) = \mu(t) + \epsilon(t), \quad \mu(t) = a_0(t) + \sum_{i=1}^{r} a_i Y(t-i)$$

Where $\mu(t)=E\{Y(t)|Y(t-1),Y(t-2),...\}$ is the conditional mean of minimum temperatureat time t; $\varepsilon(t)$ are independent identically distributed normal errors $\varepsilon(t) \sim N(0,\delta^2)$; $a_0(t)$ is thefixed trend coefficient; $a_1, a_2,\ldots,a_r$are autoregressive coefficients. We allow $a_0(t)$ to include bothseasonal and long-term effects by using a Fourier series with period, $\omega = \frac{2\pi}{366}$, and a quadratictrend:

$$a_0(t) = \{\alpha_0 + \sum_{j=1}^{k} \alpha_j \cos(j\omega t) + \beta_j \sin(j\omega t)\} \quad + \quad \{\gamma_1 t + \gamma_2 t^2\},$$

The estimation of the parameters is done by maximizing the (partial) likelihood of the data asdiscussed in [4]. The nice property of the gaussian error assumption is themaximization can be done in closed form exactly in the same way as minimizing the mean-squareerror of classical regression problem originally solved by Gauss and therefore the estimation is fastin statistical packages such as "R"(a free widely used software by statisticians and practitioners).

3. Statistical model selection

In the above we introduced several autoregressive models of: (1) various lags; (2) various seasonal complexity (number of Fourier terms); (3) various long-term trends. Therefore we need to usesome criteria to select an optimal model. The problem of model selection is an important onein statistical theory and application. Various criteria are suggested in the literature for example:AIC in [5]; BIC in [6] and AICc in [7]. Denote thelikelihood of the data by L (in this paper the "partial likelihood"), the number of covariates by pand the sample size by n. Then we have

$$AIC = 2p - 2\ln(L), \quad AICc = AIC + \frac{2k(k+1)}{n-k-1}, \quad BIC = p\log(n) - 2\ln(L).$$

Since n in our data is large compared to k, AIC and AICc are very close. When we compared themodels using these criteria, AIC and AICc give rise to the same optimal model while BIC pickeda simpler model. In Table 1 we have compared these optimal models using cross-validation errorand cross-validated correlation. The cross-validation proceeds by: (1) taking an existing data pointout; (2) fitting the model; (3) predicting the value of the point we took out (validation). Then thecross-validation error (CVE) is the mean square error of the predictions and the cross-validationcorrelation (CVR) is the correlation between the predictions and the observed. Table 1 shows thatwhile the CVE and CVR are very close for the two models, the model picked by AIC/AICc slightlyoutperforms the one picked by BIC and therefore we use that model for estimation.

Criterion	optimal Model: Z_{t-1}	CVE	CVR
AIC and AICc	$sin(\omega t), cos(\omega t), \cdots, sin(6\omega t), cos(6\omega t), Y_{t-1}, t, t^2$	**2.691**	**0.9458**
BIC	$sin(\omega t), cos(\omega t), Y_{t-1}$	2.696	0.9456

Tab 1. We compare the optimal model picked by AIC and AICc (first row) with the optimal model picked by BIC, (second row) using cross validation error and cross-validated correlation.

4. Applications in frost risk assessment

Previous section found an optimal fit to the data from which estimating the probability of anydesired (possibly complex) event is possible by performing multiple simulations. In order to find out the probability of frost in any given day during 2011-2012 we have done 10000 simulationsfrom the model for 2011-2011 and then for each day we have calculated the proportion of frostdays (number of frost days divided by 10000). The results are plotted in Figure 1.As we pointed out in the introduction because the flowering time of different varieties ofpistachios in Rafsanjan is generally between March 27th to April 20th, it is important to investigate the frost-occurrence during this period which we call the hazard period. Figure 2 shows thedistribution of the "number of frost days" during the hazard period of 2012, where the frequencyout of 10000 of any "number of frost days" is plotted. We observe that while it is most likely thatno frost occurs in that period, there is a considerable probability that there are at least one frosts.This probability turns out to be about 9 percent which is a plausible number with our experience of pistachio damages caused by frosts in the past 20 years.

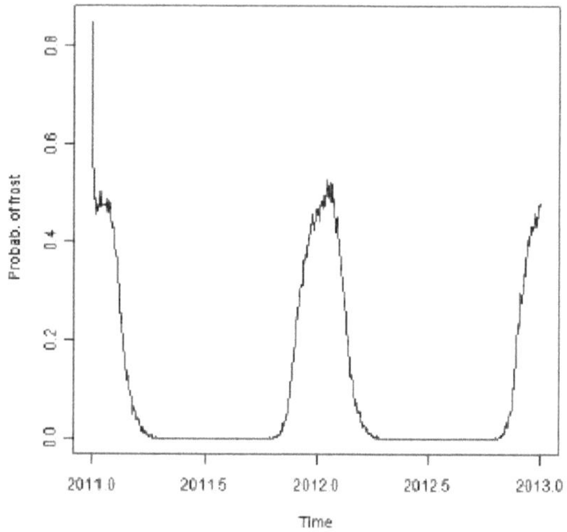

Fig 1. Estimated daily frost probability for 2011-2012 from the model, obtained using 1000 simulations of future weather.

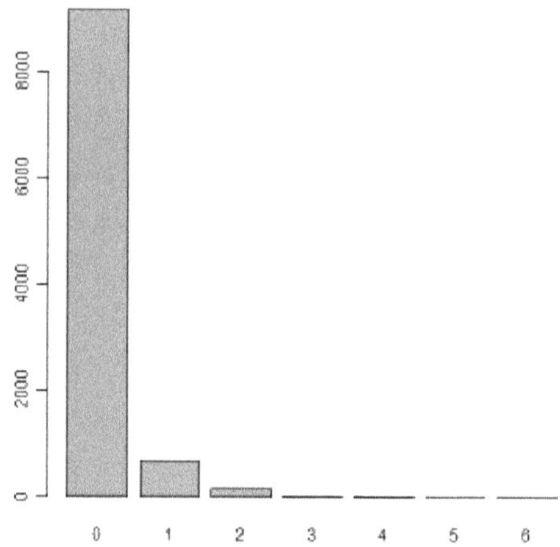

Fig 2. Distribution of frost days during the Hazard period (March 27th to April 20th 2012). This is based on 10000 simulations of the future chains. The probability of at least one frost based on this simulations is 0.0872 which is about 9 percent.

5. Summary and conclusion

This paper developed and compared several statistical models to estimate the probability of hazardfrost events for pistachio industry in Rafsanjan. Despite the importance of such risk factors, nosystematic studies and estimations of these risks are available in this region as far as we know;this paper is one of the first attempts in developing methods that can assess such risks. Assessingthe probabilities of the hazard events are useful in estimating the risk of investing in this industryfrom production to distribution and exporting. However here we have not investigated other riskfactors such as: extremely high temperature during summer; heavy short-time rain during floweringperiod; slow but long rain during the flowering time. For future studies we plan to acquire thedata for precipitation, maximum temperature and developing models that assess these other riskfactors.

Another important aspect of assessing the risk is relating the risk factors to the losses inyield or monetary values involved. For this study we relied on expert knowledge (by interviewingfarmers and agriculture engineers) to define our hazard period. However if for example data foryield per km2 becomes available for enough number of years and/or locations, one can develop astatistical model to relate the weather events to the losses in the yield in the same model.

6. Acknowledgements

We are indebted to Mr. islami for providing the data for this study.

7. References

[1] Cao, M. and Wei, J. Weather derivatives, valuation and market price of weather risk The Journal of Future Markets, 24(11):1065-1089, 2004.

[2] Richards, T. J., Manfredo, M. R. and Sanders, D. R. Pricing Weather Derivatives American Journal of Agricultural Economics, 86(4):1005-1017, 2004.

[3] Benth, F. E., and Benth, J. S. Volatility of temperature and pricing of weather derivatives Quantitative Finance, 7(5):553-561, 2007.

[4] Hosseini.R, Le.N, Zidek.J,2012. Time-Varying Markov Models for Binary Temperature Series in Agrorisk Management. Journal of Agricultural Biological and Ecological Statistics, Vol. 17,Num. 2,283-305.

[5] Kedem, B. and Fokianos, K. Regression Models for Time Series Analysis, Wiley Series in Probability and Statistics, 2002.

[6] [Akaike, H. A new look at the statistical model identi_cation. IEEE Transactions on AutomaticControl, AC- 19:716-723, 1974.

[7] Schwartz, G. Estimating the dimension of a model. Annals of Statistics, 6:461-464, 1978.

[8] Brockwell, P.J., and Davis, R.A. Time Series: Theory and Methods, 2nd edition Springer,2009.

Detection of Rifampin- and Isoniazid-Resistant Genes in *Mycobacterium tuberculosis* Clinical Isolates

Noraziah Mohamad Zin[1,*], Nor Farha Hussain[2],
Rahizan Isa[3], Mohamed Kamel A Ghani[1], Nik Marzuki Sidik[4]

1 School of Diagnostic and Applied Health, Faculty of Health Sciences, Universiti Kebangsaan Malaysia, Kuala Lumpur

2 Pathology Unit, Hospital of Tengku Anis, Pasir Puteh, Kelantan

3 Bacteriology Unit, Infectious Disease Research Centre, Institute of Medical Research, Kuala Lumpur

4 School of Biosciences and Biotechnology, Faculty of Science and Technology, University Kebangsaan Malaysia, Bangi Selangor

*Corresponding author

Abstract

Tuberculosis still remains the leading cause of death worldwide. The morbidity has been reported to decrease, but incident and prevalence of multi-drug resistance Tuberculosis is still on rise. Rifampicin and isoniazid are the first line treatment to TB patients and resistance to these drugs has been linked to mutations in genes such as *rpoB* and k*atG*. In the present study, PCR method was employed to detect three types of restricted genes associated with drug resistance tuberculosis namely *rpoB,* regulatory-*inhA* and *katG*. Sixty-two samples were obtained from different parts of Malaysia hospital which consisted of 35 pulmonary and 27 extra-pulmonary specimens. Twenty-seven specimens showed positive results as detected by duplex PCR method whereas 3 specimens positive as detected by acid fast bacilli and culture method. Out of 27 isolates, 3 isolates from culturable isolates harbored restricted genes that where associated with drug resistance tuberculosis. The mutations involved in *rpoB* genes comprised of acid amino transposition (isolate 148) and frameshift mutations (isolate 624 and 374). This study is clinically important because it focuses in molecular diagnosis and can act as an early warning on the emerging status of multidrug resistance of *M. tuberculosis* in Malaysia.

Keywords: Rifampin, Isoniazid-Resistant, M. tuberculosis, PCR

1. Introduction

Tuberculosis (TB) is an infectious bacterial disease caused by *Mycobacterium tuberculosis* which commonly affects the lungs and respiratory system. The TB morbidity was reported decreasing, but its remains alarming due to increase in incidence and prevalence of Multi Drug Resistance-TB (MDR-TB) [1]. To date, seven antimicrobial agents that have been used in the treatment of resistant TB cases were; - isoniazid, rifampicin, pyraxinamide, enthambutol, streptomycin and fluoroquinolones. MDR phenotype is defined as resistance at least to isoniazid and rifampicin which is the most effective drug recommended by World Health Organization (WHO) and being used as the first line treatment in TB patient. Therefore, resistance towards these two drugs has become major problems in the treatment of TB patients. Resistance to first line anti-TB drugs has been linked to mutations in at least 10 genes; *katG, inhA, ahpC, kasA* and *ndh* for INH resistance; *rpoB* for RIF resistance, *embB* for EMB resistance, *pncA* for PZA resistance and *rpsL* and *rrs* for STR resistance [2]. Nevertheless, nearly 95% of the RIF resistant strains possess a mutation in the *rpoB* gene encoding a DNA-dependent RNA polymerase [3]. In addition, approximately 90% of INH resistant strains have a mutation in the *inhA, katG,* and *ahpG* genes encoding enzymes that are related to mycolic acid synthesis of cell wall [3]. Rapid development of drug resistance caused by *M. tuberculosis* has led to measure resistance accurately and easily. This knowledge will certainly help us to understand how to prevent the occurrence of drug resistance as well as identifying genes associated with new drug resistance. Keeping the above facts in mind, the present, this study was carried out to detect resistance-associated mutations gene in *M. tuberculosis* clinical isolates.

2. Materials and methods

2.1. Bacterial isolates and clinical samples

Bacterial samples used in this study were collected from various hospitals in Malaysia. Prior ethical approval was obtained from the institutional ethics committee. The samples consisted of pulmonary (n=35) and non-pulmonary (n=27) specimens. The samples such as sputum, gastric lavage and urine were decontaminated with 4% NaOH for 15 minutes before being used. All samples were investigated for the presence of acid-fast bacilli by Ziehl-Neelsen and cultured on Loewenstein-Jensen medium. *M. tuberculosis* strain ATCC 27294 was grown in Ogawa medium and used as positive control. Whereas, *Bacillus subtilis* (ATCC 26633) was grown on blood agar and used as negative control.

2.2. DNA extraction, PCR amplifications and DNA sequencing analysis

Extraction of DNA from bacterial culture and clinical samples were carried out by using High Pure Viral Nucleic Acid Extraction Kit (Roche Inc. USA) according to manufacturer's recommendation. DNA from all bacterial isolates and clinical samples were subjected to Duplex-PCR using mixture of primers to amplify the IS*6110* gene and p53 gene as previous report [4] with some modifications. The amplification mixture consisted of 1 µg of template DNA, 1X final concentration of MasterMix (Eppendorf), and primers (0.4 pmole/µl). Amplification was carried out using Master Cycler Gradient Thermocycler (Eppendorf). The cycling parameters were 94 ºC for 3 min followed by 34 cycles of denaturation at 94 ºC for 15 seconds, annealing at 66 ºC for 15 seconds and extension at 72 ºC for 20 seconds. Final extension was then carried out at 72 ºC for 3 min. The PCR product was analysed by a 1.6 % agarose gel. Primers used for detection of *rpo*B, *inh*A and *kat*G resistance genes were design by Vector NTI software. The amplification mixture consisted of 1 µg of template DNA, 1X final concentration of MasterMix (Eppendorf), and primers (0.4 pmole/µl). The amplification was carried out using Master Cycler Gradient Thermocycler (Eppendorf). The cycling parameters were 95 ºC for 3 min followed by 34 cycles of denaturation at 95 ºC for 30 seconds, annealing for 30 seconds at 64, 63, and 65 ºC for *inhA, rpoB* and *katG* respectively.The extension cycle were72 ºC for 90 seconds for *inhA* and *katG* genes whereas 60 seconds for *rpoB* genes. The final extension was carried out at 72 ºC for 5 min. The PCR product was analysed by a 1.6 % agarose gel. DNA fragments of PCR products were purified using QIA-quick PCR Purification Kit (QIAGEN, Germany) according to manufacturer's recomendation. Purified DNA fragments were sent to the 1st Base Laboratories Sdn. Bhd, Petaling Jaya, Kuala Lumpur for sequencing. DNA sequences were analyzed using BLAST (http://www.ncbi.nlm.nih.gov/BLAST). Multiple sequence alignments were conducted using ClustalW (http://www.genome.jp/tools/clustalw/).

3. Result and Discussion

3.1. Detection of M. tuberculosis by PCR and culture method

In total, 27 extra-pulmonary and 35 pulmonary clinical specimens were obtained for AFB smear, culture and PCR analysis. Out of 62 isolates, only 3 isolates namely 148, 374 and 624 were culture positive as detected by AFB. The culture was then confirmed as *M. tuberculosis complex* using GenProbe System performed by Bacteriology Unit, National Public Health Laboratory at Sungai Buloh, Kuala Lumpur. Nevertheless 27 isolates were positive for IS6110 of *M. tuberculosis* as detected by DPCR. Of the 27 isolates, 12 of them were from extra-pulmonary isolates and 15 were from pulmonary isolates (Figure 1). It was interesting to note that the PCR methods were able to detect the AFB negative of the non-pulmonary specimens such as from CSF, bone biopsy, lumbar puncture, rectal biopsy and lymph nodes. A similar result was shown by Suhaila et al. (2008) [4]. Number of positive isolates detected by PCR analysis was higher compared to culture method. Time required for detection of *M. tuberculosis* from clinical specimens via PCR analysis was less than 4 hours whereas more than 8 weeks were required in culture method.

3.2. Detection of resistance genes

Out of 27 isolates positive for *M. tuberculosis* genes, only 3 culturable isolates showed amplification of all resistance genes tested (Figure 2). Low quantity of *M. tuberculosis* DNA in the clinical sample and the presence of inhibitor [5] might be the contributed factors. It was interesting to note that all the positive samples bearing the Zielhl Neelson staining scores more that 10/3L directly showed the high bacteria numbers. Approximately 104 organisms/ml are required for reliable detection with Ziehl-Neelsen stains [6] In addition, the properties of the genes of interest also affect the success of amplification. Only one specific site of *katG, rpoB* and regulatory-*inhA* genes presence in the genome of *M. tuberculosis* and may explain the limited excess to the genes. Hence, the amplification cannot be made. The primer used for detection of resistance genes have been successfully design by Vector NTI software. Amplification of resistance genes from clinical specimen have been succeeding in three isolates namely 148, 374 and 624. Amplified genes for *rpoB, katG* and *inhA* were shown in Figure 2a, 2b and 2c respectively. The product size for *rpoB, katG* and *inhA* were 442bp, 2206bp and 442bp, respectively as confirmed by sequencing analysis.

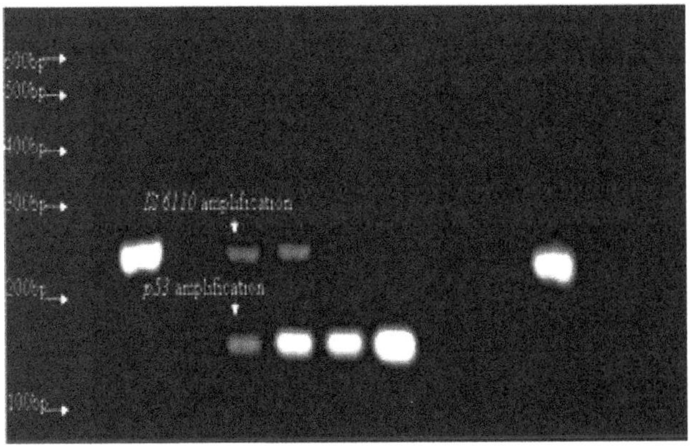

Figure 1. Agarose gel showing amplification of p53 and ISP6110. Amplifications were performed using marker 100 bp (M) and DNA from M. tuberculosis ATCC 27294 (P), various isolate (1-5), negative control, B. subtilis (N), positive control (PP).

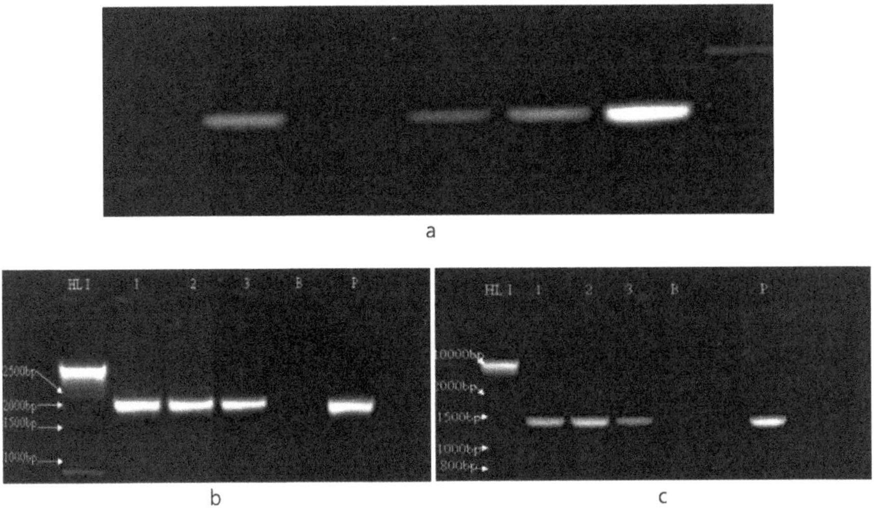

Figure 2. Amplification of rpoB gene (a), katG gene (b) and inhA gene (c) From left; Lane HL: Hyperladder maker, Lane 1, isolate 148; Lane 2, isolate 374; Lane 3, isolate 624; Lane B, negative control, Lane P, positive controlMutation analysis

PCR products amplified from regulatory-*inhA, rpoB* and *katG* genes of *M. tuberculosis* were sequenced with the same primer used in amplification of the genes. The sequences obtained were compared to the sequences available in Gen Bank Data using BlastN analysis. The DNA nucleotide sequences were translated to amino acid sequences through BlastX software available in http://www.ncbi.nlm.nih.gov. The nucleotide and amino acid changes of each isolates for respective genes were compared using ClustalW (Vector NTI). As for mutation analysis of *rpoB* genes, isolate 148 showed minor mutation where the additional, changes and transposition of certain base of amino acid Whereas, isolate 374 showed showed frameshift mutation in which total loss of A base that subsequently affects the rest of amino acid coding. It is known as.. Nevertheless, isolate 624 showed both minor and major mutation at different base position in which all amino acid code was totally changed.

4. Conclusion

Two isolates of 624 and 374 showed frameshift mutations in which all the amino acid codes were totally changed. the quantity of DNA and the present of inhibitor may affect the successful of amplification. In addition, only one specific site of *katG*, *rpoB* and *inhA* sequence in whole genome of *M. tuberculosis*.. As for conclusion, the identification of *M. tuberculosis* through amplification of *IS6110* was successfully achieved whereas amplification of restricted genes associated with drug resistance was achieved in culturable clinical specimens.

5. Acknowledgment

This research study was funded by Ministry of higher education of Malaysia UKM-NN-07-FRGS0224-2010 and UKM Research Grant University UKM-OUP-TKP-21-99/2009

6. References

[1] WHO, 2008. Global Tuberculosis Control 2008, Surveillance,Planning, Financing. *WHO Regional Office For the Western Pacific*

[2] Johnson R., Streicher E.M., Louw G.E., Warren R.M. 2008. Drug Resistance in *M. tuberculosis*. Curr. Issues Mol. Biol. 8: 97–112

[3] Sajduda, A., Brzostek A., Popławska M., Augustynowicz-Kopec E., Zwolska Z., Niemann S., Dziadek J., and Hillemann D. 2004. Molecular Characterization of Rifampin- and Isoniazid-Resistant *Mycobacterium tuberculosis* strains Isolated in Poland. *Journal of Clinical Microbiology*. 42(6): 2425–2431

[4] Suhaila, H., Rahizan, I., Nik Marzuki S. and Noraziah M. Zin, 2008. The Usefulness of PCR Amplification for Direct Detection of *M. tuberculosis* DNA from Clinical Samples. *Biotechnology* 7(1): 100-105

[5] Leven, M. and Goossens,H. 1997.Relevance of nucleic acid amplification techniques for diagnostis of respiratory transinfections in the clinical laboratory. *Clinical Microbiology Review*.10:242

[6] Wilson, M.L, Stone, B.L and Hildred, M.V. 1995. Comparison of recovery rates of mycobacteria from BACTEC 12B vials, Middlebrook 7H11 selective biplates and Lowenstein-Jensen slants in a public health mycobacteriology laboratory. Journal of Clinical Microbiology 33: 2516-2518

43

Assessment of Phylogenetic Inter-Relationships in Mud Crab Genus Scylla (Portunidae) Based on Mitochondrial DNA Sequence

Darlina Md. Naim[1,*], Hurul Adila-Aida Mohamad Rosly[2] and Siti Azizah Mohd. Nor[1,2]

[1] School of Biological Sciences, Universiti Sains Malaysia, Penang, Malaysia

[2] Centre for Marine and Coastal Studies, Universiti Sains Malaysia, Penang, Malaysia

Abstract

Mud crab of the genus *Scylla* possessed almost similar characteristics despite the established key identification which contributed to a slight confusion when inspecting them at a morphological level. Therefore, this phylogenetic study may give an insight to differentiate between species at a molecular level. This study examined 520 base pairs (bp) of the mitochondrial cytochrome *c* oxidase I (COI) gene from 60 individuals belonging to the genus *Scylla*, a group of mud crabs that inhabit over vast geographic areas ranging from south eastern and eastern Africa to South-east Asia and Indo-Pacific regions. The samples were taken from various locations throughout peninsular Malaysia, Sabah and Sarawak. The nucleotide sequences were subjected to phylogenetic analyses by using the neighbour – joining (NJ) and maximum parsimony (MP) methods. All two methods revealed the reciprocally monophyletic relationship between *Scylla olivacea*, *S. paramamosain*, *S. tranquebarica* and *S. serrata*. The clustering of all four species of *Scylla* samples into four different clades suggested that their genetic identity belongs to their respective species, thus strongly supporting their status as different species available throughout Malaysia. Overall, the present study was able to be a reference on the phylogenetic relationships and assessment of genetic structure of *Scylla* sp. in Malaysia.

Keywords: mud crab, Scylla, phylogenetics, COI

1. Introduction

Taxonomy is the foundation of traditional conservation practices [1, 2] and understanding the taxonomic details of a species is central to the development of successful management strategies for sustainable fisheries resources. Mitochondrial DNA (mtDNA) has been one of the most widely used molecular markers for phylogenetic and taxonomic studies in animals [3] due to its maternal mode of inheritance and mainly non-recombining nature [4]. Though mtDNA sequence data have proved valuable in determining phylogenetic relationships, there are considerable differences in the characteristics of different types of gene and it is crucial that the choice of gene is appropriate to the problem being tackled [5]. Cytochrome oxidase subunit I (COI) is the largest of the three mitochondria-encoded Cytochrome Oxidase subunits, and is one of the largest protein-coding genes in the metazoan mitochondrial genome which has been used as a target gene for a number of molecular phylogenetic and identification studies [6].

Mud crab of the genus Scylla belongs to the Portunidae family. They are also known as swimming crabs which characterized by their broad paddles of the flattened fifth pair of legs. The discoverer recognized the existence of only one species, Scylla serrata (Forskål, 1775), until Estampador's [7] revision made clear that there are more than one species exists. Numerous convincing arguments for the recognition of the species involved were made to minimize the confusion of the taxonomic nomenclature [8] until genetic studies have come up to support and correct the findings [9].

Several genetic methods using allozyme electrophoresis, mitochondrial genes and nuclear genes [10] havebeen carried out in attempt to justify species in this genus and to date, there are four distinct species wererecognized namely Scylla serrata, S. paramamosain, S. tranquebarica and S. olivacea [10]. However, the relationships among these species are still not clear. Therefore, it is imperative that data collected from COI gene sequence be employed for defining the phylogenetic relationships among species within the Scylla genus. In this current study, the sequence of COI gene has been employed to reappraise the taxonomic status and unravel the phylogenetic relationships among species within genus Scylla.

2. Materials and Methods

2.1. Sample collection

A total of 50 individuals were caught from eight different mangrove swamps across Malaysia, two within the east coast Peninsular (Terengganu, $n = 10$; Kelantan, $n = 11$), three within the west coast of Peninsular Malaysia (Kuala Perlis, $n = 5$; Perak, $n = 5$; Penang, $n = 5$), one within the south Peninsular (Johor, $n = 4$) and the other two within Sabah and Sarawak (Sarawak, $n = 6$; Sabah, $n = 4$) [Figure 1]. Mud crabs were caught using five to seven crab pots baited with fish laid at suitable mangrove area (spaced approximately 100 m apart) and lifted and re-baited every 24 hours for three successive days. More details on sampling method were described in Rosly et al. (in press). From each specimen, approximately 5-10 g muscle tissue was removed from a single claw and placed in 95% alcohol.

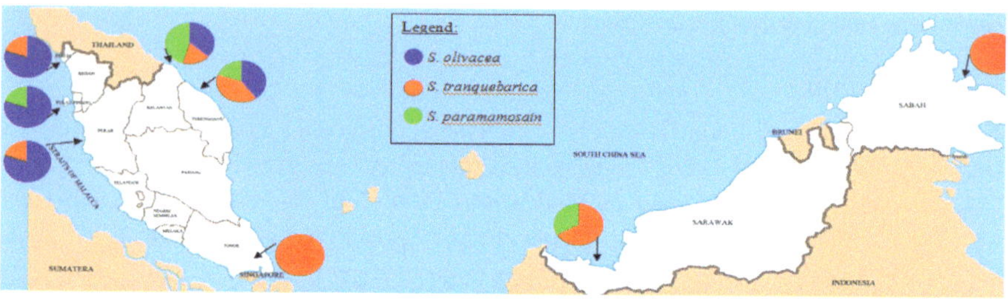

Figure 1. Sampling sites and species distribution in Malaysia. Each colour represents each species as indicated in the legend and the size of fractions corresponds to the frequency of samples obtained.

2.2. Total DNA extraction, polymerase chain reaction (PCR) and sequencing

Total genomic DNA from muscle tissue was extracted following the protocols from AquaGenomic Solution Kit (BioSyntech, USA). Polymerase chain reaction (PCR) was used to amplify the target region of the COI gene in the mtDNA genome of all sampled individuals. Heavy strand primer Mtd-10 5′- T TGA TTT TTT GGT CAT CCA GAA GT - 3′ [11] and light strand primer C/N 2769 5′- TT AAG TCC TAG AAA ATG TTG RGG GA - 3′ [16] were used to amplify a 542 bp region of the COI region. Each PCR amplification was performed in a total volume of 20 μl of PCR mixture consisting of 10x PCR buffer, 2.5 mM dNTP mixture, 5U of *i-Taq* DNA polymerase, 25mM $MgCl_2$, 0.5 pmol of each forward and reverse primer and 1.6 μl (20 ng) of DNA from each sample. Thermal cycling conditions (on a C STORM; Gene Technologies Ltd.) were 35 x [94°C for 30 s, 50°C for 30 s, 72°C for 1 min, 94°C for 3 min] and a final incubation at 72°C for 5 min.PCR products were then purified and sent to First Base Laboratories Sdn Bhd (1st BASE) for sequencing.

2.3. Alignment and sequence properties

The program MEGA ver. 4.0 [12] was used to visualize and align all sequences, including COI partial sequences of *S. serrata* obtained from Genbank (Genbank accession no: GU055497.1 – GU055506.1). The resulting sequences for each individual were then aligned using CLUSTAL W ver. 1.6 with default settings and were manually checked and trimmed in the BIOEDIT ver 7.0.9 sequence editing program; alignments were subsequently revised by eye in an effort to maximize positional homology. All aligned sequences were then imported into Basic Local Alignment Search Tool (BLAST) software to ensure the identity of the samples.

2.4. Phylogenetic analysis

The sequence characteristics of the COI region were calculated using MEGA ver. 4.0 [12]. The phylogenetic relationships among haplotypes were reconstructed using maximum parsimony (MP) and neighbour joining (NJ; [13]). We conducted MP and NJ phylogenetic analysis in MEGA ver. 4.0 [12]. Neighbour joining analysis was executed using non-parametric re-sampling procedure with 10,000 replicates. We performed MP analysis under the HKY+I model using the heuristic search algorithm with tree bisection reconnection (TBR) branch swapping and 100 random sequence addition replicates. Both trees reconstructions included 5 sequences of indigenous portunid from Genbank; *Portunus pelagicus* (Genebank accession no: GQ272560.1 – GQ272564.1). The genetic distance, G_{st} between populations was calculated using the Tamura-Nei distance [14] based on unequal base frequencies and unequal ratios of transitions to transversions (Ti:Tv) implemented in MEGA ver. 4.0 [12]. The frequency of each haplotype, haplotype diversity (*h*) (i.e., the probability that two randomly selected haplotypes are present in the sample) and the nucleotide diversity (π) within populations and geographical regions was estimated using ARLEQUIN ver. 3.11 [15].

3. Results and Discussions

Overall 50 partial sequences (see Supplementary Material for Genbank accession number for each sequence analysed) of 520 base pairs (bp) each of the mtDNA COI gene correspond to 50 individuals representing three *Scylla* species (*S. olivacea, S. paramamosain* and *S. tranquebarica*) and 10 COI partial sequences of *S. serrata* obtained from Genbank (Genbank accession no: GU055497.1 – GU055506.1) were analyzed. All the sequences show high sequence similarity (99-100%) to their respective species sequence in Genbank, indicate that misidentification of species does not occur in this study. The 520 bp *Scylla* sequences comprised 109 variable sites, of which 102 are parsimony informative sites (Table 1). This high number of parsimony-informative sites indicates that COI mtDNA is capable to be an informative locus candidate for phylogenetic studies [16] as well as for population differentiation and population genetic structure (Rosly et al., in press).

Taxa	n	Nhap	G+C (%)	Number of sites				Ti	Tv
				Variable	Conserved	Parsim-Info	Multiple substitution		
S. olivacea	20	16	42.44	20	500	11	21	19	2
S. tranquebarica	20	4	19.40	6	514	5	6	5	1
S. paramamosain	10	6	27.04	6	514	0	6	6	0
S. serrata	10	10	24.16	16	504	0	16	14	2

Table 1. Summary of number of haplotypes, nucleotide diversity (π) and haplotype diversity (h) with standard deviation (SD), nucleotide composition, transition, transversion and multiple substitution dataset for each species implemented in ARLEQUIN ver. 3.11 software [25]. The number of sites was generated by using MEGA ver. 4 [17]. n = sample size; Ti = transition; Tv = transversion.

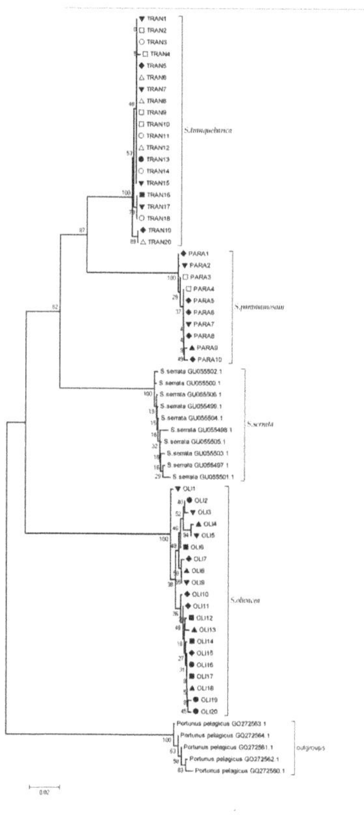

Figure 2. Neighbour-joining (NJ) phylogram showing the relationships among cytochrome c oxidase I (COI) sequences of Scylla species and 5 outgroups of genus Portunus analyzed in the present study. The number at each node represents the bootstrap percentage value based on 10000 pseudoreplications for the neighbor-joining (NJ) analysis generated using MEGA ver. 4.0 [17] software.

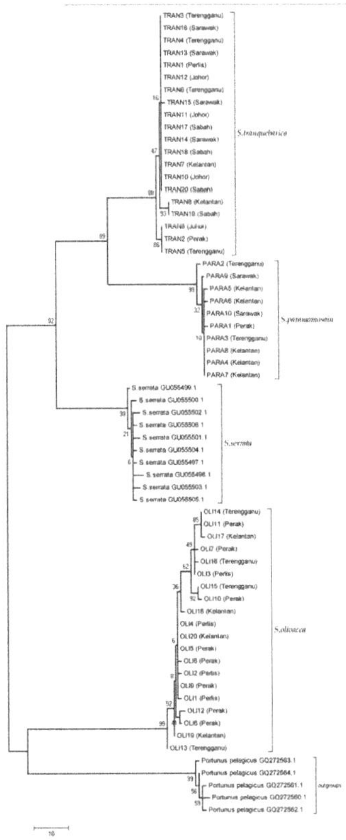

Figure 3. Maximum Parsimony (MP) phylogram showing the relationships among cytochrome c oxidase I (COI) sequences of Scylla species and 5 outgroups of genus Portunus analyzed in the present study. The number at each node represents the bootstrap percentage value based on 1000 replicates and applies the close-neighbour interchange (CNI) using the initial tree by random addition of 100 replicates for the maximum parsimony (MP) analysis generated using MEGA ver. 4.0 [17] software.

This study has highlighted the monophyly pattern of mud crab genus *Scylla* as evident by the consistent of four major clades in both phylogenetic trees (Figures 2 and 3). All samples were clustered according to their respective species; *S. tranquebarica, S. paramamosain, S. serrata* and *S. olivacea* with both trees showed that *S. tranquebarica* branched the earliest from all the species analysed in this study (Figures 2 and 3).The generic relationships revealed by our neighbour-joining (NJ) and maximum parsimony (MP) tree are highly in agreement with the previous molecular phylogeny study of *Scylla* based on COI sequence data [10]. However, our current results provide new insights into the classifications of the genus *Scylla* and presents additional evident regarding their inter-relationships that were not included and discussed in previous studies.

The construction of phylogenetic trees using NJ and MP analyses produced identical tree topologies with strong levels of support for all nodes (Figures 2 and 3). Additionally, clear relationships among mud crab species was observed in both trees and supports the placement of outgroup taxa, *Portunus pelagicus* at the base of both trees (Figures 2 and 3). The relative efficiencies of the NJ and MP in obtaining the correct topology for phylogenetic inference were studied by computer simulation. The NJ method gives a correct topology even when the distance measures used are not unbiased estimators of nucleotide substitutions, while for the MP method, both the weighted and unweighted parsimony are generally less efficient than the NJ method even in the case where the MP method gives a consistent tree. However, the NJ and MP analysis in our study returned a similar tree topology (see Figures 1 and 2), as demonstrated by several other studies at different taxa. Thus, the same tree topology (although different bootstrap values) that demonstrated by both of the phylogenetic trees has highly support the inter-taxa relationship among mud crabs genus *Scylla* sampled from Malaysia.

The relatively deep gene trees clearly show a large divergence between mud crab genus *Scylla* analysed in this study (Figures 1 and 2). This was further corroborated by the pairwise genetic distance among the four species of *Scylla* which ranges from *Gst* = 0.086-0.198 (mean±SD; 0.146±0.261) for COI gene (Table 2). Other studies, particularly in marine fishes also show a large genetic distance between species. The intraspecific and interspecific differentiation in various crustacean taxa has been reported to be ranges at *Gst* = 0.00-1.47 and *Gst* = 5.2-31.6 respectively. Accepting this threshold as valid, the genetic distances between four species of mud crab in our study falls within the range of interspecific differentiation of crustacean taxa, thus further supported the study by Fuseya and Watanabe [8] which reported that there exist at least three distinct species within the genus *Scylla* (see e.g. [10, 12]).

Taxa	S. olivacea	S. tranquebarica	S. paramamosain	S. serrata
S. olivacea				
S. tranquebarica	0.165			
S. paramamosain	0.198	0.086		
S. serrata	0.181	0.114	0.131	

Table 2. Pairwise Tamura-Nei genetic distances (Gst) for COI gene sequence among Scylla.

In conclusion, in this study we were able to provide useful insights into phylogenetic relationships and the genetic identity of *Scylla* species from mangrove areas in Malaysia. However, further studies using larger samples from other areas of its geographical distribution, sequence data from other mtDNA regions, and information based on nuclear DNA markers are required to support our findings.

4. Acknowledgements

We thank the Fisheries Research Institute of Penang for their assistance in sample collection. This project was funded by Universiti Sains Malaysia through research grant 304/PBIOLOGI/6311003.

5. References

[1] Avise, J.C., 1989. A role for molecular genetics in the recognition and conservation of endangered species. Trend in Ecology and Evolution, 4: p. 279-281.

[2] Brophy, T.R., 2004. Geographic variation and systematic in the South-east Asian turtles of the genus (Testudines: Bataguridae). Faculty Pulications and Presentations. Paper 58.

[3] Avise, J.C., 2004. Molecular markers, natural history, and evolution. 2nd Edition, Sinauer, Sunderland, MA.

[4] Avise, J.C., 1994. Molecular Markers, Natural History and Evolution. Chapman and Hall, Inc., New York.

[5] Sunnucks, P., 2000. Efficient genetic markers for population biology. Trend in Ecology and Evolution 15: p. 199-203

[6] Hebert, P.D.N., A, Cywinska, S.L. Ball, et al. 2003. Biological identifications through DNA barcodes. Proceding of Royal Society London Series B, 270: p. 313-321.

[7] Estampador, E.P., 1949. Studies on Scylla (Crustacea: Portunidae). I. A revision of the genus. Philippine Journal of Science 78: p. 95–108

[8] Fuseya, R, S. Watanabe, 1996. Genetic variability in the mud crab genus *Scylla* (Brachyura: Portunidae). Fisheries Science 62(5): p. 705-709

[9] Keenan, C. P., D. L. Mann, S. Lavery, P. Davie, 1995. Genetic relationship, morphological identification and taxonomy of mangrove crabs, genus *Scylla*, from throughout the Indo-Pacific. ACIAR Project Report, QDPI, Brisbane.

[10] Keenan, C.P., 1999. Aquaculture of the Mud Crab, Genus Scylla – Past, present and future. In: Keenan CP, Blackshaw A (eds) Mud crab aquaculture and biology. Proceedings of an international scientific forum held in Darwin, Australia, 21-24 April 1997. ACIAR Proceedings No. 78, Ferguson and Co., Brisbane Australia, p. 9-13

[11] Tamura, K, J. Dudley, M. Nei, S. Kumar, 2007. MEGA4: Molecular Evolutionary Genetics Analysis (MEGA) software version 4.0. Molecular Biology and Evolution, 24: p. 1596-1599.

[12] Thompson, J.D., D.G. Higgins, T.J. Gibson, 1994. "CLUSTAL W: improving the sensitivity of progressive multiple sequence alignment through sequence weighting, position-specific gap penalties and weight matrix choice". Nucleic Acids Research, 22: p. 4673-4680

[13] Hall, T.A., 1999. BioEdit: A user-friendly biological sequence alignment editor and analysis program for Windows 95/98/NT. Nucleic Acids Symposium Ser 41: p. 95-98

[14] Tamura, K., M. Nei, 1993. Estimation of the number of nucleotide substitutions in the control region of mitochondrial DNA in humans and chimpanzees. Molecular Biology and Evolution, 10: p. 512-526

[15] Excoffier, L., G. Laval, S. Schneider, 2005. Arlequin ver. 3.0: An integrated software package for population genetics data analysis. Evolutionary and Bioinformatics Online, 1: p. 47-50

[16] Kamarudin, K.R., R.M. Aisyah, H. U. Ridzwan, U. Gires, A. Hajar Fauzan, A. Mohd Hanafi, I. Mohd Yaman, 2011. Molecular phylogeny of *Holothuria (Mertensiothuria) leucopilota* (Brandt 1835) as inferred from cytochrome C oxidase I mitochondrial DNA gene sequences. Sains Malaysiana, 40: p. 125-133.

Influence of Congruent-Incongruent Teaching and Learning Style on Agricultural and Natural Resources Student Performance

Farzad Eskandari

Department of Agricultural Economics, Faculty of Agriculture, Kurdistan University, Sanandaj, Iran

Abstract

Based on Kolb's theory of learning and teaching a Post-test Group Design was applied to investigate the effect of congruent/ incongruent teaching and learning styles on students' performance at the faculties of Agriculture and Natural Resources at Kurdistan University, Iran in 2007. Using SPSS software descriptive statistics, descriptive statistics and T- test was employed to analyze the data. Results showed that: 1- the dominant learning styles among male and female students are Accommodator and Assimilator, respectively; 2- the main teaching style among teachers is Assimilator which corresponds to female students learning style; 3- Congruency between teachers' learning style and students' learning style has a positive significant effect on students' performance. The result could be used to improve the quality of teaching and learning at the faculties of Agriculture and Natural Resources at the University of Kurdistan, Iran.

Keywords: Higher education in agriculture, Learning-style, Teaching-style, Experiential learning, Student performance.

1. Introduction

An extensive amount of literature is available pertaining to learning styles and teaching styles and whether the match or mismatch between these two affects student performance. Of those studies which measured the effects of teaching approach on student achievement, mixed results have been reported. While a paucity of research exists in education which identifies a significant relationship between a teacher's teaching style and a student learning style and student performance [See for example 19, 16, 4, 10, 9, 3, 2 &8], there is some other that rejects this hypothesis [See for example, 7 & 18].

Although the influence of match and mismatch between teaching and learning style on student performance has received deep consideration in different countries and different parts of higher education system, the situation in Iran especially in Higher Agricultural Education System is not satisfying. Literature shows that most of the teachers in Iranian Higher Agricultural Education System (IHAES) pay less attention, if any, to their student learning style when defining their strategies to teaching. This is while most research shows a positive relationship between congruent teaching- learning styles and student achievement. Many IHAES teachers assume if their students do not show a satisfying performance on exams, it is because of variables related to students not because of quality and style of teaching.

As with agricultural education the researcher could not find any practical research testing the relationship between learning- teaching style and student performance. Currently, research on relation between learning and teaching style and student performance in Iranian colleges of agriculture is very limited. If teachers in Iranian colleges of agriculture are to recognize and appreciate differences in students' learning style and meet the challenge set forth by Anderson and Adams [1], an expansion of this research area is essential.As a result, the purposes of this study were to identify the teaching styles of agricultural education instructors and the learning styles of their students in specific field areas, to determine if a match existed between the two, and to determine if relationships existed between student success and style match or mismatch.

Research questions studied were as follows:

1. What are the teaching style profiles of the agricultural education instructors as measured by the Kolb's Teaching Styles Inventory?
2. What are the learning style profilesof students in specified agricultural classes as measured by the Kolb's Learning Styles Inventory?
3. What is the percentage of match and mismatch of teaching styles and learning styles across classes of agricultural education instructors?
4. What is the relationship between students' success as indicated by final exam scores and a match between their learning styles and the teachers' teaching style?

2. Methods

To achieve the purpose of the study a post-test group quasi-experimental design was employed. The subjects of the study were 23 agricultural teachers, 2000 students and 23 courses from faculty of agriculture at the University of Kurdistan, Iran.

All teachers at the faculty were studied. But with the student population, stratified and systematic sampling techniques were applied resulting in selecting 300 students from different areas of study at the faculty. For each teacher one course was selected.

The Kolb's Learning Styles Inventory (1999) and the Kolb's Teaching StylesInventory (1999) were used to obtain a profile of the teaching styles of the instructorsand the learning styles of their students.

Descriptive statistics and T-test were employed to analyze the data.

3. Result

Data related to learning style of the student sample is presented in table 1.

variable	frequency	percentage
Learning style of the total sample		
Accommodator	105	35
Assimilator	80	26.7
Convergent	59	19.7
Divergent	56	18.6
Learning style of male students		
Accommodator	70	51.1
Assimilator	18	13.1
Convergent	33	24.1
Divergent	16	11.7
Learning style of female students		
Accommodator	35	21.5
Assimilator	62	38
Convergent	26	16
Divergent	40	24.5

Table 1. Learning style of the sample

According to the data in table 1, the dominant learning styles among male and female students are Accommodator and Assimilator, respectively. In other words, male students at the faculty of agriculture prefer to learn through real experience and by doing while females have a preference for learning through working on abstract ideas and theoretical issues.

Profile of teaching style preferences of teachers is shown in table 2.

Teaching style	Frequency	Percentage
Assimilator	9	39.1
Convergent	6	26.1
Divergent	6	26.1
Accommodator	2	8.7

Table 2. Teaching style of teachers

As data in table 2 shows, the dominant teaching style among teachers is Assimilator (39.1 percent of the participants) which is more congruent with learning styles of female students.

Descriptive statistics for congruent / incongruent teaching-learning style at the studied faculty is demonstrated in table 3.

	N (%)	Average final exam score	S.D.
congruent	119 (39.7)	15.70	2.72
incongruent	181 (60.3)	14.65	2.83

Table 3. Descriptive statistics for congruent/ incongruent teaching-learning style

Table 3 shows that for most students (60.3%) the teaching style of the teacher does not match their learning style. In other words for almost two third of students teachers at the faculty do not consider the learning style of them when planning for teaching strategies. Table also shows that the congruent group has a higher average final exam score (15.70) compared to incongruent group. To test whether this difference in final exam score is accidental or because of the match and mismatch between teachers teaching style and student learning style, T-test was applied.

Result considering whether matching or mismatching between teaching style of teacher and learning style of student influences student performance is exhibited in table 4.

T- student	d.f	Level of significance
2.54	298	0.001

Table 4. T-test result for congruent / incongruent teaching-learning style and student performance

Based on the data exhibited in table 4, congruency between teachers' learning style and students' learning style has a positive significant effect on students' performance. So, one can conclude that the higher average final exam score gained by congruent student group is due to the match between teachers teaching styles and students learning styles.

4. Conclusions and Implications

It can be concluded from this study that a positive relationship exists between congruent teaching- learning styles and student performance hence confirming the theory established by researchers that style match will produce higher performance by students as measured by final exam scores and course grades [14, 6, 11& 15]. The findings are congruent with research by Cano et al.[5]; Welborn, [19]; Cafferty, [2]; Daniel etal, [8]; Witkin et al.[20]; and Koppleman [12] that found a match between teacher's teaching style and student's learning style will cause in a more higher significant performance by learners.

Iranian agricultural educators can use the outcomes of this study to assess the importance that their teaching styles may have to the learning of their students. This study shows that student learning will be improved when the instructor's teaching style and the students' learning style match [13 &10]. Henson and Borthwick contended that "assessing learning styles provides today's instructors with a new direction to take toward developing a more personalized form of instruction.

In this article we have discussed the significance of matching teaching and learning styles in IHAES and provided some empirical evidence to indicate that IHAES students exhibit distinctive learning style characteristics. To understand and respect individual's diverse learning styles, We suggest that teachers employ instruments to identify students' learning styles and provide diverse instructional strategies to address their differences, and that teachers plan lessons to match students' learning styles while at the same time encouraging students to diversify their learning style preferences. By doing this we can assist our students in becoming more effective learners

Like other research this study had some limitations. This study was limited to only one faculty at one university in Kurdistan province, Iran. This study should bereplicated in other higher agricultural education institutions across the country with a larger population to gain a better picture of the relationship between congruent-incongruent learning and teaching style and student performance as well as to compare withthe findings of this study and previous research.

Also, this study did not focus on extraneous variables. A similar study should be conducted which incorporates student characteristics, social variables, socioeconomic levels, and gender to determine if these variables significantly affect style match and student success.

5. References

[1] Anderson, J.A., & Adams, M. 1992. Acknowledging the learning styles of diverse student populations: Implications for instructional design. In L. L. Border & N. Van Note Chism (Eds.), New Directions for Teaching and Learning, (pp. 19-33). San Francisco: Jossey-Bass Publishers, Inc.

[2] Cafferty, J. 1980. Learning styles.Chicago, IL: National Association of Nursing Education. ERIC document, Service number: ED 2495533.

[3] Campbell, B. W. 1989. A study of the relationship between teachers' and students' learning styles and students' achievement in business communications. Unpublished doctoral dissertation, University of Illinois at Urbana-Champaign.

[4] Canfield, A. A., & Canfield, J. S. 1988. Canfield Instructional Styles Inventory (ISI) Manual. Los Angeles, CA: Western Psychological Services.

[5] Cano, J. And Garton, B. L. 1994. The Relationship Between Agriculture PreserviceTeachers' Learning Styles and Performance in a Methods of Teaching Agriculture Course. Journal of Agricultural Education. 35(2), PP. 6-10.

[6] Carthey, J. H. 1993. Relationships between learning styles and academic achievement and brain hemispheric dominance and academic performance in business and accounting courses. Unpublished master's thesis, Winona State University, Minnesota.

[7] Cupkie, L. 1980. The effects of similarity of instructor preferred teaching style and student preferred learning style on student achievement in selected courses in a metropolitan community college. Unpublished doctoral dissertation, university of Missouri-Kansas city.

[8] Danile, A., Rasmussenm C., Jackson, J. and Brenner, D. 1984. Cognitive style as a predictor of achievement: multivariate analysis. ERIC document, NO. ED248217.

[9] Dyer, J. E. and Osborne, E.W. 1996. Effects of teaching approach on achievement of agricultural education students with varying learning style. Journal of Agricultural Education, 37(4), PP. 38-45.

[10] Henson, K. T., &Borthwick, P. 1984. Matching styles: A historical look. Theory Into Practice, 23(1), 3-9

[11] Kaplan, E. J., &Kies, D. A. 1993. Together: Teaching styles and learning styles improving college instruction. College Student Journal, 27(4), 509-13.

[12] Koppleman, K. 1980. Relationship of cognitive style to teaching style. (Paper presented to the Midwest Educational Research Association). Toledo, OH. (ERIC Document Reproduction Service No. Ed 194 609)

[13] Ladd, P. D. 1993. The learning and teaching styles of Tennessee secondary business education teachers. Unpublished doctoral dissertation, The University of Tennessee.

[14] Matthews, D. B. 1995. An investigation of the learning styles of students at selected postsecondary and secondary institutions in South Carolina. Research Bulletin, 60, 1-151.

[15] Miglietti, C. L. 1994. The relationship of teaching styles, expectations of classroom environments, and learning styles of adult students at a two-year college. Unpublished doctoral dissertation, Bowling Green State University, Ohio.

[16] Miller, S. L. 1990. Effects of matching learning styles of students and instructors on students' perceptions of effective teaching behaviors in the clinical setting. Unpublished master's thesis, Bellarmine College, Louisville, KY.

[17] Simon, J. R. 1987. The preferred learning styles of two-year college students and the preferred teaching styles of two-year college instructors. Doctoral dissertation, Ohio University.

[18] Spoon, C. J. 1996. Impacts of student learning style and instructor teaching style on achievement of adult basic skills students. Doctoral Dissertation, GeorgiaStateUniversity.

[19] Welborn, R. 1986. Influences of achievement among nontraditional adult health professional students. Paper presented at the "Teaching to Potential" Regional Conference on University Teaching. (ERIC Document Reproduction Service No. ED 279 258).

[20] Witkin, H. A., Moore, C. A., Goodenough, D. R. & Cox, P. W. 1977: Field-dependent and field-independent cognitive styles and their educational implications. Review of Educational Research, 47(1), l-64.

Study on Pollen Germination and Pollen Tube Growth of Five Iranian Apricot Cultivars on in Vitro Condition

Reza Kamrani

Islamic Azad University, Bam Branch, Iran

Abstract

This research was conducted as factorial experiment based on CRD in three replication with two factors including: cultivars in 5 levels and counting time after culture in three levels (10, 24 and 48 hours). The aims was evaluation of pollen germination and pollen tube growth. Therefore, pollen grain was collected from unopened flowers of Iran apricot cultivars including : jahangiri, azghandi, royal, shahrood 49, shahrood 18. Pollens was stored in desiccators and cultured in a medium including %15 sucrose, 100 ppm boric acid and %0.5 agar. The result was showed wide range of percentage of germination and pollen tube growth between cultivars. highest and least pollen tube length were observed in shahrood 49 (1287μ m) and royal (812 μ m). there was not any significant relationship between germination percentage and pollen tube growth. The range of germination was % 20-54. and 24 hours was enough for obtain maximum germination.

Keywords: apricot, iran cultivars, pollen germination, pollen tube growth, in vitro

1. Introduction

Apricot is one of important fruit at world. and is very important in Iran too. and this tree are capable for irregular yielding and is limit for ecological adaptation and can use for researchers. Irregular yielding has some reasons and can omit proper varieties. Self incompatibility is an important factor And reduce yielding which proceed from lack of compatible pollen. (11, 24). Incompatibility report at apricot different varieties at all world (1,4,11,19). It is one of important big difficults. limiting at adaptation is other difficult at Apricot produce. every cultivar need to special condition (2, 9, 24). Apricot need to cold winter and warm summer (2, 15) ovule longevity, anther Reception, pollen tube growth in pistil, and temperature are important at apricot produce (3, 5, 8). pollen tube growth and germination speed is very important. at low speed, yielding reduce (8). Health in pollen is important too (13). For successful Zygosis, 20- 30 pollen is necessary (10). in vitro condition is best ways for this studies. for this experiment, pollen collection from flower anther Flower must be at the balloon stage. in turkey cultivars, sucrose reports: 10-15 % (18). At Euguslavia varieties reports :agar 1%, sucrose 15%, germination rate 14.9-88.7%. (6). and in armenia varieties report 54.9 – 72.65 % (11). At some Iranian varieties germination percentage reported 33.2 – 51.42 % (12). At different varieties need to different temperature (17, 23). low and high temperature is Harmful for germination (8). proper temperature for pollen germination in apricot Suggest 10 – 20 (8, 20). Hajilo suggested temperature 15 -20 for pollen Germination in apricot (12). Negation suggested sucrose 15 % agar 0.6 %and boric Acid 100 ppm (21). In other experiment use agar 1 – 6 %. Sucrose 10 – 20 %, Boric acid 0.1 % and temperature 15 – 20 %centigrade degree was considered. After (10 -24 and 48)hours measure germination percent and measure pollen tube Length after 48 hour. (8, 22, 25).

2. Material and methods

This research was done at seed and plant improvement research institute of karaj in iran. All of cultivars are Iran native. Trees were six years old. The Root stocks of this trees are wild apricot. Temperature average in this city is 13.8 Centigrade degree. The Maximum of annual temperature is 26 centigrade degree on july and the minimum of annual temperature is -12 centigrade degree on anuary. The highest Temperature was 42 centigrade degree from thirty years old and The lowest temperature was -12 centigrade degree from thirty years old. Annual rain average is 260 illimeters. The lowest annual rain is 100 millimeters On april. This research was conducted as factorial experiment based on CRD in three replication with two factors including : cultivars in 5 levels including : jahangiri, azghandi, royal, shahrood 49, shahrood 18. and counting time after culture in three levels (10, 24 and 48 hours). Data analysis with "mstatc". For this research in 5 Iran cultivars, collect pollen from flower anther and Put at Desicator and protect in refrigerator at 3-5 centigrade degree. For Pollen culture use Sterile Petridishes. Medium was included sucrose 15 %. agar 0.5 % acid boric 100 pm. and then pollens scatter on cultivation field And put in incubator at 20 centigrade degree and normal light. for estimation of pollen germination percentage at three time (10, 24, and 48) hours counted. Then get out petridishes from incubator and add some drops chloroform that Determined germination (8). at end by a microscope that connect to computer Take 15 photo from different view. and save in computer. at any view counted Total pollens and total germinated pollens that pollen tube length was equation or longer of pollen thickness. (21, 16). 45 counting use for estimate germination speed and percentage. pollens distributed low numbers in the same ways. High amount of pollen causes germination (14) and petridishes with pollen above 50 discount.

3. Summary and conclusion

Table of data analysis shows difference between apricot varieties. Germination percentage average at 5 apricot Varieties was 20 – 54 %. Pollen tube growth is different with each other.

S.V	DF	MS
variety	49.583	
Counting time	2	1.823
Time*variety	8	0.388
error	30	0.411
total	44	

Table 1.

S.V	DF	MS
Variety	4	59583.43
Error	8	11911.081
Total	14	

Table 2.

According to table :shortest pollen tube length belong to "royal" And longest pollen tube length belong to "shahrood 49" and at this experiment Pollen tube length was 812 – 1287 micro millimeter(18). Iran apricots are different with each other and can use at plant breeding and have incompatibility. yielding at incompatibility varieties need to pollination by a proper variety (21). suggestion this experiment carry out by other varieties such as turkey cultivars and other place. that's on aportunity to thanks dr. bouzary and dr. abdolbaghi and mr. tavoosi that they were my permanent guide in doing this reaserch.

4. References

[1] Anonymous. fao state database results.available on the www. fao. org. 2001

[2] Bahly. G. H. and L. F. Hough. apricots.i n:j.janic and j.n.moore.advencesin fruit breeding. purdue university press.west lefa yet Uindiana. V.S.A. 1979

[3] Burgos. L. and J. Egea. apricot embryo-sac development in relation to fruit set. J.Hort.sci. 1993

[4] Burgos. L. T. Berenguer and j.Egea.self and cross compatibility among apricot cultivars. hortscience. 1993

[5] Burgos. L. T. Berenguer and j.Egea.embryo sac development in pollination and non pollination flowers of two apricot cultivars. J. Hortscience. 1995

[6] Duric, B.pollen germination in some apricot varieties in vojvodina.CAB abstract 1991

[7] Egea, j. andL.Burgos.apricot breeding for quality and self compatibility. acta.hort.. 1999

[8] Egea,J.,burgos,L.Zoroa,N.andEgea,L..influences of temperature on the invitro germination of pollen of apricot. horticultural science. 1992

[9] Faust,M.physiology of temperate zon fruit trees.john wiley and sons,new York.USA. 1989

[10] Gerriero, R and Bartolini,s.flower biology in apricot :main aspects and problems. 1993

[11] Guleryuz,M.and I.Bolat.investigation on characteristics of apricot cultivars in Erzincan urkey. acta horticulture.1999

[12] Hajillo, j. and gerigourian.the effect of different temperature on apricot pollen germination. Iranian scientistics and technology journal. (2000)

[13] Javadi, t and arzani, k,the effect of protection duration on iranian olive pollen in vitro germination. seed and seedling journal. (1999)

[14] Kwack,b.h..the effect of calcium on pollen germination. Proceeding of the American society of horticulture science. 1965

[15] Layne, R.E.G.c.H.Bailey and L.F.Hough..apricots.in :.janick and j.n. Moore.fruit breeding. New York. 1996

[16] Loupassaki, M.Vasilakakis,M.and Androulakis, I. effect of pre-incubation humidity and temperature treatment on the on the in vitro germination of ovacado pollen grains.euphytica. 1997

[17] Luza, J. G.polito, V.S..staminate bloome date and temperature responses of pollen germination and tube growth in two walnut. American journal of botany. 1989

[18] Mahanoglu,G.Eti,s.Kaska,N.Gulcan, R. and Aksoy, U.correlations between pollen quality, pollen production and pollen tube growth of some early ripening apricot cultivars. acta horticulture. 1995

[19] Mclaren,G. F, j.A. fraser and J.E.Grant.. pollination of apricots. Orchardist of new Zealand. 1992

[20] Mellenthin,W.M.Wang, c.y. and wang S.Y.influence of temperature on pollen tube growth and initial fruit development in pear. hortscience.. 1992

[21] Nejatian, m. and ebadi, a. the studi of germination and apricot pollen tube growth in vitro. agriculture scientist journal.

[22] Obonovea, J. evaluation of physiological and morphological properties of pollen from selected apricot cultivars. CAB abstracts 1996.

[23] Polito, V. weinbaum, S.A. and Muraoka, T.T. Adaptive responses of walnut pollen germination to temperature during pollen development. journal of american society for horticulture science. 1991.

[24] Rodrigo, j. and M. herrero. evaluation of pollination as the cause of erractic fruit set in apricot. Hort science. 1996

[25] Suzuki, n.Wang, x. kataoka, I. and Inoue, H.. effect of temperature on flowering and pollen germination in Japanese apricot cv. nanko. journal japans society for horticultural science. 1993

Biological Control Potential of Parthenium Hysterophorus Against Fusarium Solani – A Cause of Fusarium Wilt in Potato

Shazia Shafique, Sobiya Shafique

Institute of Agricultural Sciences, University of the Punjab, Lahore, Pakistan

Abstract

Fusarium wilt, caused by *Fusarium solani,* is potential disease of potato in Pakistan. Recent studies have shown that fungal plant pathogens can be controlled by plant products as the plant materials are biodegradable, display structural diversity and complexity. Presently the pathogenic potential and biological control of *Fusarium solani* was studied by inoculating potato plants with three strains of *F. solani,* to screen the most virulent isolate among *F. solani* FCBP-016, *F. solani* FCBP-434 and *F. solani* FCBP-470. It was found through pathogenicity test that *F. solani* FCBP-434 was the most pathogenic among three *F. solani* strains with variation in genetic level. This disparity in genetic constitution might be a cause of its high pathogenicity. Further, antifungal bioassays were conducted to confirm mycotoxic potential of different plant parts i.e., root, shoot and leaf of *P. hysterophorus* against *F. solani* FCBP-434 using 1, 2, 3, and 4% concentrations of the aqueous, methanol and n-hexane extracts. Bioassays revealed that among the three solvents of *P. hysterophorus,* the growth of *F. solani* FCBP-434 was greatly inhibited at 1 & 2% conc. of aqueous and methanol extracts of leaf and stem while in case of n-hexane extract 3 & 4% conc. were proved more effective. Among root extracts, the higher concentrations of aqueous and n-hexane exhibited more promising results by causing reduction of 85 & 74%, respectively, whereas in methanol extract again lower conc. were more inhibitory.

Keywords: Fusarium wilt, Potato, Biocontrol, *P. hysterophorus,* Aqueous and organic solvents.

1. Introduction

Potato (*Solanum tuberosum*) is graded at third level among food after wheat and rice and ranks fifth in Pakistan for its total production [1]. Diseases are the most important factor for its low per acre production. Fusarium wilt is a fungal disease which can be caused by *Fusarium* species, particularly *F. solani.* Plant extracts and essential oils show antifungal activity against a wide range of fungi [2]. Therefore, the development of biopesticides has been focused as a viable pest control strategy in recent years. Among these, *Parthenium hysterophorus* is important. The allelopathic and antifungal potential of *P. hysterophorus* is due to release of phytotoxic substances such as caffeic, ferulic, vanillic, chlorogenic, *p*- coumaric and parthenin, *p*-hydroxybenzoic acids, ambrosin and coronopilin [3]. The antifungal activities of root and shoot extracts of two Asteraceous plant species viz. *P. hysterophorus* L and *Ageratum conyzoides* were determined against *Macrophomina phaseolina* (Tassi) Goid., the cause of charcoal rot disease of sunflower (*Helianthus annus* L.). A measured reduction in *M. phaseolina* biomass was evaluated due to aqueous extracts of different concentrations [4]. On the basis of efficacy of plant extracts, present study has been designed to find out the effective plant product for the management of *F. solani* by using *P. hysterophorus* and molecular characterization of *F. solani.*

2. Methodology

Pure cultures of *Fusarium solani* FCBP-016, *F. solani* FCBP-434 and *F. solani* FCBP-470 were taken from FCBP, Institute of Agricultural sciences, University of the Punjab, Lahore. They were maintained and subcultured on Malt extract agar medium at 4 °C monthly.

For pathogenecity test the protocol of Grogan et al. [5] was adopted. Pathogenicity test was performed by inoculating the conidial suspension @ 10-15 mL/plant containing 2×10^5 conidia mL^{-1} on 1-month-old potted potato plants and soil with three selected isolates of the fungus. Disease severity was calculated with the help of following formula.

$$\text{Disease severity} = \frac{\text{Affected area of a plant}}{\text{Total area of the plant}} \text{x } 100$$

Screening of the most pathogenic isolate was carried out on the basis of pathogenicity test. The most pathogenic species was isolated and subjected to further biocontrol assays.

The genomic DNA of three different isolates of *F. solani* was extracted by CTAB method [6] with some modifications.

Aqueous extract was primed according to Bajwa et al. [7]. While the method of Alkhail [8] was followed for preparation of extract in methanol and n-hexane. The lower conc. of 1, 2, 3 & 4% of aqueous, n-hexane and methanol extracts of leaf, stem and root were prepared by adding appropriate quantity of sterilized distilled water. To make methanol and n-hexane control, 2 mL of methanol and n-hexane were dissolved in sterilized distilled water to make final volume 100 mL, in respective flasks.

Extract bioassays were carried out in liquid medium according to Bajwa et al. [9]. Their dry weight yield was determined after 24 h oven drying at 60 °C and percentage inhibition in biomass production was calculated as:

$$\text{Biomass reduction (\%)} = \frac{\text{Biomass in Control - Biomass in extract treatment}}{\text{Biomass in Control}} \text{x } 100$$

All the data was subjected to analysis of variance (ANOVA) followed by Duncan's Multiple Range Test to delineate mean differences [10].

3. Results

3.1. Pathogenecity Test

The pathogenicity of three isolates of *F. solani* (FCBP-016, FCBP-434, FCBP-470) was assessed which enabled the reproduction of typical symptoms of the disease over the time-scale after 15 days of incubation at 25 °C. Five levels of pathogenicity were detected on potato plants (Fig. 1). *Fusarium solani* FCBP-434 was proved to be the most pathogenic. Rapid results were obtained with only two primers i.e. A-02 and B-05 and the dendrogram generated from each primer and cluster of primers by MINITAB are presented Fig. 2. The RAPD data obtained with 13 primers was evaluated to analyze the genetic parity or disparity among different genotypes. A dendrogram was constructed on the basis of genetic distances by UPGMA method and two main groups of cluster were identified in the homology tree: F. solani FCBP-016 and F. solani FCBP-470 on one side and F. solani FCBP-434 on the other side (Fig. 3). F. solani FCBP-016 and F. solani FCBP-470 are 100% similar to each other but they show only 44.34% similarity with F. solani FCBP-434. The findings of pathogenicity test and molecular analysis showed that F. solani FCBP-434 have high pathogenic potential so it was subjected to further biocontrol assays.

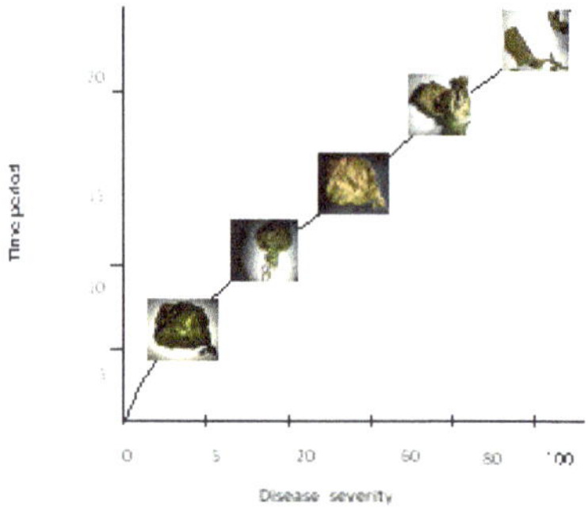

Fig 1. Periodic progression of Fusarium wilt of potato.

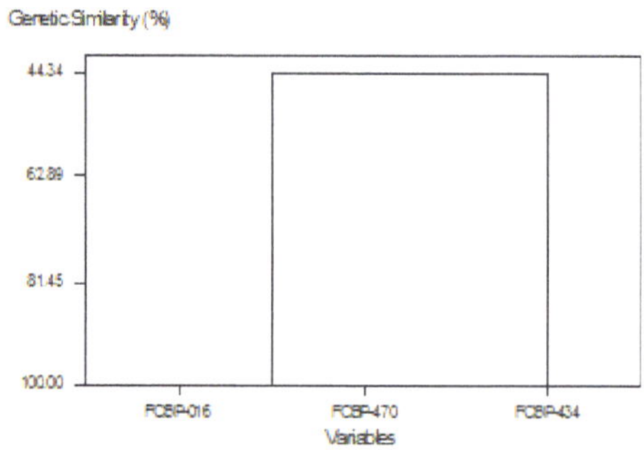

Fig 2. Homology tree constructed by UPGMA method for genetic similarity amongst genotypes of F. solani FCBP-016, 434 and 470.

3.2. Biological control through leaf extract

The data on dry biomass production in early growth phase of 7-10 days after incubation (DAI) revealed that all the conc. significantly decreased the fungal biomass as compared to control (Fig. 4). Amongst all concentrations 1% concentration exhibited the most promising results as it caused about 85% reduction in fungal biomass while 62, 52 and 41% decline in fungal biomass was depicted by 2, 3 and 4% aqueous extract, respectively. The response of *F. solani* in terms of dry biomass production was variable when grown in different concentrations of methanol extract of *P. hysterophorus*. Amongst these 2% concentration was the most effective in suppressing the biomass production up to 76% and 3% concentration was also found to decline fungal biomass production up to 67% In dry fungal biomass production with an increase in concentration of n-hexane extract revealed the maximum antifungal stress by 4% concentration causing a decline of about 66% in the biomass production of *F. solani*. It was followed by 3, 2 & 1% concentration which revealed a significant reduction, in the range of 56, 50 and 44%, respectively.

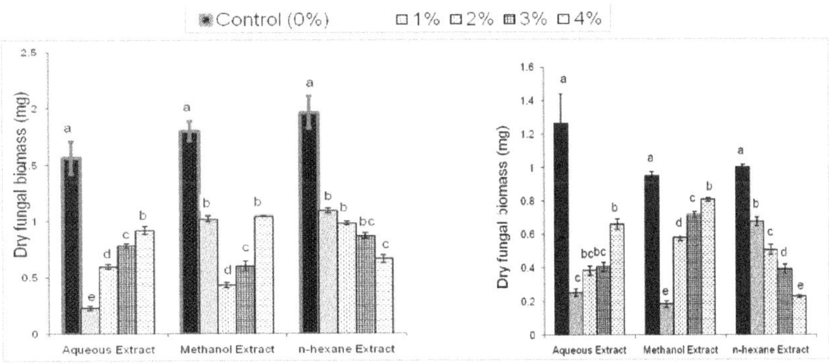

Fig 3. Effect of various conc.s of aqueous, methanol and n-hexane leaf extracts of P. hysterophorus on dry biomass production of F. solani FCBP-434.

Fig 4. Effect of various conc.s of aqueous, methanol and n-hexane stem extracts of P. hysterophorus on dry biomass production of F. solani FCBP-434.

3.3. Biological control through stem extract

All extract doses significantly inhibited the growth of the test pathogen in the same manner as depicted by aqueous leaf extract (Fig. 5). The greatest reduction in biomass production (70%) was observed in 1% extract. There was 48-70% reduction in fungal biomass production as noticed due to 1-4% concentrations of the extracts. All regimes of methanol extract of *P. hysterophorus* caused considerable inhibition in biomass production. The maximum inhibition in fungal growth was evidenced by 1% concentration which decreased biomass production upto 80%. While all the other concentrations (2-4%) caused about 15-38% reduction in biomass production, respectively. A variable response of dry biomass production in *F. solani* FCBP-434 was recorded to n-hexane extract of *P. hysterophorus* in different concentrations. *F. solani* FCBP-434 exhibited a significant reduction when exposed to different concentrations of extracts compared to control. The reduction in biomass ranged from 33 to 77%.

3.4. Biological control through root extract

All the doses of aqueous root extracts significantly retarded growth of the test fungus pathogen (Fig. 6). The fungal biomass suppressed at lower concentration (1-2%) in a range of 14-45%. While higher concentrations (3-4%) proved the most effective as these caused a reduction up to 67-85%, respectively. In methanol treatment, the lowest concentration (1%) was the most effective in reducing and suppressing the target fungal pathogen up to 77% as compared to higher concentrations. The fungal biomass increased at higher concentrations (2-4%) but still a significant inhibition in biomass production was detected with reference to control.A similar kind of suppressive effect of various concentrations was recorded against target fungal pathogen as exhibited by n-hexane leaf and stem extract. All the concentrations significantly reduced the fungal biomass production. Amongst these 4% concentration was the most effective in suppressing the biomass production up to 74%.

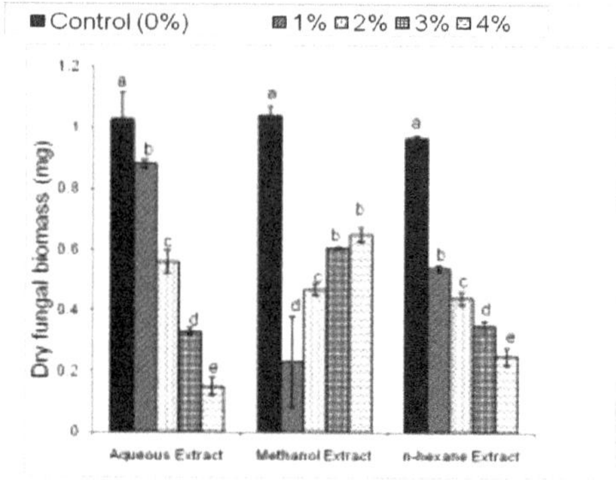

Fig 6. Effect of various conc.s of aqueous, methanol and n-hexane root extracts of P. hysterophorus on dry biomass production of F. solani FCBP-434.

4. Discussion

In the present study, pathogenicity test was carried out to assess the pathogenicity of three isolates of *Fusarium solani*. Among the isolates FCBP-434 proved to be more pathogenic and destructed the leaves on infected plants in 15 days after inoculation. Similar results were reported from various host plant - *F. oxysporum* combinations [11]. Molecular analysis of three isolates was also conducted. It was obvious from the results that *F. solani* FCBP-434 had genetic variability than other two isolates. In several studies RAPD fingerprinting technique has been employed to detect mutation, genetic relatedness and genetic variation within and between natural bacterial and human DNA and fungal populations [12, 13]. With the help of these two evidences about high pathogenic ability of *F. solani* FCBP-434, further its biological control assays were conducted.

It is obvious from the study that mycelial growth rate was significantly inhibited by antifungal compounds specifically at lower concentrations of aqueous leaf and stem extract. These results are supported from previous investigations in which leaf extracts of *Datura stramonium* have been shown to cause a decline in the development of rust pustules on wheat leaves [14]. Further increase in extract concentration exhibited significant difference in antimycotic activity as compared to 1% extract. In case of aqueous and organic root extract of *P. hysterophorus,* phytotoxins or antifungal compounds inhibited the biomass at higher concentration. The variation in antifungal activity of shoot and root extracts may be attributed to the different chemical nature of the compounds present in these parts [15]. Greater inhibition (42-76%) of growth of *F. solani* at 2-3% conc. of leaf methanolic extract was observed. While in case of stem and root methanolic extracts, lower conc (1-2%) were more effective as compared to higher conc. (3-4%). The pattern of gradually higher production of biomass in response to increasing conc. of aqueous extract was also similar to investigation of Bajwa et al. [16]. Comparative effectiveness of n-hexane extracts of selected test species revealed that higher concentrations were relatively more allelopathic thanlower concentrations. These results are supported by the work of Vir and Sharma [17] that employed 10% conc. of neem oil that exhibited 100% inhibition in *A. niger, D. rostrata* and *M. phaseolina.*

Thus it can be recommended that use of *P. hysterophorus* against *Fusarium solani* give better results as they are environmental safe alternatives and can be further exploited for formulating integrated disease management.

5. References

[1] N.J. Malik. Potatoes in Pakistan. World MatPrinters, 8A. Saharwwardy Avenue, Islamabad. 1995, Pp **3**: 27.

[2] P. Masoko, J. Picard, J.N. Eloff. The antifungal activity of twenty-four southern African *Combretum* species (Combretaceae). *South. Afr. J. Bot.* 2007, **73**:173-183.

[3] B.B. Jarvis, N.B. Pena, M.M. Rao, R.S. Comezoglu, T.F. Comezoglu, N.B. Mandava. Allelopathic agents for *Parthenium hysterophorus* and *Baccharis megapotamica*. In *The Chemistry of Allelopathy, biochemical interactions among the plants, American Chemical Society*. 1985, pp: 149–59.

[4] R. Bajwa, S. Shafique, S. Shafique. Evaluation of antifungal activity of aqueous extracts of two Asteraceous plants species. *Mycopath*, 2007a, **5**(1): 29-33.

[5] R.G. Grogan, K.A. Kimble, I. Misaghi. A stem canker disease of tomato caused by *Alternata alternata* f. sp. Lycopersici. *Phytopathol.* 1975, 65: 880-886.

[6] M.A. Saghai-Maroof, K.M. Soliman, R.A. Jorgensen, R.W. Allard. Fungal DNA Isolation. Proceedings of the National Academy of Sciences. 1984, **81**: 8014-8018.

[7] R. Bajwa, S. Shafique, S. Shafique. Appraisal of antifungal activity of *Aloe vera. Mycopath.* 2007, **5** (1): 5-9.

[8] A.A. Alkhail. Antifungal activity of some extracts against some plant pathogenic fungi. *Pak. J. Biol. Sci.* 2005, **8**: 413–417.

[9] R. Bajwa, T. Anjum, S. Shafique, S. Shafique. Evaluation of antifungal activity of *Cicer arietinum* L. *Pak. J. Bot.* 2006, **38**: 175-184.

[10] R.G.D. Steel, J.H. Torrie. *Principiles and Procedures of Statistics.* McGraw Hill Book Co., Inc, New York, USA. 1980.

[11] G.H. Kim, Hur Jae-Seoun, W. Choi, Y.J. Koh. Fusarium Wilt of Winter Daphne (*Daphne odora* Thunb.) caused by *Fusarium oxysporum. Plant Pathol. J.* 2005, **21**: 102-105.

[12] A.P. Keinath, M.W. Farnham, T.A. Zitter. Morphological, pathological, genetic differentiation of *Didymella bryoniae* and *Phoma spp.* isolated from cucurbits. *Phytopathol.* 1995, **85**: 364-369.

[13] C. Jones, A. Kortenkamp. RAPD library finger printing of bacterial and human DNA: applications in mutation detection. *Teratogen. Carcinogen. Mutagen.* 2000, **20** (2): 49-63.

[14] I. Hassan, M.A. Nasir, M.R. Haque. Effect of different plant extract on brown rust and yield of wheat. *J. Agri. Res.* 1992, **30**: 127–31.

[15] S. Afsharypuor, A. Mostajeran, R. Mokhtary. Variation of Scopolamine an Atropine in different parts of *Datura metel* during development. *Planta Med.* 1995, **61**: 383-384.

[16] R. Bajwa, S. Shafique, T. Anjum, S. Shafique. Antifungal activity of allelopathic plant extracts IV: Growth response of *Alternaria alternata, Fusarium monili-frome* and *Drechslera hawaiiensis* to aqueous extract of *Parthenium hysterophorus* L. *Int. J. Agri. Bio.* 2004, **6**: 511-516.

[17] D. Vir, R.K. Sharma. Evaluation of neem oil for control of plant pathogens. *Asian. Farm. Chem.* 1985, **1** (7-8): 23-24.

47

Antifungal Activity of Essential Oils Extracted From Clove, Cumin and Cinnamon Against Blue Mold Disease on Citrus Fruit

Tehmina Anjum* and Nosheen Akhtar

Institute of Agricultural Sciences, Quaid-e-Azam Campus, University of the Punjab, Lahore, Pakistan

* Corresponding author

Abstract

Essential oils obtained from Cumin seeds, Clove buds and Cinnamon bark was checked for their antifungal potentials against *Penicillium italicum,* causal agent of blue mold disease in citrus fruit. Selected essential oils were checked in different concentrations of 3, 6, 12, 24 and 48μl/ml for their ability to inhibit the mycelial growth of the test fungi. The *in vitro* study revealed that the essential oils of cumin and clove have the potential to inhibit mycelial growth of test fungi completely at concentrations of 12 and 48μl/ml, respectively. Essential oil of cinnamon, however failed to completely inhibit the mycelial growth even at maximum used concentration of 48μl/ml. *In vivo* assays also supported these results. Clove and cumin oils showed complete fungal inhibition at concentration of 24 and 48μl/ml, respectively when applied on citrus fruits. Whereas, cinnamon essential oil could not stop fungal infection even at its highest tested concentration. The study was extended to chemical identification of tested essential oils through GC-MS.

Keywords: essential oils, citrus, Penicillium itallicum, antifungal activity

1. Introduction

Fungi cause significant losses in almost all perishables due to post harvest rots. One of these fungi is *Penicillium italicum* Whemer (blue mold) that results in a universal post-harvest disease of almost all kinds of citrus fruit[1]. Chemicals imazalil, sodium ortho-phenyl phenate, and thiabendazole, have been widely used to control this problem [2]. However increase in public concern regarding contamination of perishables with fungicidal residues and proliferation of resistance in the pathogenic population [3] has forced the community to search for environment safe strategies. In the past few years, there has been a huge increase in the search of natural substances such as essential oils and plant extracts as potential antifungal agents [4]. The use of essential oils to control post-harvest fruit diseases have been deeply investigated and is well documented since the volatile compounds may have better applicability as fumigants for control [5]. In the present work, it has been therefore thought desirable to discover the antifungal potencies of essential oils extracted from three different plants against *P. italicum,* a dominant mycotoxin producing fungi during citrus storage.

2. Methodology

2.1. Test Fungi

The strain of *Penicillium italicum* used in this study was isolated from the decaying citrus fruit (variety: Mandarin). The fungal culture was maintained on Malt Extract Agar (MEA) medium at 4±1°C. A 7-14 days old culture of the isolate was used as the source of inoculum and for the preparation of spore suspension for various studies. The spores were removed from the surface of the culture, suspended in 5 ml of sterile distilled water containing 0.05% (v/v) Tween 80 and its concentration was adjusted to 106sp/ml using a haemocytometer for further *in vivo* studies.

2.2. Extraction of Essential Oils

Seeds of cumin (Cuminum cyminum L.), clove buds [Syzygium aromaticum (L.) Merrill & Perry], and cinnamon bark (Cinnamomum verum J. Presl) were imperiled to hydrodistillation in a modified Clevenger apparatus for 3-4 hours. Isolated EOs were stored in glass after dehydrating with anhydrous sodium sulphate and were kept in the refrigerator at 4±1oC before use. Different concentrations of plant essential oils (Eos) were prepared by adding 3, 6, 12, 24 and 48 μl of pure Eos in 1 ml of 0.05% (v/v) Tween 80 in case of in vivo and 0.5% (v/v) Tween 80 in case of in vitro experiments.

2.3. In Vitro Antifungal Assay by Agar Dilution Method

Prepared concentrations of EOs were mixed with sterile molten MEA medium. Thirty milliliters of media containing different concentrations of EOs and 0.5% Tween 80 was poured into each petri plate which was then inoculated with test fungi and incubated for 7 days at 25±1°C.

2.4. In Vitro Volatile Assay to Check Antifungal Activity

In this method the MEA plates were first inoculated with test fungi and then pure extracted EOs in quantities of 3, 6, 12, 24, and 48 μl were applied on the surface of the sterilized filter paper, which was placed in the lid of the petri plate. After inoculation the plates were incubated in inverted position for seven days at 25±1°C. The zone of inhibition was measured in two directions at right angles to each other. The percentage of mycelial growth inhibition by each essential oil concentration was calculated from the mean colony diameter (cm) on medium without essential oil amendment (control) and from the mean colony diameter (cm) on each essential oil amended plate (zone of growth).

2.5. In vivo antifungal assay

Fruits were sterilized with 6% sodium hypochlorite solution followed by immersion in sterile distilled water for two minutes and surface sterilization in 70% ethanol for another two minutes. Fruit was wounded (2-wounds per fruit) at the equatorial side with a sterile stainless steel scalpel where each wound was about 4 mm long and 2 mm deep. 15 μl of spore suspension was inoculated into each wound using a micropipette under aseptic conditions. Two hours later, each wound was inoculated with a pre-determined concentration from each plant essential oil. Control fruit was subjected to the same treatments except that sterile distilled water was used instead of essential oil. The treated fruits were labeled, placed in sterilized petri plates and incubated at 23±2 °C for two weeks to assess decay and fungal growth symptoms on daily basis.

2.6. Chemical identification of essential oils by GC-MS

Qualitative analysis of the tested essential oils was undertaken by gas chromatography-mass spectroscopy (GC-MS) using a Hewlett-Packard mass detector (model 7890) coupled with mass spectrometer selective detector 5975. Analysis was carried out using a column HP5 mass-selective detector (MSD) (30 m x 0.25 mm; 0.25 μm film thickness), the operating conditions were as follows: Helium was the carrier gas at a flow rate of 1 ml/min. diluted samples (1:100 v/v, in methanol) of 1 μl were injected manually at temperature 250°C. NIST (National Institute of Standards and Technologies) Mass Spectra Library was also used as a reference.

3. Results and Discussion

3.1. In vitro antifungal efficacy of essential oil

In vitro antifungal activity of selected oils were checked through two methods i.e., dilution method and the volatile method (Table 1). A little variation was observed in results of both the assays. In dilution method cumin oil gave best control on the mycelial growth of P. italicum as its lowest concentration of 3μl/ml gave 96% inhibition. This increased with increase in oil concentration and 12μl/ml completely inhibited the fungal growth. The results of cumin oil was followed by clove oil, whose lowest tested concentration i.e., 3μl/ml showed 84% mycelial growth inhibition. However, the complete inhibition of fungal growth was recorded at maximum tested concentration i.e., 48 μl/ml. Cinnamon gave the lowest effect among the three tested EOs.

In volatile method clove oil gave best control of P. italicum instead of cumin that showed highest inhibition of mycelial growth in dilution method perhaps because of high volatility of phenols that are abundantly present in clove essential oil. Clove oil in 3, 6 and 12μl/ml concentrations gave 97% control on tested fungi. Increase in oil concentration of clove oil gave similar results as were recorded when the same oil was checked through dilution method. Complete inhibition of mycelial growth was observed at maximum tested concentration. In case of cumin the lowest tested concentration of 3μl/ml could only inhibit *P. italicum* up to 32% in contrast to its significant control of 96% recorded in dilution method. Concentration increased the inhibition percentage up to 98-99%. Romagnoli et al.,[6] reported a strong antifungal activity against dermatophytes and phytopathogens including fungi and yeast. They also found cumin aldehyde, pinenes, and *p*-cymene, and a fraction of oxygenate compounds such as alcohol and epoxides as the most active ingredients of cumin essential oil. Cinnamon oil also depicted better control on tested fungi in volatile method when compared to the dilution method perhaps because it is also rich in volatile phenols live clove essential oil. However, still the inhibition effect of cinnamon essential oil failed to match with the controlling capacity of other two tested oils.

Essential Oils	Concentrations (μl/ml)	Average colony diameter of *P. italicum* (cm)		Inhibition of mycelial growth of *P. italicum* (%)	
		Dilution Method	Volatile Method	Dilution Method	Volatile Method
Control	0	6.666±1.7a	5.66±0.03a	0.0±0.0g	0.0±0.0g
Cumin	3	0.266±0.03de	3.83±0.09bc	96±8.3a	32.33±7.1e
	6	0.1±0.07e	1.76±0.03e	98.49±13a	68.9±9.4b
	12	0.0±0.0g	0.066±0.02g	100±0a	98.83±13a
	24	0.0±0.0g	0.1±0.07f	100±0a	98.23±7.5a
	48	0.0±0.0g	0.066±0.03g	100±0a	98.83±3.9a
Clove	3	1.066±0.05c	0.166±0.08f	84±6.2b	97.06±19a
	6	0.733±0.09d	0.166±0.05f	89±11b	97.06±3.7a
	12	0.333±0.04d	0.166±0.09f	95±8.6ab	97.06±6.7a
	24	0.066±0.03f	0.1±0.06f	99±14a	98.23±9.1a
	48	0.0±0.0g	0.0±0.0h	100±0a	100±0.0a
Cinnamon	3	6.666±.74a	4.83±0.59b	0.0±0.0g	14.66±1.5f
	6	6.333±1.04a	3.33±0.37cd	4.995±0.03f	41.16±9.5cd
	12	6.166±0.07ab	3.5±0.83c	7.5±0.14e	38.16±6.2d
	24	5.833±0.43b	3.66±0.51bc	12.496±0.37d	35.33±6.8de
	48	4.5±0.21bc	3.166±0.09d	32.49±2.9c	44.06±5.7c

* Values are means (n=3). Mean values followed by different letters within the column are significantly different according to Duncan Multiple Range Test (P<0.05).

Table 1. In vitro antifungal activity of plant essential oils, used at various concentrations, on mycelial growth of Penicillium italicum*

3.2. In Vivo Antifungal Efficacy of selected Essential Oils

The results of in vivo antifungal efficacy indicated that all the three EOs had a good inhibitory effect on mycelial growth of P. italicum when tested on the surface of the citrus fruit. Yahyazadeh et al.,[7] revealed that essential oil can result in loss of pigmentation in fungal conidia as they became hyaline that may affect virulence of the pathogen; hence a decrease in the incidence of the infection.

Figure 1A, shows that clove essential oil exhibited most pronounced antifungal potentials against P. italicum as it completely inhibited the mycelial growth at concentration of 24μl/ml after 15 days of incubation. At concentration of 3μl/ml the inhibition of P. italicum was 62% and growth started at 4th day of inoculation. Increase in concentration to 6 and 12μl/ml increased fungal inhibition up to 79 and 93% respectively, and also the fungal growth on citrus fruit was delayed till 7th and 11th day.

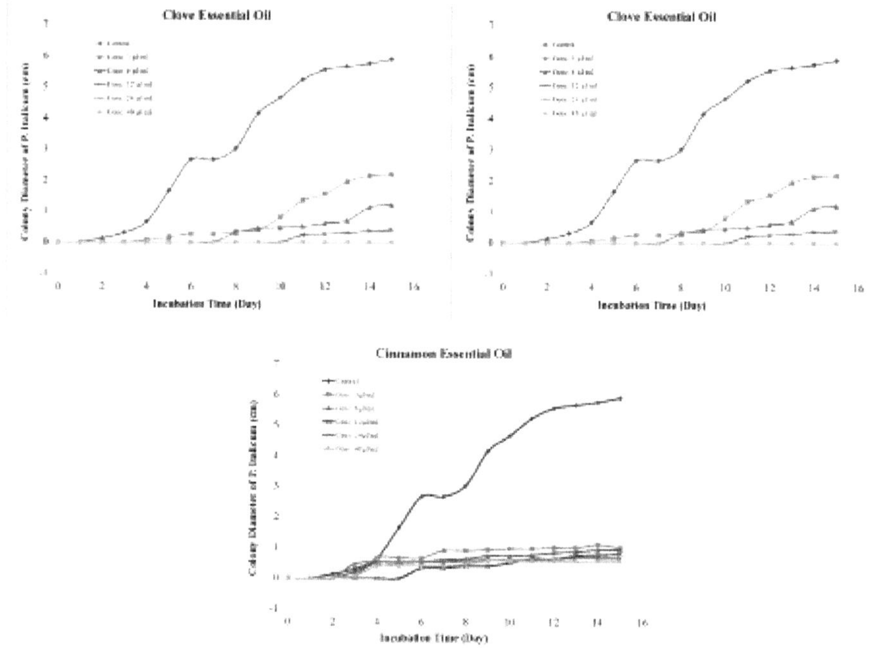

Fig 1. *In vivo* antifungal activity of **A:** Clove; **B:** Cumin; **C:** Cinnamon essential oils, used at various concentrations, on mycelial growth of *Penicillium italicum*

Cumin oil showed complete inhibitory effect at concentration of 48µl/ml while cinnamon oil did not completely inhibit fungal progression even at its highest concentration when applied on citrus fruit. The minimal inhibitory effect of cumin and cinnamon essential oils at lowest tested concentration (3µl/ml) was 88.27 and 83.05 % respectively. Whereas, no significant difference was recorded in inhibitory effects caused by both the cumin and cinnamon essential oils used in higher concentrations of 6 – 48µl/ml.

The data presented in Fig. 1B indicated that cumin oil in concentrations of 3 and 6µl/ml resulted in the inhibition of 88.71 and 89.27% respectively and growth started from the 5th day of the inoculation. However the concentrations of 12 and 24µl/ml delayed the mycelial growth in citrus fruit till 6th and 7th day. While the data presented in Fig. 1C shows that the mycelial development of tested fungus was started from 3rd day of inoculation at all tested cinnamon oil concentrations (3-48µl/ml) except the concentration of 24µl/ml that delayed the growth up to 5 days. These results depicted that the efficacy of clove essential oil during in vivo assays followed by that of cumin, whereas cinnamon oil showed least inhibitory effect.

In general, the results obtained from GC-MS analysis of the essential oils used were in accordance to the previous literature. Clove oil shows the presence of eugenol, alpha-terpineol, Iso-eugenol and beta-terpinene as its major components. Eugenol has been reported by different workers to be the most effective component of the clove and cinnamon EOs against various pathogens[8]. Vazquez et al.,[9] reported complete inhibition of P. citrinum by 2000 ppm of eugenol in a liquid medium.

The major components found in cinnamon oil were eugenol and cinnamaldehyde, whereas cumin oil revealed the presence of gamma-terpinen, cuminaldehyde and 4-carvomenthenol. Singh and Upadhayay[10] showed antifungal activity of Cuminaldehyde against Aspergillus flavus and Aspergillus niger. In a recent study Romagnoli et al.,[6] found cumin aldehyde, pinenes, and p-cymene, and a fraction of oxygenate compounds such as alcohol and epoxides as the most active ingredients of cumin essential oil.

4. References

[1] D. Prusky, J.L. McEvovy, R. Saftner, W.S. Conway, and R. Jones. Relationship between host acidification and virulence of *Penicillium* spp. on apple and citrus fruit. *Biotech. Cell. Biol.* 2004. 94: 44-51.

[2] Eckert, J. W.; Eaks, I. L. Postharvest disorders and diseases of citrus fruits. In The Citrus Industry; Reuter, W., Calavan, E. C., Carman, G. E., Eds.; University of California Press:Berkeley, CA, 1989; pp 179– 260.

[3] P. Tripathi, and N.K. Dubey, Exploitation of natural products as an alternative strategy to control postharvest fungal rotting of fruit and vegetables. *Post-harvest Biol. Technol.* 2004. 32: 235-245.

[4] V.K. Bajpai, T.J. Lee, and S.C. Kang. Chemical composition and in vitro control of agricultural plat pathogens by the essential oil and various extracts of *Nandina domestica* Thunb. *J. Sci. Food Agric.* 2009. 89: 109-116.

[5] J. Kaloustian, J. Chevalier, C. Mikail, M. Martino, L. Abou and M.F. Vergnes. Etude de six huiles essentielles: composition chinique et activite antibacterienne. *Phytotherapie*. 2008. 6: 60-164.

[6] C. Romagnoli, E. Andreotti, S. Maietti, R. Mahendra and D. Mares. Antifungal activity of essential oil from fruits of Indian *Cuminum cyminum*. *Pharm. Biol.* 2010. 48(7): 834-838.

[7] M. Yahyazadeh, R. Omidbaigi, R. Zare, and H. Taheri. Effect of some essential oils on mycelial growth of *Penicillium digitatum* Sacc. *J. Microbiol. Biotechnol.* 2008. 24: 1445-1450.

[8] R. Montes-Belmont, and M. Carvajal. Control of *Aspergillus flavus* in maize with plant essential oils and their components. *J. Food Protect.* 1998. 61: 616-619.

[9] B.I. Vazquez, C. Fente, C.M. Franco, M.J. Vazquez, and A. Cepeda. Inhibitory effects of eugenol and thymol on *Penicillium citrinum* strains in culture media and cheese. *Int. J. Food Microbiol.* 2001. 67: 157- 163.

[10] G. Singh, and R.K. Upadhyay. Fungitoxic activity of cuminaldehyde, main constituent of *Cuminum cyminum* oil. *Fitoterapia*. 1991. 62: 86.

Clinical, Serological, Hormonal, Bacteriological and Molecular Detection of Brucellosis in Aborted Cows and Buffalos

Jabbar A.A. AL-Sa'aidi, Mohsen A. Al-Rodh, Ali Anok Najum

Dept. Physiol. & Pharmacol., College of Vet. Med., Al-Qadisiya Univ., Iraq

Dept. Vet. Med., College of Vet. Med., Al-Qadisiya Univ., Iraq

Dept. Microbiol., College of Science, Al-Muthana Univ., Iraq

Abstract

Abortion is the most obvious manifestation of *Brucella* infection. In this study, 59 aborted buffalos and 91 aborted cows were included. Diagnosis of *Brucella* infection in these abortions was based on clinical, serological, bacteriological, hormonal, and molecular assays. Serological studies included the use of RB and ELISA tests as screening tests for infection. Argumentative differences between RB and ELISA results have been shown. *Brucella* isolated and identified from aborted fetuses, vaginal discharge and milk samples were 7 isolates from aborted cows and 3 from aborted buffalos. *Brucella* isolates revealed amplification of a 223-bp fragment with B4 and B5 primers. Hormonal assessment in both, brucella infected cows and buffalos, registered significant decrease in progesterone and P/E ratio in comparison with that aborted due to other causes. Immunohistochemical study revealed down expression level of 3BHSD enzyme in placentas of *Brucella* positive animals. ELISA technique was the valuable serological test to confirm the diagnosis of brucellosis. In conclusion, both RB and ELISA are necessary to be performed together as screening tests in diagnosis of brucellosis, whereas serum hormonal, placental immunohistochemical, and molecular (PCR) assessments have an efficient diagnostic values which can be included for confirmation of brucellosis.

Keywords: *B. abortus*, *B. melitensis*, Brucellosis, ELISA, RB, PCR.

1. Introduction

Brucellosis, is a major infectious disease afflicting humans and a wide range of domesticated animals and wildlife. It is known to be a worldwide problem and one of the most important among zoonoses in the Mediterranean region, India, and Central and South America (1). Brucellosis results significant human morbidity (2). Reports from the areas where *Brucella melitensis* infection is endemic, suggest that there is an increased rate of abortion in asymptomatic pregnant women (3). Outbreaks of bovine brucellosis are associated with abortion during the last trimester of gestation, and produces weak newborn calves, and infertility in cows and bulls (4). The diagnostic method known to produce the best results in terms of specificity is the isolation of *Brucella* organisms from the suspected animal. However, this method has a limited sensitivity, is expensive and cumbersome and has the added difficulty of being unpractical to apply at a large scale in control campaigns. Accordingly, the indirect diagnosis of disease based on serological tests is of choice in the eradication programmes. The standard Rose Bengal (RB) is the main serological test used to detect antibodies against *B. abortus* and *B. melitensis* infections. This test has been used for several decades, proving to be successful for eradicating bovine brucellosis in some countries. Nevertheless, there is evidence that RB test is significantly less effective for the diagnosis of brucellosis in sheep and goats than in cattle (5). During recent years, different ELISAs have been developed using more or less purified S-LPS as the antigen and have been reported to be at least as sensitive and specific for the diagnosis of brucellosis in ruminants (6, 7, 8). The aims of the present work were: (i) to compare the diagnostic performance of serological: standard RB and ELISA, hormonal; in the blood and placenta, and molecular; PCR studies in cattle with known clinical and bacteriological status, (ii) to determine the diagnostic performance of these tests in aborted cattle positive in the classical RB test, but in which *B. abortus* and *B. melitensis* could not be isolated.

2. Materials and Methods

Materials: Bovine brucellosis kit was provided by EUROPEAN VETRINARY LABORATORY; EVL, Netherland. Gram's stain solution and modified Ziehl-Neelsen stain solution was prepared and used as described by Alton *et al* .,(9). Rose Bengal antigen was provided by Omega company, UK. Monospecific antiserum; antibrucella abortus and antibrucella melitensis were supplied by Difico, USA. It was provided as a gift from FAO. Antibiotic supplements were provided by Hi-media, India. Fetal calf serum was provided by Difico company, USA. Materials used in PCR; Wizard Genomic DNA purification kit GoTaq® Green Master Mix were provided by Promega company, USA. Primers used for diagnosis; the system used was B4/B5 (Baily) primers system; advised by Baily *et al*.,(10), and provided by Alpha DNA Company, as described below:

Primers name	Sequences	Predictive product
B4	5′-TGGCTCGGTTGCCAATATCAA-3′	223- bp region within a gene coding for 31-kDa membrane protein specific to the genus *Brucella*
B5	5′-CGCGCTTGCCTTTCAGGTCTG-3′	

Methods:

Samples collection.

1. Clinical observations: was performed on aborted cows and fetuses throughout the last few days of gestation period. At parturition, retained placenta, body temperature and bleeding status has been registered.

2. Blood samples: 150 blood samples (91 from cows and 59 from buffalos) were obtained. serum samples have been obtained for serological assessments using RB and ELISA tests.

3. Aborted fetuses: Specimens from 15 aborted fetuses, at the first and last stage of pregnancy (9 from aborted cows and 6 from aborted buffalos) were cultured in duplicated *Brucella* agar plates as described by Alton *et al* (9).

Serological assessments: were preformed according to OIE, (11).

Identification and isolation of *Brucella*: *Brucella* growth was confirmed by bacteriological and biochemical tests as suggested by Alton *et al.*, (12).

Molecular Identification by PCR-based assay:

A-DNA purification: was performed according to the manufacturer instructions.

B-The PCR amplification process: Enzymatic amplification of DNA was carried out in a final volume of 25 μl according to the recommendations of manufacture. The following reaction mixes were prepared on ice. For a 25μl reaction volume:

Component	Volume	Final Conc.
GoTaq® Green Master Mix, 2X	12.5μl	1X
upstream primer, 10μM	0.25–2.5μl	0.1–1.0μM
downstream primer, 10μM	0.25–2.5μl	0.1–1.0μM
DNA template	1–5μl	<250ng
Nuclease-Free Water to	25μl	N.A.

PCR consisted of a preheating at 95ºC for 5 min. After this initial denaturation step, the mixture was subjected to 40 amplification cycles as follow:

Loop's steps	Temperature	Time	Number of cycle
denaturation	94 ºC	1 min	40
annealing	55 ºC	1 min	
extension	72 ºC	1 min	
Final extension	72 ºC	7 min	1
Hold	4 ºC	Indefinite	1

C-Detection of PCR products by agarose electrophoresis: A horizontal slab gel electrophoresis apparatus was used. Ten μl of amplified products were mixed with 3 μl of loading buffer, analysed by electrophoresis in 2% agarose gel, and stained with 0.5 μg /ml ethidium bromide at 100 V for 1hr. in 1 X TBE buffer. Then visualized under UV light using ultraviolet transelumenater. DNA ladder (100-1000) was used and the gel was photographed when necessary by digital camera. A sample was considered positive for *Brucella* spp. when a specific fragment of 223 bp was detected in the gel (10).

Statistical Analysis: All data were analyzed using the statistical package for social science (SPSS) for Windows program on the computer. Chi-square was used to compare between the frequencies. Student *t* test was used to compare between means of groups. The significance was accepted as P value < 0.05 and <0.01.

3. Results

Clinical observation:

All positive cases for *Brucella* infection were found in late stage of pregnancy (5-9 month), some cases accompanied with abortion with retention of placenta, aborted buffalos were in herds not sporadically like cows (table 1). Pneumonia and pericarditis were the main complications shown in aborted fetuses positive for brucellosis. Placentas were, edematous and opaque with bleeding in some of cases.

Serological assessments:

ELISA, RB tests and clinical signs results are shown in tables (1 and 2) that RB test results were positive in 33 cases (22%) including 26 cases (28.5%) in cows and 7 cases (11.86%) in buffalo. ELISA results were positive in 38 cases (25.3%), including 21 cases (23%) in cows and 17 cases (28.8%) in buffalo. ELISA was positive in aborted cattle which had previous history of abortion, whereas RB test was positive in 2 cows that was negative for ELISA & bacterial isolation, also RB test was negative in 10 case of buffalo that aborted before several months, which was positive with ELISA test. According to the results of Pearson Chi-Square test, the difference in RB positive cases between groups was significant ($P<0.05$), whereas difference in RB negative cases between cows and buffalos was not significance ($P>0.05$). According to ELISA technique, statistical analysis showed insignificant difference ($P>0.05$) between groups.

No.	History & clinical signs	Pregnancy period	Samples & isolation of agent	R.B	ELISA
1	Aborted before 2hrs with edema. of placenta	7 Months	Fetal stomach (+)	++	0.362
2	Pneumonia of Fetus.	8 Months	Fetal stomach (+)	++	0.409
3	Aborted before 6hrs with opaque placenta edema.	6 Months	Fetal stomach (+)	++	0.469
4	Aborted before 12hrs.	9 Months	Uterine fluid (+)	+++	0.772
5	Aborted before 24hrs with retained placenta.	9 Months	Uterine fluid (-)	+++	0.666
6	Aborted before 24hrs with Opaque placenta	8 Months	Uterine fluid (+)	++++	1.037
7	Aborted before 5hrs with pneumonia of Fetus	7 Months	Fetal stomach & Fetal organ (+)	+++	1.00
8	Aborted before 48hrs	7 Months	Uterine fluid (+)	++++	1.08
9	Aborted before 72hrs	6 Months	Uterine fluid (-)	+++	0.846
10	Aborted before 14 d.	5 Months	Milk (-)	+++	1.00
11	Aborted before 23 days.	8 Months	Milk (+)	++	0.676
12	Abortion pneumonia of Fetus with opaque of placenta	9 Months	Fetal stomach & Fetal organ (+)	+++	0.650
13	Retained of placenta	9 Months	Fetal stomach (+)	++	0.570
14	Opaque of placenta	8 Months	Placenta &uterine fluid(-)	+++	0.840

No.	History & clinical signs	Pregnancy period	Samples & isolation of agent	R.B	ELISA
15	Aborted before 40 d with retained placenta	6 Months	Milk (-)	++	1.026
16	Aborted before 60 d. (Abortion also occurred in all herds before 5 months at late pregnancy)	9 Months	Milk (-)	_	1.016

Table 1. History, clinical signs, pregnancy period, Samples and isolation, RB and ELISA results of aborted cattle and their fetuses.

	NO	R.B			ELISA			Bacterial isolation		
		+	-	%	+	-	%	+	-	%
Cow	91	26	65	28.5	21	70	23	7	79	13.18
Buffalo	59	7	52	11.86	17	42	28.8	3	55	6.77

***Only one** case had ELISA titer 10.09 (suspected grey zone)

Table 2. Results of RB test, ELISA and bacterial isolation for aborted cows and buffalos.

Bacterial isolation & identification:

Brucella organisms first recognized in smears obtained from fetal stomach stained with modified Ziehl Nielsen stain, which appeared red clumps against blue background. *Brucella* culture recognized on the basis of colonial morphology which appeared round translucent pale honey. Routine bacteriological examination has been carried out for identify the genus *Brucella* before they submitted for *Brucella* typing tests any isolate that was differ in even one test was excluded from further consideration as member of the genus *Brucella*. So the obtained isolates were Gram-negative, coccobacilli, arranged singly, in short chain pairs, with small groups, negative for haemolysis on blood agar, and it neither grow nor perform lactose fermentation on MacConkey agar, positive for nitrate reduction oxidase, catalase and urease, negative for MR-VP, gelatinase, Citrate utilization and indol production. Out of 91 aborted cows, 7 (13.18%) were positive by culture, whereas out of 59 aborted buffalos, 3 (6.77%) were positive for culture (table 1 and 2). Out of 15 aborted fetuses from cows and buffalos 4 from aborted cow's fetuses and 2 from buffalo's aborted fetus were positive by culture. Out of 12 uterine fluid swabs 2 from aborted cows and 1 from aborted buffalo were positive by culture. Out of 122 milk samples positive 1 from aborted cows, was positive by culture. Uterine swabs and milk samples where firstly cultured on the selective media, so *Brucella* colonies were recognized first by colonial morphology then subcultured to obtain pure culture before submitted to the bacteriological and biochemical test. According to the results of Pearson Chi- Square tests, isolates from cows specimens were significantly higher ($P<0.05$) than that isolated from buffalos (figure 2).

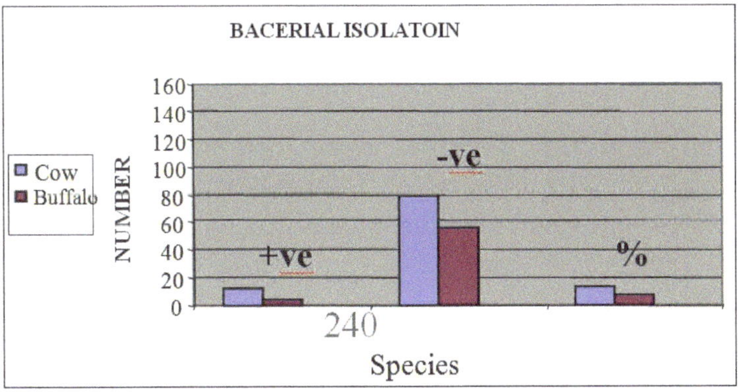

Fig 4. Illustrate the positive, negative and percentages of bacterial Isolation results for cattle.

Steroid hormones assessments:

Cows and buffalos positive for *Brucella* infection showed insignificant differences (P>0.05) of serum estradiol concentration compared with that of non infected (463.92±176.57 pg/ml for infected versus 589.66±235.67 pg/ml for non infected). Serum cortisol in infected animals revealed insignificant higher concentration (P>0.05) than non infected animals (78.36±12.14 nmol/l versus 65.48±13.38 nmol/l, respectively). on the other hand, serum progesterone concentration revealed significant decrease (P<0.05) in infected animals (0.495±0.13 ng/ml) compared with (18.468±6.26 ng/ml) in non infected group. This decrement of progesterone concentration reflected on the progesterone/ estradiol balance (P/E Ratio) which reach (0.107%) in infected group compared with (3.132%) in non infected group (figure 2).

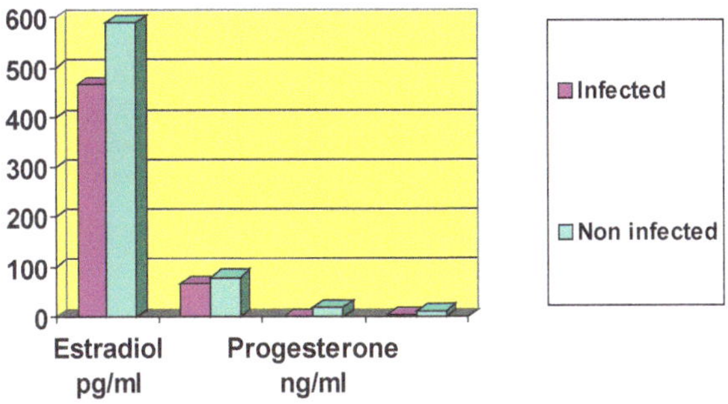

Fig 2. Mean ±SE of steroid hormones levels in aborted cattle.

Immunohistochemical assay:

Results of IHC analysis demonstrated positive staining for 3β-HSD in placentas of cattle, in which the localization was found in both chorionic villi and chorionic plate (fig. 3a and b). It has

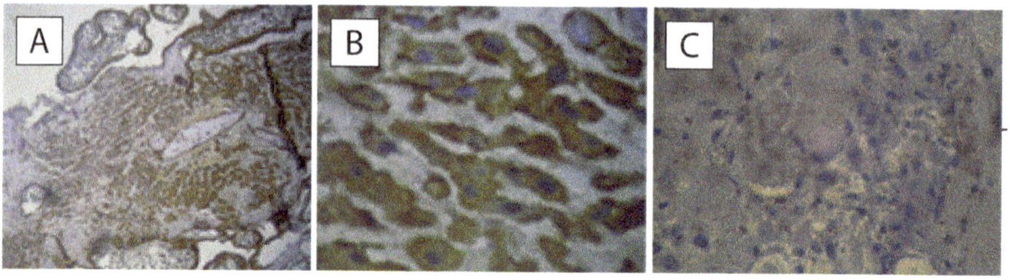

Figure 6. IHC staining results. A & B: cow placenta negative for brucellosis, specific staining of chorionic plate with DAB chromogen (brown) and counterstained with Hematoxylin (blue) (A:100x, B:400x). C: cow placenta positive for brucellosis; stained by DAB chromogen (brown) and counterstained with Hematoxylin (blue) notice, non IHC reaction for 3BHSD enzyme .100x.

Molecular detection of Brucella spp by PCR technique:

The primer pair used in this study succeeded in the amplification of a 223-bp fragment from *Brucella* isolates cultures that were studied, meanwhile, the DNA extracted from culture harboring *Brucella*'s DNA, so that they yielded predicted 223-bp fragment. All *Brucella* strains that studied with PCR have same 223-bp fragment. (figure 8).

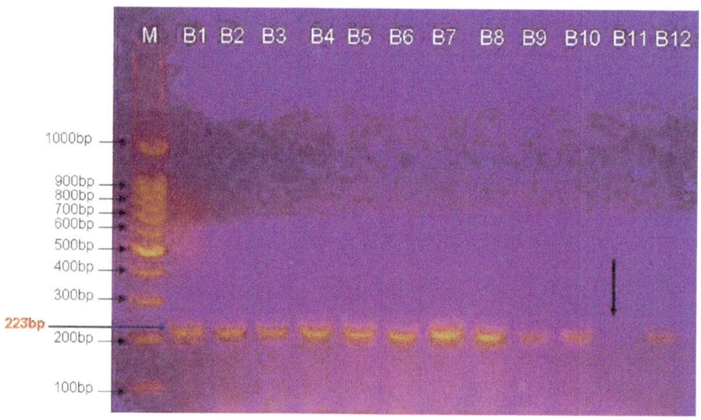

Figure 4-23. Agarose gel electrophoresis for PCR products of Brucella isolates, where 223bp PCR products appear as positive results.

4. Discussion

Clinical observation: Clinical findings observed in aborted cattle were in agreement with other researchers (13, 14). Abortion is a frequent complication of brucellosis in animals, where placental localization is believed to be associated with erythritol, a growth stimulant for *B. abortus*. Whether the rate of abortions from brucellosis exceeds rates associated with bacteremia from other bacterial causes is unclear. In any event, prompt diagnosis and treatment of brucellosis during pregnancy can be lifesaving for the fetus (1). It has been postulated that generalized suppression of adaptive immune response mainly occurred during pregnancy. This immune suppressed state prevent maternal rejection of the fetus but has unfortunate consequence of increasing maternal susceptibility to certain infectious agents (15).

Bacterial isolation and identification: *Brucella* strains isolated from cattle were obtained from aborted fetuses and vaginal discharge and were compared to that isolated from milk samples. The number of *Brucella* organisms in milk and colostrums samples was lower than that in abortion material, fetus stomach, fetal fluids and membranes, also milk samples is highly contaminated with other organisms. These results were in agreement with that mentioned by OIE (11). On the other hand, bacteriological and biochemical test of *Brucella* isolates were same as that advised by Alton *et al.* (12) and Qunin *et al.* (13).

Serological results: Serological examination performed by RB test in the present study gave positive result in some aborted animals but were negative for ELISA and bacterial isolation, also it gave negative result in buffalo aborted before 2 month or more. This diversity may attributed to the infection with microorganisms other than *brucella* spp. This infection are likely to cause cross-reactions in serological tests with smooth *Brucella* antigens and give false positive for RB. These results indicate that RB test is not confirmative test for diagnosis of brucellosis. Although RB test is known to have many false positive or negative results, but generally it is simple, rapid and can be used as screening method for infection (16). For confirmation of brucellosis, ELISA technique is the suitable and precise detectable test. It has been recorded that antibodies to *Brucella* appear in the serum within 1-2 weeks of infection. The initial response is the appearance of IgM isotype (which can be easily detected by RB) followed by a switch to IgG, after a while titers of both immunoglobulins classes increase. Distinct most of the usual serological tests, ELISA is effective in detecting all immunoglobulins (antibodies) classes and sub-classes important in diagnosis and appears to be the most sensitive serological test (17). IgM is produced soon after infection but declines quickly when production of IgG increases. IgM reacts non specifically in many serological tests and can cause high rates of false positive reactions. IgG1 is consistently produced at high levels in *Brucella*-exposed cattle sera and has a high affinity and specificity for *Brucella* antigens particularly the O-chain.

Steroid hormones interference: It has been registered, in the present study, that the results of reproductive hormones (E and P) concentrations in aborted cows and buffalos positive for brucellosis were in orchestration with that registered in immunohistochemical assays. Although estradiol concentration slightly decreased in brucella-infected aborted animals, but progesterone concentration sharply decreased. The important significant point was that related to the decrement of P/E ratio. On the other hand, immunohistochemical results revealed that 3β-HSD enzyme expression on trophoblast registered significant decrease, where 3β-HSD considered as main enzyme in progesterone biosynthesis that play important role in maintenance of pregnancy (18, 19, 20). Gorvel and Moreno (21) showed that in trophoblasts, *B. abortus* induces steroid biosynthesis and modulates the metabolism of prostaglandin precursors, favoring bacterial growth. These changes mimic to some extent what happens during parturition and are likely to contribute to the abortion.

Immunohistochemistery assay: The decrement in the expression of 3β-HSD enzyme in brucella infected animals may attribute to utilization of this enzyme by *Brucella* itself, which may results in the decrement of progesterone biosynthesis. Or due to infection of placenta with *Brucella* that may caused damage in placental tissue. This results were in agreement with that recorded by Samartino *et al.* (22). From the present findings, it can be demonstrated that *B.* abortus may traffics from a phagosome compartment towards the endoplasmic reticulum of the host cell; where the organism has an optimal environment for replication (23).

Brucella detection by PCR: All *Brucella* strains that were studied with PCR have the same 223-bp fragment. Results of PCR were the same as that obtained by Baily *et al.* (10) whom used PCR amplification contained a single pair of oligonucleotide primers designed to amplify a 223 bp product. Fekete *et al.* (24) used PCR in the diagnosis of brucellosis and described it as specific, sensitive and simple and could become a routine diagnostic test for brucellosis.

5. References

[1] WHO/OIE/FAO/CDS. (2006). Brucellosis in humans and animals. 7:1-66.

[2] Sauret, J.M., Vilissova, N., (2002). Human brucellosis review. The Journal of the American Board of Family Practice 15, 401–406.

[3] Boschiroli, M.L., Foulongne, V., O'Callaghan, D., (2001). Brucellosis: a worldwide zoonosis. Current Opinion in Microbiology 4, 58–64.

[4] Ashford, D. A., J. Lingappa, C. Woods, H. Noll, B. Neville, and B. A. Perrkins. (2004). Adverse events in human associated with accidental exposure to the livestock brucellosis vaccine RB51. Vaccine. 22: 3435-3439.

[5] Garin-Bastuji B., Blasco J.M., Grayon M., Verger J.M., Brucella melitensis infection in sheep: present and future, Vet. Res. 29 (1998) 255-274.

[6] Alonso-Urmeneta B., Marín C., Aragón V., Blasco J.M., Díaz R., Moriyón I., Evaluation of Lipopolysaccharides and Polysaccharides of different Epitope Structures in the Indirect Enzyme-Linked Immunosorbent Assay for Diagnosis of Brucellosis in Small Ruminants and Cattle, Clin. Diagn. Lab. Immunol. 5 (1998) 749-754.

[7] Cloeckaert, A.; Tibor, A. and Zygmunt. M. S.(1999). Brucella outer membrane lipoproteins share antigenic determinants with bacteria of the family Rhizobiaceae. Clin. Diagn. Lab. Immunol. 6:627-9.

[8] Marín C., Moreno E., Moriyón I., Diáz R., Blasco J.M., Performance of competitive and indirect enzyme-linked immunosorbent assays, gel immunoprecipitation with native hapten polysaccharide and standard serological tests in diagnosis of sheep brucellosis, Clin. Diagn. Lab. Immunol. 6 (1999) 269- 272.

[9] Alton, G.G.; Jones, L.M. and Pietz, D.E. (1975). Laboratory techniques in brucellosis 2nd . Ed. Geneva. Monograph. Series No. 55. ch. 2 p. 113.

[10] Baily, G.G.; Krahn, J.B.; Drasar, B.S. and Stoker, N.G. (1992). Detection of Brucella melitensis and Brucella abortus by DNA amplification. J. Tropical Med. Hyg. (95) No. 4 p. 271-275.

[11] OIE Manual of diagnostic tests and vaccines for terrestrial animals, 5th edition, (2009) part 2, section 2.3, chapter 2.3.1.

[12] Alton, G.G.; Jones, L.M.; Angus, R.D. and Verger, J.M. (1988). Techniques for the brucellosis laboratory. 1st ed. INRA Paris. Ch.2 p. 114.

[13] Quinn, P.J.; Carter, M.E.; Markey, B. and Carter, G.R. (2006). Clinical veterinary microbiology. 6th ed. Mosby an imp. Wolf, London. P: 261-267.

[14] Carvalho Neta, A.V., Steynen, A.P.R., Paixmo, T.A., Miranda, K.L., Silva, F.L., Roux, C.M., Tsolis, R.M., Everts, R.E., Lewin, H.A., Adams, L.G., Carvalho, O.A.F., Lage, A.P., Santos, R.L., (2008). Modulation of bovine trophoblastic innate immune response by Brucella abortus. Infection and Immunity 76, 1897–1907.

[15] Krishnan, L.; Guilbert, L.; Russell, A.S.; Wegmann, T.G. and Belosevic, M. (1996). Pregnancy impaire resistance C57Bl/6 mice to Lishmania major infection and cause decrease antigenic specific IFN-gamma response and increased production of T helper 2 cytokines. J.Immunol. 1156:644-651.

[16] Nicoletti, P. (1984). Diagnosis and Vaccination for the control of brucellosis in the Near East. FAO, Ani. Prod. and Hlth paper. Series No. 38, Rome.

[17] Young, E. J. (1995). An overview of human brucellosis. Clin. Infect. Dis. 21: 283-290.

[18] Enright, F.M., Walker, J.V., Jeffers, G., Deyoe, B.L., (1984). Cellular and humoral responses of Brucella abortus-infected bovine fetuses. American. Journal of Veterinary Research 45, 424–430.

[19] Enright, F.M., Samartino, L., (1994). Mechanisms of abortion in Brucella abortus infected cattle. In: Proceedings of the 98 Annual Meeting of the United States Animal Health Association. USDA, Richmon, Virginia, pp. 88–95.

[20] Anderson, T.D., Meador, V.P., Cheville, N.F., (1986b). Pathogenesis of placentitis in the goat inoculated with Brucella abortus. II. Ultrastructural studies. Veterinary Pathology 23, 227–239.

[21] Gorvel, J.P., Moreno, E.,(2002). Brucella intracellular life: from invasion to intracellular replication. Veterinary Microbiology 90, 281–297.

[22] Samartino, L.E., Traux, R.E., Enright, F., (1994). Invasion and replication of Brucella abortus in three different trophoblastic cell lines. Zentralblatt fur Veterinarmedizin B 41, 229–236.

[23] Pizarro-Cerda, J., Moreno, E., Gorvel, J.P., (2000). Invasion and intracellular trafficking of Brucella abortus in non phagocytic cells. Review Microbes and Infection 2, 829–835.

[24] Fekete, A., Bantle, J.A., Halling, S.M. (1990). Detection of Brucella by polymerase chain reaction in bovine fetal and maternal tissues. J. Vet. Diagnostic Invest., 4: 79-83.

Greenhouse Experiments of Symbiotic Effectiveness of Acid-Aluminium Tolerance Bradyrhizobium Japonicum Strains on Soybean Plant

Nisa Rachmania Mubarik [1,*], Hanum Habibah[2] and Aris Tri Wahyudi[1]

1 Department of Biology, Faculty of Mathematics and Natural Sciences, Bogor Agricultural University,

Jalan Agatis, IPB Dermaga, Bogor, Indonesia

2 Postgraduate School of Bogor Agricultural University, Bogor, Indonesia

[*] Corresponding author

Abstract

The aim of the research was to study the symbiotic effectiveness of seven strains of acid-aluminium tolerance *Bradyrhizobium japonicum* on soybean plant cultivar Slamet. The research conducted in the greenhouse and used complete randomized design with seven inoculation treatments, two controls and one reference strain which used the nutrient solution at pH 4.5. Each treatment had three replications. The all of parameters were measured at 37 days after planting (DAP). Result of the experiments showed that mutant Bj 11 (19) inoculated to soybean plant had the highest symbiotic effectiveness. The treatment of Bj 11 (19) could increase the dry weight of t upper crop (64,88%), N-uptake (190,88%), and symbiotic effectiveness (65,87%) better than treatments with and without nitrate control and the reference strain, USDA 110.

Keywords: soybean, acid-aluminium tolerance, Bradyrhizobium japonicum, symbiotic effectiveness.

1. Introduction

Availability of sufficient nitrogen is one of the keys to increase the productivity of soybean plants. Soybean plants generally take nitrogen from the air by root-nodule bacteria and then the bacteria convert nitrogen into ammonia that is needed for plant growth. *Bradyrhizobium japonicum* is one species of slow-growing nodule bacteria which is very important to uptake atmospheric nitrogen in soybean plant [1].

Acid soils usually cause problems in soybean production, such as consist of low phosphorus and high aluminium [2] that strongly inhibit the growth of symbiotic nitrogen fixation bacteria on soybean plants [3]. The failure of nodulation under acid soil conditions is common, especially in soils of pH less than 5.0 [4]. B. japonicum is more tolerant at low pH, around pH 4.0-4.5 than the fast growing nodule bacteria, such as Rhizobium [5]. Previous research by Endarini et al. [6] had managed to get acid aluminium tolerant B. japonicum from several locations in Indonesia [6]. The results showed that BJ 11 isolate has the highest tolerance on acid and had a good ability to grow on pH 4.5 media. Some of the strains showed more competitive than reference strain, USDA 110, in testing of the effectiveness symbiotic at greehouse. Furhermore, Wahyudi et al. [7] had constructed several strains of acid-aluminium tolerance B. japonicum with increased symbiotic effectiveness through transposon TN5 mutagenesis, such as Bj 11 (20) and KDR 15 (37). Some mutants showed the ability to form root nodules more than the wild-type strains viz. Bj 11 (5), Bj 11 (19), Bj 11 (20), and KDR 15 (37).

The efforts to obtain potential strains are still wide open in agricultural research. The purpose of the research was aimed to study the symbiotic effectiveness of seven strains of acid-aluminium tolerance Bradyrhizobium japonicum on soybean plant cultivar Slamet.

2. Materials and Methods

2.1. Materials

Acid tolerant isolates *B. japonicum* were used in the study viz. Bj 11 (wt), KDR 15 (wt), Bj 13 (wt), Bj 11 (5), Bj 11 (19), Bj 11(20), KDR 15 (37), and USDA 110. All bacteria were collected at IPB Culture Collection, Department of Biology, Faculty of Mathematics and Natural Sciences, Bogor Agricultural University. Soybean seeds cultivar Slamet were obtained from Research Institute for Food Crops and Genetic Resources, Bogor, Indonesia.

2.2. Methods

Experimental Design. All the data collected in greenhouse were analyzed using complete randomized block design using Statistical Analysis System (SAS) and the means at p (0.05) level of significance. The experiment was arranged into seven inoculation treatments, two controls and one standard strain which used the nutrient solution at pH 4,5. Each treatments were made in three replications. Growth parameters such as height of plant, dry-weight of upper crop, dry-weight of nodules, number of nodules, N uptake of plants, nitrogenase activity of root nodule, and symbiotic effectiveness were determined.

Medium and Inoculants Preparation. *Bradyrhizobium japonicum* isolates were grown on Yeast Mannitol Agar (YMA) for 7-8 days at room temperature. The YMA media consist of mannitol (10 g/L), K_2HPO_4 (0.5 g/L), $MgSO_4 \cdot 7H_2O$ (0.2 g/L), NaCl (0.2 g/L), yeast extract (0.5 g/L), added with 0.0025% congo red and rifampicin (50 µg/ml). The isolates were resistant to rifampicin [7] & [8]. Then they were subcultured into Yeast Mannitol Broth (YMB) and incubated for seven days with 125 rpm at room temperature.

Soybean Seed Inoculation. Soybean seeds were selected based on size and healthiness (able to shoot). Seed surface were sterilized using 95% alcohol for 10 seconds and 5% H_2O_2 for five minutes then rinsed seven times using sterilized water. The seeds are germinated in a petri dish at room temperature without light. Two days old sprouts grown in Leonard jar, pots filled with sand [7] which filled with N-free nutrient solution of pH 4.5 [9]. Each sprout was inoculated with

10^8 cell. ml^{-1} of *B. japonicum*. N-free nutrient solution and nutrient solution contained KNO_3 as control was added every second day. All treatment plants were maintained until 37 DAP.

Plant Maintenace at Greenhouse. During soybean plants growing in the greenhouse regularly added with sterile nutrient solution into the bottom of Leonard bottle. Since 10 days after planting (DAP), plants sprayed with fungicide (1g/L) once a week. Harvesting plants were done by cutting plants at the cotyledon to the former boundary separating the top and plant roots. The roots are removed from the bottle and carefully cleaned of sand and charcoal then washed.

Test of Nitrogenase Activity. Nitrogenase activity was measured by acetilen reduction using gas chromatography. Each of root and nodule soybean plant put into incubation bottle and then sealed with a rubber cover. The next stage was to capture gas from the bottle as much as 2 ml and replaced it with the injection of 2 ml acetylene gas. The bottles were incubated for 30 minutes and then 0.1 ml was taken for gas injected into the gas chromatography Shimadzu 17A. There are three triplicates for each treatment. Ethylene gas produced was calculated based on peak areas on paper chromatograms. Nitrogenase activity was defined as the total amount of ethylene formed per number of plants per hour in units of μmol [8].

Symbiotic Effectiveness Test. Symbiotic effectiveness values (SE) was obtained by formula of Gibson [10] SEN (Symbiotic Effectiveness N) = percentage of dry weight of plants inoculated test strain to dry weight of plants treated with KNO_3 and SER (Symbiotic Effectiveness R) = percentage of dry weight of plants inoculated test strain to dry weight of plants treated with reference strain, USDA 110.

Test of N Total Plant. N Total number of plant referred to the N total number of the canopy. Amount of N content was determined by Kjeldahl method [8].

3. Results and Discussion

3.1. Results

Bacterial Isolate Growth. Isolates were able to grow on YMA which were added with 0.0025% congo red and 50 μg/ml rifampicin after 7 days incubated on room temperature. Morphology of *B. japonicum* colonies were mucoid, not quite able to absorb congo red, and curve elevated (Fig.1)

Fig 1. The growth of Bradyrhizobium japonicum Bj 11 (19) on YMA media + 0.0025% congo red + 50 μg/ml rifampicin ten days after inoculation.

Number and Dry-Weight of Nodules. Inoculation of variuos strains of *B. japonicum* on soybean cultivar Slamet showed variation number of nodule between 9-21 nodule per plant. The highest number of nodule was found in the plant inoculated with mutant strain, Bj 13 (wt) (Table 1). Most of nodules were located on the secondary roots. The range of nodule dry-weight was 0.0089-0.0440 g per plant. The highest nodule dry-weight presented in soybean plant inoculated by BJ 11 (5) (Table 1).

Height of Plant and Dry-weight of Upper Crop. All treatments were inoculated with *B. japonicum* strains showed height plant higher than control N without inoculation, except Bj 11 (20) and KDR 15 (37). In general, the dry-weight of upper crop showed significantly different with control without inoculation and without added with 0.05% KNO_3.

Nitrogenase Activitity. Inoculation of various strains of *B. japonicum* on soybean cultivar Slamet were significantly influence on the activity of nitrogenase. Mutant strain Bj 11 (5) had a higher nitrogenase activity (12,79 μmol 2 $plant^{-1}$) and significantly different than the other strains except Bj 11 (wt) and reference strain, USDA 110 (Table 1).

Symbiotic Effectivenes. The highest symbiotic effectiveness was found in the plant inoculated with mutant strain Bj 11 (19) of 165,87% compared to control N effectivenes and 156.78% compared to USDA 110. While strain KDR 15 (37) showed the lowest symbiotic effectivenes only 97,30% (Table 1).

N Uptake. In general, N uptake in plants inoculated with *B. japonicum* strains were significantly different with control treatments, except strain KDR15 (37). Maximum N uptake of the plant was noticed with the strain Bj 11 (19) up to 20,10 mg N $plant^{-1}$ and significantly different with control treatments (Table 1).

3.2. Discussion

The symbiotic interaction between soybean and root nodule bacteria played an important role in increasing the plant growth of soybean plant. Effectivity of a root nodule bacteria in fixing nitrogen were affected by the compatibility between bacteria and the soybean plant [11]. Data on the effect of inoculation acid-aluminium tolerance *B. japonicum* on nodulation and vegetative growth of soybean plant (Table 1) showed that inoculation of root nodule bacteria could increase height of could increase plant, and dry weight of upper-crop up to 37 DAP. Increasing in nodule dry weight could increase N fixination and the plant growth [12]. The nodule dry weight was positively correlated with the ability of plants to fix N and dry weight of the shoot. In the study, three strains viz. Bj 11 (wt), Bj 11 (5), and Bj (19) showed the best on height of plants, dry weight of upper crop, and dry weight of nodule. The highest symbiotic effectiveness, dry-weight of upper crop, and N uptake was found in the soybean plant inoculated with Bj 11 (19) compared to plant inoculated with other strains and the reference strain, USDA 110. Bj 11 (19) was proposed to be useful isolate for soybean plant on acid soil pH 4.5. The success or failure of inoculation depends on the competitive nodulation ability against indigenous bradyrhizobia under natural conditions [4].

Treatment	Number of nodule (nodule plant $^{-1}$)	Dry-weight of nodule (g plant $^{-1}$)	Height of plant (cm)	Nitrogenase activity (µmol 2 plant $^{-1}$ hour $^{-1}$)	Dry-weight of upper crop (g plant $^{-1}$)	N uptake (mg N plant $^{-1}$)	SEN(%)	SER (%)
Bj 11 (20)	17 abc	0,0338 abc	54,4 d	10,44 b	0,7407 bc	16,03 c	137,01 bc	124,85 ab
Bj 11 (19)	16 abc	0,0377 ab	69,3 ab	10,17 b	0,9083 a	20,10 a	165,87 a	156,78 a
Bj 11 (5)	10 c	0,0440 a	67,8 ab	12,79 a	0,8317 ab	18,65 ab	151,48 ab	144,05 ab
Bj 11 (wt)	17 abc	0,0397 a	71,4 a	12,54 a	0,8447 ab	16,88 ab	155,37 ab	144,55 ab
Bj 13 (wt)	21 a	0,0262 bc	60,7 c	9,79 b	0,7963 b	16,08 bc	145,86 ab	136,60 ab
KDR 15(37)	9 c	0,0089 d	48,8 d	9,38 b	0,5342 e	7,72 d	97,30 d	92,19 b
KDR 15(wt)	20 ab	0,0373 ab	63,6 ab	10,23 b	0,8031 ab	18,12 ab	146,54 ab	140,35 ab
USDA 110	12 bc	0,0241 c	63,0 b	12,21 a	0,6164 d	13,63 c	114,92 cd	100,00 ab
Control N	0 d	0 d	46,0 d	0 c	0,5509 de	13,58 c	100,00 cd	96,26 ab
Control NO	0 d	0 d	37,3 e	0 c	0,4561 e	6,91 d	83,88 d	77,55 b

Tabel 1. Effect of inoculation of B. japonicum on soybean cultivar Slamet at 37 DAP using N-free solution at pH 4,5 + A150 µM

Numbers on the same column followed by the same letter were not significantly different based on Duncan Multiple Range Test ($\alpha = 0.05$). 0 =no detection, N:without inoculation consist of KNO_3 0,05%, N0: without inoculation and without KNO_3 0,05%, Symbiotic Effectiveness (SE) against N/R

4. Conclusions

Inoculation of acid-aluminium tolerant *Bradyrhizobium japonicum* lead to good nodulation, vegetative growth, and symbiotic effectiveness of soybean cultivar Slamet at pH 4.5. Mutant of *Bradyrhizobium japonicum* strain Bj 11 (19) could increase the dry weight of the upper crop (64,88%), the

N-uptake (190,88%), symbiotic effectiveness (65,87%) better than treatments with and without nitrate and the standard strain, USDA 110.

5. Acknowledgements

This research was funded by Incentive Programs for Applied Research the Ministry of Research and Technology, Republic of Indonesia to NRM. Previous research was supported by competitive grants, DGHE, Ministry of Education, Republic of Indonesia to ATW.

6. References

[1] R.M. Atlas, R. Bartha. *Microbial Ecology, Fundamentals and Aplication.* Ed ke-4. Menlo Park: Addison Wesley Longman, 1998.

[2] H.H. Keyser, D.N. Munns. Tolerance of rhizobia to acidity, aluminium and phosphate. *Soil. Sci. Soc. Am. J.* 1979, 43:519-523.

[3] A.E. Richardson, J.S. Richard, A.D. Michael, G.R. Barry. Expression of nodulation genes in *Rhizobium leguminosarum* biovar *trifolii* is affected by low pH and by Ca and Al ions. *Appl. Environ. Microbiol.* 1988, 10: 2541-2548.

[4] C. Appunu, B. Dhar. Symbiotic effectiveness of acid-tolerant *Bradyrhizobium* strains with soybean in low pH soil. *Afr. J. Biotechnol.* 2006, 5:842-845.

[5] R.P. Tiwari, W.G. Reeve, A.R. Glenn. Mutation conferring acid sensitivity in the acid-tolerant strains *Rhizobium meliloti* WSM419 and Rhizobium leguminosarum biovar viciae WSM710. *FEMS Microbiol. Lett.*1992, 100:107-112.

[6] T. Endarini, A.T. Wahyudi, Tedja-Imas. Selections of acid-alumunium *Bradyrhizobium japonicum* indigenous strains. (In Indonesia Languge) *Hayati* 1995, 2:74-79.

[7] A.T. Wahyudi, A. Suwanto, Tedja-Imas, A. Tjahyoleksono. Screening of acid-aluminium tolerant *Bradyrhizobium japonicum* strain analysis of marker genes and competition in planta. *Aspac. J. Mol. Biol. Biotechnol.* 1998, 6:13-20.

[8] P. Somasegaran, H.J. Hoben. *Handbook for Rhizobia. Methods in Legume-rhizobium Technology.* New York: Springer-Verlag, 1994.

[9] A.K. Alva, D.G. Edwards, B.J. Caroll, C.J. Asher, P.M. Greehoff . Nodulation and early growth of soybean mutants with increased nodulation capacity under acid soil infertility factors. *Agron. J.* 1988, 80: 836-84.

[10] A.H. Gibson. Methods for Legums in Glasshouses and Controlled Environment Cabinets. In: F.J. Bergensen (ed.). *Methods for Evaluating Biological Nitrogen Fixation.* New York: John Willey & Sons. 1980, pp. 139-184.

[11] M.T. Madigan, J.M. Martinko, J. Parker. *Biology of Microorganisms.* 9th Ed. New Jersey: Prentice Hall, 2000.

[12] D.J. Mathews, P. Hayes. Effect of root zone temperature on early growth nodulating and nitrogen fixation in soybean. *J. Agric. Sci.*1982, 96:371-376.

50

Tagetes erectus – A Tool for the Management of *Alternaria alternata* Strains of Tomato

Sobiya Shafique, Shazia Shafique

Institute of Agricultural Sciences, University of the Punjab, Lahore, Pakistan

Abstract

Tomato (*Lycopersicum esculentum* Mill.) is among the most economically valuable vegetable crops in the world. It is estimated that diseases reduce tomato production to a greater extent worldwide. Natural plants derivative compounds contribute a lot in fight against pathogens. The current study indicates the pathogenic potential of *Alternaria alternata* FCBP-573 against tomato. RAPD analysis confirmed that *A. alternata* FCBP-573 had variability in its genetic constitution with other two isolates; this disparity in genetic constitution might be a cause to stir up more pathogenicity in this isolate. Therefore, *A. alternata* FCBP-573 was selected and subjected to biological control by *Tagetes erectus* L. In antifungal bioassays plant parts of *T. erectus* with 1, 2, 3, & 4% conc. of aqueous, methanol and n-hexane extracts of each part were evaluated against *A. alternata* FCBP-573. Results revealed that growth of *A. alternata* FCBP-573 was greatly inhibited at 4% conc. of methanol extract followed by aqueous and n-hexane extract. Among different plant parts tested, root extract exhibited more promising results by causing 81-92% reduction in biomass. The study concludes that aqueous and organic extracts of ornamentals have potential to obstruct dreadful effect of pathogenic fungi by suppressing their growth. *T. erectus* conferred vital and surprisingly stable compounds having inhibitory potential against *A. alternata* FCBP-573.

Keywords: Tomato, *Alternaria alternata, Tagetes erectus,* aqueous, organic extracts, antifungal bioassays.

1. Introduction

Tomato is among the economically essential and nutritious vegetable crops in the world. The average yield of tomato in Pakistan is 10.1 tons/ha (Anonymous, 2005). There are about 200 known diseases of tomato, of which 30 are economically important (Jones et al., 1997). Early blight of tomato, stem canker, black mold rot, leaf spot and black shoulder disease is caused by the fungus *Alternaria alternata* f sp. *Lycopersici, so a single pathogen proves economically destructive*. To protect the plants from diseases and pathogens, chemical control methods are exercised. Usage of biodegradable materials as effective micro-organisms and plant extracts from different parts are being used during last few years for plant disease control (Shafique et al., 2011). Numerous studies conducted in Pakistan revealed a wide spectrum prospects of using extracts of plants for biological control of pathogenic fungi (Shafique et al., 2011). On the basis of above mentioned investigations, *Tagetes erectus* (marigold) was selected to investigate antifungal activity of its various parts against *Alternaria alternata*.

2. Materials and Methods

The pure cultures of *A. alternata* FCBP- 573, FCBP- 479 and FCBP-349 isolated from tomato plants were acquired from FCBP, Institute of Agricultural Sciences, University of the Punjab, Lahore.

Pathogenicity test was performed according to Grogan et al. (1975). Conidial suspension of 2.0 $\times 10^5$ conidia/mL of all the isolates was primed as described by Noomrio and Dahot (1992). Disease rating scale was made on the basis of disease incidence and disease severity. Disease severity was calculated by following formula and screening of the most pathogenic isolate was carried out:

$$\text{Disease severity (\%)} = \frac{\text{Infected area of plant}}{\text{Total arca of plant}} \text{ x } 100$$

The genomic DNA of isolates of *A. alternata* was mined using CTAB method (Saghai-Maroof et al., 1984). The RAPD amplification conditions were optimized by following method described by Williams et al. (1990). Five primers RAPD-6 (GATGACCGCC), RAPD-7 (TGTCTGGGTG), RAPD-8 (GTTGCCAGCC), RAPD-9 (GAACGGACTC), and RAPD-10 (TCGCCAGCCA) were used in RAPD analysis.

Fresh samples of *T. erectus* were collected from PU Lahore Aqueous extract (20% w/v) was prepared according to Bajwa *et al.* (2007). The protocol of Alkhail (2005) was used to prepare plant extracts in methanol and n-hexane. Aqueous and organic solvent extract bioassays were carried out in liquid 2% malt extract (ME) medium and incubated at 28±3 ºC for 7 days. Their dry weight yield was determined after 24 h oven drying at 60 ºC according to Bajwa et al. (2006). The rate of fungal biomass increase or decrease was determined from the dried biomass. Percentage reduction in fungal biomass was calculated as:

$$\text{Growth inhibition (\%)} = \frac{\text{Growth in control - Growth in treatment}}{\text{Growth in control}} \times 100$$

All the data was analyzed by analysis of variance (ANOVA), following this, Duncan's Multiple Range (DMR) test (Steel and Torrie, 1980) was applied to separate the treatment means.

3. Results

3.1. Pathogenicity Test

A. alternata FCBP-573 of was proved the most pathogenic (Plate 1), as it induced symptoms of dark brown to black canker with concentric zonation on stems near soil line or aboveground. The cankers became enlarge slowly and plants died. Foliar symptoms were visualized in the form of curling and pointed necrotic lesions on lowest leaflets or in later stages, complete necrosis of leaflets on sides of the midrib was noticed (Plate 1). On the basis of symptoms disease rating scale was developed that is as follows. On the basis of symptoms, *A. alternata* FCBP-573 exhibited maximum pathogenicity towards tomato plants, so was screened out as the most pathogenic strain.

3.2. Molecular Analysis

In RAPD analysis primer RAPD-7 amplified the genome of all isolates while remaining four primers (RAPD-6, RAPD-8, RAPD-9 and RAPD-10) provided no any amplification (Fig. 1). At 225 bp, a unique band was primed by A. alternata FCBP-573 and A. alternata FCBP-479. At 400-500 bp level except isolate A. alternata FCBP-573, in other both isolates amplifications were observed. Thus, A. alternata FCBP-573 was distinguished by the presence of two DNA fragments with an approximate size of 700 and 900 bp produced by the same primer which was not evident in other two. It was inferred from the analysis of amplicons that this difference in genetic framework could be a reason for more pathogenicity in A. alternata FCBP-573 (Fig. 1). The dendrogram was constructed on the basis of pattern of amplifications in RAPD analysis by using software MVSP32 3.12 version (Fig. 2). In dendrogram analysis the

Plate 1: Disease rating scale for *A. alternata* FCBP-573.

highest similarity was found to be ~ 94.11% among *A. alternata* FCBP-479B and *A. alternata* FCBP-349. While isolate *A. alternata* FCBP-573A showed 21% similarity with both the other isolates. The studies reveal that *A. alternata* FCBP-573 induced maximum infection in host plant and exhibited about 79% genetic disparity from other two strains so this most pathogenic isolate was selected for subsequent biocontrol through *T. erectus*.

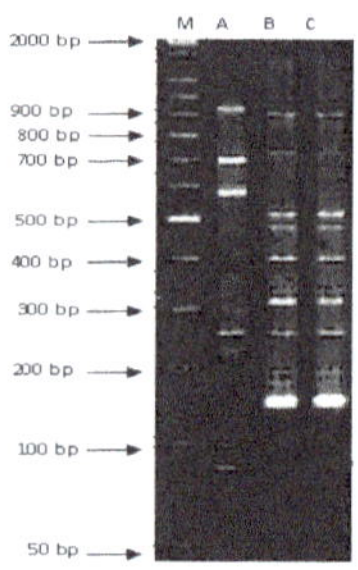

Fig 1. RAPD DNA fragment amplification with primer RAPD-7 showing genetic constitution of *A. alternata* isolates. Lane M indicates DNA marker, A: *A. alternata* FCBP-573, B: *A. alternata* FCBP-479 and C: *A. alternata* FCBP-349.

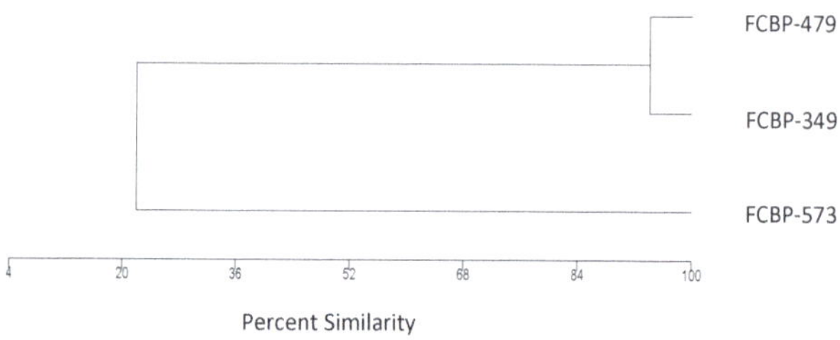

Fig 2. Dendrogram showing similarities among different isolates of *A. alternata*.

3.3. Biological Control of A. alternata FCBP-573 through Aqueous Extracts

Aqueous fractions of root extracts of *T. erectus* exhibited the most promising results in suppressing the fungal growth. A marked decrease of 90% in biomass production was recorded at 4% conc. Aqueous fractions of shoot extracts of *T. erectus* exhibited significant but less inhibitory potential in suppressing the fungal growth than aqueous root extract. However, 4% conc. caused significant reduction of about 44% in mycelial growth. The antifungal activities of aqueous flower extracts of

T. erectus in terms of growth inhibition potential were recorded highest at 3% concentration. This suppression in biomass production was in the range of 40-57%.The results of aqueous fractions conclude that root extracts of *T. erectus* exhibited more promising results as compared to aqueous shoot and flower extract as it suppressed the growth of *A. alternata* FCBP-573 up to 90% while shoot and flower extract induced only 44% and 57% biomass production, respectively.

3.4. Biological Control of A. alternata FCBP-573 through Methanolic Extracts

The maximum antifungal stress was induced by 4% root and shoot extract conc. causing 92% and 86% decline in dry biomass., respectively. Antifungal activity of methanolic flower extracts was assayed and its effect on the growth of *A. alternata* FCBP-573 is presented in Fig.4. The data analysis revealed significant reduction in growth of *A. alternata* FCBP-573 with flower extracts of *T. erectus* but the extract fractions showed significant differences in their efficacy. The reduction in biomass was ranged from 28 to 82%.It is indicated from the results that methanolic extracts of root of *T. erectus* exhibited the best potential in suppressing the fungal growth than methanolic shoot and flower extract.

3.5. Biological Control of A. alternata FCBP-573 through n-Hexane Extracts

The n-hexane extract of root showed the highest antifungal activity against A. alternata FCBP-573. At 4% concentration extract exhibited the strongest antifungal upshot by expressing 81.5% growth inhibition. In case of n-hexane shoot extract a variable pattern of antimycotic activity was observed but 4% concentration of shoot extract was the most effective in suppressing the growth of A. alternata FCBP-573 up to 64%. The data of n-hexane flower extracts revealed a steep and significant reduction in growth by 1 to 4% concentrations in comparison to control. It is indicated from the results that there was insignificant reduction in growth at 3 and 4% concentrations. Maximum arrest in biomass production was evident at 4% concentration as it induced about 57% capture at this dose. The antifungal activities of n-hexane extracts of T. erectus root exhibited more promising results in suppressing the fungal growth than n-hexane shoot and flower extract.

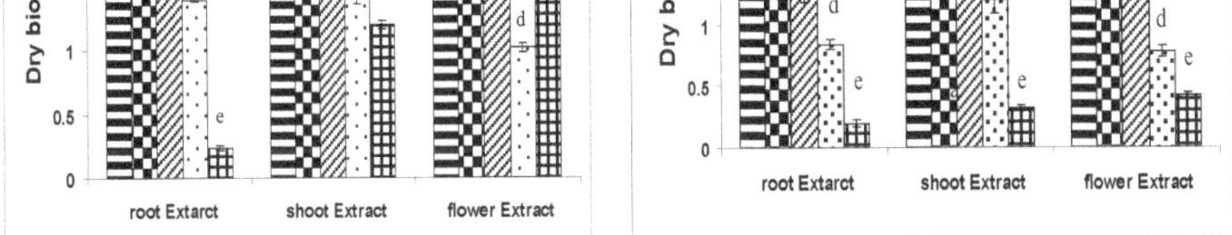

Fig 3. Effect of various conc.s of different parts of aqueous extracts of T. erectus on the biomass production of A. alternata FCBP-573.

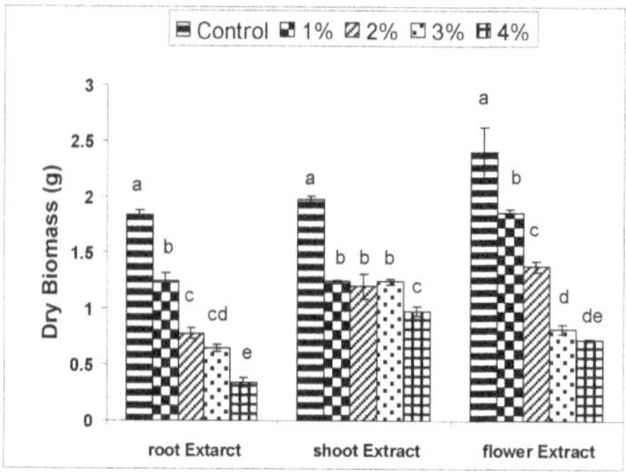

Fig 4. Effect of various conc.s of different parts of methanolic extracts of T. erectus on the biomass of A. alternata FCBP-573.

4. Discussion

In Pathogenecity test *A. alternata* FCBP-573 induced maximum characteristic symptoms. These results were found in agreement with the work conducted by Gilchrist and Grogan (1974) who reported same trend of disease development in tomato by *A. alternata*. Presently, RAPD analysis was carried out to evaluate the isolates of *A. alternata* for their genetic diversity. In dendrogram analysis, isolate *A. alternata* FCBP-479 and *A. alternata* FCBP-349 showed 79% disparity with *A. alternata* FCBP- 573. In previous study, Roberts et al. (2000) also reported a high genetic variation in *A. alternata* populations at molecular level.

Pathogenicity test and RAPD analysis confirmed the strong pathogenic potential of *A. alternata* FCBP-573 so subsequently in present study, root, shoot and flower extracts of *T. erectus* in different solvents were examined. Presently, the results indicated that aqueous fractions of root extracts of *T. erectus* effectively inhibited the growth of *A. alternata* FCBP-573. A marked decrease in biomass production (90%) was recorded at 4% conc. of aqueous root extracts because at higher concentration, thiophenes concentration also increased. It is similar to the work done by Riaz et al. (2008) who observed 54-79% suppression in biomass by employing various concentrations of aqueous fractions of *T. erectus*. In present study, Methanol extracts showed maximum antifungal stress at 4% conc. of root, shoot and flower extracts causing a decline of about 92, 86 and 82%, respectively, in biomass. Working on parallel line Bajwa et al. (2008) reported maximum antifungal activity of methanol extract of rice varieties on *M. phaseolina* and *A. rabiei*. The n-hexane root extract showed the highest antifungal activity against *A. alternata* FCBP-573. A gradual decrease in biomass production was observed with increase in extract conc. At 4% conc. root, shoot and flower extracts exhibited the strongest antifungal effect by expressing 81, 64 and 70% growth inhibition. Earlier Daoud et al. (1990) have reported good antifungal activity of *M. azedarach* against *Alternaria, Aspergillus* and *Penicillium spp.*

This study concludes that aqueous and methanolic extracts of *T. erectus* possess potential antifungal compounds against *A. alternata*, which hold strong antimycotic activity and can be used as a perfect approach for future plant disease management programs eliminating fungal spread.

5. References

[1] S.M. Ahmed, S.A.M. Abdelgaleil. Antifungal activity of extracts and sesquiterpene lactones from *Magnolia grandiflora* L. (Magnoliaceae). *Int. J. Agric. Biol.* 2005, **4**: 638–642.

[2] A.A. Alkhail. Antifungal activity of some extracts against some plant pathogenic fungi. *Pak. J. Biol. Sci.* 2005, **8** (3): 413-417.

[3] Anonymous. Agricultural statistics of Pakistan 2005-2006. Govt. of Pakistan, Ministry of Food, Agriculture and Livestock (Economic Wing), Islamabad 2005.

[4] R. Bajwa, S. Shafique, S. Shafique. Appraisal of antifungal activity of *Aloe vera. Mycopath.* 2007, **5** (1): 5-9.

[5] R. Bajwa, T. Anjum, S. Shafique, S. Shafique. Evaluation of antifungal activity of *Cicer arietinum* L. *Pak. J. Bot.* 2006, **38**: 175- 184.

[6] R. Bajwa, A. Javaid, S. Shafique, A. Javaid, K. Jabeen, S. Shafique. Fungistatic activity of aqueous and organic solvent extracts of rice varieties on phytophathogenic fungi. *Allelopathy J.* 2008, **22**: 363-370.

[7] C.H. Bock, P.H. Thrall, C.L. Brubaker, J.J. Burdon. Detection of genetic variation in *Alternaria brassicicola* using AFLP fingerprinting. *Mycol. Res.* 2002, **106**: 428-434.

[8] F.G. Braga, L.M. Maria, R.L. Bouzada, Fabri, M.O. Matos, F.O. Moreira, E. Scio, E.S. Coimbra. Antileishmanial and antifungal activity of plants used in traditional medicine in Brazil. *J. Ethnopharmacol.* 2007, **111** (2): 396-402.

[9] A.S. Daoud, N.A. Qasim, Al-Mallah. Comparison study on the effect of some plant extracts and pesticides on some phytopathogenic fungi. *Mesopotamia J. Agric.* 1990, **22**: 227-235.

[10] D.G. Gilchrist, R.G. Grogan. A phytotoxin produced by an Alternaria species pathogenic on tomato. 66th Annual Meeting American Phytopathological Society, Vancouver, Canada. Aug. 1974,11-15, Abs: 64.

[11] R.G. Grogan, K.A. Kimble, I. Misaghi. A stem canker disease of tomato caused by *Alternata alternata* f. sp. Lycopersici. *Phytopathol.* 1975, 65: 880-886.

[12] J.B. Jones, J.P. Jones, R.E. Stall, T.A. Zitter. Compendium of tomato diseases. The American phytopath, Society, 3340. Pilot Knob Road, 1997. Minnesota.

[13] M.H. Noomrio, M.U. Dahot. Comparativestudy on the production of cellulases by *Aspergillus niger* using agricultural wastes as a carbon source. Proceeding of all Pakistan Science Conference. 1992, **1**: 95-98.

[14] T. Riaz, S.N. Khan, A. Javaid. Antifungal activity of plant extracts against *Fusarium oxysporum*–the cause of corm-rot disease of Gladiolus. *Mycopath,* 2008, **6**(1&2): 13-15.

[15] R.G. Roberts, S.T. Reymond, B. Andersen. RAPD fragment pattern analysis and morphological segregation of small-spored *Alternaria* species and species groups. *Myco. Res.* 2000, **104**: 151-160.

[16] M.A. Saghai-Maroof, K.M. Soliman, R.A. Jorgensen, R.W. Allard. Fungal DNA Isolation. *Proceedings of the National Academy of Sciences.* 1984, **81**: 8014-8018.

[17] S. Shafique, S. Shafique, R. Bajwa, N. Akhtar, S. Hanif. Fungitoxic activity of aqueous and organic solvent extracts of *Tagetes erectus* on phytopathogenic fungus-*Ascochyta rabiei*. *Pak. J. Bot.* 2011, **43** (1): 59-64.

[18] R.G.D. Steel, J.H. Torrie *Principiles and Procedures of Statistics'* McGraw Hill Book Co., Inc, New York, 1980, USA.

[19] J.G.K. Williams, A.R. Kubelik, K.J. Lival, J.A. Rafalski, S.V. Tingey. DNA polymorphisms amplified by arbitrary primers are useful as genetic markers. *Nucleic. Acids. Res.* 1990, **18**: 6531-6535.

Competitive Interaction of Common Lambsquarters (*Chenopodium album* L.) and Maize (*Zea mays* L.) at Different Time of Emergence and Density

Vahid Sarabi*, Mehdi Nassiri Mahallati, Ahmad Nezami and Mohammad Hasan Rashed Mohassel

Department of Agronomy, Faculty of Agriculture, Ferdowsi University of Mashhad, Mashhad, Iran

Abstract

Common lambsquarters (*Chenopodium album* L.) is an important weed of maize fields in Iran. In order to study the effect of common lambsquarters relative time of emergence on single-cross 704 maize (*Zea mays* L.), an experiment was designed with different density levels of the weed. The experimental design was a split plot based on a randomized completed block design with three replications. The emergence time of lambsquarters was considered as main plots with three levels: emergence of the weed 14 days earlier, 7 days earlier than maize and simultaneously with maize. Density of weed was the subplot treatment with six levels: 0, 4, 8, 12, 16, and 20 plants per m^2. The results showed that height and leaf area index of maize decrease with earlier emergence time and increasing density of lambsquarters, so that 14 days earlier emergence of weed at high densities (16 and 20 plants per m^2) led to maximum reduction. In contrast, height and LAI of lambsquarters increased with earlier emergence time especially at high densities. Common lambsquarters was a stronger competitor when emerged 14 days earlier than maize. Maize yield decreased more than 70% in the 7 and 14 days earlier emergence of lambsquarters at high density. Totally, it can be state that relative time of weed emergence compare to its density had maximum effect on maize growth. In addition, controlling lambsquarters before maize emerging at any densities is recommend preventing of maize growth and yield loss.

Keywords: Corn, density, emergence time, interference, leaf area, yield loss

1. Introduction

Common lambsquarters is the most prevalent weed of soybean [*Glycine max* (L.) Merr.] and maize (*Zea mays* L.) cropping systems in the upper Midwest of the USA [3, 6, 9] and it is one of the most competitive annual broadleaf weeds in maize production in Ontario, Canada [15]. Common lambsquarters causes significant yield damage in many cropping systems because of its rapid growth characteristics, competition for nutrients [12, 14], prolific seed production [6], and seed germination, under a wide range of environmental conditions [10, 11]. Weed density is a crucial factor in crop-weed competition. Crook and Renner [5] observed a 20% reduction in soybean yield where lambsquarters was present throughout the entire growing season at a density of four plants m^{-2}. Sibuga and Bandeen [13] found that maize yield reduced by as much as 58% when lambsquarters density reached 277 plants m^{-2}.

The timing of weed emergence relative to crop is important to crop growth and yield. Yield losses are usually high when weeds emerge earlier or at the same time as the crop [1, 2, 7]. Common lambsquarters is known as one of the most troublesome weeds in maize field of Iran. The majority of studies record effects of weeds on yield in the simultaneous emergence time or times after crop emergence, whereas weeds usually cause to yield loss when emerge before crops and interference with crops during the growing season. Therefore, the objectives of this study were to (1) evaluate the importance and influence of earlier emergence time of weed at different densities on maize yield and leaf area index and (2) determine the competitiveness of lambsquarters and its effects on maize yield when emerged at different times and densities.

2. Materials and methods

An experiment was conducted in 2006 growing season at the Agricultural Faculty of Ferdowsi University of Mashhad, Iran. The experiment was a split plot based on randomized complete block design with three replications. Emergence time of lambsquarters as E_1, E_2 and E_3 (emergence of the weeds 14 or 7 days earlier and simultaneously with maize emergence, respectively) were assigned to main plots and density of weed at 6 levels of D_0, D_1, D_2, D_3, D_4 and D_5 (0, 4, 8, 12, 16 and 20 plants m^{-2}) as subplot. Maize seed were planted 1.5 to 2 cm deep in 70-cm between and 20-cm in-row spacing at density of 7 plants m^{-2}. The field was moldboard-plowed, harrowed and cultipacked in the spring to be prepared for planting. Starter fertilizer was broadcasted at a rate of 300-200-200 kg ha^{-1} N-P-K before planting based on local recommendations and was incorporated by cultivation and shallow disking for seedbed preparation. Nitrogen fertilizer was applied three times, 50% before planting, 25% at the six-to eight- leaf stage, and 25% at the ten-to twelve-leaf stage of maize. Based on a pre-test it was determined that emergence took 15 days for lambsquarters and 7 days for maize. We adjusted sowing dates of lambsquarters to achieve emergence 14 and 7 days prior to maize emergence and simultaneously with maize. Common lambsquarters emergence times were based on visual estimates of 50% emerged plants. At the initiation of lambsquarters senescence, five-tagged maize and five plants of lambsquarters near to them were clipped at the soil surface, sectioned, and placed in cloth bags. After measuring plants leaf area with leaf area meter (Delta-T Devices, Cambridge, England), plants were dried to constant weight in an oven for 70 h at 75°C, and dry weights were recorded.

The relationship between lambsquarters density and maize leaf area index was analyzed separately for each time of emergence by using a nonlinear hyperbolic model described by Cousens [4]. Equation 1 was used to describe relationship between lambsquarters density and maize leaf area index:

$$LAI = LAI_{WF}\left[1-\frac{D}{100\left(1+\frac{D}{A}\right)}\right] \qquad (1)$$

Where LAI is maize leaf area, LAI_{WF} is estimated leaf area index in weed-free plots, D is weed density (plants m^{-2}), I is the percent leaf area index loss per unit weed density as D approaches zero, and A is the percent leaf area index loss as D approaches infinity.

The relative leaf area of lambsquarters at the beginning of senescence was determined from Equation 2:

$$L_W = \frac{LAI_W}{LAI_W + LAI_C} \qquad (2)$$

Where L_W is lambsquarters relative leaf area, LAI_W is the leaf area index of weed and LAI_C is the leaf area index of maize. The relationship between maize yield loss and weed relative leaf area was determined using Equation 3 [8]:

$$Y_L = \frac{qL_W}{1+(q-1)L_W} \qquad (3)$$

Where Y_L is the predicted proportional maize yield loss, q is the damage coefficient associated with weed and L_W is the relative leaf area of weed. Another version of the model was derived from the empirical model introduced by Cousens [4]. This model includes an extra parameter for the maximum yield loss caused by weed (m) (Equation 4):

$$Y_L = \frac{qL_W}{1+\left(\frac{q}{m}-1\right)L_W} \qquad (4)$$

3. Results and discussion

The result of this research showed that earlier emergence time and density of lambsquarters has remarkable effect on maize leaf area index. Maximum maize leaf area index (3) was obtained in the same emerging at low density of weed except of weed free check. In contrast, the lowest LAI (0.4 and 0.3) was observed when maize emergence was delayed 14 days at 16 and 20 plants m^{-2}, respectively (Figure 1). Based on the coefficients of Equation 1, reduction in maize leaf area index per unit weed density as D approaches zero at 14 and 7 days earlier emergence of lambsquarters (I) was 17.63 and 9.09 percent respectively, while, in simultaneous emergence of weed was 2.71 percent.

These coefficients show that first plant of lambsquarters impose more competition pressure to maize at the earlier emergence time especially in 14 days earlier emergence of weed. Also, maximum reduction in maize leaf area index was obtained at high densities in 14 and 7 days earlier emergence of weed than maize, so that maize leaf area index reduced by almost 100 percent (Table 1). Hence, the time of weed emergence is very important in competition between crop-weed, as earlier emergence of weed would be led to reduction of leaf area index of crop close to 100 percent and consequently the maximum yield loss.

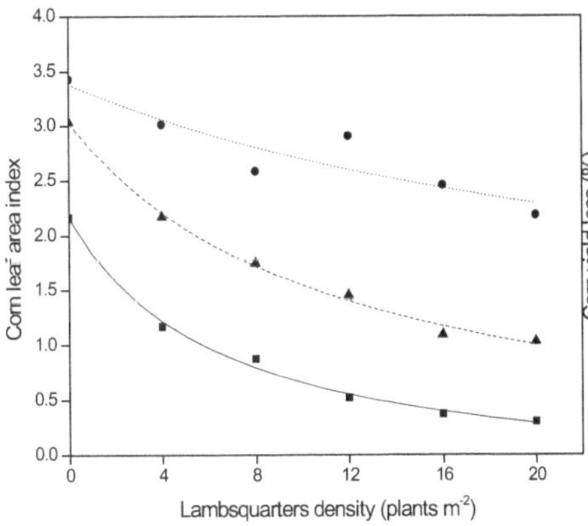

Figure 1. Observed and estimated maize leaf area index at the beginning of lambsquarters senescence as a function of weed density at the 14 days (■), 7 days (▲) earlier emergence of weed and simultaneous (•) emergence time. Regression lines were fitted using Equation 1.

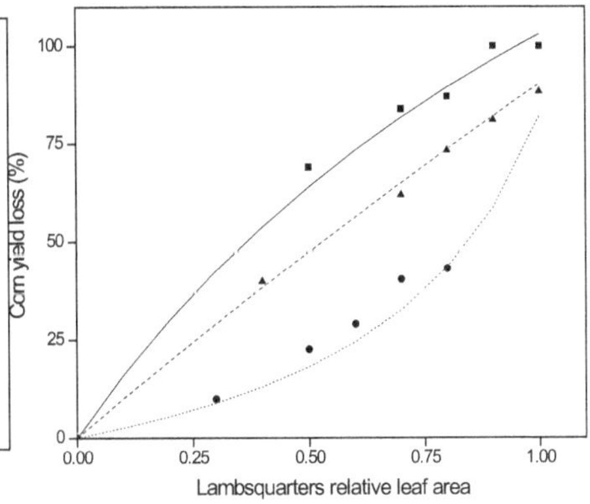

Figure 2. Relation between estimated yield loss in maize and relative leaf area of lambsquarters at the initiation of weed senescence for the 14 days (■), 7 days (▲) earlier emergence of weed and simultaneous (•) emergence time. Common lambsquarters densities were zero, 4, 8, 12, 16 and 20 plants m^{-2}.

Lambsquarters relative time of emergence	Weed-free maize leaf area index	Parameter estimates [a]			
		LAI_{WF}	I	A	R^2
	—	—	%	%	—
E_{-14} [b]	2.16	2.15 ± 0.05	17.63 ± 2.23	100 ± 7	0.99
E_{-7}	3.03	3.02 ± 0.06	9.09 ± 1.11	100 ± 9.44	0.99
E_0	3.42	3.38 ± 0.21	2.71 ± 2.31	78 ± 10.09	0.84

[a] **LAI_{WF}**, represents predicted weed-free leaf area index ± SE of maize; I, represents leaf area index loss ± SE as weed density approaches zero; A, represents leaf area index loss ± SE at high weed densities. The coefficient of determination (R^2) describes the fit of each model to the observed data.

[b] **E_{-14}**, E_{-7} are 14 and 7 days earlier emergence and E_0 is the same emergence of this weed with maize, respectively.

Table 1. Observed weed-free maize leaf area index and rectangular hyperbola parameter estimates for the first, second and third lambsquarters emergence times as a function of weed density.

Relative damage coefficient indicated that lambsquarters in 14 days earlier emergence was a stronger competitor than maize, so that at this time of emergence, q was larger than one (1.64 and 1.3 for one and two parameter model, respectively) and a convex curve is found above the diagonal line. In contrast, this coefficient was smaller than one (0.28 and 0.04) for both parameter model and a concave curve is found under the diagonal line (Table 2 and figure 2). Hence, lambsquarters was not good competitor when emerged simultaneously with maize and crop was the stronger competitor. Coefficients also showed that maize and weed had equals competition ability at the 7 days earlier emergence, but maximum yield loss (m) was obtained at high densities showing that emerging lambsquarters in this time and 14 days earlier than maize was led to yield reduction by 75 and 79 percent, respectively (Table 2).

These results show that crop yield would be reduce more than 70%, when weed emerged earlier than crop even though competitiveness of crop is the same as weed. Hence, this weed should be controlled in the earlier emergence time to prevent greater yield loss and in the same emerging time to prevent more plants of weed reaching reproductive growth and developing seed bank.

Model	Parameter estimates [a]			
	Lambsquarters relative time of emergence	q	m	R^2
One-parameter model	E_{-14} [b]	1.64 ± 0.27	—	0.99
	E_{-7}	1.1 ± 0.62	—	0.92
	E_0	0.28 ± 0.15	—	0.93
Two-parameter model	E_{-14}	1.3 ± 0.14	0.79 ± 0.89	0.99
	E_{-7}	0.83 ± 0.3	0.75 ± 0.27	0.92
	E_0	0.04 ± 0.23	0.15 ± 0.83	0.93

[a] **q,** represents the damage coefficient associated with lambsquarters; m, represents the maximum yield loss caused by weed. The coefficient of determination (R^2) describes the fit of each model to the observed data.

[b] $\mathbf{E_{-14}}$, E_{-7} are 14 and 7 days earlier emergence and E_0 is the same emergence of this weed with maize, respectively.

Table 2. Estimated parameters obtained from maize yield loss as a function of the relative leaf area of the lambsquarters for the first, second and third emergence times.

4. References

[1] R. J. Aldrich. Predicting crop yield reduction from weeds. *Weed Technol.* 1987, **1**: 199–206.

[2] R. E. Blackshaw, E. H. Stoble, and A. R. W. Sturko. Effects of seeding dates and densities of green foxtail (*Setaria viridis*) on the growth and productivity of spring wheat (*Triticum aestivum*). *Weed Sci.* 1981, **29**: 212–217.

[3] C. Boerboom. Ready to tackle lambsquarters? University of Wisconsin Extension. 2009.http://ipcm.wisc.edu/WCMNews/tabid/53/EntryId/686/Ready-to-Tackle-Lambsquarters.aspx. Accessed: February 3, 2010.

[4] R. Cousens. An empirical model relating crop yield to weed and crop density and a statistical comparison with other models. *J. Agr. Sci.* 1985, **105**: 513-521.

[5] T. M. Crook, and K. A. Renner. Common lambsquarters (*Chenopodium album*) competition and time of removal in soybeans (*Glycine max*). *Weed Sci.* 1990, **38**: 358–364.

[6] F. Forcella, R. G. Wilson, K. A. Renner, J. Dekker, R. G. Harvey, D. A. Alm, D. D. Buhler, and J. Cardina. Weed seed banks of the U.S. Maize Belt: magnitude, variation, emergence, and application. *Weed Sci.* 1992, **40**: 636–644

[7] S. K. Harrison. Interference and seed production by common lambsquarters (*Chenopodium album*) in soybean (*Glycine max*). *Weed Sci.* 1990, **38**: 113–118.

[8] A. Jakstaite. Weed harmfulness critical period on spring barley. Scientific works of Lithuania Institute of Agriculture. 1988, **36**: 108–124 (in Lithuanian).

[9] [8] M. J. Kropff, and L. A. P. Lotz. Systems approaches to quantify crop-weed interactions and their application in weed management. *Agr. Syst.* 1992, **40**: 265-282.

[10] Kruger, G. R., W. G. Johnson, S. C. Weller, M. D. K. Owen, D. R. Shaw, J. W. Wilcut, D. L. Jordan, R. G. Wilson, M. L. Bernards, and B. G. Young. U.S. grower views on problematic weeds and changes in weed pressure in glyphosate-resistant maize, cotton, and soybean cropping systems. *Weed Technol.* 2009, **23**: 162–166.

[11] D. Mulugeta, and D. E. Stoltenberg. Influence of cohorts on *Chenopodium album* demography. *Weed Sci.* 1998, **46**: 65–70

[12] A. G. Ogg, and J. H. Dawson. Time of emergence of eight weed species. *Weed Sci.* 1984, **32**: 327–335.

[13] H. N. Pandy, K. C. Misra, and K. L. Mukheriee. Phosphate uptake and its incorporation in some crop plants and their associated weeds. *Ann. Bot.* 1971, **35**: 367–372.

[14] K. P. Sibuga, and J. D. Bandeen. Effects of green foxtail and lambsquarters interference in field maize. *Can. J. Plant Sci.* 1980, **60**: 1419–1425.

[15] J. Vengris. Plant nutrient competition between weeds and maize. *Agron. J.* 1955, **47**: 213–215.

[16] S. E. Weaver. Impact of lambsquarters, common ragweed and green foxtail on yield of maize and soybean in Ontario. *Can. J. Plant Sci.* 2001, **81**: 821–828.

52

Environmental and Economic Evaluation for the Breeding of Grass Carp in Egypt's Water Channels

Ahmed Mohammed Ahmed[1], Dr. Abdelbaky Mousa Abdelbaky Elshaib[1], Abdelmagid Hassan Abdelmagid[2]

[1] Department of Agricultural Economics, Tanta University, Tanta, Egypt

[2] National Water Research Center, Cairo, Egypt

Abstract

Research aims to evaluate the biological control of weeds for the culture of fingerlings fish grass carp, by highlighting the economic, environmental and social, based on this research to the economic analysis descriptive and quantitative to estimate and explain the phenomena related to the subject matter on the basis of measurements studies feasibility and evaluation of economic projects, and noted the results of research to the return of the production of fingerlings of about 1.02, 1.36, 2.635, million pounds in cases production current expectancy and full capacity respectively. The yield indirect has been estimated at about 46.28, 61.7, 1119.55, one million pounds for the cases of three respectively were estimated payback period for the project in case of return of direct 45, 18.5 4, the year of the three cases respectively while the estimated internal rate of return of 2% , 5%, 25% of the cases of three The yield indirect has increase than 100% and recommended that the results of research that denominated biological led to raising the efficiency of use of water resources, increase production fish, increase the agricultural area, to increase agricultural production, reduce water loss 0.71 at about 1.033 billion cubic meter annually as equivalent to 1.9% of Egypt's share of the Nile River where led to provide the amount of irrigation water in the light of these results, the study recommends the need to expand the dynamic resistance of aquatic weeds, especially fish farming Congratulations and grasses.

Keywords: the internal rate of return for the project (IRR),the percentage of benefits tocosts (B / C Ratio), Net present value(NPV), payback period(PBP). Water losses by evaporation transpiration (E.T):

1. Introduction

Aquatic weeds are considered one of the most dangerous problems that weaken the efficiency of irrigation and drainage networks because of the resulting obstruction of the flow of water and it does not reach to the ends of waterways in addition to the loss of large amounts of water by evaporation transpiration. The more spread aquatic weeds, the more snails spread disease-causing deadly epidemic of health are common in many developing countries.The issue of environment protection and preservation of various types of pollution is considered one of the most important contemporary issues facing both the developed and developing countries alike. Therefore, it was necessary to look for better ways to weed and maintain the environment at the same time. It had been found that both of the manual and mechanical resistances lead to environmental pollution due to dumping products of removing weeds on the bridges of waterways that are de-weed them which is considered as a favorable environment for mice and insects especially mosquitoes epidemics and diseases harmful to the human health. Also, the mechanical resistance causes collapse for bridges waterways. Therefore, employing the biological resistance through using fish grass carp is the best option among these methods. Breeding grass carp has two important dimensions, where it increases fish production by converting useless aquatic weeds to an animal protein and then reducing the food gap, as well as reducing the wastage of fresh water and use that water in agriculture and reclamation of new agricultural lands.Research problem: The risks of aquatic weeds have been escalated recently in many parts of the world. The water is one of the most important sources of life, aquatic weeds affecting it perniciously by hampering flow of water in the canals as well as raise the water by pumps in irrigation projects, also caused damage to power plants hydroelectric, and lead to loosing water evaporation transpiration. Objective of this research: The research aims to assess the biological control of weeds farming fish grass carp fingerlings through indicating the economic, environmental and social effects of that method.

2. Research results:

2.1. Economic Analysis of the total costs and revenues of the hatchery:

Establishment of hatcheries for the production of industrial fish requires large amount of capital varies according to the used method of the hatchery, the types of fish that are spawned, hatchery's size and quality of the used raw materials and their prices. This section indicates the economic analysis of hatchery works for fish grass carp in waterfall of Aswan. Production costs of any project reflect the values paid for factors of production, or to ensure the continuation of production process during a period of time. In the short run, the total production costs are divided to fixed and variable costs. The fixed cost is the cost that paid annually as a fixed amount paid whether there was a production or not, such as the annual depreciation, insurance and taxes. The variable cost includes several items such as the price of used eggs, salaries, the cost of organic and chemicals fertilizers, feed costs, fuel costs, electricity, maintenance of equipment, and other costs related to the hatchery process.

2.1.1. Production costs of grass carp fish hatchery:

2.1.1.1. Fixed costs:

Table (1) indicates the hatchery's fixed costs. It reveals that the total fixed costs were estimated at nearly 5.3 million pounds.

fixed costs			operating costs		
Type	Cost (in Egyptian pound)	%	Type	cost (in Egyptian pound)	%
Land	2100000	39.7	Concentrated feed	352000	38.9

fixed costs			operating costs		
Basin's set up	831000	15.7	Green feed	3800	0.4
Buildings	560000	10.6	Pituitary gland, drags, chemicals, cotton, gauze and syringes	32500	3.6
Pumps	970000	18.3	Fertilizers	22700	2.5
Filters	300000	5.7	Employment	420000	46.5
Laboratories Tools	173500	3.3	Cleaning basins	4000	0.4
Mothers and future mothers	112250	2.1	Electricity, water, telephone	20000	2.2
Generator	12000	2.3	Maintenance of machinery and buildings	49000	5.4
machines and mixers for making feed	50000	0.9			
Refrigerator to store feed and raw materials	60000	1.1			
Pumping air blowers	15000	0.3			
Total	5291750	100	Total	904000	100

Source: Ministry of Irrigation and Water Resources, General Directorate of the Aswan irrigation, unpublished data

Table 1. fixed costs and operating costs of Aswan's hatchery works

2.1.1.2 Second, the variable costs:

Table 1 indicates the value of variable costs. It reveals that the total value of variable costs amounted to 904 thousand pounds per year.

2.2. Economic feasibility's Criteria

Table (2) shown that the discount and non discount criteria of economic feasibility for three cases, of Aswan's hatchery current, hoped, and full capacity for indirect returns and direct returns, direct returns include the sale of fingerlings but indirect returns include the saving water leaded to reclamation of new land in addition to providing employment opportunities, and food security

2.3. Indicators of social and environmental feasibility of the project

2.3.1. First, Indicators of social feasibility:

This project is considered one of the most important projects that have a vital role in the process of economic development. It leads to increasing the efficiency of one of the most important agricultural resources, which is the water, as well as increasing fish production which leads to narrowing of the food gap of fish. Also, these projects contribute to the increase in the agricultural area by adding new land to the agricultural area of old, leading to increased agricultural production and increase agricultural exports and reduce imports of agricultural and adjust the balance of trade in favor of Egypt, all of these social returns is working to improve living standards with increasing the efficiency of the Egyptian agricultural economy

Items	Full capacity	production can be hoped	the current production
In the case of calculating the direct return (The price of fingerlings)			
Standard non-discounted			
payback period of capital years	4.9	16.2	45.6
the internal rate of return for the project (IRR) of around %	20.3	6.2	2.2
Standard discounted			
the internal rate of return project (IRR)	25	5.4	2
Net present value NPV (million pounds)	5.81	2.08	3.33-
the percentage of benefits to costs (B / C Ratio)	131.37	85.4	74.4
Capital recovery period	4	18.53	50
In the case of calculating the return indirect			
Standard non-discounted			
payback period of capital years	first year	first year	first year
Internal rate of return IRR%	2039.2	1104.86	857.454
Standard –discounted			
Rate of return on invested capital	more than 100%	more than 100%	more than 100%
Net present value NPV (million pounds)	1066.6	519.3	407.3

Items	Full capacity	production can be hoped	the current production
the percentage of benefits to costs (B / C Ratio)	5858.9	3742.9	3230
payback period	first year	first year	first year

Source: Calculated from data tables No. (1), (3)

Table 2. Indicators of the financial evaluation of Aswan's hatchery.

2.4. Impact of the project implementation on food security:

The project is interested in saving water, which leads to increasing the cultivated area of Egypt, where the amount of water available per hectare is enough to plant about 0.34 acres in the year, which is used to produce the most important field crops, vegetable crops and crops, medicinal and aromatic plants and fruit crops, which enjoys cultivated in Egypt, and thus the project is able to raise levels of food security, increase area of wheat, which leads to increased self-sufficiency in wheat in addition to the production of oil crops, of which the main soybean export crops and most important of which onions and green beans as well as medicinal and aromatic plants and is characterized by a high demand it can increase exports Egyptian them greatly, as the project will add large amounts of agricultural production to the all.

In addition to the above, the implementation of the project will result in increasing fish production significantly and at the lowest possible cost, with support fish production here on the conversion aquatic weeds harmful to animal protein high nutritional value as the result as nearly 5.76 million kg in the year, which contributes to the reduce the food gap of fish at low cost and without the need for large capital investments or the cost of production, especially if we know that this type of fish up to large sizes and it is a kind of good taste, resulting in reduced imports of fish and reduce the burden on the balance of payments.

2.5. Impact of the project implementation on the redistribution of income:

The project provided a number of jobs and generating incomes for small-and medium-workers both in the field of agricultural production is estimated about 40 workers per acre per year was estimated at wages of about 1200 pounds per acre in the year or in the field of fishing as well as reducing the price of fish which would result in access poor and middle classes on the cheap animal protein which increases the real incomes of the poor and middle class, and it turns out that such projects be of a positive effect on economic well-being of society.

2.6. The feasibility of environmental indicators:

Became the issue of environment protection and preservation of various types of pollution and one of the most important contemporary issues faced by both developed and developing countries alike, especially after the increasing problem of climate change and global warming, and play a environmental impacts of a prominent role in the implementation of investment projects, particularly agricultural projects and water and that the most important projects of land reclamation projects and the disposal of agricultural waste, grasses and agricultural water due to the impact on the surrounding environment and public health for humans and animals. Get rid wrong from agricultural residues, aquatic weeds leads to the spread of diseases and epidemics, which affect the overall health of humans and animals which would result in lower productivity, leading to lower per capita incomes of the citizens. And this effect is reciprocal between agricultural projects and projects of the disposal of agricultural waste, water and the environment as the production of agricultural crops, and aquatic weeds heavily influenced by environmental factors in terms of temperature, humidity and soil type and quality of irrigation water and agricultural drainage water and other factors surrounding environment at the same time are affected by environmental agricultural practices and through the use of chemical fertilizers and pesticides as well as agricultural waste, water and how to get rid put them both on the shores of waterways or burn them and drainage. Hence the importance of environmental studies to identify the positive and negative effects of the implementation of the project as well as procedures for mitigating the negative effects of the implementation of the project

2.6.1. The positive environmental effects expected in the implementation of the project:

1. that the implementation of the project amounts to provide Evaporative water loss in transpiration rate of 2263 cubic meter per hectare.
2. Use of residues of harmful aquatic weeds for the production of animal protein, high nutritional value.
3. increase the cultivated area, which causes an increase of agricultural production and livestock production, fisheries and which leads to increased food security and improve the lives of human and non human exposure to diseases, especially diseases, anemia and diseases of food shortages.
4. Increase employment and the eradication of unemployment relative to where the project is working to provide about 40 workers per acre is farmed.
5. not throwing grass on the shores of waterways, which was the result in the spread of diseases and epidemics and affects the overall health of humans and animals which would result in lower productivity, low per capita incomes of the citizens.
6. bridges prevents the flow that occurs when using mechanical methods.
7. provides much of the cost of removing aquatic weeds where they are providing about 5255 pounds, compared to mechanical resistance of the grass and about 3125 pounds compared with manual resistance of the grass.

2.7. Water losses by evaporation transpiration (E.T):

Losses water is important factor that affect the efficiency of irrigation and drainage network and the River Nile. We have different views in Egypt, where the value of these losses showed the results of some research and studies vary widely in estimating the value of these losses, but these studies were not based on actual experiences. Given the importance of this subject, the Ministry of Irrigation and Water Resources conducted a series of field studies and laboratory experiments to estimate the losses have concluded from these studies indicate that the average loss from operations of evaporation transpiration of water weeds about 0.62 cm3 / cm 2 / day annual average.(2)

E.T(Million cubic meter / year Evaporative transpiration losses)	the total area affected for more than cut (acres)	areas infected for more than a desolation of the different types of grass			Channel water
		floating	Gravis	sub-merged	
		Acre	Acre	Acre	
71	7468	4893	2233	342	River Nile
578.5	60863	11165	14935	34763	Irrigation network
383.2	40314	17125	14235	8954	Drainage network
1032.7	108645	33183	31403	44059	Total

Source: Departments of Irrigation and Drainage, unpublished data, 2009.

Table 3. the amount of water loss from evaporation transpiration for each of the River Nile and the irrigation and drainage networks in million cubic meters in 2009

Previous table shows the amount of evaporation loss from transpiration of the Nile River and irrigation and drainage networks in 2009. And it is clear gravity of the loss amounts, which amounted to approximately 1.033 billion cubic meters / year, equivalent to about 1.9% of Egypt's quota of Nile water. Where the amount of the loss of the Nile River as a result of evaporation transpiration of water weeds by about 0.071 billion cubic meters, while the estimated loss from the irrigation network of about 0.579 billion cubic meters, the total amount of the loss of a network of drainage about 0.383 billion cubic meters, and multiple methods of resistance of aquatic weeds in Egypt between manual methods, mechanical methods, chemical methods, biological methods.

3. Important Informatoion

The research aims to assess the biological control of weeds farming fish grass carp fingerlings through indicating the economic, environmental and social effects of that method. This will be assessed through evaluation of hatchery fish in Aswan, which produces fish grass carp fingerlings.

4. Acknowledgements

The study depends on collecting the secondary data to achieve its objectives. The data are extracted from General Authority for Fish Resources Development (GAFRD), Egypt's Ministry of Agriculture and Land Reclamation, Central Agency for Public Mobilization and Statistics (CAPMS), and Ministry of Irrigation and Water Resources.

5. References

[1] Prof. Dr .Ahmed Mohammed Ahemd. Investment Feasibility enterprise economics, 2005

[2] Central Agency for Public Mobilization and Statistics, Bulletin of Irrigation and Water Resources, 2010.

[3] Ministry of Irrigation and Water Resources, General Directorate of the Aswan irrigation, unpublished data

[4] General Directorate of the Aswan irrigation, unpublished data, Aswan, Egypt ,2009. Cairo, Egypt, 2009.

[5] Departments of Irrigation and Drainage

53

Factors Influencing Consumers' Willingness to Pay for Agricultural Organic Products (AOP)

Mahtab Pouratashi

Department of Agricultural Extension and Education, Faculty of Agricultural Economics and Development, University of Tehran, Karaj, Iran

Abstract

The main purpose of this study was to investigate factors influencing consumers' willingness to pay for agricultural organic products (AOP). This article has done with reviewing and analyzing various researches in different countries. The findings revealed that the premiums that consumers were willing to pay vary from country to country. Consumers' knowledge and awareness about AOP differs. Some had more knowledge about AOP and some were un-knowledgeable about AOP. Some factors influencing consumers' willingness to purchase AOP were consumers' socio-economic characteristics and AOP characteristics. Due to the importance of consumers' knowledge about AOP for purchase decisions, it is recommended to publish materials about AOP and deliver them to families.

Keywords: Agricultural organic product, willingness, purchase, knowledge

1. Introduction

With population growth, agriculture will need to produce enough supply of food to feed an expected more than eight milliard people by 2030 (1). Farmers depend mainly on chemical technologies to manage pests and to produce adequate food (2). However, despite their many benefits, it is now clear that pesticides may also have unfortunate consequences to human health and the environment (3). Hence, an interest in organic products is increasing throughout the world due to response to concerns about conventional agricultural practices, human health, and environmental safety. The role of organic agriculture in providing food is now gaining wider recognition (4).

Stobbelaar et al. (5) stated that "an organic product is food produced without artificial fertilizer or chemical pesticides, nor containing artificial coloring, flavoring or aromatic substances, preservatives, or genetically modified ingredients". Organic foods are perceived as healthier than conventional alternatives (6). At the same time, these products are perceived as rather expensive (7). Hence, it is essential to investigate if consumers are willing to pay for AOP. According to the studies done in the field of organic products in different countries, this study attempts to investigate how much consumers are willing to pay for organic products. Also, which factors influence they willingness to pay for these products.

1.1. Purpose and Objectives

The main purpose of the study was to investigate Factors influencing consumers' willingness to pay for agricultural organic products (AOP). The special objectives of the study were:

Consumers' willingness to purchase AOP;

Effective factors on consumers' willingness to purchase AOP.

2. Consumers' willingness to purchase AOP

Different studies have been done in regard to consumers' willingness to purchase AOP. For example, Skuras and Vakrou (8) found that 60 to 70 percent of respondents willing to pay a premium for quality food products. Meier-Ploeger and Woodward (9) in German reported that 52 percent of the respondents were willing to pay more for organic fruits and vegetables and 39 percent for grain products. Davis et al. (10) found that men would pay more at a percentage of 41 percent compared to 44 percent of women.

Gil et al. (11) showed that in Spain only likely and actual organic consumers were willing to pay a premium of 15-25 percent for organic food. Fotopoulos and Krystallis (12) stated that in Greece, organic consumers expected to pay from 19 to 63 percent for food products.

Corsi and Novelli (13) in Italy found that only organic consumers who could remember the price of conventional minute steak beef and roast beef were willing to pay 52 and 58 percent respectively above regular prices for the organic type of these products. Millock (14) found that 35% of the respondents in Denmark were willing to pay more for any type of organic products. In contrast, 18% of consumers were not willing to pay for all kind of products.

Canavari et al. (15) found that the proposed premium price for organic peaches and apples was accepted by 65.8 percent of the Italian respondents of their survey. Angulo et al. (16) found that although the Spanish consumers were concerned with the issue of food safety, 72.5 percent of them were not willing to pay a premium for a labeled food with a traceability certificate.

Aryal et al. (17) revealed that all respondents were willing to pay price premium, but the level of acceptability varied considerably. A total of 58% of the consumers were willing to pay 6- 20% price premium, whereas 13% were willing to pay up to 50% premium. The average premium was estimated about 30%. Asadi et al. (18) found that in Iran, the majority of respondents were not willing to pay a price premium higher than 20%.

According to the findings, it is revealed that the premiums that consumers are willing to pay vary. This difference can be explained by the consumers' knowledge and awareness about agricultural organic products.

3. Effective factors on consumers' willingness to purchase AOP

Attention to studies in the field of organic agriculture has found different factors influencing purchase of AOP. Krissoff (19) found that consumers' perception about food safety of organic products - these products are safer, healthier and more environmentally friendly than conventionally products- affected consumers to pay for AOP. Govindasamy and Italia (20) showed that gender, age, income and education, are among the most important factors influencing willingness to purchase AOP.

Makatouni (6) reported that product characteristics such as nutritive value, freshness, taste, and food safety influence consumer' willingness to purchase AOP. Bonti-Ankomah and Yiridoe (2006) also reported that respondents were willing to pay more for organics products because of food safety, taste and nutritive value.

Angulo et al. (16) stated that among the factors that affect willingness to pay for organic foods were consumers' use of food labels, experience with the product, and the prices consumers actually pay. Krystallis and Chryssohoidis (22) found that Consumers purchased organic products because they perceived these products as higher quality, safer foods that they could trust more than their conventional counterparts.

Aryal et al. (17) found that lack of information available to consumers, higher prices over those of conventional foods, and the limited and erratic domestic supply were factors influenced consumers' willingness to purchase AOP. Rajabi et al. (23) found that the consumers' knowledge of organic products was at medium level. Also, consumers' attitude toward using organic products was at neutral and favorable levels. The researchers found that four most highly ranked factors influencing the adoption of organic products were educational, access, improving product characteristics and supportive services

According to the findings, effective factors on consumers' willingnessto purchase AOP can be shown by Fig. 1.

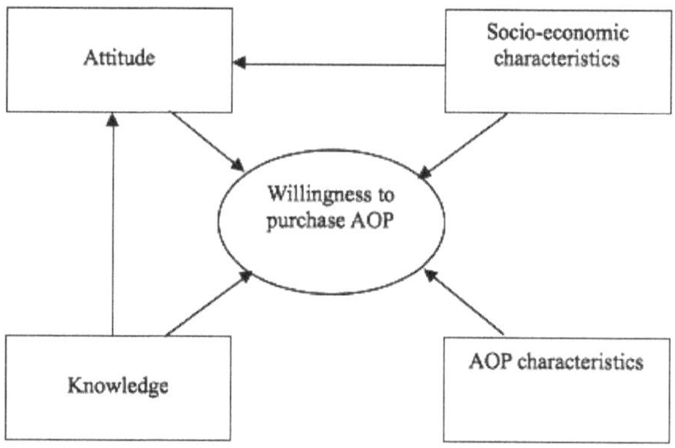

Fig. 1- Factors influencing consumers' willingness to purchase AOP

4. Summary and conclusion

According to the importance of AOP, this study investigated factors influencing consumers' willingness to purchase AOP. Attitude, knowledge, age, income, characteristics of agricultural products such as tests, color, nutritive value were among the factors affected consumers' decision making about purchase.

If consumers are not knowledgeable about organic products, they are unwilling to pay anything more than standard prices. Knowledge and awareness are perceived prerequisites in the adoption process (24). Bhatta et al. (25) found that a majority of the respondents knew about organic agriculture. Aryal et al. (17) found that nearly all of the respondents had heard about the organic products. But, they were not sure which products were organic and which were not. Rajabi et al. (23) found that the consumers' knowledge of organic products was at medium level. Therefore, consumers must be knowledgeable about organic products and their benefits including human health. Rajabi et al. (23) found that uses TV the most. After that, there were books, Newspapers and Journals. It is recommended to increase consumers' knowledge about AOP via different delivery methods such as radio and TV programs, CD and DVD. Zhou and Chen (26) found that 56 percent of the consumers had heard about organic products from TV, 47 percent learned from magazines, 23 percent through internet, 16 percent obtain the information from supermarket, 10 percent from friends and 5 percent had obtained the organic food information from other channels.

High prices and deficiencies in distribution channels are obstacles to the purchase of AOP (11). Hence, it is recommended that policy makers indicate distinct places for selling of these products.

5. Acknowledgement

The author gratefully acknowledges the National Elite Foundation for their support.

6. References

[1] GHolifar, E., Ali Asadi, A., Akbari, M. and Pouratashi, M. 2010. Effective factors in apple waste in Islamic republic of Iran:a comparative study. Journal of Human Ecology, 32(1): 47-53.

[2] Mauceri, M. 2004. Adoption of integrated pest management technologies: A case study of potato farmers in Carchi, Ecuador. Master of Science in Agricultural and Applied Economics, Virginia Polytechnic Institute and State University, Blacksburg, Virginia.

[3] Pimentel, D. 2005. Environmental and economic costs of the application of pesticides primarily in the United States. Environment and Development, 7, 229–252

[4] Van Elzakker, B., Parrott, N. Chola Chonya, M. and Adimdao, S. 2007. Organic Farming in Africa in H. Willer and M. Yusseffi (Eds.). The World of Organic Agriculture: Statistics and Emerging Trends. Bonn, IFOAM, pp.96-106.

[5] Stobbelaar, D.J., Casimir, G., Borghuis, J., Marks, I., Meijer, L., and Zebeda, S. 2006. Adolescents' attitudes towards organic food: a survey of 15 to 16 year old school children. International Journal of Consumer Studies. Blackwell Publishing Ltd.

[6] Makatouni, A., 2002. What motivates consumers to buy organic food in UK? Results from a qualitative study. British Food Journal. 104(3/4/5): 345-352.

[7] Radman, M. 2005. Consumer consumption and perception of organic products in Croatia. British Food Journal, 107(4), 263-273.

[8] Skuras, D. and Vakrou, A. 1999. Willingness to pay for origin-labeled products: a case study of Greek wine consumers. 67th EAAE Seminar Proceedings, Le Mans, 28-30 October.

[9] Meier-Ploeger, A. and Woodward, L. 1999. Trends between countries. Ecology and Farming, 20, January-April, 15.

[10] Davis, A., Titterington, A.J. and Cochrane, C. 1995. Who buys organic food? A profile of the purchasers of organic food in N. Ireland. British Food Journal, 97(10), 17-23.

[11] Gil, J.M., Gracia, A. and Sanchez, M. 2000. Market segmentation and willingness to pay for organic products in Spain. International Food and Agribusiness Management Review, 3, 207-26.

[12] Fotopoulos, C. and Krystallis, A. 2001. Defining the organic consumer and his willingness to pay for selected food products in Greece. paper presented at the 51st International Atlantic Economic Society Conference, Athens, March 13-20.

[13] Corsi, A. and Novelli, S. 2002. Consumers' willingness to pay a price for organic beef meat. Paper presented at the Xth EAAE Congress, Zaragoza, 28-31 August

[14] Millock, K., Hansen, L.G. Wier, M. and Anderson, L.M. 2002. Willingness to Pay for Organic products: A Comparison between Survey Data and Panel Data from Denmark, AKF Denmark.

[15] Canavari, M., Nocella, G. and Scarpa, R. 2003. Stated willingness to pay for environment-friendly production of apples and peaches: web-based versus in-person surveys. Paper presented at the 83rd EAAE Seminar, Chania, 4-6 September.

[16] Angulo, A.M., Gil, J.M. and Tamburo, L. 2003. Food safety and consumers' willingness to pay for labeled beef in Spain. paper presented at the 83rd EAAE Seminar, Chania, 4-6 September.

[17] Aryal, K.P., Chaudhary, P., Pandit, S. & Sharma, G. 2009. Consumers' willingness to pay for organic products: a case from kathmandu valley. The Journal of Agriculture and Environment, 10, Jun.2009, 12-22.

[18] Asadi, A., Akbari, M. Sharifazadeh, A. and Hashemi, S.M. 2009. Analysis of Factors Affecting Agricultural Organic Products Diffusion among Consumers: Perception of Extension Workers. World Applied Science Journal. 6 (3): 331-338.

[19] Krissoff, B. 1998. Emergence of U.S. organic agriculture - can we compete? American Journal of Agricultural Economics. 80(5): 1130-1133.

[20] Govindasamy, R. and Italia, J. 1999. Predicting willingness to pay a premium for organically grown fresh produce. Journal of Food Distribution Research, 30(2), 44-53.

[21] Bonti-Ankomah S. and Yiridoe, E.K. 2006. Organic and conventional food: A literature review of the economics of consumers' perceptions and preference. Final Report. Organic Agriculture Centre of Canada. Nova Scotia Agricultural College, truro, Nova Scotia, Canada.

[22] Krystallis, A. and Chryssohoidis, G. 2005. Consumers' willingness to pay for organic food. British Food Journal, 107(5),320-343.

[23] Rajabi, A., Pouratashi, M. and ShabanAli Fami, H. 2011. An Analysis of consumers' knowledge and willingness to pay for organic products. First National Congress on New Technologies in Agriculture, 10-12 Sep., Zanjan University, Iran.

[24] Rogers E. 1995. Diffusion of innovations. 4th ed. New York: Free Press.

[25] Bhatta, G.D., Doppler, W. KC, K.B. and Ranabhat, A. 2008. Potentials of Organic Agriculture in Nepal. In P. Chaudhary; K. Aryal and D. Tharu (ed.), Proceedings of International Workshop

on Opportunities and Challenges of Organic Production and Marketing in South Asia, NPG, Kathmandu, Nepal, pp.34-46.

[26] Zhou, L. and Chen, T. 2007. Consumer Perception of Organic Food in Urumqi. Contributed Paper prepared for presentation at the 5th Seminar 'International Marketing and International Trade of Quality Food Products' Bologna, Italy, March 8-10.

54

A Gravity Model Analysis of Egypt's Trade and Some Economic Blocks

El-Sayed M. A.

Department of Economic & Rural Development, Suez Canal University, El-Arish, Egypt

Abstract

This paper aims to study the economic effects of trade flows between Egypt and some economic blocs, where study confine to AFTA agreement particularly a Arab interface business and agreements of each bloc COMESA and EU generally because of strength opportunities for these blocs by using descriptive analysis and Gravity Model (GM). The major results confirmed the efficiency of the model in explaining Egyptian trade flow for the three previous blocs. Based on the above results, the study recommends continuing to increase the volume of foreign trade, in addition to reducing the constraints faced by Intra-Arab trade.

Keywords: Economic Effects of Trade Flows, Economic Blocks, Gravity model approach

1. Introduction

Trade among Arab countries (ACs) has been consistently weak in spite of several efforts to engage into different forms of regional economic integration. The most important attempts to achieve Arab economic integration were the agreement of 1953 on Transit Trade, the Common Market attempt of 1964, and the agreement of 1981 on the facilitation and development of trade, all signed under the auspices of the Arab League. These attempts, in addition to about 135 bilateral trade-related agreements, were not capable of stretching inter-trade beyond its peak of 10 percent of the total trade of ACs [1].

In 1994, the Intra-trade of ACs as percentage of their total exports was around 8.3 percent. This rate compares unfavorably with the corresponding rates of many regional groupings from both Developed and Developing countries. The latter ratios were 69.9 percent for APEC, 61.7 percent for EU, 47.6 percent for NAFTA, and 11.6 percent for EFTA. For regional groupings from the Developing countries these rates were 18.2 percent for MERCOSUR (Latin America), 12.0 percent for UEMOA (West Africa), and 21.2 percent for ASEAN (South-East Asia).(1)

These rates are not strictly comparable across groupings. Difference in the degree of development, size, and weight in international trade of the different countries of the groupings, explain to a great extent the observed variation between these regional groups. This can be said, however, the extent of Intra-Arab trade is arguably weaker than what it should have been given the common historical, religious, social, cultural, and language characteristics shared by these countries.

Many factors were presented to explain the weakness of Intra-Arab trade and the obvious failure of previous Arab regional agreements to stimulate trade among Arab countries. These factors

1 (1) UNCTAD (1997): Handbook of International Trade and Development Statistics.

range from mere economic factors, such as difference in economic systems, similarity of production structure and traded goods, lack of adequate transportation infrastructure compounded by distance, overprotection, heavy reliance on trade taxes, the lack of convertibility of Arab currencies, lack of market information, weakness of marketing strategies, and poor competitiveness of products.

Other factors are of an institutional nature. These include colonial links or the moral commitment to a well established partner (North African Arab countries to Europe, and Middle Eastern countries to the U.S. and Europe), the poor preparation of and lack of commitment to the regional agreements, the lack of adequate trade financing schemes at the regional level, the low quality of bureaucracy, and lengthy trade-related procedures. Last but not least, trade among Arab countries is very sensitive to political events and relationships among these countries.

These impediment factors notwithstanding, Arab leaders have unanimously taken the decision, during their summit of June 1996 in Cairo, to revive the 1981 agreements and create an Arab Free Trade Area (AFTA). The unanimous decision bears witness to the commitment of Arab countries to reinforce trade among each other as a means of facing the fierce competition in international markets entailed by a rampant globalization.

2. Research problem and objective

The agreements of AFTA which came into effect January 1, 1998, entailed the elimination of non-tariff barriers and the reduction of tariff rates on goods traded among ACs by an average of 10 percent a year, over a period of ten years. Although, the general feeling is that AFTA has been well prepared for and unambiguously committed to, many suspect that most of the factors that are behind the failure of previous agreements are still present. Arguably, this minimizes the chance of a breakthrough in Intra-Arab trade.

This paper presents an objective evaluation of the potential for success in AFTA agreements based on a model accounting for most of the dimensions involved in explaining Intra-Arab trade flows.

3. Methodology

Gravity models, which were originally proposed by [2] and [3], have become one of the most commonly used workhorse models to analyze patterns in international trade. By analogy with Newton's theory of gravitation, these models express bilateral trade as a function of two key variables: the economic size of the two countries engaged in trade and the distance between them. Accordingly, in their most general form these models suggest that the magnitude of trade between two countries depends on the supply conditions in the source country, the demand conditions in the host country (and other factors which may stimulate or hinder bilateral trade); they are consistent with standard models of international trade (see [4] [5] [6]).

As regards other factors, four variables are commonly added (see, e.g. [7]: Firstly, it is likely that countries sharing the same language trade more with each other than otherwise. This may be partly related to historically established trade ties. A common language dummy could for instance explain the relatively high levels of Spain's trade with its former colonies in Latin America. Secondly, if two countries were part of the same territory (such as the countries of former Yugoslavia or the former Soviet Union), they may still have closer trade ties than otherwise (history matters). Thirdly, if countries share a common border, transaction costs may be reduced beyond the mere distance factor, translating into a higher bilateral trade. Finally, the accession to a free trade arrangement may stimulate trade among the constituent countries, as the rise of Spanish trade with other euro area countries in the second half of the 1980s indicates.

In view of their simplicity and high explanatory power, gravity models have been applied to the particular case of CEE countries in several studies. [8] and [9], two of the most influential early studies in the field, showed that trade of the CEE countries with developed countries has been only a fraction of potential trade. [9] suggested that actual trade with the EU12 was up to 5 times smaller than potential trade for Bulgaria and former Czechoslovakia in 1989. Some CEE transition economies were found to be much closer to equilibrium (this is the case of Hungary, with a ratio of potential to actual trade of 1.8), while countries like Romania and Albania, which did not participate in the Council of the Mutual Economic Assistance, started trade liberalization with regional trade structures closer to the gravity predictions estimated by Baldwin. [10] found that Romania achieved a significantly higher actual share of trade with the EU than predicted by the model in 1996. [11] and) [12] suggest a rapid convergence towards trade potential levels in non-EU countries which have a trade agreement with the EU (so-called associated countries). [13] and [14]found that trade between the EU15 and the CEE countries was close to the predicted level at the end of the1990s. Meanwhile, [15] found that trade potentials were largely exhausted by the end of the decade (in the longer-run however, trade flows could increase in larger proportions once all structural adjustment is completed).

After the theoretical foundation of gravity model had been established, in the 1990s, further studies concentrated on its empirical application. [16] Formulated a more complex and advanced form of gravity equation where he particularly emphasized the role of geographical factors, such as distance, landlockedness and population, as determinants of bilateral trade flows. He also included regional trading blocs, such as APEC, NAFTA and Mercosur, in his gravity equation in order to estimate the impact of regional integration on bilateral trade flows. In a similar way, [17] tried to analyze the impact of economic integration, as embodied by the LAIA, the Andean Pact, and CACM, in Latin American countries' Intra-regional trade flows, based on the gravity-type equation.

Furthermore, they do not present a broader comparison with other regions. Such comparison would bolster the results and put them in perspective. However, with the increasing importance of geographical factors in international trade theory, the gravity model started to attract a re-awakening interest in the 1980s [18].

Works by [19], [20], [4] and [21] greatly contributed to the establishment of a theoretical foundation for the gravity model by showing that the gravity equation can be derived from a number of different international trade models.

There are two competing models of international trade that provide theoretical justification for the gravity model. They are the Differentiated Products Model and the Heckscher-Ohlin Model. [5] and [19] tried to identify the relationship between the bilateral trade flows and the product of two countries' GDPs by utilizing the Differentiated Products Model. According to Krugman

& Helpman, under the imperfect substitute model, where each firm produces a product that is an imperfect substitute for another product and has monopoly power in its own product, consumers show preference for variety. When the size of the domestic economy (or population) doubles, consumers increase their utility, not in the form of greater quantity but of greater variety. International trade can provide the same effect by increasing consumers' opportunity for even greater variety. Therefore, when two countries have similar technologies and preferences, they will naturally trade more with each other in order to expand the number of choices available for consumption.

In this paper, we use the log-linear form to estimate bilateral import flows among some economic blocs over average 2008-2010, to give an order of magnitude for the impact of the main factors often advanced in explaining the weakness of Intra-Arab trade.(2) Thus, the gravity equation to be estimated is as follows: (3)

Log(Mij)= ββ1Log(GNPi*GNPj) + β1Log(GNPi*GNPj) +β3Log(DISTANCEij) + β4Log(INEQGNPC) + β5PARTNER + β6COSINEij + β7POLFACT +β8Log(XRCi) + β9Log(Mji) + β10ATFD81 + β11BORDER + CONSTANT

Mij: Flow of imports of country j from country i, in millions of U. S. dollars.

GNP: Gross National Product of country i or j in millions of U.S. dollars.

GNPC: GNP per capita in millions of U. S. Dollars.

DISTANCE: Distance in kilometers between the capitals of countries i and j. BORDER: Dummy variable taking the value of unity if i and j share a common border and zero otherwise.(4)

PARTNER: Dummy variable taking the value of unity if i and j are members of GCC or AMU and zero otherwise.

INEQGNPC: Measure of GNP per capita inequality between countries i and j.(5)

POLFACT: Dummy variable taking the value of unity in case of border closing, political disagreement or event affecting normal diplomatic and commercial relations between countries i and j.

COSINE: Measure of trade correspondence between the export structure of country i and the import structure of country j.(6)

---

(2) Arab countries, excluding Palestine, are: Algeria, Bahrain, Comoros, Djibouti, Egypt, Iraq, Jordan, Kuwait, Lebanon Libya, Mauritania, Morocco, Oman, Qatar, Saudi Arabia, Somalia, Sudan, Syria, Tunisia, U.A.E. &Yemen.

(3) Frankel et al. (1995) and Cyrus (1996) followed a closer specification.

(4) GCC stands for the Gulf Cooperation Council formed in 1981 grouping Bahrain, Kuwait, Qatar, Oman, Saudi Arabia, and the United Arab Emirates. AMU stands for the Arab Maghreb Union formed in 1989 and grouping Algeria, Libya, Mauritania, Morocco, and Tunisia.

(5) Balassa (1986), Balassa and Bauwens (1987), and Bergstrand (1990) used this measure of inequality. For a given variable x it is given by: 1+[xlog(x)+(l-x)log(l-x)l/log(2). It has the advantage of depicting relative rather than absolute inequality and falls between zero and one.

(6) The cosine measure indicates the cosine of the angle between the export vector of country i and the import vector of country j and is given by:

$$CONSINEij = \sum_k E_k M_k \sqrt{\sum_k E^2{}_k \sum_k M^2{}_k}$$

where E_{ik} stands for exports of commodity k by country i and M_{jk} for imports of commodity k by country j.

XCR: Export Concentration Ratio of country i measured as the share of the three most important commodities in the total value of its exports.

ATFD81: Dummy variable taking the value one if both countries, i and j, have signed the Arab Trade Facilitation and Development agreement of 1981, and zero otherwise.

Mji: Flows of imports of country i from country j, in millions of U.S. dollars.

4. Results & Discussion

I- Description of GM variables:

The country was proved (i) Egypt, with changes of countries (j = 1, 2, ... n) in gravity equation. Also was used all variables in the form of the natural logarithm without dummy variable.

(1) Multiplying GDP: multiplying GDP represent the economic size of the two countries, also was as representative of the productive possibility and market size, so the larger countries - with large production possibility - are the most likely to reach economies of scale leading to increased exports of competitive advantage. Also have large domestic markets possibility to absorb more imports, so the increase multiplied GDP is possibility to lead to increase the volume of bilateral trade and it is expected that the coefficient is estimated to be greater than zero. Since the GDP of Egypt remained constant, multiplying GDP depends upon GDP of the partner countries (j) and this included the lack of impact or influence of the domestic market neutral (internal).

(2) Multiplying GDP Per Capita: the variable GDP Per Capita represent the level of income and/ or the purchasing power of imported and exported countries, with the installation of GDP per capita in Egypt, will this variable to explain what If the flow of Egyptian trade depends upon the level of income of trading partners.

(3) Geographical Distance: the distance is a variable resistant to a trade where represent trade barriers such as transportation costs, time, cultural differences and barriers to market access...etc. And the distance used in this study is (Great Circle Distance) between the Egyptian capital and the capitals of countries (trading partners). It is worth noting that many previous studies have been translated coefficient of geographical dimension, it's flexibility of trade in respect of the absolute level of the geographical dimension, where is the volume of trade greater between pairs of countries that are far from the rest of the world [22]. And coefficient distance measures the relative distances of the States, where the lack of distance coefficient refer to trade with nations far more than compared with the more closer. While the increase refer to trade with the more convergence faster than those away. And it is noted that the notion of relative distance (relative dimension) is significant in the case of States (N×N). whereas, in the case of the model used in this study. (N×1) measured all distances in absolute terms for Egypt, so it is expected that the coefficient of the distance be less than zero.

(4) Border: it's a dummy variable illustrated that the whether the countries share the same or contiguous borders, and this variable takes the picture the correct one if two countries share a common border and zero otherwise.

Table (1), illustrated that the GM variable definition.

Variable	Definition
In Tij Aggregate trade balance	Aggregate trade balance of source country (Ti) to & from (Tj) host one.
In Country GDP	Logarithm (In) of GDP (Ti) × GDP (Tj)
Ln Country GDP Per Capita	Logarithm (In) of Per Capita (Ti) × Per Capita (Tj)
Ln Distancc	Logarithm (In) of distance between Source country and the host one
Border	Dummy variable for Border between Source country and host one (1,0)

Table 1. GM Variables Definition

II- Descriptive Statistics for the Basic Sample:

While, table (2) showed that the Descriptive Statistics for the Basic Sample: Average 2007-2009.

Variable	Units	Obs.	Mean	Std. Dev.	Min	Max
Ln (Y) Tij	000 $	21	12.188	2.451	4.007	15.287
Ln (X_1) GDP	Million $	21	22.373	1.770	18.276	25.057
In (X_2) Per Capita	Mill. $/000 per.	21	2.283	1.462	0.214	4.961
(X_3) Distance	Kilometers	21	2137.667	1057.122	680	4599
(X_4) Border	Dummy variable	21	0.190	0.402	0	1

Source: Compute from Unified Arab Economic Report 2010.

Table 2. Descriptive Statistics for the Basic Sample: Average 2007-2009

III- Empirical Results of GM:

The empirical results showed that in table (3). The first column in the table represents estimates for coefficient. Results obtained from running T-Statistics are re-ported in columns two.

The aggregate performance of the model in this table looks good value for the coefficient of determination was estimated at about 0.86 in the case of the agreement with AFTA, which means that GM, efficient in explaining the flow of bilateral trade to Egypt.

Results indicated that the GM to study the most important variables affecting on the volume of Egyptian bilateral trade in framework AFTA.

	AFTA		COMESA		EU	
	Coeff.	T-stat.	Coeff.	T-stat.	Coeff.	T-stat.
Constant	-7.697	-2.055	-8.781	-2.021	-6.478	-1.436
	(3.746)		(4.344)		(4.511)	
In (x_1) GDP	0.991	5.681**	1.098	5.468	0.918	4.267
	(0.174)		(0.201)		(0.215)	

	AFTA		COMESA		EU	
	Coeff.	T-stat.	Coeff.	T-stat.	Coeff.	T-stat.
ln (x_2) GDP Per Capita	-0.038	-0.185	0.781	2.664	-0.794	-1.880
	(0.204)		(0.293)		(0.422)	
ln (x_3) Distance	-0.001	-4.389**	-0.001	-5.229	-0.00008	-0.224
	(0.0002)		(0.0003)		(0.0004)	
Ln (x_4) Border	-0.277	-0.478	-2.147	-1.207	0	65535
	(0.578)		(1.778)		(0.000)	
No. of Observation	21		22		27	
R-squared	0.89		0.82		0.45	
R-Adjusted	0.86		0.78		0.34	
F- test	31.381		19.469		6.284	

** Significant at level 0.01 * Significant at level 0.05" ( ) refers to standard Error

Source: Compute from Unified Arab Economic Report 2010.

Table 3. Results of GM for some Economics blocs: Average 2007-2009

(1) The Egyptian bilateral trade with AFTA was increased about 1% by increased of GDP of the partner countries by about 1% (equivalent). While the trade volume bilateral Egyptian with COMESA and EU were estimated about 1.1, 1% respectively, indicating the importance of EU as major trade partner for Egypt.

(2) The variable GDP per capita is not significant in the case of the three agreements, which refer to the flow of Egyptian trade does not depend upon the income level of trading partners, but depends on the economic size of trading partners even more.

(3) The distance variable or geographic dimension was statistically significant and corresponding with economic logic in influencing the volume of Egyptian trade bilateral in framework AFTA. But it is not significant in the case of the agreement with COMESA and EU, which requires the development of land, sea and air transportation networks between ACs, including more than the speed of passage of goods and reduce transport costs, where the results indicated that increasing the distance between Egypt and its trading partners within the framework of AFTA was estimated about 1% would lead to a reduction of the flow of the bilateral trade by about 1.9%.

(4) The border variable was not statistically significant which led to decline from important of common border between in framework AFTA as factor, determined of trade exchange.

5. Summary and conclusion

This paper has reviewed most recent factors in Egyptian Trade with some Economic blocks. It was discussed that GDP & Distance only approach that maximizes the trade between Arab countries. It is yet anticipated future works are required to find the statistical significant of AFTA sample size using the other approaches to achieve this on the rest of Economic Blocs. Furthermore, the current approach may be extended for the rest of other factors of Egyptian trade such as GDP per capita, Border, and Language.

6. References

[1] A. Al-Shibani, "Arab trade liberalization: The Arab free trade area," *the General Secretariat of the General Union of Chambers of Commerce, Industry, and Agriculture in the Arab Countries*, 1997, pp. 67-114.

[2] S.B. Linder, "An Essay on Trade and Transformation," *New York: John Wiley and Sons*, 1961.

[3] H. Linnemann, "An Econometric Study of International Trade Flows," *North-Holland*, 1966.

[4] A.V. Deardorff, "Determinants of Bilateral Trade: Does Gravity Work in a Neoclassical World?," *NBER Working Paper Series #5377*, 1995.

[5] J.E. Anderson, "A theoretical foundation for the gravity equation," *American Economic Review*, vol. 69 (1) no. 1, 1979, pp. 106-116.

[6] J.E. Anderson, and E.V. Wincoop, "Gravity with gravitas: a solution to the border puzzle," *American Economic Review*, vol. 93 (1), 2003.

[7] H. Cheng, and H. Wall J., "Controlling for Heterogeneity in Gravity Models of Trade and Integration," *January/February Review of Federal Reserve Bank of Saint Louis*, vol. 87(1), 2005, pp. 49-63.

[8] [8] C.B. Hamilton, Winters, Alan L. , "Trade with Eastern Europe," *Economic Policy*, vol. April, 1992.

[9] R.E. Baldwin, "Towards an Integrated Europe," *Centre for Economic Policy Research*, 1994.

[10] O.a.H.M.A.-A. Hvrylyshyn, "Opening up and geographic diversification of trade in transition economies," *working paper No. 98/22, Washington DC: International Monetary Fund*, 1998.

[11] B. Kaminski, Z. K. Wang and L. A. Winters "Export performance in transition economies," *Economic Policy*, no. October, 1996, pp. 421-442.

[12] Z. Jakab, M. A. Kovács and A. Oszlay "How far has trade integration advanced? An analysis of the actual and potential trade of three Central and Eastern European countries," *Journal of comparative Economics*, vol. 29 (2), 2001, pp. 276-292.

[13] P. Egger, "An econometric view on the estimation of gravity models and the calculation of trade potentials," *The World Economy*, vol. 25 (2), 2003, pp. 297-312.

[14] J. Fidrmuc, Fidrmuc, Jan., "Disintegration and Trade," *CEPR Discussion*, vol. Paper No. 2641, no. Centre for Economic Policy Research, December., 2000.

[15] L. Fontagné, "Foreign Direct Investment and International Trade: Complements or Substitutes?," *STI Working Papers*, vol. 1999-3, no. OECD-DSTI, 1999.

[16] [16] J. Frankel, D. Romer, and T. Cyrus, "Regional Trading Blocs in the World Economic System," *Institute for International Economics, Washington, DC.*, 1997.

[17] G. Garman, and G. Debora, "Economic Integration in The Americas:1975-1992," *Journal of Applied Business Research*, vol. Vol.14 #3, 1999.

[18] C. Sohn, "A Gravity Model Analysis of Korea's Trade Patterns and the Effects of a Regional Trading Arrangement," *The International Centre for the Study of East Asian Development, Kitakyushu*, 2001.

[19] E.a.P.K. Helpman, "Market Structure and Foreign Trade: Increasing Returns, Imperfect Competition and the International Economy," *MIT Press, Cambridge, MA.*, 1985.

[20] J.H. Bergstrand, "The Generalized Gravity Equation, Monopolistic Competition, and the Factor-Proportions Theory in International Trade," *The Review of Economics and Statistics*, vol. LXXI (1), 1989, pp. 143-153.

[21] S. Evenett, and K. Wolfgang, "On Theories Explaining the Success of the Gravity Equation," *National Bureau of Economic Research*, vol. Working Paper #6529., 1998.

[22] J. Harrigan, "Specialization and The Volume of Trade, in E. Kwan.," *Handbook of International Trade, Blackwell Publishers.*, 2003.

Effect of Priming on Dormancy Breaking and Seedling Establishment of Caper (*Capparis spinosa* L.)

Saeed Khaninejad*, Iman Hessam Arefi, and Mohammad Kafi

Department of Agronomy, Ferdowsi University of Mashhad, Iran

Abstract

Caper *(Capparis spinosa)* has deep roots, drought tolerant species, and produces a satisfactory vegetative cover which protects soils from erosion; it can be highly useful for the prevention of land degradation. Domestication of capers as medicinal, vegetable or soil surface coverage plant is complicated by limited and variable seed germination under artificial conditions. In order to examine the role of different levels of KNO_3 (0, 500, 1000, 2000, 4000 and 8000 ppm) and gibberellic acid (GA_3) (0, 50,100, 250, 500, 1000 and 2000 ppm) and durations (3, 12, 24 and 48 hr) on germination of Iranian caper seeds. In general 2000 mg/l gibberellic acid treatment resulted in more vigorous seed germination (42%) at any duration compared to any other concentration of the gibberellic acid. The highest seed germination of 26% was achieved when the seeds were treated 24 hour with 4000 ppm KNO_3 solution but it was decreased in 8000 mg/l. The highest germination percentage (72%) was observed in seeds placed in filter papers wetted with in 250 ppm gibberellic acid after treatment with 8000 mg/l KNO_3 for 24 hour (this duration was the best time span in two previous experiments). It seems that GA_3 and KNO_3 can replace partly to improve seed germination of caper. The highest seedling dry weight was achieved as seeds were treated in 100 ppm gibberellic acid plus 1000 ppm potassium nitrate. Therefore, it can be concluded that for best germination percentage of caper seeds, 250 ppm GA_3 and 8000 ppm KNO_3 and for the strongest seedling 100 ppm GA_3 plus 1000 ppm KNO_3 could be recommended.

Keywords: Caper, Dormancy, GA3, KNO3, Medicinal Plant

1. Introduction

Caper (*Capparis spinosa* L.) plant is typical of the tropical Mediterranean areas as well as central Asia, Europe (Spain, Italy, Greece and Turkey) as well as North Africa and Middle East countries including Iran [9,11]. It is a perennial shrub, covering soil surface widely, and produces one of the deepest root systems [14]. The long roots and wide ecological amplitude allow it to withstand harsh environments. The species thus appears to be a suitable candidate for the protection of degraded areas in arid areas [14]. Apart from its roles in soil conservation, different parts of the caper plant is used in Iranian traditional medicine as a liver provoking agent, treatment of vessel clogs, anti-rheumatism and a diuretic agent [13]. Seed dormancy is an adaptive mechanism in many species particularly wild species to protect seedlings from freeze damage during the winter, or from drought stress in water shortage conditions [15]. Caper species is endangered in its natural habitats in Iran and many other countries due to climate change and overutilization.

Domestication of capers as medicinal, vegetable or soil surface coverage plant is complicated by limited and variable seed germination under artificial conditions. According to some researchers, there are germination obstacles in the caper seeds and; thus, there are propagation difficulties of caper seedlings [11]. To ensure high plantation and viability, a reliable and high germination percentage is required. Plant growth regulators such as gibberellic acid (GA_3) [6] and chemicals such as KNO_3 [5] have been recommended to break seed dormancy and enhance germination. Gibberellins promote growth by increasing extendibility of the cell wall followed by the hydrolysis of starch to sugars which reduces the potential in the cell, resulting in the entry of water into the cell causing elongation [2]. KNO_3 is the most widely used chemical for germination promoting [11]. Information on seed germination of capers is still limited. Therefore, it was thought that treatment of the seeds with plant growth regulators may influence rapid germination and root formation. The present study was conducted to examine the role of KNO_3 and GA_3 that might affect germination of Iranian capers seeds and study the possible advantages of in vitro germination over direct sowing of seeds in the soil.

1.1. Materials and Methods

In order to evaluate germination properties and effect of different treatments on dormancy breaking of *caparis spinosa,* these experiments were conducted at Physiological Research Laboratory, Ferdowsi University of Mashhad, Iran, in 2010. The seeds used in this study were collected from Shothern Khorasan province, Iran, where capers grows abundantly. The seeds were separated from the fruit material, rinsed in tap water, dried in shade and kept at room temperature in linen sacks until sowing.

The germination test conducted to find out germination properties of caper seeds in H_2O as control, concentrated gibberellic acid, potassium nitrate, gibberellic acid + potassium nitrate. A sample of 25 randomly selected seeds was used with four replications. Experiments were carried out in 90mm diameter Petri dishes using 90mm diameter Whatman No. 1 filter papers in the bottom and top of the petri dishes to cover the seeds. Experiment was continued for 21 days and every day the procedure of germination was monitored.

Similar to previous experiment, samples of twenty five randomly selected seeds with four replications were soaked in different doses (0, 50,100, 250, 500, 1000 and 2000 ppm) and durations (3, 12, 24 and 48 hr) of gibberellic acid solution. No further water was added during the experiment period. For control treatment, Filter papers were moistened with distilled water.

Another 21 days experiment was conducted to find out potassium nitrate solution on germination. Seeds soaked in different doses (0, 500, 1000, 2000, 4000 and 8000 ppm) (Amri, 2010) and durations (3, 12, 24 and 48 hr) of potassium nitrate. Samples and germinated seeds were counted every day until the end of the experiment.

The seeds were placed in filter papers were soaked in 0, 50,100, 250, 500, 1000 and 2000 ppm gibberellic acid solution after applying different doses (0, 500, 1000, 2000, 4000 and 8000 ppm) of potassium nitrate solution for 24 h. For control treatment, filter papers were moistened with distilled water. This experiment also continued for 21 days and seed germination was monitored every day.

The combined experiment of gibberellic acid and potassium nitrate was arranged in the polyethylene pots filled with growing medium composed of clay, sand and manure (1:1:1). Pots were kept in open air conditions after sowing. Seven different gibberellic acid treatment (0, 50, 100,

250, 500, 1000 and 2000 ppm) and six different potassium nitrate (0, 500, 1000, 2000, 4000 and 8000 ppm) treatment were applied on the seeds in the spring (March) of 2010. The experimental design was a factorial based on randomized complete block with three replications (35 pots for each block) for every treatment.

Analysis of variance performed using SAS 9.1 (SAS Institute, Cary, NC) and general linear models (PROC GLM) procedure. Multiple comparisons were conducted for significant effects with the least significant difference (LSD) test at $\alpha = 0.05$ to determine the rate of seed germination for each patch and seed treatment.

2. Result and Discussions

There was no any symptom of germination of caper seeds in distilled water, Olmez *et al.*, (2004) reported only 3.67% germination in this plant using distilled water. The seed infection was started form the 4th day of the experiment and after 21 days, there were some damaged seeds in each petri dish.

2.1. Gibberellic acid treatment

The highest seed germination of 42% was achieved after 24 h of soaking in 2000 mg/l GA_3 (Table 1). However, 38% seed germination was achieved in the same duration at 1000 mg/l GA_3 (Table 1). In general 2000 mg/l GA_3 treatment resulted in more vigorous seed germination at any duration compared to any other concentration of the gibberellic acid. This indicates that the regulation of endogenous GA_3 levels after seed imbibitions is a crucial factor in determining seed germination. Duration of exposure of seeds to GA_3 is also important; for instance the germination percentage at 1000 mg/l of GA_3 for 6 hours was only 8% while at the same concentration of GA_3 it was 38% after 24 hours of GA_3 treatment. Germination percentage was reduced beyond 24 hours of soaking in gibberellic acid at all concentrations. Gibberellic acid is known to play an essential role in seed germination, leaf expansion, stem elongation, flowering and flower development [16]. Our results are in agreement with that of Negbi et al. [10] for *Hirschfeldia incana* and Orphanos [12] and Olmez *et al.*, [11] for *Capparis spinosa*, who found that the seed dormancy is mainly due to the seed coat that prevents germination and GA_3 has a positive effect on germination. Mayer and Shahin [8] observed that the gibberellins reduces oxygen requirement for germination. The caper germination percentage obtained in this experiment is higher than previously reported e.i. 27.4 % in Olmez *et al.*, [11] experiment.

Soaking time (hours)	GA_3 dose (mg/l)					
	50	100	250	500	1000	2000
6	6j*	6j	6j	10hi	8i	10hi
12	8i	8i	10hi	12gh	12gh	13fgh
24	24d	30c	30c	32c	38b	42a
48	14fg	16f	16f	24d	20e	22de

*Values followed by different letters are significantly different at 0.05 level using LSD. There was no germination in distilled water (zero GA_3)

Table 1. Effect of different duration and concentration of GA_3 on seed germination percentage of *C. spinosa*

2.2. Potassium nitrate treatment

Treatment with exogenous KNO_3 stimulated of germination percentage of caper seeds. The highest seed germination of 26% was achieved when the seeds were treated 24 hour with 4000 mg/l KNO_3 solution but it was decreased in 8000 mg/l. Moreover, germination percentage of caper seeds were dependence to concentrations up to 4000 mg/l and duration dependence up to 24 hours (Table 2). Several workers have reported that KNO_3 improved the seed germination of many plants seed [3,11]. Potassium nitrate was found to be effective in breaking dormancy of many species [1]. Use of KNO_3 has been an important seed treatment in seed-testing laboratories for many years without a good explanation for its action [5].

Soaking time	KNO_3 dose (mg/l)				
	500	1000	2000	4000	8000
6 hour	6c*	6c	8b	12c	10d
12 hour	12b	14fg	20cd	24ab	14fg
24 hour	16ef	18de	20cd	26a	22bc
48 hour	18de	16ef	18de	22bc	18de

*Values followed by different letters are significantly different at 0.05 level using LSD. There was no germination in distilled water (zero KNO_3)

Table 2. Effect of different duration and concentration of KNO_3 on seed germination percentage of *C. spinosa*

2.3. Potassium Nitrate + Gibberellic acid treatment:

The highest germination percentage (72%) was observed in seeds placed in filter papers wetted with in 250 ppm gibberellic acid after treatment with 8000 mg/l KNO_3 for 24 hour (this duration was the best time span in two previous experiment), however there was no any significant difference between this germination percentage and that of 1000 ppm gibberellic acid after treatment with 4000 mg/lit KNO_3 (Table 3). It seems that GA_3 and KNO_3 can replace partly to improve seed germination of caper. No germination in case of control was possibly due to the seed coat of the capers that forms mucilage on soaking in water [12]. The mucilage surrounding the seed is supposed to inhibit diffusion of oxygen to the embryos and prevent germination. Khan and Ungar [7] believed that vegetative hormones can break embryo dormancy and neutralize prevention role of Abscissas acid (ABA) directly or indirectly. But we could not find the synergistic effects of GA_3 and KNO_3 on germination of caper and other species in the literature.

GA_3 dose (mg/l)	KNO_3 dose (mg/l)					
	0	500	1000	2000	4000	8000
0	0	16u*	18tu	20st	22rs	22rs
50	24qr	22rs	26q	36lm	36lm	44hi
100	30op	30op	32no	42ij	42ij	64bc

GA_3 dose (mg/l)	KNO_3 dose (mg/l)					
	0	500	1000	2000	4000	8000
250	30op	40jk	34mn	48fg	52e	72a
500	32no	44hi	42ij	50ef	66b	46gh
1000	38kl	46gh	44hi	62c	70a	48fg
2000	42ij	46gh	56d	46gh	50ef	46gh

*Values with in a column followed by different letters are significantly different at 0.05 level using LSD.

Table 3. Effect of different concentration of KNO3 + GA_3 on seed germination percentage of *C. spinosa*

2.4. Seedling establishment:

Significant effect of seed priming treatment was observed in the final seeding establishment. The highest seedling dry weight of 25 mg/plant was achieved when the seeds were treated with 100 ppm gibberellic acid treatment + 1000 ppm Potassium Nitrate (Table4). The germination phase of planted seeds is critical because it directly determines the density of a crop stand especially under arid conditions. Where dry soil may impair imbibitions of water and high temperature may affect seed viability and eventual density of a crop stand [4]. Hadas and Russo [4] further observed that a good stand can be ensured by a complete and fast germination and if germination seed is slow in taking up water, emergence is impaired and consequently the final stand is reduced.

GA_3 dose (mg/l)	KNO3 dose (mg/l)					
	0	500	1000	2000	4000	8000
0		17ghi	18fgh	19efg	19efg	18fgh
50	18fgh*	19efg	19efg	20def	20def	19efg
100	20def	21cde	22bcd	23abc	23abc	22bcd
250	20def	23abc	25a	24ab	23abc	22bcd
500	19efg	20def	22bcd	21cde	20def	18fgh
1000	16hij	18fgh	18fgh	19efg	17ghi	15ijk
2000	12l	13kl	14jkl	14jkl	13kl	12l

*Values followed by different letters are significantly different at 0.05 level using LSD.

Table 4. Effect of KNO3 + GA3 on seedling dry weight (mg) of *C. spinosa*

In general to improve caper seed germination, 2000 mg/l gibberellic acid treatment resulted in more vigorous seed germination (42%) at any duration compared to other concentrations. While, the highest seed germination of 26% was achieved when the seeds were treated 24 hour with 4000 ppm KNO_3 solution but it was decreased in 8000 mg/l. The highest germination percentage (72%) was observed in seeds placed in filter papers wetted with in 250 ppm gibberellic acid after treatment with 8000 mg/l KNO_3 for 24 hour (this duration was the best time span in two previous experiments). It seems that GA_3 and KNO_3 have a synergist effect and can replace partly to improve seed germination of caper. The highest seedling dry weight was achieved as seeds were treated in 100 ppm gibberellic acid + 1000 ppm potassium nitrate. Therefore, it can be concluded that for best germination percentage of caper seeds, 250 ppm GA_3 and 8000 ppm KNO_3 and for the strongest caper seedling 100 ppm GA_3 + 1000 ppm KNO_3 could be recommended for dormancy breaking of caper seeds.

3. Reference

[1] Agrawal, P., K. and M. Dadlani. Techniques in seed science and technology. Second Edition. South Asian Publishers, New Delhi International Book Company Absecon Highlands: 1995. pp. 109–113.

[2] Arteca, R., N. Plant growth substances principles and applications. Chapter 3: Chemistry, Biological Effects and Mechanism of Action, 1996. p: 66. Chapman & Hall 115 Fifty Avenue New York, NY 10003.

[3] Cirak, C., A. K. Ayan and K. Kevseroglu. 2004. The effect of light and some presoaking treatments on germination rate of St. John's worth (*Hypericum perforatum* L.) seeds. Pakistan J. Biol. Sci., 7(2): 182-186.

[4] Hadas, A. and D. Russo. 1974. Water uptake by seeds as affected by water stress, capillary condectivity and seed-soil water contact. Agron. J., 66: 643-647.

[5] Hartmann H. T., D. E. Kester, F. Jr. Davies, and R. L. Geneve. 1997. Plant propagation principles and practices. Sixth Edition. New Jersey, Prentice Hall.

[6] Iglesias, R. G., and M. J. Babiano. 1997. Endogenous abscisic acid during the germination of chickpea seed. Physiol. Planta., 100: 500–504.

[7] Khan, M. A, and I. A. Ungar. 1985. The role of hormones in regulating the germination of polymorphic seeds and early seedling growth of *Atriplex triangularis* under saline conditions. Physiol. Planta., 63: 109–113.

[8] Mayer, A. M. and Y. Shahin. 1974. Control of seed germination. Annual Rev. Plant Physiol., 25: 167–93.

[9] Movafeghi A., Gh. Habibi and M. Aliasgarpoor. 2008. Plant regeneration of *Capparis spinosa* L. using hypocotyl explants. Iranian J. Biology, 21(2):289-297.

[10] Negbi, M., E. Rushkin and D. Koller. 1966. Dynamic aspects of water relations in germination of *Hirschfeldia incana* seeds. Plant Cell Physiol., 7: 363–76.

[11] Olmez, Z., Z. Yahyaoglu and A. Ö. Üçler. 2004. Effects of H_2SO_4, KNO_3 and GA3 treatments on germination of caper (*Capparis ovata* Desf.) seeds. Pakistan Biol. Sci., 7 (6): 879-882.

[12] Orphanos, P. I. 1983. Germination of caper (*Capparis spinosa* L.) seeds. J. Hort. Sci., 58: 267–70.

[13] Ramazani M., M. Taghvaei, M. Masoudi, A. Riahi and N. Behbahani. 2009. The evaluation of drought and salinity effects on germination and seedling growth of Caper (*Capparis spinosa* L.). Iran. J. Rangeland, 2(4): 411-420.

[14] Sakcali M. S., H. Bahadir and M. Ozturk. 2008. Eco-physiology of *Capparis spinosa* l. : a plant suitable for combating desertification Pak. J. Bot., 40(4): 1481-1486.

[15] Shirazi, A. M. 2003. Standardizing methods for evaluating the chilling requirements to break dormancy in seeds and buds (including geophytes): introduction to the workshop. HortScience, 38: 334–335.

[16] Yanmaguchi, S. and Y. Kamiya. 2002. Gibberellins and light stimulated seed germination. J. Plant Growth Regul., 20: 369–76.

Allelopathy an Environmentally Friendly Method for Weed Control

Hamid Sodaeizadeh[1], Zahra Hosseini[2]

1 Faculty of Natural Resources and Desert Study, Yazd University, Iran

2 Management and Planning Organization, Yazd County, Iran

Abstract

Biological weed management is a system that incorporates the use of diverse biological organisms and biologically-based approaches including allelopathy, crop competition, and other cultural practices to significantly reduce weed densities in a manner that is similar to use of chemical herbicides alone. Interest in developing effective biological weed management systems continues to increase because of a growing awareness of problems associated with the constant and intensive use of chemical herbicides, which include surface- and groundwater contamination, detrimental impacts on nontarget organisms, development of weeds resistant to herbicides, and consumer concerns for residues on food. Among different biological methods of weed control, allelopathy could lead to reduced labour costs and increased efficiency, without any adverse effects on the environment. Many of the compounds produced by green plants that are not involved in primary plant metabolism are observed to function as chemical warfare agents against competing plants and pests. Many such natural compounds have the potential to be exploited as herbicides or as leads for discovery of new herbicides. The paper highlights the different concepts of using allelopathy for eco-friendly control of weeds.

Keywords: allelopathy, allelochemicals, natural compounds, weeds control

1. Introduction

The phenomenon of allelopathy, whereby a plant species chemically interferes with the germination, growth or development of other plant species has been known and documented for over 2000 years.

The term allelopathy, however, was first coined in 1937 by the Austrian Professor Hans Molisch from two Greek words: allelon 'of each other 'and pathos 'to suffer'and means the "injurious effect of one organism upon the other" [16]. Today, the term is generally accepted to cover both inhibitory and stimulatory effects of one plant on another plant [16]. In 1996, the International Allelopathy Society defined allelopathy as follows: "The science that studies any process involving secondary metabolites produced by plants, micro-organisms, viruses, and fungi that influence growth and development of agricultural and biological systems (excluding animals) "[24]. Nowadays, allelopathy has a significant role in research involving sustainable agriculture, like biological weed and pest control [3]. The current trend is to find a biological solution to minimize the perceived hazardous impacts from herbicides and insecticides in agriculture production. In this regards, the harmful impact of allelopathy can be exploited for pest and weed control [7, 20].

The chemicals responsible for the phenomenon of allelopathy are generally referred to as allelochemicals or phytotoxins [8]. Allelochemicals are usually classified as secondary metabolites and are produced as offshoots in the primary metabolic pathways of plants [9]. Many such natural compounds have the potential to induce a wide array of biological effects and can provide great benefits to agriculture and weed management [3, 10].

2. Allelopathy for Weed Management

The word weed means any wild plant that grows at an unwanted place for example in fields and interferes with the growth of cultivated plants [17]. Farmers must contend with approximately 30,000 plant species identified as weeds. Among them, 250 are really important and about 80 are known to reduce crop yield [8].

Weeds have substantially adapted characteristics (e.g. produce an abundance of seed, rapid seedling growth, quick maturation, dual modes of reproduction, environmental plasticity) that enable them to grow, flourish, invade and dominate an important part of natural and agricultural ecosystems [8, 25]. In agro-ecosystems, weeds compete with crop plants for resources, interfere in crop handling, reduce crop yield and deteriorate their quality, and thus result in huge financial losses [8] . Degree of loss depends on crop species present, timing and duration of competitive interactions, and resource availability [1]. Oerke et al. [12] reported that weeds, pathogens and animal pests cause a loss of around 13.2, 13.3 and 15.6% (totally 42.1%), respectively, in the eight most important food and cash crops, even when they are intensively controlled. However, if no physical, biological or chemical measures were used to protect crops, yield losses would be around 69.8%. So, losses prevented by crop protection measures are about 27.6% of attainable production. The only basis on which it is possible to calculate an overall figure for crop losses in all crops is financial one. In US agriculture, weeds cause an overall reduction of 12% in crop yields, and this represents approximately $32 billion in lost crop production each year (USCB 2007). In addition to the direct losses, approximately $4 billion is spent each year on herbicides used to control pest weeds. Thus, the total annual cost of introduced weeds to US agricultural economy is about $36 billion [13].

In light of these characteristics of weeds and their hazards, it becomes imperative to control them. Several techniques (e.g. mechanical and chemicals) are used for weed control. These techniques attempt to achieve a balance between cost of control and crop yield loss [8, 20]. Mechanical methods, such as hand weeding require enormous labour and time input. Nowadays, chemical method provides an effective strategy for weed control. Since their discovery in the 1950s, synthetic herbicides have developed as a major tool for weed management. Herbicides have helped farmers to increase yields while reducing labour. Indeed, without herbicides, labour would be a major cost of crop production in developed countries. Nevertheless, the indiscriminate use of herbicides has provoked an increasing incidence of resistance in weeds to some herbicides, changes in weed population to species more related to the crop, environmental pollution, and potential health hazards [10]. Overuse of synthetic chemicals for weed control worsens the quality of soil, water, other life support systems, human health and food [21]. Fast-developing herbicide-resistant ecotypes of weeds are also posing serious threats to agricultural production. So far, at least 334 weed-resistant biotypes belonging to 190 species (113 dicots and 77 monocots) toward herbicides have been identified (www. weedscience.com). Because of all these problems, efforts are being made to find out alternative low-input strategies for weed management. In this regard, much attention has been focused on the use of allelopathic plants and their products for managing weeds in a sustainable manner [21]. Natural products release from allelopathic plants may help to reduce the use of synthetic herbicides for weed management and therefore, cause less pollution, safer agricultural products as well as alleviate human health concerns [6]. So, it is worthwhile to explore the potential of plants with strong allelopathic activity for the management of agricultural weeds.

The use of allelopathy for controlling weeds could be either through directly utilizing natural allelopathic interactions, particularly of crop plants, or by using allelochemicals as natural herbicides. In the former case, a number of crop plants with allelopathic potential can be used as cover, smother, and green manure crops for managing weeds by making desired manipulations in the cultural practices and cropping patterns. These can be suitably rotated or intercropped with main crops to manage the target weeds (including parasitic ones) selectively. Even the crop mulch/ residues can also give desirable benefits.

3. Allelochemicals as natural herbicides

There is increasing evidence that allelochemicals or natural plant products derived from higher plants/microbes can be ideal agrochemicals. Initially, the reason why plants devote resources to the production of these compounds was not understood as they were regarded as functionless waste products. It is now increasingly accepted, however, that these compounds function as defensive agents against pathogens, insects and neighbouring plants [11]. Many such natural compounds have the potential to induce a wide array of biological effects and can provide great benefits to agriculture and weed management [3, 10]. Evidence showed that higher plants release a diversity of allelochemicals into the environment. Despite so much chemical diversity, allelochemicals can be broadly characterized into phenolics and terpenoids. They are released by volatilization, root exudation, death and decay of plants, and leachation from living or decaying residues [1, 18]. After release, allelochemicals are involved in a variety of metabolic processes [18]. Several factors determine their toxicity such as concentration, flux rate, age and metabolic state of plant, and prevailing climatic and environmental conditions [18]. Their amount and production varies in quality and quantity with age, cultivar, plant organ, and time of the year. Einhellig [5] mentioned that both abiotic (temperature, nutrient amount, and moisture deficit) and biotic (disease and insect damage and interaction of plans with herbivory) factors enhance the amount and biosynthesis of allelochemicals in plants.

These allelopathic chemicals are produced by a 'donor' and transmitted to a 'receiver' that can either be 'injured' or 'stimulated'.

Allelochemicals act through direct interference with physiological functions of 'receiver' such as seed germination, root growth, shoot growth, stem growth, symbiotic effectiveness or act indirectly through additive or synergistic impact along with pathological infections, insect injury and/or environmental stress. Though many of these allelochemicals exhibit inhibitory response on various morpho-physiological functions of receiver plants and such responses being observed to be dose dependant in a linear fashion, their concentrations required for control of weeds on a field scale are impracticably higher.

4. Discussion

Despite herbicidal activity of allelopathic plants, to attain significant weed reduction under field conditions a large quantity of plant materials or pellets is required. This needs heavy field work. Therefore, the possibility of its periodical application for greater weed control should be further examined [14]. The various combinations of allelopathic plants and herbicides to reduce dependence on synthetic herbicides should be tested [15]. In addition, a combination of different allelopathic plant species with strong weed-suppressing ability, may be capable of controlling more weed species than a single allelopathic plant species. Another alternative to reduce field work is to select allelochemicals from various sources, such as plants or microorganisms, and use them as herbicides in place of synthetic chemicals [15] . This procedure can have desirable results, because most natural products are broken down rather rapidly by common microorganisms and thus are not persistent pollutants in the environment, as are many of the synthetic herbicides. Among the plant products as herbicides, juglone, isolated from walnut tree has been found effective against redroot pigweed, velvetleaf and barnyard grass [22, 23]. Caffeine derived from coffee showed considerable selectivity in inhibiting germination of *Amaranthus spinosus* L. at a concentration that has no effect on blackgram [17]. Strigol, a sesquiterpenoid derivative from cotton roots is a potent germination stimulant of witchweed (Striga asiatica L. Kuntz), an obligate parasite of maize, sorghum [2] and Orobanche minor [22]. Dhurrin (sorghum); gallic acid (spurge); Phlorizin (apple root); trimethylxanthene (coffee) and cinch (eucalyptus) are some other important plant products having promising herbicidal activity. I this regard continuous study on isolation and identification of allelopathic compounds in plants and rhizosphere should be conducted. Although many biologically active compounds have been found, we still need to explore new compounds from plants and microorganisms.

5. Conclusions

Increasing attention has been given to the role and potential of allelopathy as a management strategy for crop protection against weeds and other pests. Incorporating allelopathy into natural and agricultural management systems may reduce the use of herbicides, insecticides, and other pesticides, reducing environment/soil pollution and diminish autotoxicity hazards. There is a great demand for compounds with selective toxicity that can be readily degraded by either the plant or by the soil microorganisms. In addition, plant, microorganisms, other soil organisms and insects can produce allelochemicals which provide new strategies for maintaining and increasing agricultural production in the future.

6. References

[1] Anaya, A.L., 2006. Allelopathic organisms and molecouls : Promising bioregulators for the control of plant diseases, weeds and other pests. In: Inderjit, & Mukerji, K. G., (Eds.). *Allelochemicals: biological control of plant pathogens and diseases.* Springer. Dordrecht, The Netherlands.

[2] Cock, C.E., Whichard, L.P., Turner, B., Well, M.E. and Egley, G.H. 1966. Germination of witchweed (*Striga lutea* Lour) : Isolation and properties of potent stimulant. *Science*, 154: p. 1189-1190.

[3] Duke, S.O., and J. Lydon, 1987. Herbicides from natural compounds. *Weed Technology*, 1: p. 122-128.

[4] Economou, G., O. Tzakou, A. Gani, A. Yannitsaros, and D. Bilalis, 2002. Allelopathic effect of *Conyza albida* on *Avena sativa* and *Spirodela polyrhiza*. *Journal of Agronomy and Crop Science*, 188, 248.

[5] Einhellig, F. A, 1996. Interactions involving allelopathy in cropping systems. *Agron. J*, 88: p. 886–893.

[6] Khanh, T.D., A.A. Elzaawely,. I.M. Chung, J.K. Ahn, S.Tawata, and T.D. Xuan, 2007. Role of allelochemical for weed management in rice. *Allelopathy Journal*, 19: p. 85-96.

[7] Kohli, R. K., D. Batish, and H. P. Singh, 1998. Allelopathy and its implications in agroecosystems. *J. Crop Prod*, 1: p. 169–202.

[8] Kholi, R.K., H.P. Singh, and D.R. Batish, 2004. Allelopathy in agroecosystems. Food Products Press. New York, USA.

[9] Kruse, M., M. Strandberg, and B.Strandberg, 2000. Ecological effects of allelopathic plants: a review. Silkeborg, Denmark, National Environmental Research Institute.

[10] Macías, F.A., N. Chinchilla, R.M. Varela, and J.M.G. Molinillo, 2006. Bioactive steroids from *Oryza sativa* L. *Steroids*, 71: p. 603-608.

[11] Mattner, S.W. 2006. The impact of pathogens on plant interference and allelopathy In: Inderjit, & Mukerji, K. G., (Eds.). *Allelochemicals: Biological control of plant pathogens and diseases.* Springer. Dordrecht, The Netherlands.

[12] Oerke, E.-C., H.-w. Dehne, F. Schonbeck, and A. Weber, 1995. Crop production and crop protection: estimated losses in major food and cash crops. Elsevier Science B.V. Amesterdam.

[13] Pimentel, D. 2009. Invasive plants: their role in species extinctions and economic losses to agriculture in the USA. In: Inderjit, (Ed.). *Management of invasive weeds.* Springer Verlag, New York.

[14] Rice, E.L. 1984. Allelopathy. Academic Press,Orlando.

[15] Rice, E.L. 1995. Biological Control of Weeds and Plant Diseases : Advances in Applied Allelopathy. University of Oklahoma Press

[16] Rizvi, S.J.H., and V. Rizvi, 1992. Allelopathy: basic and applied aspects. Champan & Hall. London SE1 8HN.

[17] Rizvi, S.H.J., Mukerjee, D. and Mathur, S.N. 1981. Selective phytotoxicity of 1, 3, 7-trimethylaxanthine between *Phaseolus mungo* and some weeds. *Agricultural Biological Chemistry*, 45 : p. 1255-1256

[18] Singh, H.P., D.R. Batish, and R.K. Kohli, 1999. Autotoxicity: Concept, Organisms, and Ecological Significance. *Critical Reviews in Plant Sciences*, 18: p. 257-272.

[19] Singh, H.P., D.R. Batish, and R.K. Kohli, 2003. Allelopathic interactions and allelochemicals: new possibilities for sustainable weed management. *Critical Reviews in Plant Sciences*, 22, 239.

[20] Sodaeizadeh, H., M. Rafieiolhossaini, J. Havlík, and P. Van Damme, 2009. Allelopathic activity of different plant parts of *Peganum harmala* L. and identification of their growth inhibitors substances. *Plant Growth Regulation*, 59: p. 227–236

[21] Sodaeizadeh, H., M. Rafieiolhossaini, and P. Van Damme, 2010. Herbicidal activity of a medicinal plant, *Peganum harmala* L., and decomposition dynamics of its phytotoxins in the soil. *Industrial Crops and Products*:31: p. 385- 394.

[22] Spelce, D.L. and Muselman, L.J. 1981. Orobanche minor germination with strigol and G.R. Compounds. *Zeitschrift Pflanzenphysiologie* 104: p.281-283.

[23] Weston, L.A., Burke, B.A. and Putnam, A.R. 1987. Isolation, characterization and activity of phytotoxic compounds from quackgrass (Agropyron repens (L.) Beauv). *Journal of Chemical Ecology*, 13: p. 304-431.

[24] Xuan, T.D., S. Tawata, T.D. Khanh,.and I.M. Chung, 2005. Decomposition of allelopathic plants in soil. *Journal of Agronomy and Crop Science*, 191: p. 162-171.

[25] Zimdahl, R.L. 1999. Fundamentals of Weed Science Academic Press. New York, USA.

Investigating Importance and Effects of Climate Changes in Agriculture in South Khorasan Province and Recognizing Appropriate Extension Education Activities in Confronting Them

Farhood Golmohammadi[1], Mohsen Arazmjoo[2] , Seyed Hamid Razavi[3]

1 Islamic Azad University, Birjand Branch – Birjand, Iran

2 Agricultural Research- Education and Extension Organization_ Birjand Branch – Birjand, Iran

3 Oloome Entezami University- Birjand, Iran

Abstract

In this article, author state a brief to drought management as an essential approach for regional development and maintaining employment in South Khorasan province in Iran. Rainfall is the ultimate source of water, affecting production of crops and other biomass by direct falling on the fields as well as supporting surface and ground water irrigation. However, possibilities of drought occurrence in Iran vary from once in 20 years. The frequency and intensity of extreme weather events like droughts, floods, heat/cold waves, cyclones, delayed or early onset, long dry spells, early withdrawal, and floods in drought frequented areas and droughts in flood afflicted areas have increased during the last two decades due to global warming. Since drought is defined by deviation from the normal rainfall, it can happen in all regions. Assessment and management of drought is complex due to its gradual appearance and long lasting impact or recoveries. Characteristics and impact of drought vary from region to region and year to year. Drought affects human, livestock, wildlife, bio-diversity and degrades the quality of natural resource base. This article is part of a research project titled: ((Investigating importance and effects of climate changes in agriculture in south khorasan province and recognizing appropriate extension education activities in confronting them)) that by author has been done in the Agricultural Research, Education and Extension Organization_ Birjand Branch – Birjand, South khorasan province of Iran.

Keywords: Drought, Impact, desert, south Khorasan province, Iran.

1. Introduction

Drought is universal phenomenon that can occur everywhere and can cause harmful impacts on human beings and natural ecosystems. Growing public awareness of the issue of global climate change has raised enormous concerns regarding its potential impacts and consequences. Although there are inconclusive findings on the specific impacts of climate change on regional water resources, many scientists have suggested that climate change is likely to increase the frequency and intensity of extreme climate events such as drought. Drought based on frequency of occurrence, severity, affected area, economic damages, environmental and social affects and severe long term impacts is very important and dangerous phenomenon compared to other disasters. Drought is one of the most important natural disasters which could be defined as: less than average annual rainfall and discordant distribution of rainfall in the region. With lack of rainfall for a long period of time; farms, gardens, pastures, and forests which their required water resources are provided from the atmosphere rainfall are damaged directly. Particularly, agriculture which has an important role in national economy and is a set of activities that its aim is supplying food needs of community and produce raw materials for other sectors including industry. Iran, with a rainfall average about 252 mm in year is among the dry regions in the world. Low rainfall, irregular distribution of rainfalls and climate warming, causes economic, environmental, political and social crisis in different areas. In recent years, impacts of drought were high on water sources, agriculture, livestock production, pastures, migration, rebellion of pests and disease. Drought can also reduce water quality, because lower water flows reduce dilution of pollutants and increase contamination of remaining water sources. Studies indicate that drought has priority to other natural disasters in the frequency of occurrence, duration, extent, loss of life, economic and social impacts and severe effects in the long run. Damages of drought will affect economic, environmental and social status of communities. Drought includes a set of negative effects which not only affect economic and social activities of farmers and related industries, but also affect those who are not actually employed in agriculture but are living in agricultural regions. Despite the relatively small share of agriculture in Australian GDP, the drought has reduced GDP by 1.6%, and has contributed to a decline in unemployment and to a worsening of the balance of trade. In a surveyed environmental, economic and social effects of drought and effect of solutions applied in order to reducing its effects in Sistan Province (Iran) and concluded that between the effects of drought (environmental, economic and socio- psychology) the economic effects are more than others, then environmental impacts and at last the least impacts were socio-psychology effects (Golmohammadi, 2006,2007,2011). Rezavi et al. (2011) surveyed economic, social, environmental, and ecological impacts of drought in Zanjan province and concluded that these impacts were economic, environmental, social and ecological respectively. Nuri and Bazrafshan (1996-2004) investigated direct and indirect effects of drought on rural economy of Sistan, and stated that direct effects include damage to crops, horticulture and animal husbandry and indirect effects include an increase in the population covered by the support relief organizations, an increase in migration from rural regions, reducing price of agricultural lands and orchards and also change rural economic structuring. South Khorasan Province in -East of Iran is one of the Provinces that in recent years have caught with very strong drought and water shortage. This problem has created many limitations and negative impacts for farmers. Drought and water shortage has more effect on this region. Reliable statistics indicate that South Khorasan Province in -East of Iran will face with severe water shortage and drought in the future and these problems, more than anything, would affect Barberry and Saffron production.

Figs 1., 2. Barberry as one of main agricultural product that are resistant to drought conditions in South Khorasan Province in -East of Iran (by author, 2010).

Figs 3., 4. Saffron as one of main agricultural product that are resistant to drought conditions in South Khorasan Province in -East of Iran (by author, 2010).

Figs 5., 6. seasonal floods that not properly storing as a main source of water for resistant and living people in drought conditions in South Khorasan Province in -East of Iran (and author Jun 2012).

Figs 7., 8. Some locally and medium dam projects that recently building for gathering and storing seasonal floods as a main source of water for resistant and living people in drought conditions in South Khorasan Province in -East of Iran (and author Jun 2011).

2. Materials and Methods

This article is part of a research project titled: ((Investigating importance and effects of climate changes in agriculture in south khorasan province and recognizing appropriate extension education activities in confronting them)) that by author has been done in the Agricultural Research, Education and Extension Organization_ Birjand Branch – Birjand, South khorasan province of Iran.

3. Results

Drought is one of the most important natural disasters which affect on economic, environmental and social conditions of communities. Also it might occur even in all humid and semi-humid areas, although the details and rates can be different from one region to another. south Khorasan province in -East of Iran is one of the high deserted provinces of Iran that in recent years caught with drought and water shortage. This problem, in addition reduce yield of barberry and saffron in this province, leaded to other direct and indirect environmental and socio-economical problems for barberry and saffron farmers. The results of this research indicated that between environmental, economic and social impacts of drought, the economic effects were more than others, then environmental impacts and at last were social effects. This result is consistent with the finding of shokri (2005) and Rezayi et al (2011).

In viewpoint of barberry and saffron farmers, in economic part, drought leaded to; increase in costs labor and eradicating weeds, increase in costs for water supply, decrease in purchasing power, decrease in savings, non-payment of bank loans and obligations, increase in the false financial relationship, decrease in price of crops due to reduction of quality, decrease in income due to reduction of cultivation, decrease in land price, decrease in income from side jobs, respectively.

In environmental part, drought resulted in; Decrease in rivers flow, groundwater levels, Decrease in surface water reservoirs and ponds, Increase in weeds growing in fields, Increase in mortality of fish and other aquatic in ponds, Decrease in water quality, Increase in pest attack,

Increase in plant diseases, Increase in soil erosion, Increase in amount and intensity of fires, Decrease in diversity of plant species respectively.

In social part, it resulted in increase in frustration, anxiety and emotional problems, feeling of poverty and decrease in life level, decrease in recreational activities , increase in local divisions to supply water, weakened position of institutions and cooperative unions, weakened traditions of cooperation, increase in tend to migrate, decrease in social ceremonies, decrease in the level of education in children and juveniles, disintegrate of consistency and continuity in family system respectively. We also recommend following solutions:

- To accept the risk of assigning the responsibilities to people for management drought conditions in South Khorasan Province in order to empower them;
- Constructive interaction with the offices, plays the most critical role to achieve for objectives of management drought conditions in South Khorasan Province in order to empower them;
- Formation of management drought conditions is the most important factor for participation of local communities towards sustainable rural development;
- Multidisciplinary and integrated planning in addition to bottom-up approach decision making are the most important factor in the success of management drought conditions and realization of sustainable rural development;

4. References

[1] F. Golmohammadi, M.K. Motamed, A viewpoint toward farm management and importance of barberry in sustainable rural livelihood in desert regions in east of Iran. African Journal of Plant Science Vol. 6(7), pp. 213-221, 6 April, 2012 . Available online at http://www.academicjournals.org/AJPS DOI: 10.5897/AJPS11.073 ISSN 1996-0824 ©2012 Academic Journals

[2] F. Golmohammadi, Study of Education and Extension Methods of Indigenous Knowledge of Rural Architecture (Case Study: Systan and Balouchestan Province). In: JOURNAL OF Housing And Rural Environment. Journal of Housing Foundation of Islamic Revolution. TEHRAN. IRAN. ISSN: 2008-4994. No. 136, Winter, 2012. 2, 2012; 30 (136) :79-94.URL http://www.jhre.ir/browse.php?a_code=A-10-177-3&slc_lang=fa&sid=1

[3] F. Golmohammadi, Preparing Systematic Functional Pattern of Extension to Attaining Sustainable Development of Agriculture. Title of my Dissertation for Ph.D. IN AGRIULTURAL EXTENSION AND EDUCATION . FROM : ISLAMIC AZAD UNIVERSITY_ SCIENCE AND RESEARCH BRANCH –TEHRAN. IRAN.2002-2007.

[4] F. Golmohammadi, The Role of Extension Education in Participating Rural People toward Establishing and Maintaining Rural Parks in Isfahan Province. IRAN. M.SC. Thesis IN AGRIULTURAL EXTENSION AND EDUCATION. TARBIAT MODARRES UNIVERSITY(T.M.U.). IRAN.1995-1998.

[5] F. Golmohammadi, Role of farmers' education and Extension Endeavours in Productivity and water consumption management). Poster presentation. Published in the Proceedings of the Regional Congress of Sustainable Management Science – based in Agriculture and Natural Resources. Gorgan Univ. of Agric. Sci. & Natur. Resour. Page:31. In the:((Regional Congress of Sustainable Management Science – based in Agriculture and Natural Resources)). Gorgan Univ. of Agric. Sci. & Natur. Resour. Gorgan. Iran. 21-22 May 2011.

[6] H. Nuri, and J. Bazratshan, Effects of droughts 1998-2003 on rural economies of Sistan and solutions to deal with it. Journal of Geography and Development. P. 53-72. 2005.

[7] K. Peter, Impact of drought on the social well-being of rural communities and farm families. Submission from the Department of Justice/ Office of the Emergency Services Commissioner. Submission number 143. From http: //www.daff.gov.au/data/assets/pdf. 2008.

[8] R. Rezayi, M. Hosseini, and L.Safa, Studding and identifying the economic, social, environmental, and ecological impacts of drought in Zanjan township. The first national conference of drought and climate change. PP 686-693. 2011.

Effect of Fungal Growth Inhibition from Pomegranate Flower and Peel Extracts

Mahsa Shafighi [1], Leila Amjad[2], Mahboubeh Madani[3]

1 Young Researchers Club, Falavarjan Branch, Islamic Azad University, Isfahan, Iran

2 Departments of Biology, Falavarjan Branch, Islamic Azad University, Isfahan, Iran

Departments of Microbiology, Falavarjan Branch, Islamic Azad University, Isfahan, Iran

Abstract

Candida species are now one of the most common organisms isolated from hospitalized patients. The range of antifungal agents available is limited, and some of the most effective agents are also toxic. In the other hand plants have been used for thousands of years to flavor and conserve food, to treat health disorders and to prevent diseases including epidemics. The aim of this study was to investigate the antifungal effect of petroleumether, ethyl acetate and n-butanol fractions isolated from pomegranate pericarp and flower against Candida albicans (ATCC 3153). The maximum inhibition zone of Candida albicans was obtained by peel's n-butanol fraction, 35 mm. Petroleumether fractions had no any antifungal activities.

Keywords: Punica granatum L., peel, flower, petroleumether, ethyl acetate, n-butanol

1. Introduction

The *Punica granatum L.*, is one of the oldest known edible fruits. It is native to Persia and from there it spreads into Asia, North Africa and Mediterranean Europe, including Turkey [1, 2]. According to Qur'an, Bible, Torah and Babylonian Talmud, pomegranate is a gif and heavenly fruit of God.

The extracts of traditional herbs have been shown to exert biological activity in vitro and in vivo. *Punica granatum* is employed in man medicine for the treatment of various diseases such as skin diseases, and wound healing, ulcers, fever, diarrhoea, and microbial infection. In the recently years, the *Punica granatum* has been the subject of much scientific research which have showed its antimicrobial, antioxidant and anti-cancer effects [3, 4].

The different types of phytochemical that have been showed identified from pomegranate pericarp (peel, rind) and pomegranate flower. Pomegranate pericarp's constituents are Luteolin, kaempferol, EA glycosides, EA, Pedunculagin, punicalin, phenolic punicalagins, Gallic acid and other fatty acids, catechin, EGCG, quercetin, rutin and other flavonols, flavones, flavonones, anthocyanidins. Pomegranate flower's constituents are Polyphones, punicalagins punicalin, EA, Gallic acid, ursolic acid, triterpenoids, including maslinic and Asiatic acid [1, 5- 7].

Candida species are harmless saprophyte yeasts, a normal component of the human biota in the gastrointestinal tract and oral and vaginal mucosae. These yeasts can cause superficial infections such as thrush and vaginitis; however, if the immune defences of the host become compromised, they can cause severe systemic infections. Risk factors for patients include infection by the human immunodeficiency virus (HIV), anticancer therapy, organ transplantation, abdominal surgery, catheters, diabetes and the use of broad-spectrum antibiotics [4, 11].

The aim of this study was to investigate the antifungal effect of some fractions isolated from pomegranate pericarp and flower against *Candida albicans.*

2. Materials and Methods

Samples of *P. Granatum* flowers and peels were collected and identification in June and September 2011 respectively from the Agricultural Research Centre in Isfahan, Iran.

The flowers and peels were air-dried in a low light at room temperature for 1 weak. The material was thereafter ground in an electric grinder to produce a powder separately.

Peels and flowers powder extracted with 100% petroleumether (F1) at room temperature for 24 hours respectively, the extracts were filtered. Then 20 gram of remains powders were extracted with 75% Ethanol by Soxhlet extraction for 8 hours.

The residues were dried over night and then evaporated by using a rotary evaporator. The dried extracts were suspended in distilled water (F2) and partition with ethyl acetate (F3) followed by n-butanol (F4). All fractions were frigid at -20°C until each experiment.

The yeast strain *C. albicans* (ATCC 3153) were used in this study. It obtained from Institute of Scientific and Industrial Researches, Tehran, Iran. At the first, it has been maintained at 4°C on Sabouraud dextrose agar (SDA) plates and sub cultured at 25°C in Sabouraud dextrose broth (SDB) before each experiment to ensure viability and purity.

Petri dishes contained 20 ml of SDA have been used for well-diffusion assay. Wells have been prepared in the SDA plates. In agar well diffusion, 10, 50, 100 and 200 µl of each fractions have been inoculated to each well separately. Then 100 ml of 10^6 CFU/ml yeast suspension was spread uniformly onto the agar plate using cotton swabs. Diameters (in mm) of growth inhibition zones were measured after incubation at 25°C for 24- 48 hours.

3. Results and Discussion

Candida species are now the fourth most common organism recovered from the blood of hospitalized patients. Notwithstanding the increasing need for effective therapy, the range of antifungal agents available is limited, and some of the most effective agents are also toxic. In addition, although azoles have been used successfully for the treatment of Candida infections, numerous reports of treatment failures are now appearing in the literature [4]. In view of the lack of new classes of drugs or different molecular targets, drug combinations might be considered a viable strategy for therapy, considering the multiplicity of fungal targets against which current agents are effective.

The antimicrobial and antifungal effects of alcoholic extracts of different parts of pomegranate tree were previously studied [1, 8-11]. Nevertheless, few studies have showed the antifungal effect of different pomegranate fractions. This study showed the antifungal activity of 4 pomegranate fractions. Antifungal effects of peel and flower fractions by well-diffusion assay method have been shown in Figure 1 and Figure 2 respectively.

The maximum inhibition zones of peel fractions against C. albicans ATCC 3153 were obtained in 200 μl concentrations by n- butanol fraction, water fraction and ethyl acetate fraction respectively.

C. albicans ATCC 3153 only was sensitive to flower n- butanol fraction. In addition C. Albicans ATCC 3153 was resistant to peel and flower Petroleumether fractions.

The fractions had no any antifungal activity in 10 μl, and there was direct relationship between concentration and inhibition zone.

Also, n- butanol, ethyl acetate, water and petroleumether had no any antifungal activities. In the other words, the extracted compounds of peel and flower were effective [4].

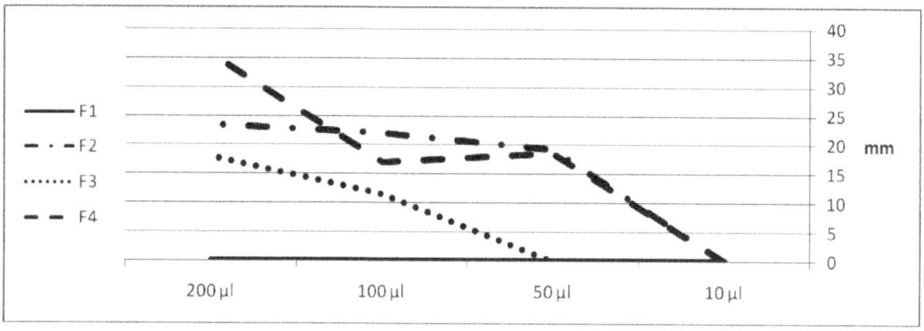

Figure 1. The inhibitory zons (mm) of peel's fractions in 4 concentrations against *C. albicans* (ATCC 3153). F1: Petroleumether Fraction, F2: Water Fraction, F3: Ethyl acetate Fraction, F4: n-butanol Frcation.

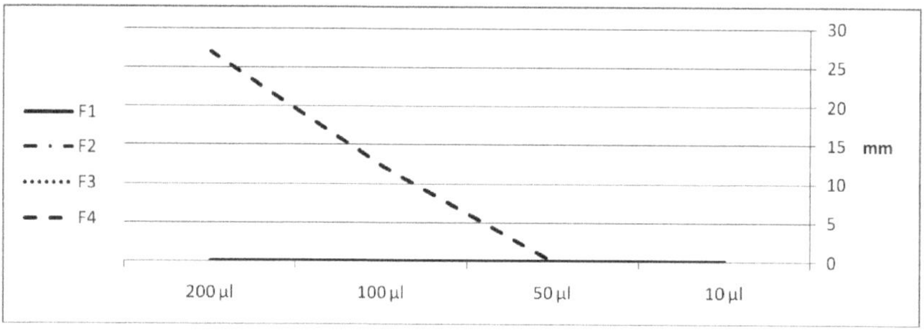

Figure 2. The inhibitory zones (mm) of flower's fractions in 4 concentrations against C. albicans (ATCC 3153). F1: Petroleumether Fraction, F2: Water Fraction, F3: Ethyl acetate Fraction, F4: n-butanol Frcation.

Acknowledgment

This work was supported by Young Researchers Club, Falavarjan Branch, Islamic Azad University. The authors also thank Dr. Ranjbar, Dr. Khatabakhsh and Mrs. Shahsar for their kindly aid.

4. References

[1] Saad Sabbar Dahham, Mir Naiman Ali, Hajera Tabassum and Mazharuddin Khan, 2010. Studies on Antibacterial and Antifungal Activity of Pomegranate (*Punica granatum* L.). American-Eurasian J. Agric. & Environ. Sci., 9 (3): 273-281.

[2] Ahmet D. Duman, Mustafa Ozgen, Kenan S. Dayisoylu, Nurcan Erbil and Coskun Durgac, 2009. Antimicrobial Activity of Six Pomegranate (*Punica granatum L.*) Varieties and Their Relation to Some of Their Pomological and Phytonutrient characteristics. Molecules, 14: 1808-1817.

[3] Kannan Ponnusamya, Chelladurai Petchiammal, Ramasamy Mohankumar, Waheeta Hopper, 2010. In vitro antifungal activity of indirubin isolated from a South Indian ethnomedicinal plant Wrightia tinctoria R. Br. Journal of Ethnopharmacology, 132: 349–354.

[4] Eliana Harue Endo, Dio´genes Aparı´cio Garcia Cortez, Ta^nia Ueda-Nakamura, Celso Vataru Nakamura, Benedito Prado Dias Filho, 2010. Potent antifungal activity of extracts and pure compound isolated from pomegranate peels and synergism with fluconazole against *Candida albicans*. Research in Microbiology, 161: 534-540.

[5] M. Viuda-Martos, J. Fern´andez-L´opez, and J.A. P′erez-´Alvarez, 2010. Pomegranate and its Many Functional Components as Related to Human Health: A Review. ComprehensiveReviews inFoodScienceandFoodSafety, 9: 635-654.

[6] J. Pongpuntaruk, 2010. Chemical constituents from the stem of *Punica granatum* and the root of *mechelia alba,* 2010. Songkla University.

[7] Ephraim P. Lansky, Robert A. Newmana, 2007. *Punica granatum* (pomegranate) and its potential for prevention and treatment of inflammation and cancer. Journal of Ethnopharmacology, 109: 177–206.

[8] L. VASCONCELOS, F. C. SAMPAIO, M. C. SAMPAIO, M. S. Vieira PEREIRA, J. S. HIGINO, M. P. PEIXOTO, 2006. Minimum Inhibitory Concentration of Adherence of *Punica granatum Linn* (pomegranate) Gel against *S. mutans, S. mitis and C. albicans.* Braz Dent J, 17(3): 223-227.

[9] Silva NCC, Fernandes Júnior, 2010. Biological properties of medicinal plants: a review of their antimicrobial activity. The Journal of Venomous Animals and Toxins including Tropical Diseases, 3(16): 402-413.

[10] Abdollah Ghasemi Pirbalouti, Mahmood Bahmani, Majid Avijgan, 2009. Anti-*Candida* Activity of Some of the Iranian Medicinal Plants. Electronic Journal of Biology, 5(4): 85-88.

[11] Maria G. Miguel, Maria A. Neves and Maria D. Antunes, 2010. Pomegranate (Punica granatum L.): A medicinal plant with myriad biological properties - A short review. Journal of Medicinal Plants Research, 4(25), 2836-2847.

Effect of Heat on Egg White Proteins

Zoubida Akkouche*, Lyes Aissat, Khodir Madani

Faculty of Nature and Life Sciences, 3BS Laboratory A. Mira University, Bejaia, Algeria

*Corresponding author

Abstract

Thermal treatments applied to the egg white cause undesirable modifications of their physico-chemical and functional properties. The goal of this study was to acquire some knowledge on the effect of heat on some physico-chemical properties of the albumen while applying different scales (temperature / time). The results show: a sigmoïdale evolution with negative slope of the transmittance, an irreversible loss of solubility. Native-Page of protein's albumen shows the apparition of several bands what denotes diversity of proteins in egg white. The comparison of these profiles dealt to different temperatures /20 min with those unprocessed, shows apparition of new bands and different relative mobility due to the formation of aggregates. Coagulation and/or precipitation of these proteins have been observed around 74°C/20 min.

Keywords: Proteins, egg white, heat denaturation.

1. Introduction

Man has used bird eggs as food since prehistoric times [1, 2]. This reserve of proteins, lipids, vitamins and minerals is remarkable for varieties of nutrients it contains and their high nutritional value, due to a perfect balance of its constituents. Egg is always recognized as a food of high nutritional quality for humans [3, 4, 5, 6, 7]. Beyond their nutritional value, their primary function, namely the formation of a new individual, suggests that they contain many compounds essential to life, and therefore they constitute a major potential source of biological active molecules which interest the pharmaceutical, cosmetic and food [8, 9, 10].The food industry currently uses largely hen egg (especially egg white) as an ingredient, despite competition from increasingly strong dairy products. The origin of this interest is many functional properties of this latter (foaming and gelling properties in particular) [11].Egg white protein's are rich in essential amino acids, and possess excellent nutritional value, thus they have long been considered as reference proteins [12, 13, 14]. Nevertheless, the white gotten from breaking eggs is sensitive to contaminations of the environment [15, 16].To fight against these infections and protect the consumer, the heat treatment of food egg is often required to ensure microbial safety or to obtain desirable organoleptic attributes [17].The egg white is a hydrated medium, viscous, rich on protein, heat treatment would cause a distortion of its components, the non-enzymatic browning and coagulation of proteins [18, 19]. Polypeptide chains are unfolded and can form a gel or aggregates by the association of these molecules [20]. This aggregation leads to significant changes in the physico-chemical and functional properties of egg proteins [20, 21, 22, 23, 24].It is therefore very important that manufacturers in the industry provide an egg product that combines functionality and safety [25].To limit the damaging effects of heat treatment, some researches propose the addition of additives to increase protein stability to heat. Plant's extracts rich in polyphenols as caffeic acid, helps to increase the thermal stability of milk proteins to the heat [26]. Chemical structure of these substances makes them a highly developed ability to bind to all sorts of molecules, primarily proteins [27].

The purpose of this study was to gain knowledge on the effect of heat on some properties of egg white proteins. This new knowledge will be useful in improving methods of pasteurization and sterilization to reduce the negative impact of heat on technological properties of egg white.

2. Materials and methods

2.1. Preparation of egg white

After washing and cleaning of eggs bought in trade, they are carefully broken and the white and yolk separated, taking the precaution of removing chalazae. The pH of the albumen is measured immediately using a pH meter (Hanna Instruments, Microprocessor pH211 pH Meter); it is equal to 9.45 ± 0.02.

The volume of egg white is measured using a test-tube and then adjusted with the buffer solution Tris-HCl (20 mM pH 8.2) to obtain a dilution solution of 1:200.

A gentle stirring for 10 minutes is done on the samples, followed by filtration with a strip agase. Identical volumes were transferred into tubes. After heat treatment in a water bath thermostated (Memertt) under continuous agitation at temperatures of 56 ° C to 89 ° C with a step of 3 ° C for 3 min ,10 min and 20 min, the tubes were immediately immersed in an ice bath for 1 minute to stop the denaturation process [16,24].

2.2. Measure of transmittance

It consists in measuring the transmittance of the egg white processed relative to that of untreated; the latter is used to bring back the turbidity at 0 (which corresponds to a transmittance of 100%). This measure is performed at a wavelength of 650 nm [24], immediately after heat treatment and after 24 hours of storage at 4 ° C.

2.3. Determination of protein solubility

The treated samples were centrifuged at 20000 g for 20 min at 4 ° C [24]. The proteins of the supernatant were measured by the Bradford method [28]. Solubility S (%) of proteins is expressed relative to an untreated white using the following formula:

$$S\% = \frac{\text{Concentration of proteins in supernatant treated}}{\text{Concentration of proteins in supernatant untreated}}$$

The measurements are carried out immediately after the heat and after 24 hours of storage at 4 ° C.

2.4. Polyacrylamide gel electrophoresis under native conditions (NATIVE-PAGE)

The equipment used is the mini-plate vertical electrophoresis on polyacrylamide gel described by Laemmli (1970) [29]. The protein electrophoresis under non denaturing conditions is done to separate proteins of egg white as a result of an electric field on a separating gel with 7.5% of acrylamide and a staking gel with 5% of acrylamide. The rate of separation depends on the mass and charge of each protein [30].

3. Results and Discussion

3.1. Turbidity

Turbidity is inversely proportional to the transmittance. Figures 1 show a negative slope sigmoidal evolution of protein denaturation of albumen to increasing temperatures ranging from 56 ° C to 89 ° C during 3 min, 10 min and 20 min respectively, it has been measured immediately after heat treatment at 650 nm.

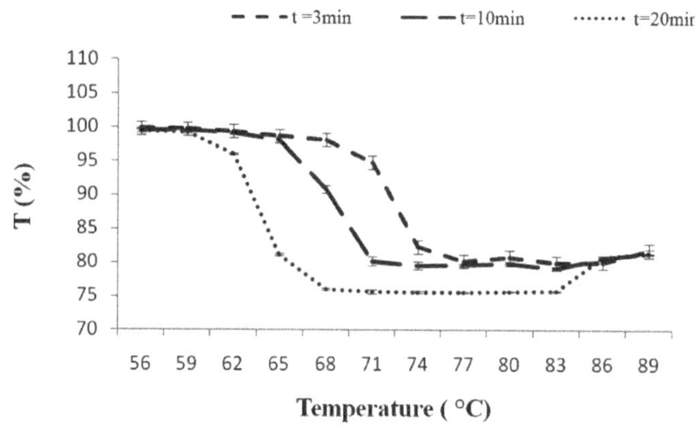

Fig 1. Effect of heat time on egg white turbidity, measures done immediately after heat.

This sigmoid with negative slope includes four distinct parts:

- Stationary Phase 1: The transmittance is steady according to the temperature. This phase could be considered as the phase of denaturation of proteins.
- Downward Phase: Corresponds to a negative evolution of the transmittance according to temperature. It is about the transition phase between the stage of denaturation and the stage of aggregation. The slope of this part informs us on the speed of transition between the stage of denaturation and the one of aggregation.
- Stationary Phase 2: This phase could correspond to the phase of aggregation of proteins.
- Ascending Phase: All curves include a positive phase from 86°C. This reduction of trouble could be associated to the sedimentation of the aggregates formed or to the formation of the S-Ovalbumine of which the ray of Stock is weaker than the one of native protein of a factor of 2, what drives to the formation of small size aggregates.

Statistical analysis reveals that the percentage of transmittance measured immediately and after 24 hours storage at 4 ° C of heat change over time was significant ($P<0.05$) and that the disorder is less important at 3 min than 10 min and 20 min respectively. So while extending heat, the trouble appears rapidly (lower temperature) ;for transmittance measured immediately after treatment, the trouble appears from 59°C at 20 min. whereas at 3 min and 10 min, it appears respectively at 68°C and 65°C.

Student test indicates no significant differences between the transmittance measured immediately after heat and after 24 hours storage at 4 ° C treated samples. So the phenomenon of protein denaturation in albumen; is irreversible and interactions occurring between them are covalent (disulfide bridge) and non covalent, (hydrophobic and electrostatic interactions).Statistical analysis reveals that the evolution of transmittance as a function of temperature is significantly different ($P <0.05$). In fixing the time of heat treatment, the temperature is increasing turbidity increases. This could be explained by protein aggregation most resistant to heat such as ovalbumin, which represents 54% of proteins in egg white [16, 24]. At temperatures understood between 55°C and 65°C, transmittance decrease could be explained by denaturation and aggregation of the most sensitive proteins to heat as the ovotransferrin (13% of proteins in white) and at temperatures more elevated (≥74C°), the fall of the transmittance could be due to the denaturation and aggregation of the most heat-resistant proteins such as ovalbumin .

3.2. Solubility

According to figure 2, the heat applied to solutions of egg white lead to irreversible loss of solubility for the three times studied as a function of temperature.

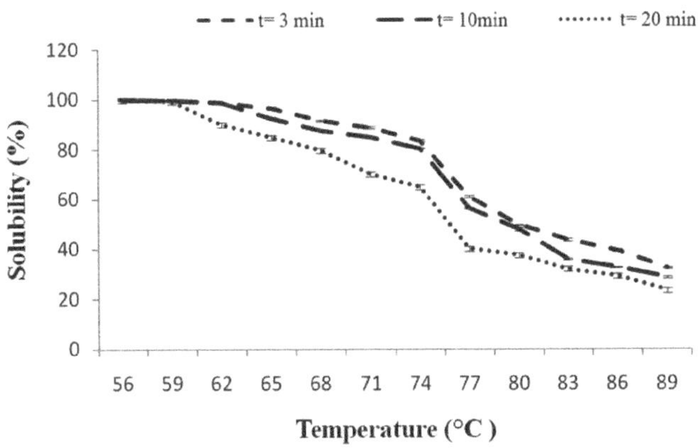

Fig 2. Effect of heat time on egg white solubility, measures done immediately after heat.

After heat during 20 min at 89 °C , only (23.44 ± 0.21) % soluble proteins remained, while after 10 min and 3 min at 89 °C (28.64 ± 0.4)% and (32.43 ± 0.27)%, respectively remained soluble. The strong decrease in solubility at higher temperatures (74-89 °C) is probably due to denaturation and aggregation of heat-resistant proteins such as ovalbumin [31]. While the weak loss in solubility in the range of lower temperatures (56-71° C), could be due to denaturation and aggregation of egg white protein more sensitive to heat, such as ovotransferrin [16, 24, 31].

3.3. Native - PAGE

Figures 3 and 4 show the electrophoretic patterns on polyacrylamide gel (7.5%) in native conditions of egg white proteins after heat during 20 min.

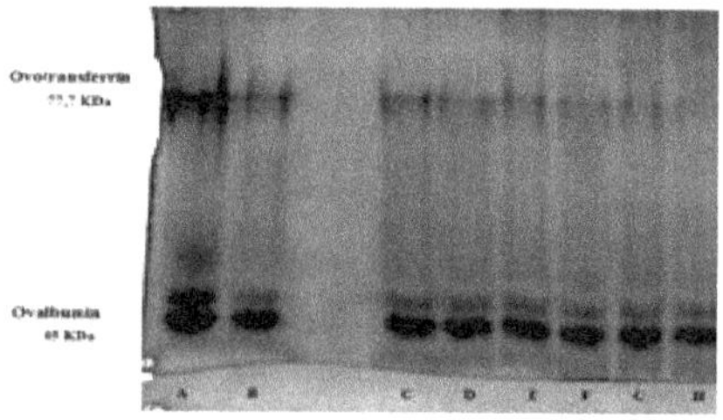

Fig 3. Native-Page of albumen heated during 20 min. A: control (Oval, Ovot, Lys) 1mg/ml, B: unheated white, C: 56°C, D: 59°C, E: 62°C, F: 65°C, G: 68°C, H: 71°C.

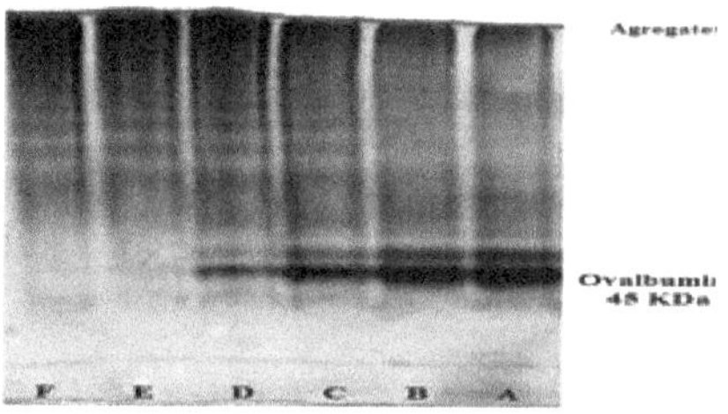

Fig 4. Native-Page of albumen heated during 20 min. A: 74°C, B: 77°C, C: 80°C, D: 83°C, E: 86°C, F: 89°C.

On the first gel (figure 3), all samples are similar to the control; but ovotransferrin band intensity decreases from 59°C, which could be due to its gelation due to its heat-sensitivity. As the highest temperature here is 71°C, we can conclude that its effect on the structure of the egg white proteins is minimal-or at least it does not cause major changes in its structure; also denaturation protein phase cannot be observed on a Blue Native gel. In the second gel (figure 4),, high temperatures are clearly a major effect because the proteins all migrated to a higher position, suggesting an aggregation-resistant electrophoresis conditions. The comparison between the two gels suggests that under the chosen conditions, coagulation and /or precipitation begins to be observed around 74°C.

4. Conclusion

Egg products are used as ingredients in many food sectors for their techno-unique feature. Stabilization treatments applied to egg white provide good control of hygiene but also induce an alteration of functional properties. Indeed, when egg white is subjected to heat, its globular proteins are prone to changes in structure and conformation. Depending on the extent of the temperature and duration of the treatment, these changes can range from denaturation at the gelation or coagulation. Our study shows effect of heat on some physico-chemical properties of egg white proteins such a sigmoidal evolution of transmittance and an irreversible loss of solubility. Analysis of electrophoretic profiles of native proteins heated at different temperatures, shows appearance of several bands, indicating diversity of proteins in egg white. The comparison of these profiles with those who did not undergo any heat shows appearance of new bands and different relative mobility due to the formation of protein aggregates. Coagulation and / or precipitation of egg white proteins around 74°C/20 min were observed.

5. References

[1] C.Allemeersch. The ovoproduits in France. Veterinary doctorate thesis. Veterinary National school Of Aflort.1983; 70:1-151.

[2] M .Hirose .Globular proteins. In: Progress Biotechnology .ISSN 0921-0423. Published by Elsevier Science B.V. 2003; 23.

[3] D .Lorient, E. Matringe, F .Vey, O .Bourgeois-Adragna. Functional Properties, sensory and nutritional of the ovoproducts. In: Egg and ovoproducts, Paris, Tec *and Doc. Lavoisier*. ISBN: 2-85206-903-2. 1994: 368.

[4] J .Adrian, J .Potus, A .Poiffait. Introduction to the nutritional analysis of the food commodities. *Tech ed And Doc,Lavoisier*. Paris .1998; 254-260.

[5] C .Alais, G. Linden. Food biochemistry Abstract. *Masson ed*. Paris. 1997: 248-250.

[6] Y .Nys, B .Sauveur. Nutritional Value of eggs. INRA Prod. Anim. 2004; 17 (5): 385-393.

[7] Ph .Lafon, F .Lafon .The egg and the ovoproducts. Technique of the engineer. 2009; F7010: 1-21.

[8] S .Rehault, M .Anton, F .Nau, J .Gautron, Y .Nys .The biologic activities of egg. INRA Prod. Anim, 2007; 20 (4): 337-348.

[9] R .Huopalahti , R .Lopez-Fandino, M .Anton, R .Schade .Bioactive Egg Compounds.ISBN-13 :978-3-540-37883-9. Springer Berlin Heidelberg New York. 2007.

[10] F .Nau, A .Marc, N .Yves .Egg of hen: a mine of molecules whith biological activities. Fifth Day of the Poultry Research, Tours.2003.

[11] F .Nau, L .Gestin, J .Protais, A .Awade , JL .Thapon .Comparative study of the functional properties and the composition of thick and liquid fractions of the hen's egg white. ENSAR Agro-Food Department, Reindeer. 1995: 113.

[12] JC .Cheftel, JL .Cuq, D .Lorient. Food Proteins: biochemistry, functional properties, nutritional value. *Tec ed and Doc. Lavoisier* ,Paris. 1985: 306-308.

[13] Y .Mine .Recent advances in the understanding of egg white protein fonctionality. Trends in Food Science and Technology. 1995; 6: 225-232.

[14] Y .Mine .Egg Bioscience and Biotechnology. Department of Food Science. University Guelph of. Willy - Interscience. ISBN: 978-0-470-03998-4. 2008.

[15] B .Sauveur .Structure,composition and nutritional value of hen egg. In: Reproduction of poultries and production of egg. Edition: INRA. ISBN: 2853409619. 1988: 347-436.

[16] I .Van der Plancken, A .Van Loey, ME .Hendrickx .Combined effect of high pressure and temperature one selected properties of egg white proteins. Innovate Food Science and Ennerging Technology. 2005; 6:11-20.

[17] JL .Thapon, JC .Bourgeois .Egg and ovoproducts. In: Collection science and agroalimentary techniques. Edition: technique and documentation, Lavoisier. ISBN: 2852069032. 1994: 1-344.

[18] L .Campbell, V .Raikos, SR .Euston .Modification of functional properties of egg-white proteins. *Nahrung/Food*. 2003; 47(6): 369-376.

[19] C .Sanchez, S .Fremont .Consequences of the thermal treatments and formulation on the structure and the allergenic of the food proteins, Direct Science. 2003; 43:13-20.

[20] M-S .Appavou. Survey of the influence of the temperature and the pressure on the structure and the dynamics of the inhibitor of the bovine pancreatic trypsine. A survey by diffusion of neutrons. Thesis of doctorate. University of Paris XI Orsay. Specialty: Biophysical. 2005; 1-228.

[21] H .Neurath, HB .Bull .Denaturation and hydration of proteins. From the Division of Agricultural Biochemistry, University Minnesota,St, Paul. 1936: 519-527.

[22] D .Lorient. Biochemical Modifications of the constituent food. Techniques of engineer.2009; 3400: 1-20.

[23] ISL .Navizet. Modelling and analysis of the mechanical properties of the proteins. Thesis of doctorate. Paris University 6 Pierre & Marie Curie. 2004.

[24] I .Van der Plancken , A .Van Loey, ME .Hendrickx .Effect of heat-treatment one the physico-chemical properties of egg white proteins: A kinetic study. Newspaper of Food Engineering. 2006; 75: 316-326.

[25] V .Lechevalier. Ciaisaillement, creation of interfaces and thermal treatment. Thesis of physico-chemistry Doctorate and quality of the byproducts. School of life-Agro-health doctorate. 2005.

[26] M-M .Filion. Improvement of the thermal stability of milk by modulation of the potential of oxydoréduction. Faculty of the Sciences of agriculture and food. University Laval, Quebec. 2006.

[27] E .Hagerman, L G .Butler .Protein precipitation Method to Quantitative and Determination of tannins. Newspaper of Agriculture. *Food Chem*. 1978; 26(4): 809-812.

[28] MM .Bradford. A rapid and sensory method to the quantification of microgram quantities of protein utilizing the principle of protein-dye biding. *Anal Bioch*. 1976; 72: 248-254.

[29] UK .Laemmli. Cleavage of structural proteins during the assembly of the head of T4 bacteriophage. *Nature*. 1970; 227(259): 680-685.

[30] GM .Walker .The Protein Protocols Handbook (second edition). SD Polyacrylmide Frost Electrophoresis of proteins. University of Hertfordshire, Hatfield, UK. 2002: 1-1103.

[31] C .Desert, C .Guérin-Dubiard, F .Nau, G .Jan, F .Valley, J .Mallard. Comparison of Defers Electophoretic Separation of Hen Egg White Proteins., School National Superior Agronomic of Rennes, *Food Chem*. 2001 ; 49 :4553-4561.

Dietary Manipulations for Enhancing Cardio-Protective Fatty Acids in the Milk of Dairy Cows

Muhmmad Subhan Qureshi, Tawheed Ali Azeemi

Faculty of Animal Husbandry and Veterinary Sciences, Agricultural University, Peshawar, Pakistan

Abstract

The ruminants milk contains a higher proportions of saturated fatty acid (SFA), which is a risk factor related to cardiovascular disease. The mono and polyunsaturated fatty acid (MUFA, PUFAs), decreasing the risk of heart disease, are low in milk fat. The crossbred cows have been a major source of milk for human consumption. This study was conducted to investigate the effect of protected palm fats feeding on milk fatty acids profiles of crossbred cows. A total of 15 of Crossbred and 15 of Holstein Friesian cows were selected and protected palm fats were supplemented as: PF-0, PF-25, PF-50, PF-100 and; PF-150; the number representing the quantity (g) of fats/day. Milk sample were collected, analyzed and the study continued for 8 weeks. SFA was significantly ($P<0.05$) decreased from 70.80 to 67.45 g/100g while MUFA and PUFA increased with the increasing supplementation. It appears that hypercholestermic properties of the milk were reduced and cardio-protective properties were enhanced by feeding protected palm fats. It was also associated with increased milk yield and progesterone level reflecting better fertility and productivity. In early lactation 150 g/day palm protected fat may be supplemented for maximum yield, better reproductive performance and healthier milk.

Keywords: Hypercholestermic; cardio-protective; milk; dairy; nutrition; cattle; diet

1. Introduction

Crossbreed cows are kept for milk production by the peri urban dairy farmer and small scale rural farmer in Pakistan. These animals have emerged through genetic improvement of local non-discriptive cows leading to gradual improvement in milk yield. Resultantly they require improved feeding and management practices [1]. The improvement of ruminant milk quality has in many ways, been made possible by dietary manipulation [2]. Researchers have attempted to obtain milk fat with healthier properties increasing its content in polyunsaturated fatty acids (PUFAs) which have unquestionable beneficial effects on human lipid metabolism [3]. Many attempts have been made, therefore, to alter the fatty acid composition of milk fat from lactating cows in order to improve its nutritional value. From these studies, it has been deduced that the fat used should be protected from ruminal microbial actions so that PUFAs reach the small intestine and their potentially toxic effect on rumen micro-organisms is minimized [2, 4]. Cows have high energy demands in early lactation to sustain milk secretion; hence one logical strategy for sustaining milk production is maximizing energy intake by increasing energy density of diet. Flaked, prilled fatty acids, calcium salts, formaldehyde and many other treated protected fatty acids are insoluble in the rumen [2].

The present study was conducted to explore response the changes in milk fatty acid profiles with protected fats diets in various cattle breeds.

2. Materials and methods

Selection of Animals and treatments: Thirty lactating cows (Holstein Friesian and Crossbred) during 3rd week of lactation were selected at University Dairy Farm. The animals were housed in open paddock with free access to water from a tank. Milking was practiced at 3 AM and 3 PM and concentrate feed was provided during the milking time. Experimental period was for 8 weeks, with 3 days of adaptation. The cows were allotted randomly to the five dietary treatments as follows: PF0 (concentrate mixture, 1kg/ 3L milk); PF25, PF100, PF150 (concentrate mixture + Protected fats 25, 50, 100, 150g/day). The fats was protected through the method described as by Strohmaier et al. [5].

Milk recording, sampling and analysis: A total of 80 samples of milk were collected in each week and daily milk yield was recorded. Weekly milk samples were collected at the rate of 50 ml and stored at -20ºC until analyzed. The milk fat was separated and analysed for fatty acids [6].

Statistical Analysis: The data were analysed through SPSS 11 for Windows XP. Analysis of variance was used for means comparison through general liner model procedures. Correlation analysis was used for detecting changes in protected fats levels affected milk yield and fatty acids. Means were subsequently ranked using Duncan Multiple Range Test.

3. Results

Milk Fatty acids concentrations: The mean values of milk yield and milk fatty acids and milk progesterone are reported in Table 1. Saturated fatty acid (SFA) showed the highest concentration out of the total milk fatty acids with an mean ± SE of 68.72±0.41 g/100 ranging from 62.3 to 78.4 g/100g. Within SFA the highest concentration was recorded for C16:0 (26.87 g/100g) followed by C14:0 (13.96 g/100 g). The sum of three hypercholestremic fatty acids (C12:0, C14:0 and C16:0) was 44.0 g/100g. Average concentration of unsaturated fatty acids (UFA) was 30.27 g/100g out of which the highest concentration was recorded for C18:1 (23.89 g/100). The concentration of monounsaturated fatty acids (MUFA) was 26.87 and polyunsaturated fatty acids (PUFA) were 3.41 g/100 g in the total milk fatty acid.

Effect of protected palm fats: In Holstein dairy cows fats supplementation affected concentration of C14:0 and C16:0 significantly decreasing with the increasing levels of fat intake from PF25 to PF150 /day (Table 2). Polyunsaturated fatty acids were higher in PF100 and PF150 groups. The highest concentration was recorded for C18:3 ($P<0.05$). In crossbred dairy cows fat supplementation had no significant effect on overall SFA; however with increasing fat intake from PF0 to PF100 a decreasing trend was observed in SFA from 71.36 to 66.50 g/100 g (Table 3). Concentration of C14:0 significantly decreased with the increasing the level of dietary protected fats. In polyunsaturated fatty acids C18:3 was significantly increased (P=0.042) in PF100 and PF150 supplemented groups, while C18:2 was not effected significantly however, highest concentration was observed in PF150 group.

Correlation analysis showed that changes in C8:0, C10:0 and C12:0 were positively correlated with the PUFA. Milk progesterone was negatively correlated with the saturated fatty acids but showed strongest correlation with the C12:0 (r = -0.379). Daily milk yield correlated negatively with C12:0 and positively with C19:0 (-0.224 and 0.250 respectively, P<0.05).

Figure 1 shows constant decrease in the level of SFA from 71 to 67% with increasing intake of protected fats from PF0 to PF100, however, further increased was not effective. The increasing level of protected fat from PF0 to PF50 did not effect the unsaturated fatty acid. Further increased up to PF150 g/day increased the later from 32.7 g/100g all these changes were significant.

Effect of protected fats on milk yield and progesterone: Figure 2 shows changes for progesterone and milk yield with different supplemented levels of protected palm fats. Highest milk yield (13.31±0.81 kg/day) was recorded in PF150 group, followed by 12.66±0.79 kg/day in PF100 supplemented group (P<0.05).

4. Discussion

Effect of Protected Palm fats on Milk fatty acids: Supplementation of palm protected fats to dairy cows

Parameters	Minimum	Maximum	Mean ±SE
Milk Yield kg/day	6	18	12.12±0.49
Progesterone ng/ml	0.22	2.78	1.80±0.06
Milk Fatty acid (g/100g)			
C12:0	0.8	9.3	3.25±0.17
C14:0	10	18.9	13.96±0.28
C16:0	22.8	32	26.87±0.24
C18:0	13	21.7	16.41±0.19
C14:1	0.1	1.6	0.415±0.03
C16:1	0.1	3.7	0.79±0.08
C18:1 *trans*	0.2	6.2	1.77±0.14
C18:1	14.3	33	23.89±0.51
C18:2	0.3	6.3	0.415±0.03
C18:2 *trans*	0.4	1.2	0.79±0.08
C18:3	0.2	2.8	1.77±0.14

Parameters	Minimum	Maximum	Mean ±SE
SFA	62.3	78.4	68.72±0.41
UFA	17.4	45.1	30.27±0.70
MUFA	15.9	37.5	26.87±0.59
PUFA	0.5	8.1	3.41±0.17

Table 1. Descriptive statistics for various parameters in dairy cows (mean ± SE)

SFA (saturated fatty acid), MUFA (monounsaturated Fatty acid), PUFA (polyunsaturated fatty acid)

Milk fatty acids	Protected palm fats					
	PF0	PF25	PF50	PF100	PF150	P vaule
C14:0	15.81±0.34	15.46±0.53	14.10±0.60	11.40±0.45	11.82±0.7	0.000
C16:0	24.79±0.87	25.49±0.64	26.69±0.92	27.82±0.40	28.57±0.4	0.002
C18:3	0.83±0.15	1.06±0.18	0.83±0.14	1.53±0.22	1.74±0.2	0.003
SFA	70.22±0.93	70.21±1.30	68.39±1.60	66.20±0.76	68.47±1.3	0.129
MUSFA	25.77±1.31	24.20±1.92	25.62±1.17	27.42±2.25	29.47±2.8	0.332
PUSFA	3.04±0.54	2.74±0.54	2.69±0.31	4.32±0.85	4.36±0.61	0.089
USFA	28.81±1.55	26.93±2.10	28.03±1.41	31.74±2.80	33.82±2.72	0.191

Table 2. Effect of protected fats intake on milk fatty acids (g/100g) in Holstein dairy cows (Mean±SE)

Milk fatty acids	Protected palm fats					
	PF0	PF25	PF50	PF100	PF150	P value
C14:0	15.85±0.60	15.36±0.71	16.17±0.52	12.56±0.52	11.42±0.4	0.000
C18:2	1.82±0.40	1.46±0.23	1.87±0.26	2.05±0.27	2.41±0.3	0.546
C18:2 *trans*	0.279±0.03	0.24±0.49	0.16±0.04	0.19±0.36	0.24±0.7	0.720
C18:3	0.94±0.24	1.18±0.20	1.01±0.10	1.51±0.14	1.65±0.1	0.042
SFA	71.40±0.96	69.37±1.0	68.11±1.50	68.37±0.9	67.49±0.8	0.085
MUSFA	26.16±1.46	27.80±2.6	26.68±2.08	28.07±1.93	28.45±1.2	0.952
PUSFA	3.07±0.3	2.89±0.45	3.05±0.38	3.71±0.42	4.30±0.35	0.162
USFA	29.04±1.08	30.77±2.96	29.74±2.43	31.75±2.19	32.7±1.50	0.903

Table 3. Effect of Protected Palm fats intake level on Milk fatty acids profile (g/100g) in Crossbred dairy cows (Mean± SE)

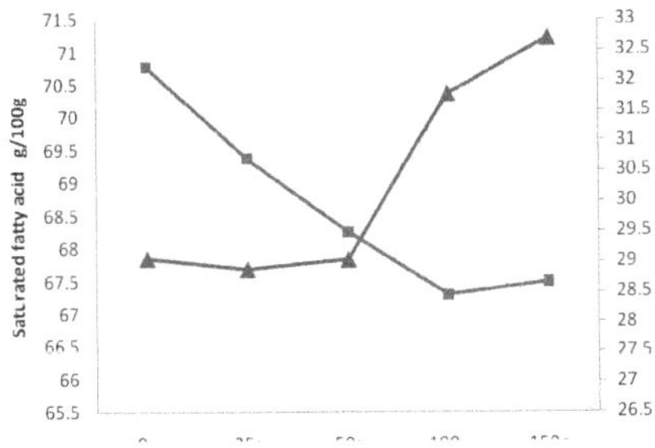

Figure 1. Effect of protected-palm-fats feeding (g/day) on unsaturated (▲) and saturated fatty acids (g/100g, ■) in the milk of dairy cows.

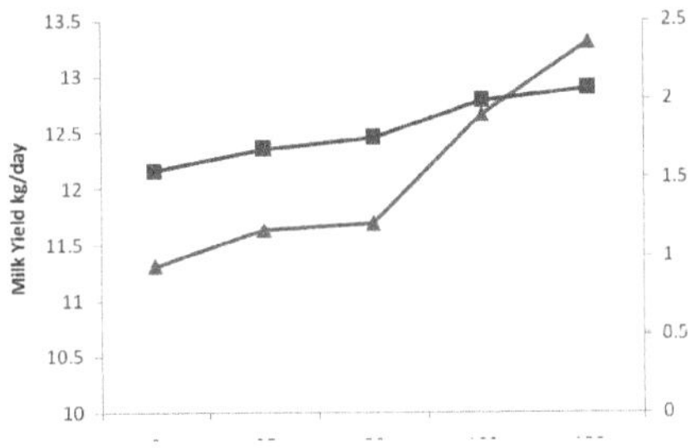

Figure 2. Effect of different levels of protected palm fats (g/day) on milk yield (kg/day, ▲) and progesterone concentrations (ng/ml, ■).

significantly ($P<0.002$) decreased SFA of milk from 70.78 to 67.49 (g/100g). Within the SFA the caproic (C6:0) capylic (C8:0) and capric (C10:0), lauric (C12:0), myristic (C14:0) acids decreased while palmitic (C16:0) was significantly ($P<0.001$) increased. In polyunsaturated fatty acids linolenic acid (C18:3) was significantly ($P<0.05$) enhanced by increasing supplement fats up to 150 g/day. Similar finding was observed by Purushothaman et al. [7] who fed 200 g palm protected fats to crossbred cows, and observed that with the supplementation the proportion of caproic (C6:0), capylic (C8:0) and capric (C10:0) acid decreased significantly ($p<0.01$) while the palmitic (C16:0), polyunsaturated fatty acid: oleic (C18:1), linoeic (C18:2) and linolenic (18:3) acids concentration in increased in milk fats of crossbred cow.

Lauric acid (12:0), myristic acid (C14:0), and palmitic acid (C16:0) are to be the serum total and LDL-cholesterol raising SFA [8]. Myristic acid C14:0 is more hypercholesterolemic than palmitic acid C16:0 [9]. The increased concentration of palmitic acid in milk fat was due to palm oil, rich in palmitic acid. The decrease in C14:0 and C12:0 by 27.94% and 38% respectively can be related to the inhibition of these fatty acid with the supplementation, rich in PUFA. The present study is supported by West and Hill [10] reporting that supplementation of protected fats decreased the percentage of short-chain fatty acids and increased the long-chain fatty acids in milk fat. Similarly, Pantoja et al. [11] in Holstein dairy cows, found an increase in C16:0 and C18:0 fatty acids and decrease in C8 to C14 fatty acids. McDonald and Scott [12] reported that cows fed protected fat containing polyunsaturated oil had marked increased linoleic acid of milk fat.

Effect of Protected fats on Milk yield: In our study milk yield is increased by the supplemented levels of protected palm fats in both crossbred and Holstein dairy cows. Similar findings were reported by Sajith Purushothaman et al. [7] about increased milk yield (by 1.9 kg/day) when fed 200 g protected palm fats. An increase in the production of milk has been also reported by Sarwar et al [13] and Palmquist and Jenkins. [14] in cows supplemented up to 300 g of rumen-protected. Supplementation of calcium salt of long chain fatty acids increased the milk production significantly ($P<0.05$) as reported by Maeng et al. [15]. Staples et al., [16] reviewed the effects of feeding fats on reproduction in dairy cows and also summarized their effects on milk yield, addition of calcium salt of palm oil increased milk yield by 2.4 kg/day.

Effect of protected fats on milk progesterone: Milk progesterone was affected by intake of protected palm fats (P<0.033) changing favourably with increasing level of fats supplementation. Lowest progesterone concentration (1.54 ng/ml) was found in controlled group while highest 2.07 ng/ml of progesterone in 150g fat supplemented group. A similar study carried by Lopes [17], observed 1.81 ng/ml of progesterone on supplemented fed of 200 g of polyunsaturated fatty acid to *Bos indicus* beef cows. Espinoza et al., [18] supplemented 125g of protected palm oil to Herford and Angus and found greater than >1ng/ml of progesterone in cyclic cows. The increase in milk progesterone may be due to combating negative energy balance as already reported [19].

5. Summary and Conclusion

This paper investigated the effect of feeding protected palm fats on milk fatty acids profiles. The results suggest that hypercholestermic properties of the milk were reduced and cardio-protective properties were enhanced by feeding protected palm fats. It was also associated with increased milk yield and progesterone level reflecting better fertility and productivity. In early lactation 150 g/day palm protected fat may be supplemented for maximum yield, better reproductive performance and healthier milk.

6. References

[1] Qureshi, M.S., A. Khan, K.B Mirbahar and Usamo. 2000. Productive and reproductive and their interaction in cross breed cattle under field condition, District Bano, Pak Vet Journal 20:31-34.

[2] Grummer, R.R. 1991. Effect of feed on the composition of milk fat. J. Dairy Sci. 74:3244–3257.

[3] Brisson, G.J. 1986 . Dietary fat and human health. In:Harsign, W.,Cole, D.J.A. (Eds.), Advances in Animal Nutrition Butterworths, London, pp.3-24

[4] Kin, Y.K., D.J. Schingoethe, D.P. Casper and F.C. ludens. 1993. Supplemental dietary fat from extruded soybeans and calcium soaps of fatty acids for lactating dairy cows. J. Dairy Sci. 75, 197-204

[5] Strohmaier, J.K. (2001). Calcium salts – saponification of polyunsaturated oils. United States Patent Number US 2004 015 8087 AI.

[6] Qureshi, MS, A Mushtaq, S Khan, G Habib, ZA Swati, 2010. Variation in milk fatty acids composition with body condition in dairy buffaloes *(Bubalus bubalis). Asian Aust J Anim Sci,* 23: 340-345.

[7] Sajith Purushothaman1, Anil Kumar and D. P. Tiwari. 2008. Effect of Feeding Calcium Salts of Palm Oil Fatty Acids on Performance of Lactating Crossbred Cows Asian-Aust. J. Anim. Sci. Vol. 21, No. 3 : 376 - 385

[8] Keys, A., J. T. Anderson, and F. Grande. 1965. Serum cholesterol response to changes in the diet. IV. Particular saturated fatty acids in the diet. Metabolism 14:776.

[9] Zock, P. L., J. H. M. deVries, and M. B. Katan. 1994. Impact of myristic acid versus palmitic acid on serum lipid and lipoprotein levels in healthy men and women. Arterioscler. Thromb. 14:567.

[10] West, J.W. and G.M. Hill. 1990. Effect of a Protected Fat Product on Productivity of Lactating Holstein and Jersey Cows, J Dairy Sci 73:3200-3207

[11] Pantoja, J., J. L. Firkins, M. L. Eastridge and B. L. Hull. 1996. Fatty acid digestion in lactating diary cows fed fats varying in degree of saturation and different fiber sources. J. Dairy Sci. 79:575-584.

[12] McDonald, I.W. and T.W. Scott. 1977. Foods of ruminant origin with elevated content of polyunsaturated fatty acids. World Rev Nutr Diet, 26: 144–207.

[13] Sarwar, M., A. Sohaib, M. A. Khan and Mahr-Us-Nisa. 2003. Effect of feeding saturated fat on milk production and composition in crossbred dairy cows. Asian-Aust. J. Anim. Sci. 16(2):204-210.

[14] Palmquist. D. L. and T. C. Jenkins. 1980. Fat in lactation rations: review. J. Dairy Sci. 63:l

[15] Maeng, W. J., J. H. Lim and S. R. Lee. 1993. Effects of calcium salts of long chain fatty acids on ruminal digestibility, microbial protein yield and lactation performance. Asian-Aust. J. Anim. Sci. 6(3):395-400.

[16] Staples, C. R., J. M. Burke and W.W. Thatcher. 1998. Influence of supplemental fats on reproductive tissues and performance of lactating cows. J. Dairy Sci. 81:856–871.

[17] Lopes, C. N., A. B. Scarpa, B. I. Cappellozza, R. F. Cooke and J. L. M. Vasconcelos. 2009 Effects of rumen-protected polyunsaturated fatty acid supplementation on reproductive performance of Bos indicus beef cows. J.Anim. Sci.87:3935–3943

[18] Espinoza, J.L.; Ramírez-Godínez, J.A.; Jiménez, J.A. & Flores, A. (1995). Effects of calcium soaps of fatty acids on postpartum reproductive activity in beef cows and growth of calves. Journal of Animal Science, 73:2888-2892.

[19] Villa-Godoy, A., T.L. Hughes, R.S. Emery, L.T. Chapin and R.L. Fogwell. 1988. Association between energy balance and luteal function in lactating Holstein cows. Journal of Dairy Science 71: 1063-1072.

Information Technology and E-Commerce Reflexes on Total and Agricultural Trade in Egypt

Mahmoud M .Fawaz[1], Abdelbaky M. Elshaib[2], Roshdy Sh. El Adwy[1]

1 Department of Agricultural Economics, kafr elsheikh University, kafr elsheikh, Egypt

2 Department.of Agricultural Economics, Tanta University, Tanta, Egypt

Abstract

Nowadays, E-commerce has been developing quickly and it brings great impact on economy of all the countries. We insist that the E-commerce development in developing countries and economies in transition will have fine perspectives if the government, corporations and all the related people make great effort together. These papers address the issue of understanding the components of the information economy and E-commerce (profitability of .ese Experiment, this research aimed to identification, entity, characteristics and divisions and to figure out its profitability in Egypt. This research aimed also to pinpoint the measures necessary to incorporate the E-commerce between Egypt and the others commentating countries. We therefore propose a framework consisting of five components. First, concept and entity of E-commerce and its importance in marketing area, secondly, information and communication technology in developing and developed countries, third, the effect of internet on the Egyptian trade, forth, the effect of Internet on the Egypt Agric. Trade

Keywords: (Ec) Electronic commerce, (ICT) Information and Communication Technology, (ITU) international Telecommunication Union, (PIS) Primary Information sector, (SIS) Secondary Information Sector, (ET) Electronic Trade,(BtoB) Business to Business, and (BtoC) Business to consumer.

1. Introduction

Nowadays, E-commerce has been developing quickly and it brings great impact on economy of all the countries, and play important role in foreign agriculture trade in both developed and developing countries. Research Problem The research problem confined to important question; that is how far profitability would Egypt gain from ET, although it regarded as an user rather than producer of information technology. In addition, nowadays, the assessment of ET amount encountered a practical difficulties; this attributable to the absence of local or international system or rules could control such trade.Furthermore, the formal statistical institutions could not monitor the precise or actual amount of ET and the scientific studies and literatures pertaining such trade regarded very few either at local or international scale. It can be said, that the agricultural marketing via ET in Egypt still limited although there are a continuous increase in number of agricultural ET sites in Egypt. Consensus. This research aimed to: Determination of the direct positive impacts of ET on the national income, by supporting and boosting the external trade. Assessment of the effect of transportation means of technology (Internet) on the Egyptian Agric.

Trade. Assessment of Relationship between information technology and GDP in Egypt. This research relied on qualitative and quantitative indications to display the economic impacts of ET and to address the gab between Egypt and the other countries.. The study comprised the data gathered between 1994 and 2009; that would be divided into 2 terms, the first between 1994 and 1998 (the period before establishment of internet handling). The second term was from 1999 to 2009 (period after launching of internet dealing in Egypt) in the aim of determination the impact of information revolution (internet dealing) on the agricultural marketing.

2. Information and Communication Technology in developing and developed countries

2.1. Internet activity worldwide

The equation (1) describes that there is a significant increase of the internet host count from 1981-2009 according to the world wide internet domain survey. By added the factor of time to the previous equation, the results of analysis statistics was as the following:

$$\text{Log } y_t = 0.465 + 5.789 \log x_t$$
$$(0.562) \quad (17.912)^{**} \qquad (1)$$
$$R^2 = 0.925 \qquad F = 320.844^{**}$$

Where: Y_t = Internet host count according to the world internet domain survey, X_t = time (1, 2, 3...)

2.1.1. Internet usage statistics by world Region

Table (1) illustrates the internet usage by region during period 2000-2009, the highest growth percentage of internet usage was achieved by middle East which recorded 267 Percent, while Latin America and Caribbean occupied the second rank with 211% Africa, Asia, Europe, North America, and Australia occupied the rest ranks of growth with percents 198,165,152,105,114, respectively.

World Regions	Population (2009)	Population % of world	Internet usage latest data	Usage Growth 2000-2009	Percent ration % population	World users %
Africa	915.210.928	14.1	13,468,600	198	1.5	1.5
Asia	3.667.774.066	56.4	302,257,003	165	8.4	34
Europe	730,991,138	11.4	259,653,144	152	35.5	29.3
Middle East	259,499,779	4	19,370,700	267	7.5	2.3
North America	328,387,059	5.1	221,437,647	105	67.4	25
Latin and Caribbean	546,917,142	8.5	56,224,957	211	10.3	9.3
Australia	33,443,448	0.5	16,269,080	114	48.6	1.9
Total world	**6,499.697.060**	**100**	**888,681,131**	**164**	**13.9**	**100**

Table 1. Average internet usage statistics by world Region, 2000-2009

Source: world internet usage and population statistics

2.1.2. Information and Communication Technology (ICT)

(ICT) play a key role transforming of economic structures. The preceded studies confirmed that the economical impacts of (ICT) occurred on macro. And micro- economics levels. On the macroeconomics level, the studies mentioned that the rapid progress in (ICT) will lead to increase in international Trade cause minimization cost, advertising information exchange exhibitions, information availability in real time, more accessibility to the international market. And on microeconomic level, the studies confirmed that the (ICT) will help small business to access to international market, optimizing the resource allocation and reducing transaction cost. Examining the e-commerce important question is what impact (ICT) and Internet have on the productivity growth. It reduces transaction costs, allocates recourses better, increases economies of scale, improves the competitiveness of business in general, increase efficiency generates important changes in the management and production processes of business.

2.1.3. Distribution of Market Information Technology (IT)

(IT) encompasses every thing that allows us to electronically gather, generate, store, analyze, distribute or otherwise use information. Although the internet has received the most attention, information Technology (IT)involves other computer Technologies such as microchips, monitors, hard- drives and software. It also includes more traditional telecommunication Technologies shushes cell phones or fax machines anything related to the electronic of information. Table (2) represents the Electronic commerce by type Business to Business and Business to consumer by region during the period 2005-2009, data showed that the global of (B to B) to increases with growth Percentage reached by 257.9% annually. The average of (B to B) reached 609.25Millar dollars. While (B to C) the average 60.5Millar dollars and growth Percentage reaches by 169.1% annually. And the average B to B/ Total 88.75%.

E- commerce	2000	2003	2006	2009	Average	Growth%
B To B	43	251	843	1300	609.25	257.9
B to C	8	33	76	125	60.5	169.1
Total	51	284	919	1425	669.75	245.2
% B to B/total	84	88	92	91	88.75	

Table 2. Electronic commerce by type (BtoB) from 1998 to 2009 (Millar dollars)

Source: Computed from the data cited from world Resource

2.2. Technology gap between developing and developed countries

When compares between developing and developed countries Shown that the Digital divide (technology gab) in 2009:

a. Internet: developing countries 34% while developed countries 66%.
b. Computer: developing countries (27%) while developed countries (73%).
c. Mobile Phone: developing countries 46% while developed countries (54%).

d. Land line: developing countries (45%) while developed countries (55%).

But population in developing countries 81% and 19% in developed countries.

2.3. Egypt Internet usage

Egypt Internet usage and population compared with the Rest of the Africa in 2009. The Egypt has 5 million users for the international information net. The Percent of internet user with represent as the indirect indicator fore – commerce in Egypt is about 21.3%. But this Percent in Rest of Africa 78.7%. The population in Egypt about 8.5% in Africa. Some studies mentioned that may be Egypt become the biggest African countries for the number of internet users even more than south Africa, which is the highest percent in Africa.

Table (3) illustrates Internet users in Egypt during the period 1998-2008. the number of Internet users increased from about 36.4 (1000 persons) in 1998 to 1.355.0 (1000 persons) in year 2008. And the index number increased from 100% in year 1998 to 3722.527. % in years 2008.

Year	Egypt (1000 persons)	Index number
1998	36.4	100
1999	51.8	142.3077
2000	72.2	198.3516
2001	92.5	254.1209
2002	282.3	775.5495
2003	437.1	1200.824
2004	591.9	1626.099
2005	782.0	2148.352
2006	996.6	2737.912
2007	1.354.8	3721.978
2008	1.355.0	3722.527

Table 3. Number of Internet users in Egypt, 1998-2008.

Source: world internet usage and population statistics. 2009.

3. The effect of transportation means of technology (internet) on the Egyptian trade

3.1. The effect of transportation means of technology (internet) on the Egyptian export

It illustrates the table export in Egypt before and after using internet, the results of the statistical analysis that there is a significant increase in the table exports during the period after the internet comparing to the period before. The study used the dummy variable to measure the effect of information revolution in Egypt, the study divided the period (1992-2009). The first period was from (1992-1998) and represented the period before the revolution of information in Egypt (internet). The second period was after that, from (1999-2009) this dummy variable represented as Dt. Average Total exports before using Internet about 11438.4 (million LE) .(S.T DEV) about 1081.042 and after using internet (47554.82 million LE) .(S.T DEV) about 31764.58

By added the factor of time to the previous equation, the results of analysis statistics was as the following:

$$Et = 11438.414 + 36116.404\ Dt$$

$$(1.205) \qquad (2.974)^{**} \qquad (2)$$

$$R^2 = 0.356 \qquad F = 8.842^{**}$$

Where: Et = Total export in Egypt during the period (1997-2009), Dt = Dummy variable equal (1) after using internet, equal (0) before using internet.

3.2. The effect of transportation means of technology (internet) on the Egyptian imports

It illustrates the total Imports in Egypt before and after using internet, the results of the statistical analysis that there is significant increase in the total imports during the period after the internet comparing to the period before. (Average) total imports in Egypt before using internet about 38955.4 million LE.(S.T DEV) about 10448.87 and after using internet about 113052.4 million LE.(S.T DEV) about 74178.37 The study used the dummy variable to measure the effect of information revolution in Egypt. By added the factor of time to the previous equation, the results of analysis statistics was the following:

$$It = 38955.386 + 74096.992\ DT$$

$$(1.747) \qquad (2.598)^{*} \qquad (3)$$

$$R^2 = 0.297 \qquad F = 6.749^{*}$$

Where: It = Total imports in Egypt during the period (1999-2009), DT: Dummy variable equal (0) for the period from (1992-1998) and (1) for the period from (1999-2009)

Years	Total Export	Total Import	Trade Balance	Agriculture Export	Agriculture Imports	Agriculture Trade balance
1992	10171.2	27656.1	-17485	1262	7900	-6638
1993	10464.5	27550.4	-17086	1115	7500	-6385
1994	11757.5	32460.6	-20703	1740	7200	-5460
1995	11703.8	39890.9	-28187	1641	10800	-9159-
1996	12006.1	44217.9	-32212	1689	12700	-11011
1997	13281.0	44885.7	-31605	1488	13400	-11912
1998	10684.8	56026	-45341	1760	15900	-14140
Average	**11438.4**	**38955.4**	**-275170**	**1527.86**	**10771.429**	**-9243.57**

Years	Total Export	Total Import	Trade Balance	Agriculture Export	Agriculture Imports	Agriculture Trade balance
1999	12052	54399	-42347	1804	12400	-10596
2000	16351	48645	-32294	2100	12700	-10600
2001	16491	50660	-34169	2400	7700	-5300
2002	21144	56480	-35336	3300	9700	-6400
2003	36712	65083	-28271	4600	10100	-5500
2004	38653	57408	-18755	5284	9900	-4615
2005	44083	121905	-77822	5995	10010	-4015
2006	67379	152504	-85125	4748	11746	-6998
2007	79955	163475	-83520	7403	14659	-7256
2008	97656	224569	-126913	8980	17881	-8901
2009	92627	248448	-155821	11571	19236	-7665
Average	**47554.82**	**113052.4**	**-65488.5**	**5289.545**	**12366.55**	**-7076.91**

Table 4. Total and Agricultural export and total and imports Agricultural imports Agricultural in Egypt Before and after using internet. (Million LE)

Source: www.Egypt.gov.eg

3.3. Relationship between total export and internet in Egypt

The equation (4) describes that there is significant increase relationship between total export in Egypt and internet user, the result of analysis statistics was as the following:

$$Et = 12426.84 + 471.334\eta 1$$
$$(5.344)^{***} \qquad (7.467)^{***} \tag{4}$$
$$R^2 = 0.918 \qquad F = 55.75$$

Where: Et: Total export in Egypt., $\eta 1$: Internet users in Egypt., T: Time (1, 2, 3 … 14)

4. The effect of transport in means of technology (Internet) on the Egyptian Agric. trade

4.1. The effect of transport in means of technology (Internet) on the Egyptian Agric. Exports

It illustrates the total Agric. Export in Egypt before and after using internet, the results of the statistical analysis that there is significant increase in the Agric. Exports during the period after the internet comparing to the period before. Average Agric. Exports in Egypt before using Internet about 1527.86 Million (LE) .(S.T DEV) about 251.79, and after using internet about 5289.545 Million (LE). .(S.T DEV) about 3056.783 The study used the dummy variable to measure the effect of information revolution in Egypt, the results of analysis statistics was the following:

$$E^{At}=1527.857+2112.514\ Dt$$

$$(3.402)^{**} \qquad (3.326)^{**} \qquad (5)$$

$$R\text{-}2=0.48 \qquad F=11.061^{**}$$

Where:E^{AT}: Agric. Export in Egypt, Dt: Dummy variable equal (0) for the Period from 1992-1998 and (1) for the Period from 1999-2009.

4.2. the effect of transport in means of technology (Internet) on the Egyptian Agric. Imports

The equation (6) illustrates the total Agric. Imports in Egypt before and after using internet, the results of the statistical analysis that there is non significant decrease in the Agric. Exports during the period after the internet comparing to the period before. Average Agric. imparts in Egypt 10771.429 Million (LE) .(S.T DEV)3382.166, and After using internet about 12366.55 Million (LE) .(S.T DEV)3591.836. The study used the dummy variable to measure the effect of information revolution in Egypt, the results of analysis statistics was the following.

$$y= 10771.429 - 1595.117\ Dt$$

$$(8.108) \qquad (0.939) \qquad (6)$$

$$R^2= 0.052 \qquad F= 0.881$$

Where: IAt= Agric. Imports in Egypt, Dt= Dummy variable equal (0) for the period from 1992-1998 and (1) far the period from 1999-2009

5. Relationship between information technology and GDP,ICT in Egypt

The equation (7) describes that there is significant increase relationship between Total Agricultural Trade and both (GDP) Gross domestic product in Egypt,(ICT) information communication technology in Egypt, the results of analysis statistics was as the following:

$$Y_{TAT}= 3.412 + 0.315\ GDP +0.731\ ICT$$

$$(3.91)^{**} \quad (6.151)^{**} \quad (4.735)^{**} \qquad (7)$$

$$R^2 = 0.723 \qquad F\ 126.3^{**}$$

Where:Y_{TAT}: Total Agricultural Trade,GDP: Gross domestic product, ICT: information communication technology.

The equation (8) describes that there is significant increase relationship between Total Agricultural Trade, GDP, Internet host in Egypt. Internet user in Egypt Population ,the results of analysis statistics was as the following:

$$Y_{TAT} = -2.634 + 0.326\ GDP +0.029\eta 1 +0.0739\ \eta 2+ 0.0234P$$

$$(-5.091)^{**} \ (4.932)^{**} \ (1.092) \quad (3.947)^{**} \quad (0.413) \qquad (8)$$

$$R^2=0.691 \qquad F=103.2^{**}$$

Where:Y_{TAT}: Total Agricultural Trade .GDP: Agric growth domestic product in Egypt.$\eta 1$: Internet user in Egypt. $\eta 2$: Internet host in Egypt and P=Population describes that there is significant increase relationship between Total Agricultural Trade, GDP, Internet user in Egypt and Internet host in Egypt. Population The equation describes that there is significant increase relationship between Total Agricultural Trade, gross domestic product GDP and internet user, n2: Internet host, Population in Egypt.:

6. Conclusion

The result of study shown that, there is a significant impact for information technology also E-Commerce applied on foreign trade sector generally, in addition to total trade balance export, import was also positive impact on Agriculture import, export Egyptian Agriculture trade balance. Through the study of the information economy and E-commerce, we can make conclusion as followings:

1. The main gab between developed countries and development of E-commerce.
2. The number of internet user around the world has been steadily growing and this growth has provided the impetus and the opportunities for global and regional E-commerce.
3. The main gab between developed countries and developing countries in information and communication technology (IT).
4. About 2.3% only Africa internet usage compared the rest of the world in 2009. While 36.5% population in Asia usage internet.
5. Egypt becomes the biggest African countries for the number of internet user even more than South Africa. Which is the highest percent in Africa?
6. The results of the statistical analysis that there is a significant increase in the total exports and imports in Egypt and during there period after the internet comparing to the before.
7. The results of the statistical analysis that there is a significant increase in the agric. Exports and agric. Imports in Egypt during the period and after the internet comparing to the before.

7. References

[1] Lavouras, S., 'Trading on the Internet Challenges and Benefits for Buyers. In; proceeding of E-commerce for agribusiness', London, 2009.

[2] Mcfarlane, D., "Internet Adoption and use of E-commerce Strategies by Agribusiness Firms in

[3] Alabama"southern agricultural economic association annual meeting, mobile, Alabama, Feb, 2003.

[4] Nixon. C., "E-commerce and its Projected Impacts on Agriculture", NZ institute of economic research (INC) ministry of agriculture and forestry, New eland Jun 2000.

[5] www.Egypt.gov.eg

[6] world internet usage and population statistics. 2009.

[7] data cited from world Resource

Assessment of Environmental Quality of Coastal Fishpond Areas Using Macrobenthic Structure: Multivariate and Graphical Approaches

Sapto Purnomo Putro and Riche Hariyati

Department of Biology, Faculty of Mathematic and Natural Sciences Diponegoro University, Jl. Prof. Soedharto, SH, Tembalang Campus, Semarang, Indonesia

Abstract

Environmental degradation that results in decreased quantity of farmed fish production is an issue that often arises in rapid aquaculture industry. This study aims to develop the method of environmental quality assessment of aquaculture using macrobenthic structure to ensure the sustainability of its activity. The research was conducted at three fish farming sites along coastal of Sayung, Demak Regency, Central Java, i.e. milkfish ponds, shrimp ponds, and mixture pond. Determination of the environmental quality of the farms was done by analyzing data and environmental parameters and macrobenthic abundance and biomass using multivariate and the graphical methods. The results of Principle Component Analysis (PCA) projected from an aquatic environment parameters showed no signs of grouping based on three types of ponds, but there are signs of grouping by time sampling, indicating fluctuations in physico-chemical conditions of waters over time. Based on the macrobenthic abundance, study sites were dominated by gastropods (97%), the rest of bivalves (2%) and polychaetes (1%). Results from ordination analysis, ABC curves and k-dominance showed no signs of clustering by types of pond, but between sampling times. This implies that multivariate and graphical methods can sensitively detect any environmental change, particularly changes in macrobenthic community, water quality and sediment over time.

Keywords: environmental quality, multivariate, graphical method, macrobenthic structure, fish farming.

1. Introduction

Macrobenthic animals are invertebrate animals that are relatively small and retained on sieve size of 500 μ m and stayed at the bottom habitat by digging a hole in the substrate or sediment, either have home tube (tubicolous) and do not have the tube [1]. The animal has an important role in the formation of sedimentary habitats. These organisms can stimulate and improve the process of mineralization of organic matter and improve the exchange of particles in the boundary layer between water and sediments [2]. They play an important role in the food chain through the transfer of organic carbon back to the pelagic ecosystem [3]. Through mechanisms such as increasing the N cycle throughout nitrification and denitrification, increase the rate of oxidation of sediments, benthic organisms are very responsive to eutrophication and hipoxia [4], and can therefore be used as organic enrichment bioindicator [5]. Vertical distribution of the processes performed by microbes in sediments influenced *infauna* animals through activities, including eating, digging holes, and the formation of the tube home [6]. The presence *bioturbation* activity, such as mixing or stirring and relocation of sediments is one of the most important factors in controlling the processes that occur around the oxic-anoxic sediment layer [7].

Several studies on the effects of fish farming activities on water quality and sediment have been conducted, among others, the existence of organic enrichment, *eutrophication*, sediment anoxic (without oxygen), decrease in redox potential, oxygen consumption increased in the sediment, increasing the total organic carbon, sulfite, nitrogen components, and phosphate. However, the results of these studies generally varied and inconsistent, indicating that *abiotic* environmental variable alone is not sufficient to determine the quality of the environment more comprehensively. In the last decade, analysis of macrobenthic *infauna* has been applied as one of the main criteria in determining the quality of the environment for aquaculture management in various countries. Efforts to determine the response of macrobenthic community to environmental changes can be done by using a multivariate analysis, and graphical method. Multivariate analysis and graphical method have been applied by several researchers to determine the rate of changes in the number of species, abundance, and biomass in response to environmental changes, especially by organic enrichment. Approach to community analysis using these methods is considered to be more accurately assessed in determining the level of disturbance, given the response at the individual/ specific taxa to environmental changes are often specific and highly variable [8, 9, 10].

2. Methods

Measurement of physical-chemical water carried out *in situ* using a Horiba U-10 Multiprobe Tester. Parameters measured include pH, dissolved oxygen (DO), temperature, turbidity, salinity, conductivity, and depth. Measurements were made around the bottom (adjacent to the sediment), and in surface water. Turbidity was measured using a Secchi disk. Sediment sampling was carried out by using the Eckman Grab, operated directly by using the hand strap. Determination of the quality of sediments was carried by several physico-chemical parameters of sediments, including the composition of the physics/particles of sediment and organic matter content.

Sediment samples were taken from Eckman Grab put in 4% formalin solution and stored in plastic jars. Sediments were filtered through a 1 mm mesh-size sieve to retain macrobenthic animals. Organisms retained on the sieve were put into a solution of 70% ethanol for further analysis, which includes sorting, counting, identification, counting the number of species, density, and classification of taxa, and biomass. Determination of dry weight after drying of each animal taxa macrobenthos in an oven at a temperature of 60°C for 24 hours.

Principal Component Analysis (PCA) using the Euclidean distance were performed to determine differences in environmental variability between sampling areas. Graphical method of Abundance / Biomass Comparison (ABC) was used to determine the extent of changing patterns of macrobenthic structure based on biomass and abundance of each sampling station in space and time [11, 12]. The *k*-dominance curves for both biomass and abundance will be projected in the same graph. Multivariate analysis using the ordination and cluster by the method of Non-metric Multi Dimensional Scaling (NMDS) of Bray-Curtis similarity was conducted to assess differences in macrobenthic structure between sampling stations in two dimensions [13]. Multivariate analysis and graphical methods performed using the software Primer Version 6.1.5 [12].

3. Results

3.1. Physico-chemical quality of waters

The results of measurements of physical-chemical parameters of waters at fishpond areas of Sayung, Demak show a normal range. The average measurements of chemical physics waters at the location of the entry of water (*inlet*) indicated low variability across the fishpond type, especially the measurement of conductivity, turbidity and brightness. Turbidity indicates the optical properties of water, resulting in the refraction of light into the water. Turbidity limits the entry of light into the water. Turbidity is influenced by the floating material, and the breakdown of certain substances, such as organic materials, microorganisms, sludge, clay and other fine floating objects. The more turbid of the water, electrical conductivity and solid will be higher [14]. Conductivity values are generally closely related to the salinity of the waters, while the brightness is influenced by both organic solute particles (the excess of animal feed and feces of farmed fish) and inorganic (particles of silt/soil). Highest brightness was located in the milkfish ponds. Brightness is also affected by the microscopic aquatic organisms, especially microalgae (phytoplankton). When the population of microorganism in the waters increases, the brightness of the waters will be decreased. Results from Principle Component Analysis (PCA), which is projected from water environment parameters of each type of fishpond showed no obvious grouping (Fig. 1.A). This means that in general the condition of waters in the three locations are relatively similar, with little variability of the value. However, if the PCA is projected based on the sampling time, it appears that stations of sampling I occurred grouping to the left and sampling II clustered to the right (sampling II) of the ordinate (Fig. 1.B). This clustering indicates that the values of physico-chemical parameters of waters fluctuated from time to time, although no extreme values were recorded. Based on water physico-chemical measurements, all three locations of the fishpond are relatively conducive to the cultivation area.

3.2. Macrobenthic Structure: spatial and temporal

Taxa of macrobenthos identified at the study site were dominated by gastropods (97%), the rest of the class consists of bivalves (2%) and the polychaetes (1%). Member of gastropods inhabiting fishpond areas consist of 27 species. Species dominated in the area of milkfish ponds, shrimp ponds, and mixed pond was *Syrmilasma venustula* (Thiaridae) with a mean abundance of 67.5 individuals/grab, followed by *Cerithidea quadrata* (Potamididae) with a mean abundance of 9.9 individuals/grab and *Thiara sermilla* (Thiaridae) with a mean abundance of 6.3 individuals/grab. In contrast to gastropods, bivalves found relatively few, i.e. 7 species with a relatively low in abundance. No dominant species was found at any sampling station. Similarly, members of polychaetes were 5 species with relatively low in abundance.

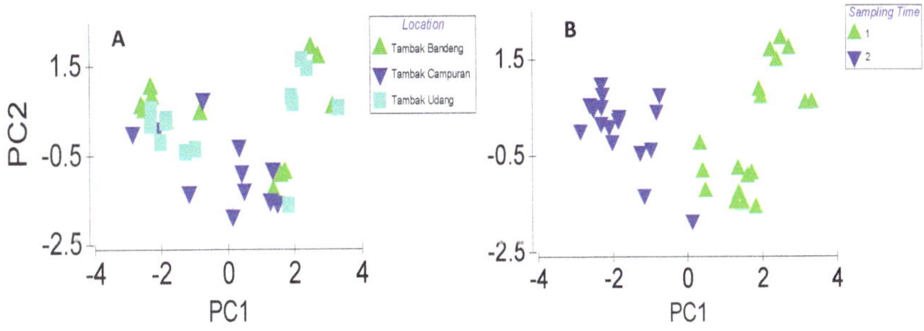

Fig 1. Principle Component Analysis (PCA) generated from the water parameter data, projected by: (A) Type of fishpond, (B) The time of sampling.

Macrobenthic abundance varied between stations (spatially) and sampling time (temporally). Results from ordination of Non-Metric Multidimensional Scaling (NMDS) projected based on transformed data of log (x +1) of macrobenthic abundance showed no grouping if it was projected by fishpond types (Fig. 3.a). The absence of grouping patterns among the three types of farms indicates that the abundance and number of species in all three types of ponds are relatively similar, except for stations TU02SD2B (shrimp ponds; outlet; sampling II) and TC01SD1B (Pond mixture; inlet; sampling I), which are projected separately from the other group of stations. Projection of station to the right of the ordinate indicates a decline in abundance and number of species that can be caused by environmental disturbance [11].

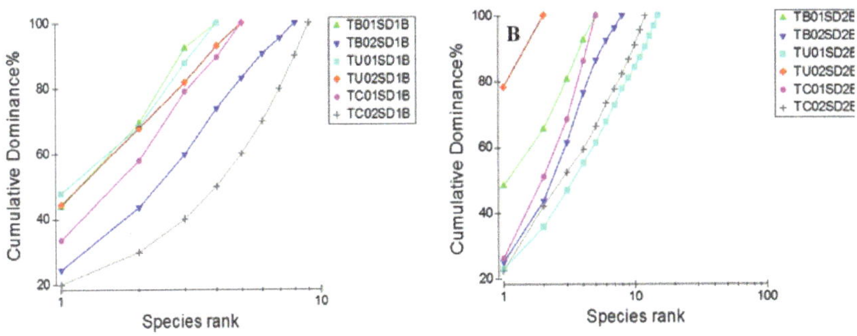

Fig 2. Cumulative k-dominance curves projected for each station based on the time of sampling: (A). Sampling I (July 2009), and (B). Sampling II (October 2009)

On the contrary, the ordination based on the sampling time indicates grouping of sampling stations I and II sampling (Fig. 3.b).These groupings may be due to differences in composition and abundance of species found between the two sampling time.

Based on sampling time, results from cumulative *k*-dominance analysis projected for each station stations placed curve of site TC02SD1B (mixed fishpond; outlet; sampling I) at the bottom of the graph. This indicates the highest evenness and species diversity among other stations (Figure 4.A). Whilst curve of site TU02SD2B (shrimp ponds; outlet; sampling II) was positioned at the top (Figure 4.b). This indicates the lowest diversity and evenness among the other stations [12]. Level of environmental disturbance can be determined by comparing the abundance and biomass of macrobenthic community. This method is known as an Abundance-Biomass Comparison (ABC) [15]. In stable environmental conditions or interference levels are considerably low, macrobenthic community will be dominated by conservative species, i.e. species that have a life strategy "*K*-selection", large body size, relatively long life span, dominant in biomass but low in the number of species. Under conditions of disturbed areas, macrobenthic communities will be dominated by organisms that have a strategy of "*k*-selection" in his life, or so-called opportunistic species, characterized by a relatively small body size, short life span, dominant in the number, but low in biomass, has potentially high reproductive rate and early maturation [16]. Depending on the level of disturbance, biomass curves can be positioned above or below the abundance curve, or the two curves can be similarly shaped adjacent and parallel to each other or intersect each other once or several times along the curve [11, 17].

Results from the analysis of abundance and biomass of macrobenthos projected as ABC curve showed variability between stations (Fig. 3 and Fig. 4). Based on the criteria proposed by Clarke & Warwick (2001) [11], all the sampling stations on the sampling time I (July 2009) is categorized as a disturbed/ polluted area, except station TC01SD1B (mixed ponds; inlet) and station TC02SD1B (mixed pond; outlet), as shown in Fig. 3. This area is used as mixtures to obtain the flow of water directly from the River Ronggolawe. The river is still affected directly by the tidal wave, so the quality of water used is still relatively good. While the input water for shrimp and milkfish ponds are from mixed pond, so that water quality has been affected by the activity of farming in the mixed pond.

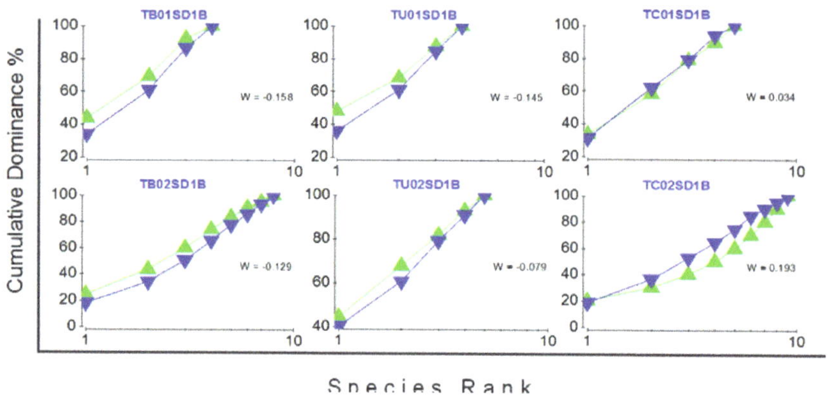

Fig 3. Abundance Biomass Curves Curve (ABC) is projected based on data transformation log (X +1) of abundance (▲) and biomass (▼) macrobenthos in the first sampling time .

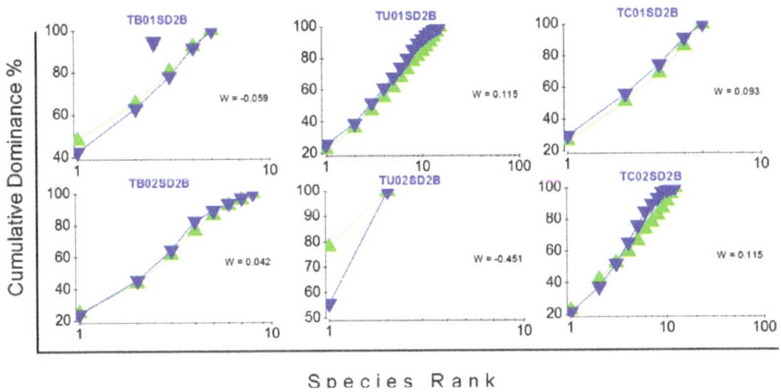

Fig 4. Abundance Biomass Curves Curve (ABC) is projected based on data transformation log (X +1) of abundance (▲) and biomass (▼) macrobenthos in the second sampling time.

Unlike the sampling time I, the curve shown for each station on the sampling II (October 2009) is categorised as undisturbed areas (unpolluted), except station TB01SD2B (milkfish ponds; outlet) and station TU02SD2B (shrimp ponds; outlet) (Fig. 4). This indicates that the two stations above have been disturbed in October's samples. Based on the results of interviews with the owner/ manager of shrimp farms (personal communication), the condition of shrimp pond a week before sampling II (early October 2009) has been environmentally disturbed, resulting in most of the stocked shrimp died (more than 70% of total population). It is not yet known the cause of the disturbance, but is expected related to the quality of feed and less aeration in the pond. Poor feed quality will reduce or even eliminate appetite of the animals. This can cause the accumulation of feed into body water and partially decomposed into sediment. The presence of high organic matter can trigger the growth of toxic microalgae and pathogenic bacteria in the waters, so it can result in lowered resistance of the cultured animals against the diseases.

4. Conclusion

Macrobenthic animals are very sensitive to changes in the environment, particularly water quality and sediment. In general, the disturbance or physical-chemical changes in the environment will respond to changes in the composition and abundance the animals. Approach to multivariate and graphical methods, especially using ordination analysis MNDS, ABC curves and *k*-dominance, can sensitively detect any environmental change, particularly changes in macrobenthic community, water quality and sediment over time.

5. Acknowledgments

The authors would like to thank to the DP2M-DIKTI through Fundamental Grant 2010 ('Hibah Fundamental') that has funded the implementation of research for the study on biomonitoring in aquatic environment.

6. References

[1] Snelgrove, P. V. R. 1999. Getting to the bottom of marine biodiversity: sedimentary habitats. *BioScience*, **49**: p.129-135.

[2] Heilskov, A. and M. Holmer, 2001. Effects of benthic fauna on organic matter mineralization in fish-farm sediments: importance of size and abundance. *ICES J. Mar. Sc., 58*:p123-130.

[3] Snelgrove, P.V.R. and C.A. Butman, 1994. Animal-sediment relationship revisited: cause versus effect. *Ocean. Mar. Biol. Ann. Rev.*, **32**: p111-117.

[4] De Roach, R. J., A. W. Rate, B. Knott, and P.M. Davies, 2002. Dennitrification activity in sediment surrounding polychaete (*Ceratonereis aequisetis*) burrows. *Mar. Fresh. Res.*, **53**: p35-41.

[5] Grall, J and L. Chauvaud, 2002. Marine eutrophication and benthos: the need for new approaches and concepts. *Glob. Chang. Biol.*, **8**: p813-820.

[6] Gerino, M., G. Stora, F. Poydenot and M. Bourcier, 1995. Benthic fauna and bioturbation on the Mediterranean continental slope: Toulon Canyon Cont. *Shelf Res.*, **15**: p1483-1488.

[7] Kristensen, E., 2000. Organic matter diagenesis at the oxic/anoxic interface in coastal marine sediments, with emphasis on the role of burrowing animals. *Hydrobiologia*, p. 426: 1-7.

[8] Yokoyama, H. 2002. Effects of fish farming on macroinvertebrates: comparison of three localities suffering from hypoxia (UJNR Technical Report No.24), http://nsgl.gso.uri.edu/source/tamuw95003/ptamuw95003_part_95001.pdf.

[9] Putro, S.P., I. Svane, , and J. Tanner. 2006. Effects of fallowing on macrobenthic assemblages in sediments adjacent to southern bluefin tuna cages. *Final report of Aquafin CRC-Southern bluefin tuna aquaculture: evaluation of waste composition and waste mitigation*. SARDI Aquatic Science, Adelaide,.

[10] Putro, S.P. 2007. Spatial and temporal patterns of the macrobenthic assemblages in relation to environmental variables. *J. Coast. Dev.*, **10**: p.15-22.

[11] Clarke, K.R. and R.M. Warwick, 2001. Change in marine comunities: *an approach to statistical analysis and interpretation PRIMER-E Ltd, Playmouth*, p.180-190.

[12] Clarke, K.R. and R.N. Gorley, 2006. *Primer v6: User Manual/Tutorial*. PRIMER-E, Plymouth, p.190

[13] Clarke, K. R., P.J. Somerfield and M. Chapman, 2006. On resemblance measures for ecological studies, including taxonomic dissimilarities and a zero-adjusted Bray-Curtis coefficient for denuded assemblages. J. *Exp. Mar. Biol. Ecol.*, **338**: 179-185.

[14] Kristanto, P., 2002. Ekologi Industri. Andi offset, Yogyakarta,p.155-162.

[15] Clarke, K.R. and R.M. Warwick, 1994. Similarity-based testing for community pattern: the two-way layout with no replication. *Mar. Biol.*, **118**: p.167-172.

[16] Diaz, R. J. and R. Rosenberg, 1995. Marine benthic hypoxia: a review of its ecological effects and the behavioural responses of benthic macrofauna. *Ocean. Mar. Biol. Ann. Rev.*, **33**: 245-251.

[17] Putro, S.P., 2009. Response of trophic groups of macrobenthic fauna to environmental disturbance caused by fish farming. *J. Coast. Dev.*, **12**: p.146-152.

The Effect of Hydro-Alcoholic Extract of Fenugreek Seeds on Female Reproductive Hormones in Mice

Mehrdad Modaresi[1], Behnaz Mahdian[2], Alireza Jalalizand[3]

1 Department of Agriculture, Khorasgan Branch, Islamic Azad University, Isfahan, IRAN

2 Payam e Noor University, Isfahan Center, Isfahan, IRAN

3 Department of Plant protection, Khorasgan Branch, Islamic Azad University, Isfahan, IRAN

Abstract

Fenugreek is a plant from Fabeaceae family which is used for many medical plans. This study was conducted to study the effect of fenugreek seed's extract on reproductive system of female *Balb/C* mice .To this, mice's were divided in five groups: control, placebo, and three experimental groups. Each group had ten mice which their estrous cycles were synchronized for starting the study. Control group did not receive any drug. Placebo group received normal saline, and experimental groups were injected in peritoneum by 50,100, and 200 mg/kg extract every day until 20 days. After finishing injection, blood samples were taken. Hormone measuring (including FSH, LH, estradiol, and progesterone) was done using one way variance analysis of SPSS program at 95% probability level. Ovary slides were prepared and were studied using optical microscope. Obtained results, showed significant reduction in FSH, and LH levels and also significant increase in stradiol level in all experimental groups, but progesterone level was increased only in second experimental group. Histology results of ovary showed significant reduction in folliculogenesis of all three experimental groups. Also, increase in number of corpus luteum was highly significant. Furthermore, destruction of ovary tissues was observed in second experimental group. According to results, the extract of fenugreek seed stopped folliculogenesis trend and destroyed ovary tissue which shows its anti fertility effect in female mice.

Keywords: Fenugreek, reproductive hormones, mice

1. Introduction

Traditional medicine is a nature based science which is inherited from ancestors and includes plants, minerals and animal matters which are used as drugs [1] . In traditional medicine, pharmaceutical plants have a special place. Medicinal plant is a plant which has specific effective matters, is used in prevention or cure of illness, and has been mentioned in one of national valid Pharmacopoeia [2]. The importance of pharmaceutical plants is more obvious today and scientists of various countries are trying to identify medicinal plants, their properties and effective matters. Fenugreek is one of the old plants which have a wide range of medicinal properties: reducer of blood sugar and fat, anti diabetes, pain reliever, and anti cancer, increase in sex ability, and increase in milk production and worm killer [3] . Scientific name of fenugreek is *Trigonella foenum -graceum* L. which *trigonou* is derived from Latin and means triangle (because of triangle leaves), and *foenum-graceum* which means Greek hay is because of its different uses in ancient Greek. Also, because two seed pods are produced from main stem's nodes, these plants has been called "bull's horn" or "Goat horn " [4] . This plant is an annual, bush plant with 10 to 50 cm length which is sown in various regions of world like small Asia, Iran, Egypt, Algeria, India, Morocco, Italy, and Spain. The region of this plant has been known west of Asia [5] . Economical products of fenugreek are seeds and leaves, and reproduction of this plant is done by seeds which are sown in clay, calcareous lands in September. Seeds are sown together with clover seeds. Harvesting time is from June to July after harvesting, stems are being cut from the bottom and being dried [6] . This plant is one of the oldest pharmaceutical plants.

It was used in old Egypt as incense, and for mummifying corpses and also for easy confinement and increase in milk production. Even nowadays Egyptian woman use this plant for curing menstrual pain, as a tea for stomach problems of tourists, and also as a complement matter for wheat and corn flours for baking breads and confectionaries [7] . In ancient Chinese drugs, seeds of fenugreek were used as strengthen drug [8] . The nature of plant and its seeds is dry and warm. Seeds are used as sterilizer, mild laxative, diuretic, in bronchial inflammation treatment, leprosy treatment, the treatment of hemorrhoids and mouth deodorant [9] . Also, it is laxative, anti inflammatory, joint pain reliever, pulmonary and bone tuberculosis. Meanwhile, it is used for increase in weight [6] . Seeds of fenugreek have constant oils, essence, alkaloids, flavonoids, saponin, sapogenin, mucilage, free amino acids, carbohydrate, fiber, phosphorus, sulfur, lecithin, iron, calcium, magnesium, potassium, sodium, coumarin, tannin, resin, pectin, niacin, and carotenodic compounds.

Fenugreek has many pharmacological effects: results of researches show that diabetic mice (induces by stereptozotocine) which were cured by seed extract of fenugreek seeds, had an increase in weight and a reduction in ratio of kidney weight to body weight [10] .Also, reduction in blood fats was ascribed low carbohydrate absorption and fat absorption was ascribed to active presence of fibers [11]. Results of studies on wild mice showed that fenugreek increased excreting of acids and neutral stroles, and then saved cholesterol of body was decreased [12]. The effect of fenugreek on fat index of diabetic patients with high cholesterol showed that this plant reduced fats significantly [13]. Significant changes were observed in total cholesterol, LDL and triglyceride but no in HDL level. Flavonoides of fenugreek seeds have anti oxidant activities which play their roles via hydrogen reduction and deletion single oxygen. A lot of fenugreek seed poly phenoles prevent oxidative hemolysis and peroxidation of induced fats (by hydrogen peroxide in laboratory) of human's red cells. This study was conducted to evaluate the effect of hydroalcoholic extract of fenugreek on reproduction physiology of female mice.

2. Materials and Methods

2.1. Extraction

To prepare the extract of seeds, they were grinded completely and 30 g of obtained powder was poured in a sterilized erlen, 40 cc of physiological serum was added to it, and was located in a cool place. After 24 hours, erlen contents were mixed completely using a shaker for five minutes. Then, after filtering the solution by filtrative paper and calculating extract residual in solution, concentration of extract in base solution was determined and doses were prepared.

2.2. Animals

Female mice (*Balb/C*) in weight range of 25-40 g were taken from animal division of Isfahan Medical University. Samples were kept in similar conditions of water, food, light, temperature and moisture to environment adaption. These similar conditions were continued in injection time too.

2.3. Study treatments

Samples were divided to five groups with 10 mice in each group: one control group, one placebo group, and three treatment groups. Mean weight of all groups were 30 ±5 g and each group was kept in a separate cage. Prepared extract was injected in three concentrations according to body weight:

Control group: non injected

Placebo group: 9% normal saline

group1: 50 mg/kg extract of body weight

group2: 100 mg/kg extract of body weight

group3: 200 mg/kg extract of body weight

Ten Injections were done between 8-10^{am} in a twenty days period (every other day) . One day after progesterone injection, extract injection was started and one day after the last injection, blood sampling and autopsy were done. For studying drug effect on samples, all mice must be in similar estrous cycle. To this, 0.5 micro gram of cloprostenol was injected in peritoneum and three microgram of progesterone was injected under skin of all samples. One day after that, extract injection was started and one day after last injection, blood samples of heart were prepared to study variation of sexual hormones, and also autopsy was done to ovary histology of samples.

2.4. Statistical analysis

Obtained data were analyzed using one way variance analysis of *SPSS 11.5* program, and mean comparison was done using Duncan test.

3. Results

3.1. Number of graafian follicles

Counting graafian follicles of prepared tissue slides and mean comparison of groups using Duncan test ($p \leq 0.05$) showed that all experimental groups had significant reduction in proportion to control group and third group had the least.

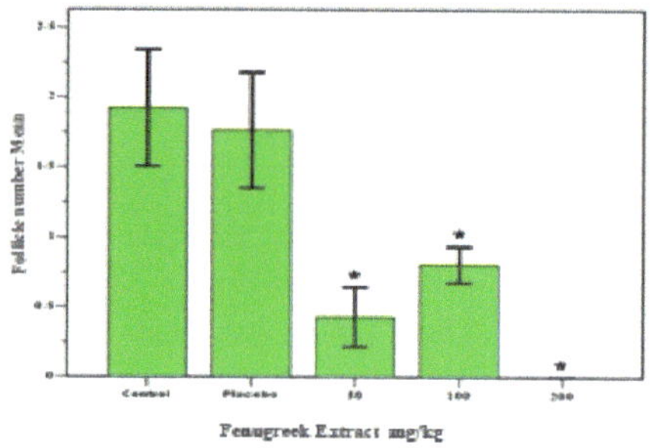

Graph 1. The number of graafian follicles in various groups

Results of counting corpus leuteum of tissue slides showed significant increase in second experimental group (100 mg/kg) and significant reduction in third group (200 mg/kg) in proportion to control but there was no significant difference between first group (50 mg/kg) and control.

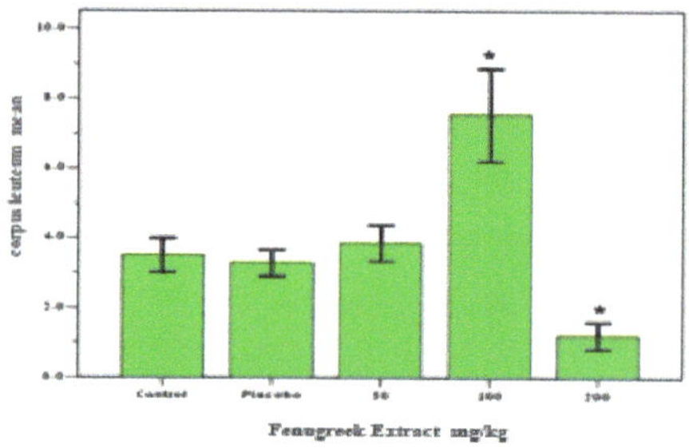

Graph 2. The number of corpus leuteum in various groups

3.2. FSH amount

Comparing FSH level (mIu/ml) in blood serum of experimental and control groups using Duncan test ($p \leq 0.05$) showed significant reduction all three experimental groups in proportion to control.

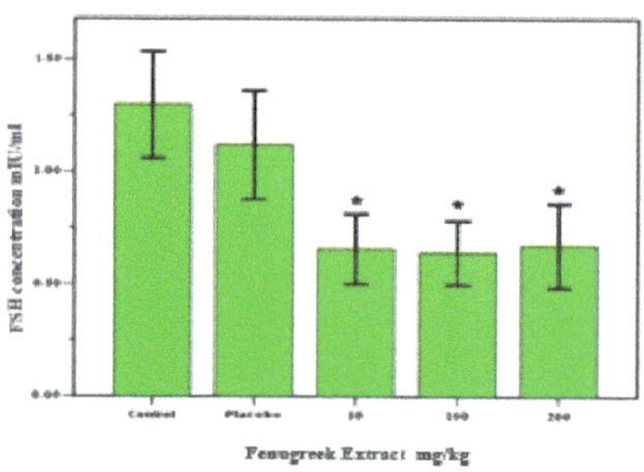

Graph 3. FSH amount of various groups

3.3. LH amount

There was significant reduction in LH level (mIu/ml) of all experimental groups according to mean comparison results.

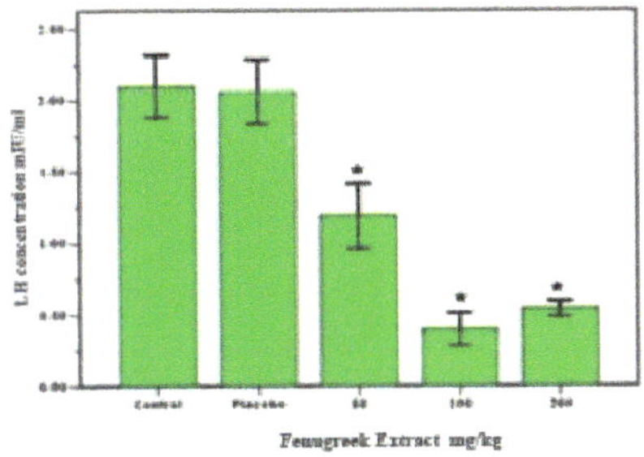

Graph 4. LH amount of various groups

3.4. Estrogen amount

Mean comparison of estrogen level in blood serum of experimental and control groups using Duncan test ($p \leq 0.05$) showed significant increase in hormone level of all three experimental groups and second group (100mg/kg) had the highest hormone level.

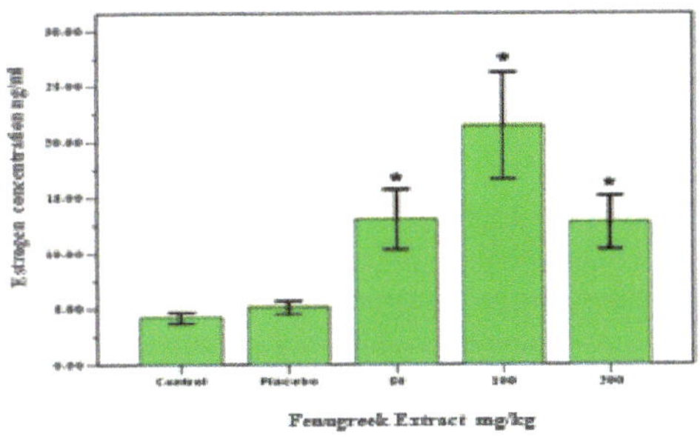

Graph 5. Amount of Estrogen hormone in various groups

3.5. Progesterone amount

Mean comparison results of progesterone level in blood serum of experimental and control groups showed increases in first group (50 mg/kg) and second group (100 mg/kg) which this increase was significant only for second group. Also, third group (200 mg/kg) showed significant reduction in proportion to control group.

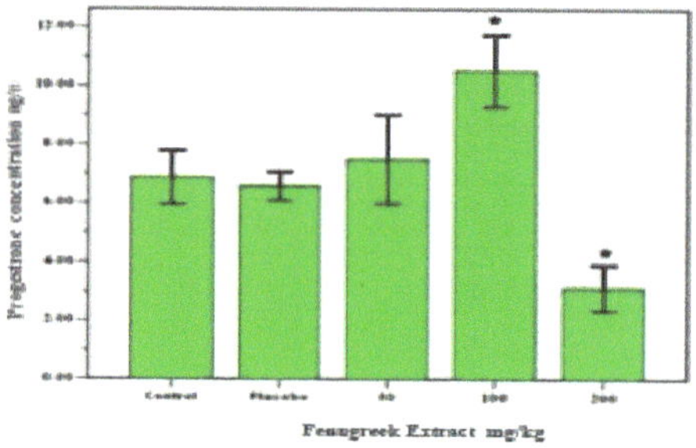

Graph 6. Amount of progesterone hormone in various groups

4. Summary and conclusion

Obtained results show significant reduction in number of graafian follicles in all three experimental groups (50,100, and 200 mg/kg) . This reduction is more obvious in third experimental group which is in agreement with previous results [3] . In their study, because of some compounds like sapogenin and diosgenin extant in fenugreek seeds, which are precursor of progesterone and testosterone reducer, sperm production was decreased [3].

By decreasing FSH level in follicle liquid (one of the follicle growth affecting factors) $IGFBP_{4.5}$ will be increased and action of proteases will be prevented, then, FSH antagonists will be increased and follicle will be evolved with atresia. GnRH stimulates also $IGFBP_{4.5}$ production in granulosa cells of follicle and reduces $IGFBP_{4.5}$ proteases, then, causes follicle atresia. Probably, secreting Aromatase preventing protein from dominant follicle is effective on the other follicles and cause growth stop and atresia. Furthermore, low concentration of leptin in follicle liquid may have negative effect on growth and ripening of ovules. Results of studies show that Nitric oxide (NO) prevents growth of follicles and repining of ovocytes via inducing apoptosis [14] . According to results, amount of FSH hormone decreased significantly in all experimental groups. High concentration of progesterone and low concentration of estrogen prevent follicle stimulating hormone. Meanwhile, it seems that opioid peptides of brain are mediums of this negative feedback [15] . On the other hand, inhibin hormone of dominant follicle prevents FSH secreting from pituitary, in very specific way. considering the results, LH hormone was reduced significantly in all experimental groups which can be because of :

Linoleic acid (CLA) has reducing effect on LH amount via decreasing leptine. Due to quite definite and significant relationship between leptine and nitric oxide in LH releasing from pituitary, Leptine reduction will be led to reduction in nitric oxide and then GNRH releasing [14] .

Fenugreek increases prolactin via affecting serotonergic system which will be led to prevention in GnRH releasing and then reduction in LH[14] .according to results, estrogen hormone was increased in all experimental groups. Seed extract of this plant has Gitogenin, Trigonelline, and Quercitin which have estrogen making activity. It seems that these three compounds play important roles in increasing estrogen by their biological activities. Considering the results, amount of progesterone hormone was increased significantly in second experimental group and was decreased significantly in third group. These increase and decreases are similar to increase and decrease in corpus leuteum of second and third groups. On the other hand, progesterone increase can be ascribed to diosgenin compounds of fenugreek which are progesterone precursor.

5. References

[1] Modaresi M. Gholchobian H. (2012), Effect of hydro alcoholic extract of nettle on Immune system in mice, *Asian J. Chemistry* 24(5) ; 2339-2341

[2] Davazdah Emami, S.2003. Application of medicinal plants. Isfahan Agricultural Researches organization press. Page 10.12-14 .

[3] Mokhtari, M. Shariati,M, Ghahramani,R.2007. The effect of fenugreek (Triqonella foenum-greeum L .) on hormone variation of testpterone and spermatogenesis of rat . Medicinal plants magazine. 7th year. no: 25(winter).pages 12-20.

[4] Zargari.A. 2007 .Medicinal plants. Tehran university press.2nd volume. page 608. 637-642.

[5] Samsamshariat. H.2003 medicinal plant production. Maani press. Page 307-308

[6] Samsamshariat, H. Moattar, F.2008 .Medicinal plants and natural plants (medical materia). Mashaal press. Page 79,333,336-337.

[7] Ashish Toppo F, Akhand R, Pathak A. K. 2009.Pharmacological actions and potential uses of *Trigonella foenum-graecum*: A review, Asian Journal of pharmaceutical and clinical Research, vol. 2, Issue 4,

[8] Yoshikawa M, Murakami T, Komatsu H, et al. . 2001 .Medicinal foodstuffs. IV, Fenugreek seed. (1): structures of trigoneosides Ia, Ib, IIb, IIIa, and IIIb, new furosranol saponins from the seeds of Indian *Trigonella foenum-graecum* (Fenugreek) seed extract as antineoplastic Agent, phytother. Res; 15: 257-9.

[9] Bhatia K, Kaur M., Atif F., Ali M., Rchman H., Rahman S., Raisuddin S., 2005; Aqueous extract of T. fenum-graecum l. Ameliorates additives urotoxicity of buthionine sulfoximine and cyclophosphamide in mice. Food and Chemical Toxicology 44 ;1477-1750

[10] Wan- Li xue, Xuan- she Li, Jian Zhang, Young- Hui Liu, Zhi- lun wang and Ruijuan Zhang. 2007 Effect of Trigonella foenum- graecum (Fenugreek) extract on blood glucose blood lipid and hemorheological properties in streptozotocin induced diabetic mice. Asia PC J clin Nutr; 16 (suppi): 422-426 .

[11] Hannana JMA, Rokeya B, Faruque O, 2003. Effect of soluble dietary fibre fraction of Trigonella foenum-graecum on glycemic, Insulinemic, lipidemic and platelet aggregation status of Type2 diabetic model mice. J Ethnopharmacol; 88: 73-77.

[12] Sharma R.D. 1986 . An evaluation of hypocholesterolemic. Factor of fenugreek seeds (T. foenum-graecum) in mice. Nutr. Rep. int.; 33: 669-77.

[13] Awal MA, Rashid MU, Ahmad KW, Asadi ZS, Islam K. 1999. Effect of karela and fenugreek on lipid profile in hypocholesterolemic diabetic patients . Bangladesh J physiol pharmacol.; 15: 6-8.

[14] Tammanini C, Basini G, Grasselli F, Tirlli M. 2003 Nitric oxide and the ovary. J Anim SCi.; 81(E. Suppl.2): E1-E7.

[15] Zamiri, M. 2006. Reporoduction physiology. Haghshenas press. Pages 112-114,127.

The Issue of Food Subsidies in Egypt Following the Revolution of January 25

Dr.Roshdy Sh. El Adwy

Department of Agricultural Economics, kafr Elsheikh University, kafr Elsheikh, Egypt

Abstract

Program is to support food a key component of the program of direct government support, is also on the other hand one of the basic mechanisms to mitigate the negative effects of the application of economic reform policies of Egypt, and most important of which high levels of domestic prices and the increasing rate of inflation in Egypt, especially in recent years, means support display and provide some basic goods and services at prices below actual cost and the real have to ensure that low-income and poor them in proportion to their income, has resulted in the rise in the cost of the program of support for food to the high costs of programs of direct government support has grown to support program food as more than one- third the cost of the program of direct government support

Keywords: program of support, Forecasting system nutritional support, Per capita food support,

1. Introduction

As a result of the large number of events and the controversy over government support program and the extent of its ability to fulfill its obligations towards the citizens and individuals in the community, it gave evidence for economists to study that program to assess and evaluate and develop with the guidance to ensure the destination to take advantage of it as possible. Problem of the study Despite efforts by policy makers to reduce the incidence of poverty through a program of government support and the provision of basic needs of the members of the major commodities, but this interest has led to the high cost of the invoice system of government support as a direct result of the increasing population as well as rising prices and higher inflation rate with the following open system to support programs, in addition to that the support has become a financial burden, increasing the state budget, and is one of the reasons for lack of domestic production and low production efficiency with a lower benefit the people of rural support for the people of the cities of any non-arrival of support to those who deserve it in addition to the large number of the controversy about the possibility of converting support in-kind to cash support, Government has made numerous efforts to develop this program so issued many of the procedures and developed from the system of goods offered by the program of support to citizens in terms of the types and quantities of each citizen, but he is still the same problems the previous continuous and present themselves on the scene, hence the need to study. Objectives of the studThis study aimed primarily to achieve a set of key objectives that can be highlighted in thefollowing two points:1 - shed light on the structure and the terms of the program of government support, especially the Egyptian food subsidy program.2 - shed light on the efficiency of the current support program in Egypt. Research method and data sourceThe study relied on the use of methods of analysis of quantitative and qualitative, Where the function was used in the quadratic forms, linear, and Cubic, as it relied on the coefficient of determination and the amount and rate of annual change, were obtained secondary data from published and unpublished necessary to conduct this study of the views of various government, which included both the National Information Center of the Central Agency for Public Mobilization and Statistics, records of the Ministry of Social Solidarity, records of the Statistics Department of the central administration Sector of Economic Affairs of the Ministry of Agriculture and Land Reclamation, the cabinet of Ministers (information and decision support Centre), and obtained the data from some of the websites of some bodies like the World Bank, National Bank of Egypt, Ministry of Finance, the Central Agency for Public Mobilization and Statistics.

2. The Current Status of the Food Support in Egypt

Given the importance accorded to the support system of food in Egypt and in particular, for each member of the working classes and low-income, where is the issue of support to one of the most important issues affecting the economies of the countries according to their political, economic and social development, so it was important to shed light on the evolution of per capita terms of the support structure food in Egypt.

It became clear that the average per capita food support during the study period 91/1992-09/2010 - Table (1) ranges from a minimum of about 25.44 pounds in 99/2000 and a maximum of about 71.93 pounds in 07/2008 and an average annual rate of about 42.18 pounds, while showing that the average per capita bread subsidies ranging from a minimum of about 16.58 pounds in 91/1992 and a maximum of about 66.59 pounds in 07/2008 and an average annual rate of about 29.83 pounds, while the per capita support sugar ranges from a minimum of about 3.06 pounds in 07/2008, and a maximum of about 12.48 pounds in 91/1992, and an average annual rate of about 6.17 pounds, it turns out that the average per capita support of edible oils from a minimum of about 4.57 pounds in 98/1999, and a maximum of about 10.83 pounds in 91/1992, and an average annual rate of about 7.19 pounds, and it became clear that the per capita support for wheat ranges between a minimum of about 14.14 pounds in 98/1999, and a maximum of about 72.20 pounds in 08/2009, and an average annual rate of about 29.93 pounds.

And estimate the equation of the trend overall time of the evolution of per capita share of each item of food subsidies in Egypt in real terms during the study period - Table (2), it was found that the share of individual from each of the food support and bread subsidies and support for edible oils and wheat subsidies and minutes taken a general trend increasing by amounted to about 1.3, 1.8, 0.38, 2.02 pounds per year may be attributed the rise to double the world price of wheat in one year where the price of import tons of wheat four times in 2007, in addition to the decrease of oil production in return for higher domestic consumption growing and growing food gap, while Per capita sugar support have taken a general trend by Decreasing amounted to about 0.34 pounds per year decline was attributed to higher average prices in the community (inflation) with high rates of population growth, which reduces the real value of the support which will reflect negatively on the purchasing power of individuals.

3. Forecasting System nutritional Support in Egypt in the future

The nutritional support of government is a political system that works to protect citizens from higher commodity prices, allowing a minimum of decent living for citizens and fill their basic needs basket of subsidized goods, under the government›s efforts to develop such a system has dealt with many of the presidents in their electoral programs of several aspect was to increase the number of subsidized items or increase the quantities of subsidized goods to individuals or increase the value of support so it was necessary to shed light on the current situation and the situation is expected after five years, the values of per capita terms of nutritional support

Table (3) that per capita food support has increased from about 85.9 pounds in 2012 to reach in the future to 136.02 pounds in 2017, as well as similarly to the per capita support of bread and wheat subsidies and accurate has increased from about 49.93, 52.14 pounds on the arrangement currently informed of in the future about 59.07, 62.28 pounds respectively in 2017, while per capita support of sugar will decrease from about 2.42 pounds currently amounts to about 0.71 pounds in the future, as showen that the per capita support for wheat and minutes has increased from 52.14 pounds 2012, to 62.28 pounds in 2017.

Years	per capita food subsidy	per capita support of bread	per capita support of sugar	per capita support of edible oil	per capita wheat subsidies and accurate
1991/1992	61.09	16.58	12.48	10.83	16.56
1992/1993	41.31	23.17	10.06	8.43	23.15
1993/1994	31.94	19.18	7.23	7.33	19.18
1994/1995	29.65	19.28	8.07	6.77	19.28
1995/1996	32.52	31.45	6.64	6.21	31.44
1996/1997	36.22	23.49	7.74	6.34	23.96
1997/1998	27.59	21.97	6.09	5.84	21.03
1998/1999	25.91	16.06	6.28	4.57	14.14
1999/2000	25.44	17.21	6.21	7.22	14.43
2000/2001	33.61	20.27	6.29	8.51	17.46
2001/2002	39.56	19.15	5.72	7.06	19.72
2002/2003	34.57	24.86	5.31	5.16	23.86
2003/2004	33.02	40.68	3.21	6.02	40.22

Years	per capita food subsidy	per capita support of bread	per capita support of sugar	per capita support of edible oil	per capita wheat subsidies and accurate
2004/2005	59.88	41.52	4.15	8.41	39.03
2005/2006	50.35	36.52	4.25	9.43	33.54
2006/2007	50.39	42.8	3.97	8.46	40.42
2007/2008	71.93	66.59	3.06	6.71	62.19
2008/2009	61.13	45.58	4.84	6.78	72.2
2009/2010	55.48	40.56	5.64	6.59	36.84
Average	42.18	29.83	6.17	7.19	29.92
Standard deviation	14.13	13.56	2.31	1.51	15.88
Coefficient of variation	33.51	45.44	37.51	21.11	53.08

Table 1. Average annual per capita from each of the nutritional support and support for the bread during the period 1991- 2010

Source : Ministry of Social Solidarity, the cabinet of ministers(information and decision Support Center

variable	values	Num. of function	type of model	equation	average	R^2	F	quantity Ex. yearly	Ex. rate yearly
subsidy foods	True value	1	quadratic	$\dot{Y}t= 51.755-5.157xt+0.323x^2t$ (7.753)** (-5.2)** (4.328)**	42.18	.664	15.784	1.303	3.09
Subsidy breads	True value	2	linear	$\dot{Y}t = 11.571+1.827xt$ (2.662)* (4.792)**	42.18	0.575	22.963	1.83	6.12
Subsidy sugar	True value	3	linear	$\dot{Y}t = 9.583-0.341xt$ (15.087)*** (-6.125)**	6.17	.0.688	37.362	-0.34	-5.53
Subsidy food's oils	True value	4	cubic	$\dot{Y}t= 12.410-2.239xt+0.236xt^2 -0.007x^3$ (9.914)**(-4.239)** (3.898)** (-3.568)**	7.19	0.559	6.332	0.381	5.29
Subsidy wheat &flour	True value	5	linear	$\dot{Y}t = 9.560+2.028xt$ (1.776) (4.256)**	29.92	0.516	18.114	2.028	6.78

Table 2. equations of general time trend of the average per capita from each of the nutritional support and support for the bread , Sugar and edible oils and wheat subsidies during the period (1991- 2010)

Source: Calculated from Table (1).

variable	status value(2012)	expected value(2017)
Per capita food support	85.9	136.02
Per capita support of bread	49.93	59.07
Per capita sugar support	2.42	0.71
Per capita food oils support	4.64	-9.3
Per capita wheat subsidies and accurate	52.14	62.28

Table 3. Statement of the status quo and the expected value of per

Source: Calculated from Table (2)

4. Recommendations

In light of the outcome of the study of the results it can be recommended as follows:

1. dissemination of information on the size of government food support in the state budget so that the science to the public expenses of that program.
2. Taking into account the orientations of food subsidies in the future in light of ongoing global increasing in the prices of food commodities.
3. Need to work on the restructuring of the support system in the manner which ensures the desired arrival to the extent appropriate and efficient to target groups of low-income groups.

5. References:

[1] Ministry of Social Solidarity, the Cabinet of ministers(Information and Decision Support Center.

[2] Amira Mostafa Mohammad Hamza, the Economic Study of Government Support for Sood Commodities in Egypt, Master Thesis, Faculty of Agriculture, Tanta University, June 2012

[3] Ministry of Social Solidarity,Cairo Egypt.2009

[4] The Cabinat of Ministers (Information Decision Support Center)

[5] Central Agency for Public Mobilization and Statistics, 2010.

[6] Selma Fared Policy support in Egypt historical background may 2006.

Determining Morphological Traits and Genetic Diversity of Rose Aphids Using RAPD and RFLP-PCR Molecular Markers

Ali Reza Jalalizand[1], Azadeh Karimi[1], Mehrdad Modaresi[2], Esmaeil Mahmoodi[1]

1 Department of Plant Protection, Khorasgan (Isfahan) Branch, Islamic Azad University, Isfahan, IRAN

2 Department of Animal Science, Khorasgan (Isfahan) Branch, Islamic Azad University, Isfahan, IRAN

Abstract

Rose is one of the most beautiful attractive flowers of world which is important in landscaping because of its unique botanical specifics. One of the most important pests of this plant is aphid. In this study, 135 aphid samples were collected from various regions of Isfahan landscapes. Morphological traits and mitochondrial gene sequencing were used for identifying them. According to morphological traits, these samples were belonging to *Macrosiphum rosae,aphis gossipy,* and *metopolophium dirhodum* species which diversity in their morphological traits was obvious. Studying the genetic diversity of 16 selected samples was done by RAPD-PCR molecular marker using three primers and RFLP-PCR using restriction enzyme *Rsa*I. Results showed high genetic diversity in studying population. Samples grouping were done better by RFLP marker than RAPD marker. So, all samples which were located in three groups by this method had also high relations in morphological traits. On the other hand, genetic differences were shown better by RAPD for insects of a group which didn't have similar morphological traits. Then, this method can be used to observe the highest diversity level of population.

Keywords: Rose , RAPD , RFLP-PCR

1. Introduction

Aphids, as an important group of insects which are belonged to Hemiptera, are very successful creatures with the most species diversity in temperate regions and worldwide distribution. There is few plant species in this area without any specific aphid [1] . They may cause loss of plants directly or indirectly. Plus the direct loss which is made by heavy feeding from sap and includes weakness of plant and finally reduction in yield, they cause indirect loss by honeydew secreting on leaves and branches which absorb dusts and also mold will start to grow and finally photosynthesis and yield will be reduced. Furthermore, aphids are very important economically because of transferring plant viruses and their related diseases [2 , 3] There have been identified more than 4000 insects and this number is increasing daily. Aphids are a little group in proportion to other insects groups but their diversity is very high because of polymorphism existence and creating new biological types [1]. Classification of aphids is according to their morphological traits like other insects. It means that main differences or main similarities of samples are being compared and they are being located in their own location [2] . Phylogenic classification is based on evaluation history of species, genera, and families and classifies them according to common ancestors and hosts. In this type of classification, apomorphic traits are used [4]. Primary classifications were not based on phylogenic specifics and were according to personal tastes because the importance of phylogenic specifics was not known. Aphid classification is being discussed very much today especially in one main category: their family numbers[4] . .There have been many morphological anatomical studies conducted on aphids which have prepared background for systematic studies about them [5] . Aphids have high ability for adaption and changing and their morphology is being affected by environmental factors. Many ecological physiological factors affect morphological form of aphids [6] . Considering rich vegetation coverage in Iran and high diversity in roses, many aphids' species are not collected yet and probably this diversity in vegetation coverage and weather conditions have been led to high genetic diversity in aphid population of Iran. Molecular methods are appropriate way for responding main basic questions about genetic diversity and more systematic relationships about live organisms. because of high inter species and intra species diversity in populations of some insects like tripses, white flies, and aphids, morphological tool is not efficient for dividing them, then using molecular markers can be very effective [7, 8] . In high taxonomic levels, to study phylogenic relations and also creating and determining classification systems, molecular markers are efficient resources. In molecular level, they help highly in expanding the concept of species. In population level, they explain direct relationship between heredity patterns, distribution, and colonization with temporal geographical distribution. Molecular markers have been used in last years to basic and applied studies of various organisms. So that discovering various molecular markers have caused high progresses in genetic studies [9] . DNA molecule is the base of genetic differences between two live organism and DNA fingerprint is one of the current methods for determining biological identity of live organisms. By comparing electrophoresis profile of DNA we can realize their differences. DNA polymorphism is the base of many genetic studies [10] . Some of PCR based DNA markers are: RAPD, PCR-RFLP, and AFLP [11] . This markers have been used for various purposes like: creating genetic maps, map of traits in diversing populations, saturating genome places with marker , people fingerprints, germplasm analysis, measuring the genetic distance of people, and evaluating parents portions in back cross. This method is a valuable tool in molecular genetics science which is used easily in map creating and fingerprint application [12] . There has not reported any research about morphological and genetic differences of colorful biotypes of rose aphids and then many researches can be done in this category. These studies can be a good introduction for next studies about resistance of rose varieties to these biotypes, their biological differences and their geographical distribution [13] . These types of researches cause more knowledge about pests and then we can find better ways for fighting against pests.

2. Materials and Methods

2.1. Sampling and counting aphids

In order to determine the number of rose aphids, sampling was done from Isfahan province. Because that aphid colonies take place mostly in 10 -15 cm end of branches, about 15 cm of twigs were cut and locate in plastic bags. Sampling was done from young branches because aphids are interested mainly to this type of branches. Samples were transferred to laboratory and their insects were swept by brush and were stored in 70% alcohol.

2.2. Morphological characterization of aphids

At first, microscopic slides were prepared from healthy samples using boiling with Canadabalsam method. Samples identification was done using different identification keys extant for Iran aphid fauna.

2.3. DNA extraction of roses

DNA extraction was conducted using Kawasaki method (2005) with a little change: aphids were washed by distilled water and located in 1.5 ml micro tubes plus 100 micro liter extraction buffer. Then samples were grinded for one minute to obtain a homogenized suspension. After that another 200 micro liter extraction buffer and 30 micro liter of protease K were added to each tube and were incubated at 60°C for one hour. Then 300 micro liter phenol - chloroform was added to each micro tube and tubes were centrifuged for five minutes in 1200g. Upper part was collected, was transferred to new micro tube with 300 micro liter of phenol – chloroform and then centrifuged again. Then, upper part was mixed with 300 micro liter of Isopropanol and 30 micro liter of sodium – acetate (3M) and was incubated for ten minutes at -20°_{C} Then samples were centrifuged for ten minutes at 4°_{C} in 2000g. Finally 300 micro liter of 70% ethylic alcohol was added to depsit and were centrifuged in 2000g for ten minutes at 4°_{C}. After eliminating alcohol, micro tubes were dried and 25 micro liter of twice sterilized distilled water was added to them and were kept at -20°_{C}.

2.4. RAPD-PCR reaction for evaluating genetic diversity of aphids

Polymerase chain reaction was conducted using prepared mixture made by Amplicon Company (Japan). Substances of reaction were Perimx: 12.5 micro liters, primer: 0.5 ml, 7 micro liters of twice distilled water, pattern DNA: 1 micro liter. Three primers were used in this study: UBC90 (5 -GGGGGTTAGG), R108 (5 -GTATTGCCCT), and R157 (5 -GCTGTAGTGT). Thermal plan of RAPD reaction was: six minutes at95°c, 40 cycles with 45 seconds at 94°c, 90 seconds at 32°c, 90 seconds at 72°c, and final amplification with 8 minutes at 72°c.

Evaluating of PCR product was done on Agarose gel (1.2%). Nine micro liters of each reactions product plus one micro litter loading buffer (6X from 0.25%Boromo phenol blue and 40% w/v of sucrose) were poured in gel wells and electrophoresis was done at 2.5 volts/cm. After electrophoresis, gel was colored in methyl bromide solution (1μg/ml) for 10-15 minutes and then was decolorized in distilled water for 5-7 minutes. RAPD bands were observed under UV lamps and were shot using Uvtech machine. Evaluating of PCR product was done on Agarose gel (1.2%). Execution buffer of electrophoresis machine was TAE (0.04M Tris-acetate and 0.001 M Na_2EDTA).

2.5. Data analysis of RAPD-PCR

To data analysis of electrophoresis results, existence or none existence of each band was recorded in Excel as one and zero, respectively. Cluster analysis of isolates was conducted using UPGMA method and *NTSYS V 2.2* program.

2.6. RFLP-PCR reaction

One of the molecular markers which are used for diversity studying is pattern of amplified genome segments or RFLP. To this, lepF: 5 -ATTCAACCAATCATAAAGATATGGG-3 and lepR: 5 - TAAACTTCTGGATGTCCAAAAAATCA-3 primer pairs were used in polymerase chain reaction.

PCR substances were: PCR buffer, dNTP (200MM), $Mgcl_2$ (2MM), primers (25 pichomole of each), DNA polymerase Taq (1.25) and pattern DNA (100 nanog.). Primary denaturing of PCR was done for 6 minutes at 95°_{C}, amplification in 35 cycles as 0.75 minute at $94^{\circ}_{C,}$ 1.5 minutes at $55^{\circ}_{C,}$ 1.5 minutes at 72°_{C} and final amplification 5 minutes at 72°_{C}.then,amplified segments were cut in enzymatic digestion reaction by *Rsa*I enzyme at 37°_{C} for three hours and cutting pattern of genomic segments was observed by electrophoresis on Agarose gel (1.5%).To data analysis of polymorphism results, existence or none existence of each band was recorded as one and zero, respectively. Cluster analysis was conducted using UPGMA and Jacquard coefficients and *NT-SYS V 2.2* program and then its cladogram was drawn.

3. Results

3.1. Morphological traits of rose aphids

Isfahan has its own aphid's fauna for roses like every other place. Wide sampling of all Isfahan places prepared 135 samples totally. According to morphological traits, three aphid species were collected from Isfahan roses. The most abundant species was *Macrosiphum rosae* which? samples were belonged to this species. This species which is known as rose aphid too has been sawn from average to big sizes, with long or spindle shaped bodies in green, yellow, pink and red – brown colors. According to previous reports, this species is distributed all over Iran and can be collected all seasons except summer. The most important morphological characteristics of this species are ?. The second species which was collected less in this study was *Metropolophium dirhodum*. Wingless members of this species are all long spindle – shaped with green color or green – yellow and an obvious bright green back stripe. Antennas are bright and the end of third and fifth part and appendix of final part were dark to black. Antenna length is about 0.6 of total length, and third part has 1 to 3 secondary * in its bottom. The others have a green abdomen without sclerotium spots. Hair formula of tarsus first part was 3-3-3 and tail had 9-12 hairs too. The length of wingless members bodies were from 1.6 to 3.3 mm. This characteristic is completely in accordance to Black man and Eastop (2000) reports. The main host of this insect is rose too and has been reported from many places of Iran. The third species was *aphis gossypii*. This egg shaped insect is about 1.8 mm and is sawn in many colors. Same are green or yellowish green and the other are grey to green. It has been reported from all places of Iran and various hosts. We must announce that some samples were not completely in accordance to identification Key and were ascribed to these species because of showing the most important characteristics. One of important specifics of aphids is similarity of their morphological traits which make their identification difficult [14] .

3.2. RAPD-PCR reaction

In this study 16 aphid samples which had high morphological differences were used to study their genetic diversity. All used primers of this study showed good polymorphism in studied rose aphids. The size of obtained bands was estimated about 100-2000 bp. Among used primers, UBC 90 and R 108 showed the highest polymorphism with 25 and 24 amplified segments respectively (figure1). According to cluster analysis rose aphids divided into three groups: the first group (A) consisted of 5 aphid samples from various regions of Isfahan. Second group (B) consisted of nine samples which were different in color. Members of this group were black and green and were belonging to *Aphis gossypii* and *Macrosiphum rosae*. Two samples of rose aphids (C2 and V1) were located in C group which C2 was green to grey colored and V'/ was brown colored and both were located in *Macrosiphum rosae* species. Results of this study showed that dividing rose aphids by RAPD-PCR has a close relation with their morphological traits but not with their color (figure2).

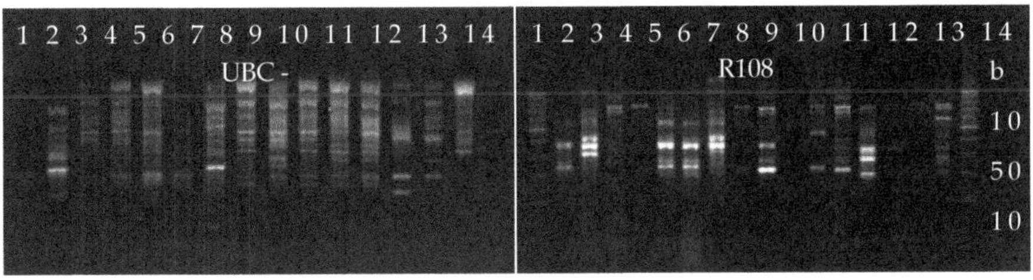

Figure 1. Band figures of rose aphids amplified segments using UBC-90, and R-108 primers in Agarose gel 1.2%

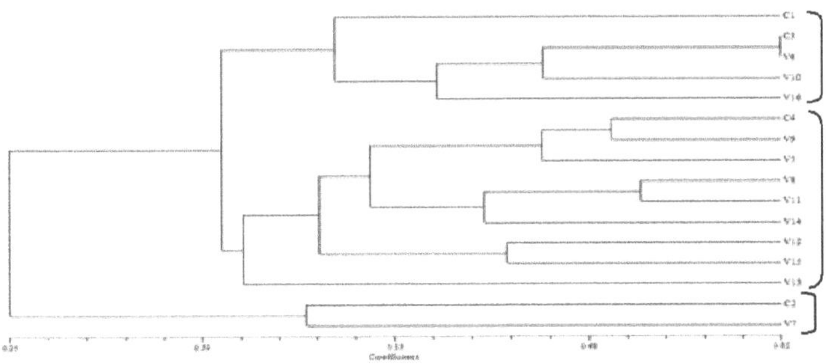

Figure 2. Cluster analysis of rose aphids using data of RAPD-PCR. Group A:, group B:, and group C:

3.3. Genetic diversity of rose aphids using RFLP-PCR

To this, PCR product of mitochondrial gene was digested by *Rsa*I restriction enzyme. The cutting distribution pattern of aphid mitochondrial gene is presented in figure 3. The enzyme could show seven band pattern of polymorphism on mitochondrial genome of aphid samples. Cluster analysis of this polymorphism was done using *NTSYS V 2.2* program (Figure4). As it was shown in cladogram, aphids are dividing in three groups according to RFLP-PCR. Group A were red aphids and were very similar morphologically. In this grouping many of black and green aphids (7 samples) was located in group B and the rest (4 samples) were located in group C. This grouping has a high accordance to morphological results of aphid population. For instance, C2, C6, and C9 which were related to numbers 109,105 and 108 (studied samples) respectively were located in one group. In morphological traits all these aphids are red. These results show efficiency of this marker for for classifying and identifying the aphid population, along with their phenotypic characteristics. In second group also samples were located which were very similar morphologically and were different only in color and some other characteristics.

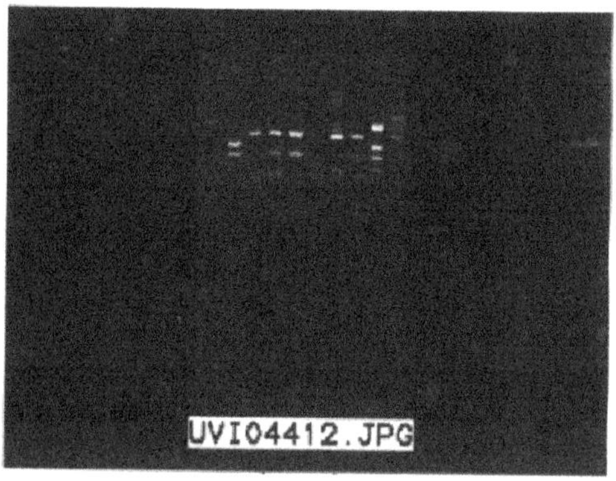

Figure 3. Polymorphism (restriction pattern) of mitochondrial gene of rose aphid in enzyme digesting using Rsal restriction enzyme on Agarose gel 1.5% in TBE buffer

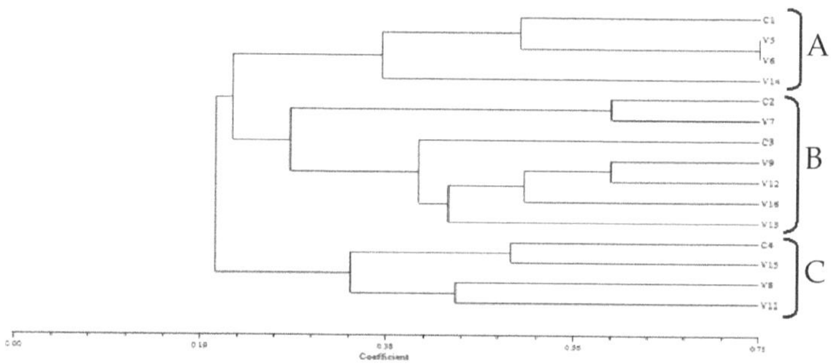

Figure 4. Cladogram of cluster analysis of aphids' polymorphism in RFLP-PCR using NT SYS V2.2 program

4. Summary and conclusion

Population structure of each organism is dependent to amount and distribution of within and between populations genetic diversity. Genetic structure of a population shows evolution history and potential of it for evolution and adaption to environment [5] . Every agricultural ecosystem is facing to environment changes like cultivating resistant cultivars, fungicides, pesticides, fertilizers, irrigation and crop rotation, then pests and plant pathogens are constantly evolving and changing to adapt to these changes. So, having knowledge about power of pest populations in evolution and adaption to environment and host is very important for completing effective controlling strategies [15] . Then, one of the other goals of studying genetic diversity and population structure is identifying the factor which plays more important role in evolution of that population and also how that factor determines genetic structure of population and its evolution potential[5] . Developing molecular technics and using genetic markers in last decades have led to developing tools and fast, cheap, and accurate methods for identifying creatures. So, polymerase chain reaction (PCR) and molecular methods based on PCR have had an important role in developing biological sciences. In recent years, many efforts have been done for using molecular methods in entomology. For example Wagou et al. (1996) used molecular markers for identifying parasite insects and studying their biology. Genome sequencing and molecular markers is an appropriate replacing method for morphological methods in studying aphids' epidemiology [13] RAPD and AFLP molecular markers have high application as population markers in studies about population diversity of living things especially insects [7] Chen et al (2008) studied genetic diversity of cabbage aphids with molecular methods and reported that when it was not possible to divide aphid populations by morphological traits, RAPD and AFLP markers divided them well. In current study, rose aphids were studied for morphological and molecular properties. According to morphological findings, all collected samples were belonging to three species: *macrosiphum rosae, aphis gossypii,* and *metopolophium dirhodum*. Previous studies confirm it too. These three aphid species have been identified before on roses of Isfahan [16] this study was done to clarify structure of rose aphid's populations from morphological and genetic aspects and also for evaluating use of RAPD and RFLP markers for grouping these populations. UBC-90 primers showed the highest polymorphism in this study. Also, primers could present significant relationship with morphologic traits of rose aphids. Polymorphism analysis of these markers with NTSYS program showed that RFLP markers acted better than RAPD in locating close populations with similar morphological traits in similar clads. Considering high distribution of this pest on various plants, for better realizing the population structure of this pest, it is better to study its genetic diversity on other hosts in various regions. On the whole, results of this study showed that morphological and genetic diversity of this aphid is relatively high, and then effective factors in its creation and expanding must be studied.

5. References

[1] Dixon A F G. 1998. Aphid ecology. Second edition, UK, Chapman and Hall.

[2] Rezvani, A. 2001. Identification key of Iran aphids. Iran organization of agricultural researches, training and advisory.

[3] Blackman R L, Eastop V F .2000. Aphids on the World's Crops: an Identification and Information Guide, 2nd edn. England, John Wiley & Sons Ltd.

[4] Ilharco F A, Van Horten A. 1987. Systematics. In: Minks A K, Harrewijn P. Aphids Their biology, natural enemies and control. Elsevier science publishers B V, PP:51- 77.

[5] Mcdonald B. 1997. The population genetics of Fungi: tools and techniques. Phytopathology, 87: 448-453.

[6] Stroyan H L G. 1999. Tribulations of an aphid taxonomist. Atti dell Accademia Nazionale Italian di Entomologia. Anno XLVII, pp: 89-100.

[7] Fergusson, J.W.H., 2002. On the use of genetic divergence for identifying species. Biol. J. Linn. Soc. 75 (4), 509– 516.

[8] Havill N.P., Foottit R.G., Von-Dohlen C.D. 2007. Evolution of host specialization in the Adelgidae (Insecta: Hemiptera) inferred from molecular phylogenetics. Molecular Phylogenetics and Evolution 44: 357–370.

[9] Caetano-Analles G, Gresshof P M. 1997. DNA makers, protocols, three different component nuclear types from *Fusarium moniliforme*. Heterokaryon and Owerviews. Wiley Liss, Inc.489pp.

[10] Krawezac M, Schmidtke J. 1994. DNA Fingerprinting. Bioss Scientific Publisher, First ed. 698pp.

[11] Gharayazi,B.1996. Application of DNA markers in plant breeding. 4th congress of agronomy and plant breeding. Agriculture faculty. Isfahan University of Technology. Pages: 328-381.

[12] Vahhabi, Kh.2004.Genetic diversity of *Trichoderma* species relating to *Agricus bisporus* using molecular and morphological method. M.Sc. Thesis. Agriculture faculty. Isfahan University of technology.

[13] Mehrparvar, M. Mobli,M., and B. Hatami.2008.Mass variation of *Macrosiphum rosae* (L.) (Homoptera: Aphididae) on various varieties of rose and eglantine. Journal of agricultural sciences and natural resources. 12th year. 45:711-718.

[14] Bernardo U., Monti M.M., Nappo A.G., Gebiola M., Russo A., Pedata P.A., Viggiani G. 2008. Species status of two populations of Pnigalio soemius (Hymenoptera:Eulophidae) reared from two different hosts: An integrative approach. Biological Control 46: 293–303

[15] Coeur d'acier A., Jousselin E., Martin J.F., Rasplus J.Y. 2007. Phylogeny of the Genus *Aphis* Linnaeus, 1758 (Homoptera: Aphididae) inferred from mitochondrial DNA sequences. Molecular Phylogenetics and Evolution 42: 598–611.

[16] Coeur d'acier, A., Sembène, M., Audiot, P., Rasplus, J.-Y., 2004. Polymorphic microsatellites loci in the black Aphid, *Aphis fabae* Scopoli, 1763 (Hemiptera, Aphididae). Molecular Ecol. Notes 4 (2), 306–308.

Synthesis of Partially Carboxymethyl Cellulose Derived from Rice Straw and Its Utilization as Dye Adsorbent

S. Tawfik[1], S. H. Abd Elsalam[1], H. M. El-Hennawi[2],
I. Abd El-Thalouth[2], E. Adel[1]

1 Faculty of Applied Arts, Helwan University, Cairo, Egypt

2 Dyeing, Printing and textile auxiliaries Department, Textile Research Division, National Research Centre, Cairo,Egypt

Abstract

Two samples of partially carboxymethyl cellulose derivatives of different D.S. values were prepared from Egyptian rice straw via pulping followed etherification using different concentrations of monochloro acetic acid under the catalytic action of sodium hydroxide. The prepared derivatives were assessed for D.S. and evaluated as dye adsorbent for different classes of dyestuff. The results obtained indicate that, the D.S. increases from 0.09 to 0.14 by increasing monochloro acetic acid from 5 to 10 g/100g cellulose pulp. The rate of dye absorbance increases by increasing the amount of adsorbent as well as the time of adsorption. While as the dye concentration increases from 0.01 to 0.5 the percent dye absorption decrease regularly. However, the magnitude of the percent decrease in the colour depends on : (a) the nature of the dyestuff used ,(b) the D.S. of the adsorbent, and (c) on the technique applied . The magnitude of colour removal in case of using ultrasonic technique is relatively higher than the mechanical shaking irrespective of the nature of the dye used and/or the conditions of adsorbance. The percent colour removal follows the order Basic green> Basic yellow> Acid green> Acid blue respectively.

Keywords: rice straw- colour removal-ultrasonic- adsorption-carboxymethylation

1. Introduction

Rice is the largest cereal crop in the world. Rice straw represents around 45% of the volume in rice production, producing the largest quantity of crop residue. As rice straw is a marginal feed compared to other cereal grain straw and a problematic fuel source due to high ash generation, exploring more viable options to utilize rice straw is pressing, particularly as an environmental concern. With its compositions of cellulose (38.3%), hemicellulose (31.6%), lignin (11.8%) and silica (18.3%) [1] rice straw is the most available cellulose source from agricultural crop residues in the world[2] In recent years, many biological materials, such as orange bagasse [3], plant leaves [4], saw dust [5] and maize cob [6], have been applied as adsorbents to adsorb dyes from wastewaters. Researchers have been trying to find ways to take advantages of straw [7,8]. One of the promising ways to use this precious bioresource is to produce straw-based adsorbents

[9-13]. However, the adsorption capacity of unmodified straw is insignificant, since straw materials are deficient in free ionic groups, which would play an important role in removal of ionic dyes. Therefore, it could improve the adsorption capacity of straw by introducing some ionic functional groups through chemical modification [14]. Lilienfeld [15,16], was the first to affect the partial carboxymethylation of cotton. Two main methods for the preparation of partially carboxymethylated cellulose are known:(a)The aqueous carboxymethylation and (b)The non aqueous carboxymethylation. Partially carboxymethylated cellulose with a D.S. of about 0.05 to 0.15 retains the original fibrous nature and exhibits a number of potentially valuable properties. The objective of this paper is to synthesize two different partially carboxymethylated derivatives from Egyptian rice straw pulp and to investigate their suitability to be utilized these derivatives as adsorbent substrates for different reactive dyes under a variety of conditions.

2. Materials and Methods

2.1. Native rice straw supplied by Racta Co. For Paper Manufacture, Alexandria, was used

The following different dyes selected from the most dyestuffs which are used in the Egyptian Textile Industry.

- Sunzol Brilliant Violeet 5 R (C.I. Reactive Violet 5) (schheme 1).
- Sunzole Blue 19 (C.I. Reactive Blue 19) scheme 2.
- Ginacryl M alachite Green M (C.I. Basic. Green 4) scheme 3
- Ginacryl G . yellow GLE 200% (C.I. Basic yellow 28) scheme 4
- Dystar. Green BW (C.I. Acid Green 27) scheme 5
- Dystar. Sup ralan.Blue 22 R (C.I. Acid blue 225)

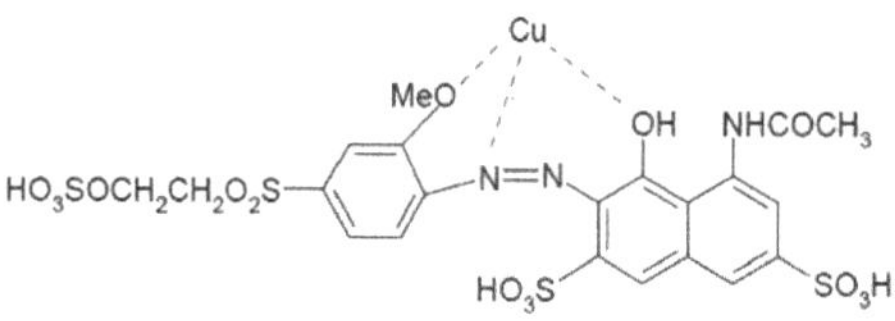

Scheme 1 -Reactive Violet 5

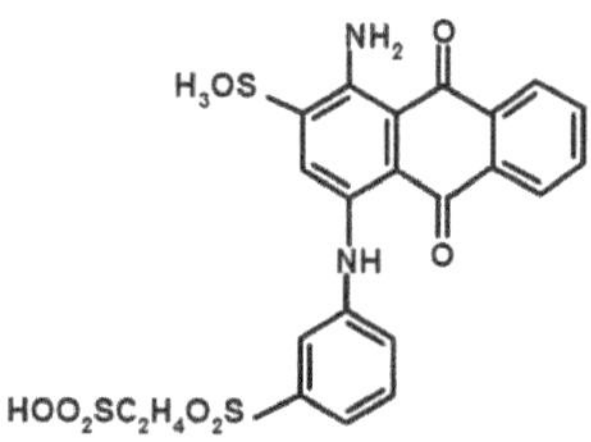

Scheme 2 -Reactive Blue 19

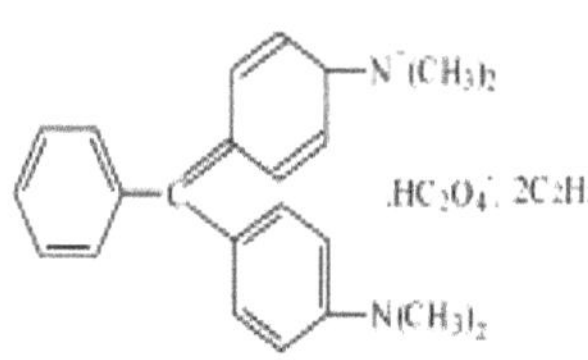

Scheme 3- Basic. Green 4

Scheme 2 -Reactive Blue 19

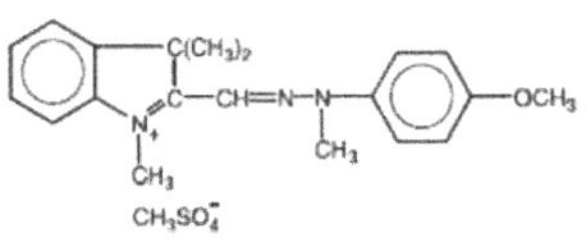

Scheme 3- Basic. Green 4

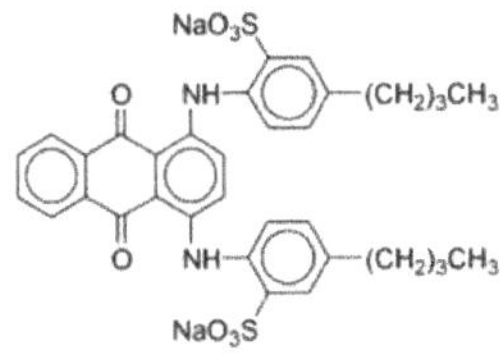

Sodium hydroxide and So dium hypochlorite both of laboratory grad e chemicalss were also used.

2.2. Methods:

2.2.1. Preparationn of bleached rice straw:

The alkali treated sample wer e subjected to sodium hypochloritte (NaOCl) bleaching (4g/l active chlorine) for two hours at room temperature, liquor ratio 10:1 foll owed by waashing thoro-oughly with running w ater and finally air dried.

2.2.2. Preparationn of partially carboxymethyl ric e straw derivatives:

Two different substituted partially carboxymethylated rice straw derivatives were prepared from bleached rice straw via using different concentrations of the etherifying agents, i.e. monochloro acetic acid and sodium hydroxide. The procedure adopted was carried out as follows:

Alkali cellulose was prepared by treating 100g of dry bleached rice straw with 200 ml of 5% aqueous sodium hydroxide solution, and mixed well, a solution of 5g monochloro acetic acid in 100ml distilled water was added gradually to the alkali cellulose with continuous agitation for two hours, and the reaction mixture was left at room temperature overnight. The excess alkali was neutralized with glacial acetic acid using phenolphthalein as indicator. At this end the product was filtrated washed well with water and finally air dried at ambient conditions. Another sample was also prepared by the same technique using 10% sodium hydroxide solution and 10g monochloro acetic acid dissolved in 100ml water

2.3. Procedure of Dye Adsorption

Different amount of different substrate (rice straw, alkali treated and rice straw pulp) were added to aqueous solutions of the selected dyes (0.01g) dissolved in 1 liter of distilled water. The suspension was treated using either mechanical shaking or ultrasonic technique for different periods of time (5, 15, 30, 45, 60 minutes)and temperatures (30, 40, 50, 60°C). At the end of the run of aliquot was centrifuged at5000 rpm for 30 min and the dye concentration in the clear solution was evaluated colourimetrically at the maximum wavelength for every dyestuff. The absorbance was measured using a double-beam spectrophotometer Thermo Electron Corporation Unican 300, England.

The percent dye absorption was calculated by equation 1 :

$$\% \text{ Colour removal} = \frac{\text{C A of original sample} - \text{C A of treated sample}}{\text{C A for the origin}} \times 100$$

Where CA is the colour absorbance

2.3.1. Determination of degree of substitution (D.S.)

The D.S. was determined according to a standard method [17]. Where The water soluble sodium carboxymethyl cellulose is converted to the insoluble acid form, purified by washing, dried and then a weight sample is reconverted to the sodium salt with a measured excess of sodium hydroxide from which the D.S was calculated.

3. Results and discussion

Pure cellulose was prepared from rice straw wastes via alkali scouring and bleaching as previously mentioned. The prepared cellulose was subjected to carboxymethylation using two different amounts of the etherifying agents i.e. monochloro acetic acid and sodium hydroxide. The caroxymethelation reaction took place according to the following reaction:

$$\text{Cell.OH} + \text{Cl.CH}_2\text{.COOH} + 2\,\text{NaOH} \circledcirc \text{Cell.O.CH}_2\text{COONa} + \text{NaCl} + 2\,\text{H}_2\text{O}$$

Table 1 represent the effect of concentration of monochloro acetic acid on the D.S. of the prepared partially carboxymethyl cellulose derivatives.

Amount of $ClCH_2COOH$/100g of cellulose	Amount of NaoH/100g of cellulose	D.S.	Solubility	
			Water	ethyl alcohol
5g	**5g**	0.09	Insoluble	Insoluble
10g	**10g**	0.14	Insoluble	Insoluble

Table 1. The effect of the amount of monochloro acetic acid on the D.S. of the prepared derivatives

It is clear from the data of Table I that the degree of substitution (D.S.) depends on the concentration of the etherfying agent i.e. monochloro acetic acid.As the latter increases from 5 to 10 g/100 pure cellulose, the D.S. increases from 0.09 to 0.14 respectively.It is also clear from Table I that the prepared partially carboxymethylated bleached rice straw is insoluble in both water and ethyl alcohol.

3.1. Effect of amount of partially carboxymethyl cellulose on dye adsorbents

The prepared two partially carboxymethylated cellulose derived from rice straw was utilized as dye adsorbent for both reactive blue19 and reactive violet5. Figure 1and 2 represent the data obtained on using the two partially carboxymethyl cellulose derivative of different D.S.values on conducting the adsorbance using mechanical shaking or ultrasonic technique.

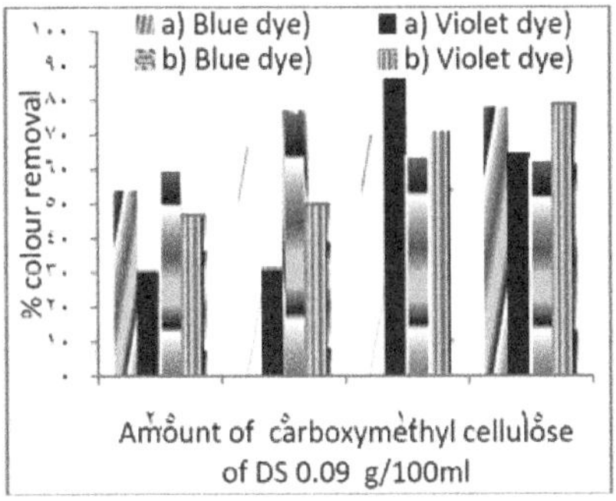

Fig 1. Effect of amount of carboxymethyl cellulose derivative of D.S. 0.09 on % colour removal using different reactive dyes reactive dyes for (a) ultrasonic and (b shaking treatment.

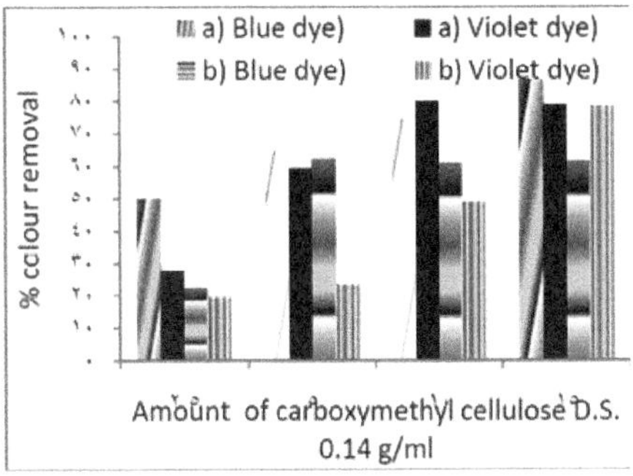

Fig 2. Effect of amount of carboxymethyl derivative of D.S. 0.14on % colour removal using different reactive dyes for (a) ultrasonic and (b shaking treatment

Generally speaking it is clear from the Figures 1 and 2 that increasing the amount of adsorbent, i.e. partially carboxymethyl derivatives, is accompanied by an increase in the % colour removal to reach to a maximum after which it either remain constant or decreases. This phenomenon holds true regardless of (a) the D.S.,(b) the nature of the reactive dye used, or (c) the technique applied. It is also clear from the figures that in case of the samples conducted via mechanical shaking the %dye adsorption of the carboxymethyl derivative of relatively low D.S. (0.09) is higher than their corresponding samples acquire relatively higher D.S. (0.14) this phenomenon is true on using either reactive violet5 or reactive blue19. While, in case of ultrasonic, there is irregularity in the results. It seems that the sample of partially carboxymethylated derivative prepared using 5g monochloro actic acid acquire the sufficient carboxymethyl groups to open the structure of cellulose and hence its absorbance reach to the maximum. Increasing the D.S. causes an increase of the COOH groups. The latter ionize in the solution into-COO^- and Na^+ which causes an increase in the negative charge on the substrate. Since the reactive dye acquire negative charge too hence the rate of dye adsorption decrease. It is also observed that ultrasonic technique increases the absorbance capacity of both carboxymethyl derivatives under investigation since they are higher than their corresponding samples conducted via mechanical stirring. For example the % colour removal for the relatively higher D.S. (0.14) derivative on using reactive blue19 was 86.6% against 62.2% for the sample conducted using ultrasonic and

mechanical shaking respectively. Furthermore, it is clear from the data that, irrespective of the amount of the adsorbent or the technique applied, Partially carboxymethylated samples acquire higher % colour removal compared with the native, alkali treated and the pure cellulose, i.e. alkali treated and bleached samples.

3.2. Effect of Treatment Time

Figures 3 and 4 represent the data obtained on studying the effect of time of adsorption of different reactive dyes, on using two carboxymethylation derivatives of different D.S. values 0.09 and 0.14 respectively.

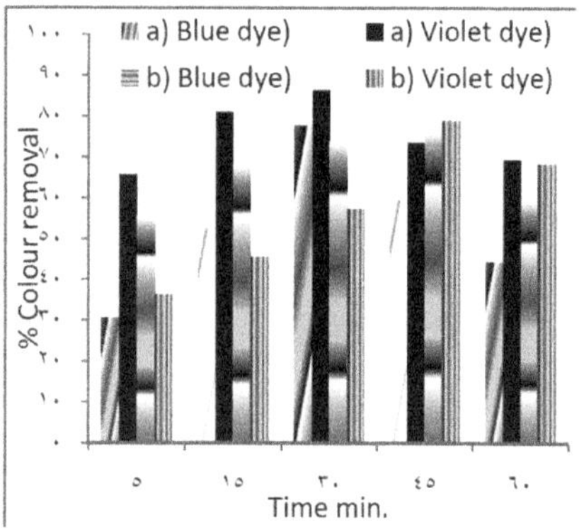

Fig 3. Effect of treatment time using carboxymethyl cellulose derivative of D.S. 0.09 on % colour removal using different reactive dyes for (a) ultrasonic and (b shaking treatment

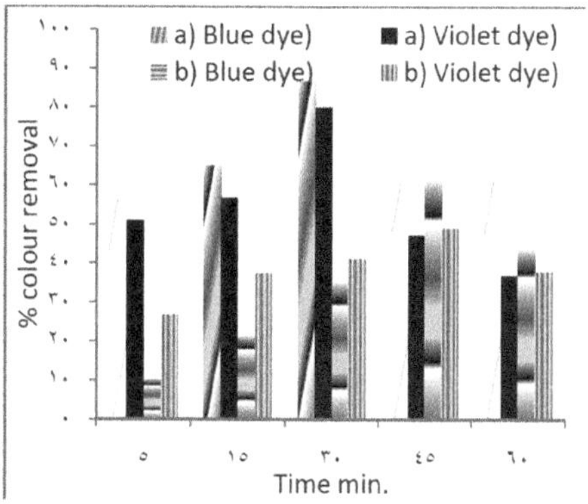

Fig 4. Effect of treatment time using carboxy- methyl cellulose derivative of D.S.0.14 on % colour removal using different reactive dyes for (a) ultrasonic and (b shaking treatment

Generally speaking, it is clear from figures 3& 4 that in all cases the rate of dye adsorption increases by increasing the time of adsorption to reach to a maximum after which it shows slight decrease or remains constant. However, the time of adsorption to reach the maximum value and magnitude of the % colour removal at the maximum adsorption depends on the nature of the colour and the adsorbent used as well as the technique applied.It is clear that for the samples conducted via mechanical shaking the maximum dye adsorption was obtained after 45minutes irrespective of the D.S. of carboxymethyl sample or the nature of the reactive dye used, where, it reaches to 78.9% and 48.9% on using reactive violet5 and 76.6% and 62.2% on using reactive blue19 for carboxymethyl samples of D.S. 0.09 and 0.14 respectively. In other words, the sample of low D.S. acquires a higher absorbance capacity than the sample which acquire relatively higher D.S. While, in case of using ultrasonic technique the maximum dye adsorption arrived at relatively lower time, i.e. 30minutes only. However, the magnitude of the maximum dye adsorption was higher for carboxymethyl derivative of relatively low D.S. value only on using reactive violet5. While in case of using reactive blue19, the opposite holds true. The decreases in the % colour removal on using carboxymethyl derivative of relatively higher D.S. value may be due to the increase in the negatively charged ($-COO^-$) groups on the polymer. The latter repel the reactive dye molecules which acquire the similar negative charge as previously explained. Hence, the numbers of the adsorbed dye molecules decreases.

3.3. Effect of Dye Concentration

Figures 5and 6 represent the results obtained on using different concentrations of the aforementioned two different reactive dyes in case of carboxymethyl derivatives of D.S. 0.09 and 0.14 respectively.

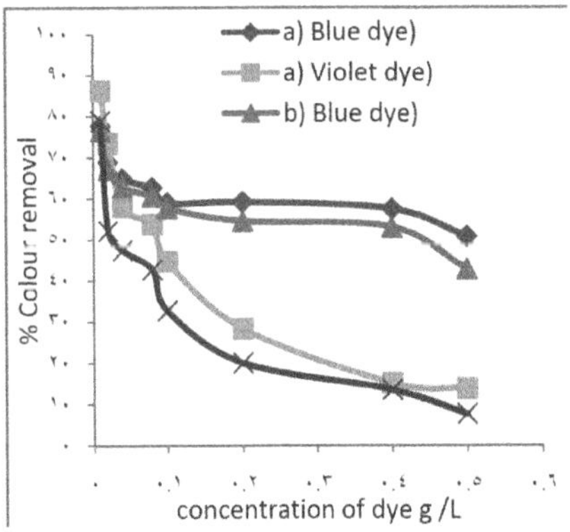

Fig 5. Effect of dye concentration on % colour removal using carboxymethyl derivative of D.S.0.09 (a) ultra-sonic and (b shaking treatment

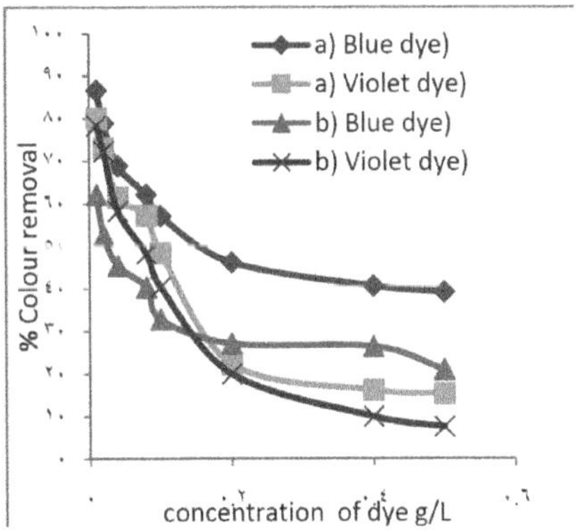

Fig 6. Effect of dye concentration on % colour removal using carboxymethyl derivative of D.S.0.14 (a) ultrasonic and (b shaking treatment

It is clear from the data that in all cases as the dye concentration increases from 0.01 to 0.5gm, % the percent dye adsorption decreases regularly. This may be due to either: (a) aggregation of the dye molecules which increases as the concentration increases, or (b) the decrease in the mobility of the dye molecules as the concentration increases.

3.4. Effect of nature of the dyestuff used

At the end, it is of great interest to investigate the effect of the nature of the dyestuffs used on the percent colour removal on using carboxy methyl cellulose derivatives derived from rice straw of D.S. 0.09 and 0.14. Hence 4different dyestuffs (two of them acid and the other are basic) were chosen and used with the mentioned carboxymethyl cellulose derivatives under identical conditions in case of either mechanical shaking or ultrasonic. The results obtained are illustrated in Figure 7 in case of using mechanical shaking and ultrasonic.

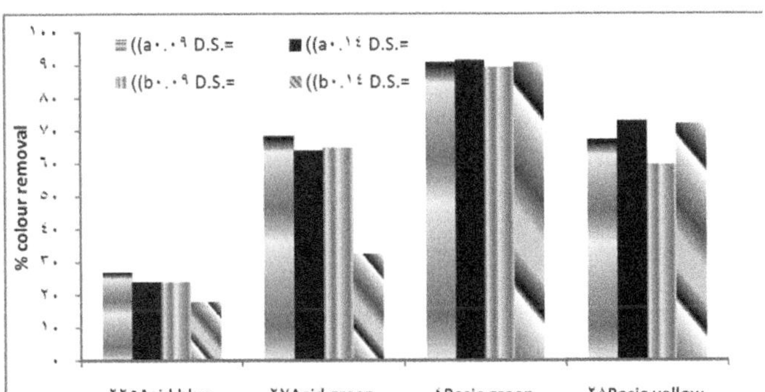

Fig 7. Effect of nature of dyestuff on the ability of carboxymethyl derivatives towards colour removal on using (a) ultrasonic and (b shaking treatment)

Generally speaking it is obvious from the figure 7 that the % decreases in colour depends on: (a) the nature of the dyestuff used, (b) the D.S. of the adsorbent and, (c) on the technique applied. It is also clear on using carboxymethyl derivative of D.S. (0.09) for the samples conducted under mechanical shaking the % colour removal in case of basic dyes is higher than that of acid dye. The percent colour removal

cellulose derivatives acquire (- ve) charges. Hence, it is expected that the capacity for dye adsorption of carboxymethyl cellulose derivatives is higher on using basic dyes than using acid dyes. It is clear from figure 7 that for basic dyestuffs used the % colour removal is higher on using carboxymethyl cellulose of relatively D.S. (0.14) values than their corresponding samples conducted using relatively D.S. (0.09) derivative. This due to the increases in the (- ve) charge of adsorbent as the D.S. increase. The same trend could be observed for the samples conducted via ultrasonic technique. Furthermore, it is clear that on using acid dyestuffs either blue or green the % colour removal decreases by increasing the D.S. of carboxymethyl derivatives from 0.09 to 0.14. As the D.S. increases the (- ve) charges increases and hence the adsorbed acid dye which acquire (- ve) charges decrease.

4. Conclusion

The percent colour removal increases by increasing the time of treatment and / or the amount of substrate. While the opposite holds true by increasing the dye concentration. In all cases the magnitude of dye adsorption in case of ultrasonic is relatively higher than mechanical shaking. Increasing D.S of carboxymethyl derivatives decrease the % colour removal of reactive dye . The percent colour removal depends on the nature of dye used and follows the order Basic green> Basic yellow> Acid green> Acid blue respectively.

5. References

[1] Hessien, M. M., Rashad, M. M., Zaky, R. R., Abdel-Aal, E. A., & El-Barawy, K. A. Controlling the synthesis conditions for silica nanosphere from semi-burned rice straw. Materials Science and Engineering B-Advanced Functional Solid-State Materials, 162(1), (2009) 14–21.

[2] P. Lu, Y.-L. Hsieh Preparation and characterization of cellulose nanocrystals from rice straw Carbohydrate Polymers 87 (2012) 564– 573

[3] L.D. Fiorentin, D.E.G. Trigueros, A.N. Modenes, F.R. Espinoza-Quinones, N.C. Pereira, S.T.D. Barros, O.A.A. Santos, Biosorption of reactive blue 5G dye onto drying orange bagasse in batch system: kinetic and equilibrium modeling, Chem. Eng. J. 163 (2010) 68–77.

[4] V. Ponnusami, K.S. Rajan, S.N. Srivastava, Application of film-pore diffusion model for methylene blue adsorption onto plant leaf powders, Chem. Eng. J.163 (2010) 236–242.

[5] C.P. Kaushik, R. Tuteja, N. Kaushik, J.K. Sharma, Minimization of organic chemical load in direct dyes effluent using low cost adsorbents, Chem. Eng. J. 155 (2009) 234–240.

[6] G.H. Sonawane, V.S. Shrivastava, Kinetics of decolourization of malachite green from aqueous medium by maize cob (zea maize): an agricultural solid waste, Desalination 247 (2009) 430–441.

[7] P. Binod, R. Sindhu, R.R. Singhania, S. Vikram, L. Devi, S. Nagalakshmi, N. Kurien, R.K. Sukumaran, A. Pandey, Bioethanol production from rice straw: an overview, Bioresour. Technol. 101 (2010) 4767–4774.

[8] F.A. Abdel-Mohdy, E.S. Abdel-Halim, Y.M. Abu-Ayana, S.M. El-Sawy, Rice straw as a new resource for some beneficial uses, Carbohydr. Polym. 75 (2009) 44–51.

[9] U. Farooq, J.A. Kozinski, M.A. Khan, M. Athar, Biosorption of heavy metal ions using wheat based biosorbents—a review of the recent literature, Bioresour. Technol. 101 (2010) 5043–5053.

[10] H. Muhamad, H. Doan, A. Lohi, Batch and continuous fixed-bed column biosorption of Cd2+ and Cu2+, Chem. Eng. J. 158 (2010) 369–377.

[11] B.C. Oei, S. Ibrahim, S. Wang, H.M. Ang, Surfactant modified barley straw for removal of acid and reactive dyes from aqueous solution, Bioresour. Technol. 100 (2009) 4292–4295.

[12] S. Ibrahim, I. Fatimah, H.M. Ang, S.B. Wang, Adsorption of anionic dyes in aqueous solution using chemically modified barley straw, Water Sci. Technol. 62 (2010) 1177–1182.

[13] S. Ibrahim, W.Z. Shuy, H.M. Ang, S.B. Wang, Preparation of bioadsorbents for effective adsorption of a reactive dye in aqueous solution, Asia-Pac. J. Chem. Eng. 5 (2010) 563–569.

[14] R.P. Han, L.J. Zhang, C. Song, M.M. Zhang, H.M. Zhu, L.J. Zhang, Characterization of modified wheat straw, kinetic and equilibrium study about copper ion and methylene blue adsorption in batch mode, Carbohydr. Polym. 79 (2010) 1140–1149.

[15] L. liliendfeld, Brit. Patent , 231, 803 (1924); C.A. 19,3600 (1925). [16] L. liliendfeld, Brit. Patent, 231, 809 (1924); C.A. 19,3592 (1925).

[16] M. El-Zawahry, Ph.D. Thesis "Utilization of some carbohydrate derivatives in dye adsorption to minimize dyestuff pollution" Faculty of science , Cairo University (1995).

Synthesis and Characterization of Arylazopyrazolopyrimidines Dyes and Studying their Antibacterial Activity

K. A. Ahmed[1], M. A. Elkashouti[1], S. T. Tawfeek[2] and Sh. S. Mohamed[1]

1 Textile division, National Research Center, Cairo, Egypt

2 Faculty of applied art, Helwan University, Cairo, Egypt

Abstract

The purpose of this research is to synthesize several new pyrazolopyrimidine containing an arylazo function containing electron with drawing groups and benzothiazole moiety, the substituted 5-arylazopyrazolopyrimidine were prepared by reaction of aryl azopyrazole with airelidene of 2- cyanomethyl benzothiazole under basic condition in boiling ethanol. The structure of arylazopyrazolopyrimidine dyes were established by their element analysis and spectral data (MS, IR and ^{1}H-NMR). The antibacterial properties of these dyes have been investigated.

Keywords: pyrazolopyrimidine – benzothiazole – antibacterial activity

1. Introduction

Innovations in azo dye based on heterocyclic systems have been made as a result of intensive studies stimulated by the mounting need for bright dyes. Generally many of heterocyclic azodyes show dramatic bathochromic shifts combined with brilliance of shade and high tinctorial strength compared with conventional anthraquinone dyes and aminobenzene azodyes[1-4]. In spite of the large number of arylazopyrazol dyes reported in literature, only very few condensed pyrazole derivatives carrying arylazo functions on the pyrazole ring have been reported. In continuation of the increasing interest in synthesis of condensed arylazopyrazole new dyestuffs [3], the present work deals with novel synthesis of condensed arylazopyrazolopyrimidines derivatives and studying their printing properties using silk screen and heat transfer printing techniques on polyester and polyamide fabrics. The antibacterial activity of these dyes was also studied, where recently in the textile industrial sector [5], there has been increasing interest in the manufacture of clothing and products with antibacterial properties. Clothing of textile can act as carrier for microorganisms such as pathogenic or odor –generating bacteria and moulds [6]. The textile material is known to be susceptible to microbial attack, in contact with the human body it offers an ideal environment for microbial growth providing oxygen, water and warmth, and nutrients from spillages and body exudates [7]. This often leads to objectionable odor, dermal infection product deterioration, allergic responses and other related diseases which necessitate the development of clothing products with antimicrobial properties [8].

2. Experlmental

2.1. Synthesis of dyes

Preparation of 3,5 diamino-4-arylzopyrzole (1)

(0.01 mole) of diazotized aniline derivatives coupled with (0.01 mole) of malono nitrite in the presence of (20 ml) ethanol and (5 gm) sodium acetate at $0\text{-}5^{\circ}C$. The precipitated solid is filtrated and dried (0.01) mole of the dried solid is dissolved in (20 ml) ethanol then (0.01 mole) of hydrazine hydrate is added drop wise, the precipitated solid of arylazopyrazole derivatives is filtrated and re-crystallized from ethanol.

Preparation of Cyanomethylarylidine benzothiazol-2-yl

Cyanomethylarylidinebenzothiazol-2-yl is prepared by the condensation reaction of (0.01 mole) of cyanomethyl benzothiol-2yl with (0.01) mole of aromatic aldehyde derivatives in (20 ml) ethanol and drops of piperidine at room temperature. The collected precipitate is filtrated and recrystallized in ethanol

General Procedure of synthesis of arylazo-pyrazolopyrimidines dyes:

To a solution of derivatives (1) (10 mmol) in ethanol (50 mL), the (2) (10 mmol) and drops of pipridine were added. The reaction mixture was refluxed for 4 h then left to cool. The formed product was filtered off, washed with ethanol, and recrystallized from ethanol to afford the corresponding arylzaopyrazolopyrimidines 4a-e. *(Scheme I)*

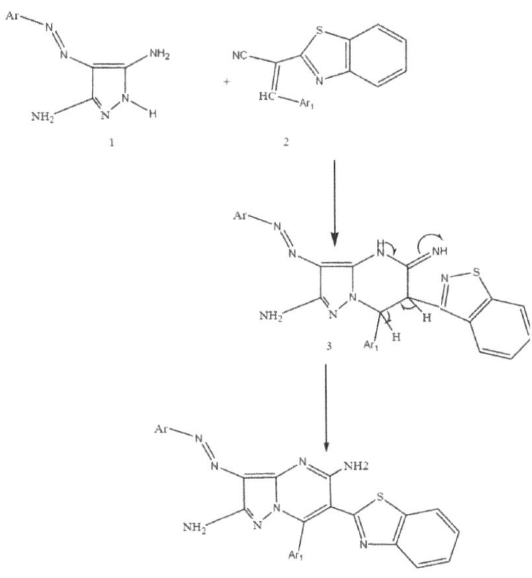

Scheme 1.

Dye number.	Ar	Ar_1
4a		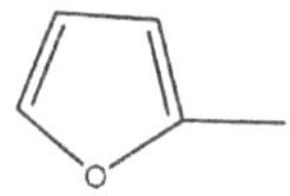
4b	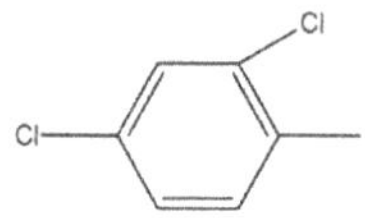	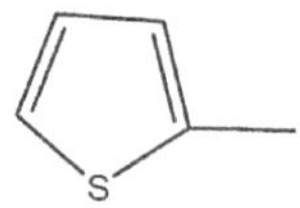
4c	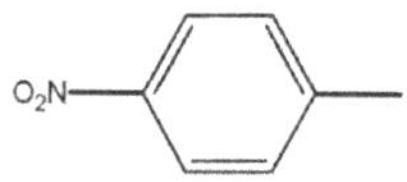	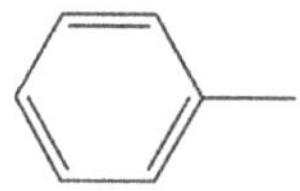
4e	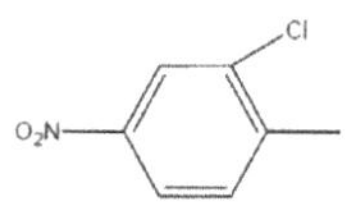	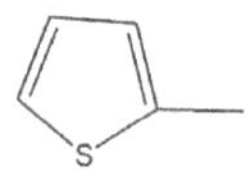
4d	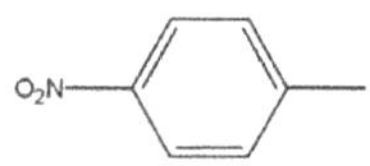	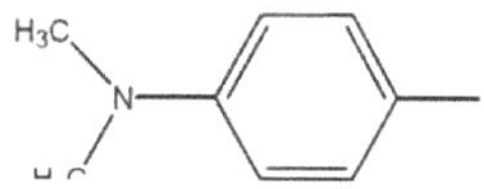

Table 1. Represents Ar and Ar1 groups

2.2. Measurements

Melting point

Melting points were measured by Electrothermal IA 9000 series digital melting point apparatus. Element analysis:

Elemental analytical data were obtained from the micro-analytical unit, National Research Center, Dokki, Giza, Cairo, Egypt.

Fourier-Transition Infrared Spectroscopy (FTIR):

Fourier- transition infrared spectroscopy (FTIR) was performed using a Pye-Unicam spectra-1000 machine to determine the functional groups on the surface of the linen samples. Potassium bromide (KBr) disc was used.

^1H-NMR spectra:

The NMR spectra were recorded on a Varian Mercury VX-300 NMR spectrometer. ^1H spectra were run at 300 MHz in deuterated dimethylsulphoxide(DMSO-d_6).

3.5 Mass spectra:

Mass spectra were measured on a Varian MAT CH-5 spectrometer (70 eV). 3.6 Spectrophotometric measurements:

The absorbance of the dyes were measured in the ultraviolet -visible region at wave length between 300-700 nm. by a UNICAM UV spectrophotometer using a 1cm. quartz cell . The dyes were dissolved in absolute ethanol at a concentration of 10^{-4} mole/l.

2.3. Antibacterial activity measurements

The newly synthesized pyrazolopyrimidines were assessed for their *in vitro* antibacterial activity in the Micro Analytical Centre of Cairo University using Kirby-bauer disc diffusion method [9].

3. Results and discussion

The formation of 2,5-di-amino-3-arylazopyrazolopyrimidine ring system (4a-e) from compounds 1 and 2 under basic condition is assumed to proceed via addition of the most basic N atom in compound 1 to the unsaturated double bond in compound 2 to give the intermediate 3 this Michel adducts is followed by neuclophilic addition of NH_2 group to CN group. The reaction was established and confirmed by studying their element analysis, and (IR –^{1}H-NMR – Mass) spectra. The following results were obtained

2,5-di-amino-3-(4-iodo)-arylazo-6-benzothiazol-2-yl-7-fur-2-yl-pyrazolopyrimidines (4a)

brown crystals, m.p. ≥300° C; yield: (85%)., IR (KBr), 3377-3328 (NH_2), cm^{-1}. , ^{1}H NMR (300 MHz, DMSO-d_6): δ 6.6-7.7. (m, 4H, Ar-H), 8.3 (s, D_2O exchangeable, 2H, NH_2).,MS (70 eV): *m/z* = 577 (M^+, 576)., Element analysis for $C_{23}H_{15}N_8SI$ molecular weight (576), Calc.: C, 47.83 ; H, 2.59; N,27.72 ; O, 10.77 ; S, 5.54. ; I, 22.01, Found: C, 47.73 ; H, 2.49 ; N, 27.62 ; O, 10.67 ; S, 5.44 ; I, 22.0 2,5-di-amino-3-(2,4-dichloro)-arylazo-6-benzothiazol-2-yl-7-thiophen-2-yl–pyrazolopyrimidine (4b)

Red crystals, m.p. ≥300°C; yield: (90%), IR (KBr), 3377-3328 (NH_2), cm^{-1}, ^{1}H NMR (300 MHz, DMSO-d_6): δ 6.6-7.7. (m, 4H, Ar-H), 8.6 (s, D_2O exchangeable, 2H, NH_2), MS (70 eV): *m/z* = 537 (M^+, 535),Element analysis for $C_{23}H_{14}N_8S_2Cl_2$ molecular weight (537), Calc.: C, 51.39 ; H, 2.60 ; N, 20.83 ; S, 11.91; Cl, 13.22., Found: C, 51.29 ; H, 2.50 ; N, 20.73 ; S, 11.81; Cl, 13.21 2,5-di-amino-3-(4-nitro)-arylazo-6-(benzothiazol-2-yl)-7-phenylpyrazolopyrimidine (4c)

yellow crystals, m.p. ≥300°C; yield: (85%), IR (KBr), 3377-3328 ($2NH_2$), cm^{-1}, ^{1} H NMR (300 MHz, DMSO-d_6): δ 6.6-7.7. (m, 4H, Ar-H), 8.3 (s, D_2O exchangeable, 2H, NH_2), MS (70 eV): *m/z* = 505 (M^+, 504), Element analysis for$C_{25}H_{17}N_9$ SO_2 molecular weight (505), Calc.: C, 59.40; H, 3.36 ; N, 24.25 ; S, 6.33; O, 6.33, Found: C, 59.30 ; H, 3.26 ; N, 24.15 ; S, 6.23; O, 6.23.

2,5-di-amino - 3-(2-chloro - 4 - nitro) - arylazo - 6 - benzothiazol - 2 - yl) - 7 - (thiophen-2yl) pyrazoloPyri midine (4d)

brown crystals, m.p. ≥300°C; yield: (85%), IR (KBr), 3377-3328 ($2NH_2$) , cm^{-1}, ^{1} H NMR (300 MHz, DMSO-d_6): δ 6.6-7.7. (m, 4H, Ar-H), 8.1 (s, D_2O exchangeable, 2H, NH_2),MS (70 eV): *m/z* = 547 (M^+, 546), Element analysis for $C_{23}H_{14}N_9S_2O_2$ Cl molecular weight (547), Calc. : C, 50.45 ; H, 2.55 ; N, 23.03 ; S,11.70. ; O, 5.85; Cl, 6.39, Found : C, 50.3 ; H, 2.45 ; N, 23.0 ; S,11.60 ; O, 5.75 ; Cl, 6.29.

2,5-di-amino-3-(4-nitro)-arylazo-6-(benzothiazol-2-yl)-7-(4-(dimethylamino)phenyl)Pyrazolo pyrimidine (4e) orange crystals, m.p. ≥300°C; yield: (85%), IR (KBr), 3377-3328 (NH_2), cm^{-1}, ^1H NMR (300 MHz, DMSO-d_6): δ 6.6 -7.7. (m, 4H, Ar-H), 8.00 (s, D_2O exchangeable, 2H, NH_2), 4.00 (s, D_2O exchangeable, 2H, NH_2) ppm., MS (70 eV): *m/z* = 577 (M^+, 499), Element analysis for $C_{27}H_{22}N_{16}SO_2$ molecular weight (577), Calc.: C, 58.99 ; H, 4.00 ; N, 25.45 ; S, 5.81.; O, 5.81.

Found: C, 58.89 ; H, 4.00 ; N, 25.35 ; S, 5.71. ; O, 5.71.

3.1. Electronic Effects and Ultraviolet-visible spectra

The electronic absorption spectra of arylazopyr- azolopyrimidine dyes 4a-e have been studied, their UV spectrum was found to be ranging between 360-440 nm.

Arylazopyrazolopyrimidine dyes have different resonating structures; the groups attached to rings play an important role in the resonance of the dyes.

It is well known that, the stabilization of these dyes are effected by the substituents originates from the charge separation through the conjugated system between different substituents. However, stabilization of different resonating structures depends on the introduction of an electron – withdrawing group in the aromatic ring and electron donating group of the other aromatic ring. deep bathchromicshifts by the presence of (N,N-dimethyl) as an electron donating group or as a rule the longer conjugation in the molecule; the deeper will be the color this is due to increase of the number of electron in the oscillation which facilities polarizations , this phenomena is clear from the bathochromic shift of dye no. 4d λ max=430

3.2. Antibacterial activity of the synthesized dyes

It is clear from the results, in table (III) that all of the newly synthesized dyes (4a- e) with different color shades possess excellent inhibition of the bacteria growth against the tested gm positive and gm negative bacteria and the inhibition zone diameters obtained are in the range of (17-21) mm which is quite good compared to the control value of 0. It appears also that dyes (4b) and (4d) possessed relatively higher inhibition zone diameter value than the other dyes and this may be attributed to its chemical structure and to the presence of Cl and NO_2 groups in its structure [10].

Dye	Inhibition zone diameter (mm)			
Sample	Bacillus Subtilis (G+)	Esherichia Coli (G-)	Staphylococcus Aureus (G+)	Pseudomonea Aeuruginous (G-)
Tetracycline	30	30	30	30
Control	0.0	0.0	0.0	0.0
4a	19	20	19	19
4b	21	21	20	21
4c	17	20	19	19
4d	19	21	21	20
4e	17	18	17	19

Table 2. Inhibition zone diameter of dyes (4a-e) against Gram positive and Gram negative bacteria

Different novel functionalized 2, 5-di-amino-3-arylazopyrazolopyrimidines derivatives (4a-e) are prepared in good yield as well as good antibacterial properties.

4. References

[1] Ahmed, K.A., El-Hennawi , H.M. Hatem A. Abdel-Aziz and Elkashouti M.A "Synthesis and biological evaluation of sulfonamide 1,3-Thiazole azo dyes and their textile printing application "J. Int. Environmental Application & Science, Vol. 5 (5): 890-900 (2010).

[2] Elgemeie,G. H. Maher H. Helal, Ahmed K. A. „Synthesis and dyeing properties of a new class of condensed carbocyclic arylazopyrazolo[1,5-a] pyrimidines" Pigment and Resin Technology, 32, No. 1 p. 10-23 (2003)

[3] Elgemeie, G.H., Helal, M.H. El-Sayed, H. M. „Recent trends in synthesis and application of nitrogen heterocyclic azo dyes" Pigment and Resin Technology, 30, No. 4 p. 210 (2001)

[4] El kashouti, M.A., Elgemie, G.H. Elmolla, M.M. .Elsayad , H.S and Ahmed. K. A. "Synthesis and application of some new disperse azo dyes derived from 2-cyanomethyl benzothiazole" Pigment and Resin Technology, 36, No. 6, p.382-391 (2007).

[5] Gupta D., Khare S. K., and LahaA., "Coloration technology" 140 p. 15 (2004).

[6] Giridev V.R., Venugopal J., Sudha S., Deepika G. Ramakrishna S. " Carbohydrate polymer" 75. P. 646 (2009).

[7] Thiry M. C. "AATCC Rev" 11, p. 11 (2001).

[8] Ranjni S. , Astha J., Shikha P., Deepti G., and Khare S. K. "Dyes and pigment", 66, p.99 (2005).

[9] Bauer, A.W., Kirby, W. M., Sherris C., Turck, M. (1966), American journal of clinical pathology, vol. 45, pp. 493-196.

[10] Mndzhoyan, S.L., Kramer, S.M., Akopyan, S.G., Grigoryan, M.M., Terzakharyan, Yu. Z., Agabayan, R.V. and Kazaryan, E.V. (1992), "Synthesis and antibacterial properties of halogen-containing 1-phenylcyclopentyl-1-penicillins and 1-cephalosporins", Pharmaceutical Chemistry Journal, Vol. 26, No. 2, pp. 169-173.

Improving Bread Wheat Productivity and Reduce Use of Mineral Nitrogen by Inoculation with *Azotobacter* and *Azospirillum* Under Arid Environment in Upper Egypt

Abd El-Lattief, E.A.

Agronomy Department, Faculty of Agriculture, South Valley University, Qena, Egypt

Abstract

The effect of integrated use of mineral N fertilizer (Urea) and biofertilizer (*Azotobacter* and *Azospirillum*) on grain yield, grain yield attributes and harvest index of wheat was assessed. Two field experiments were carried out on a sandy soil in the Experimental Farm of the Faculty of Agriculture, South Valley University at Qena Governorate, Egypt. The recommended N (230 kg N ha^{-1}) and biofertilizer (*Azotobacter* and *Azospirillum*) were applied alone and in various combinations among them. A randomized complete block design, with three replications, was used in this study.

Treatments significantly affected plant height, spike length, number of spikelets/spike, kernel weight /spike, 1000-kernel weight, grain and straw yields and harvest index. The highest values of such traits were obtained in treatment T_{11} (75% mineral N + biofertilizer with *Azotobacter* and *Azospirillum*). However, T_{12} (50% mineral N + biofertilizer with *Azotobacter* and *Azospirillum*) resulted also higher values for the above mentioned traits comparing with T_1 (100% nitrogen and uninoculated) but the differences among the two treatments almost did not attain the statistical differences.

From this study, it can be concluded that the biofertilizers (double-inoculation of *Azotobacter* and *Azospirillum*) of efficient strains could save 25 or 50 % of the recommended dose of mineral N.

Keywords: Wheat, Biofertilizers, Azotobacter, Azospirillum, Grain yield.

1. Introduction

The high cost of chemical nitrogenous fertilizers and the low purchasing power of most of the farmers restrict its use in proper amounts, hampering crop production. Besides, a substantial amount of the urea-N is lost through different mechanisms including ammonia volatilisation, denitrification and leaching losses, causing environmental pollution problems [1, 2].

The utilization of biological nitrogen fixation technology can decrease the use of urea-N, prevent the depletion of soil organic matter and reduce environmental pollution to a considerable extent [3, 4]. Also, Use of biofertilizers on Egyptian soils has decreased the pH, which had led to increased availability of trace elements that enhance plant growth. Bio-fertilizers are eco-friendly and have been proved to be effective and economical alternate of chemical fertilizers with lesser input of capital and energy [5].

Biofertilizer contains live or latent cells of efficient strains of nitrogen fixing, phosphate solubilizing or cellulolytic micro-organisms used for application to seed, soil or composting areas to accelerate microbial processes to augment the extent of availability of nutrients.

Nitrogen fixation potential of *Azotobacter* and *Azospirillum* are known. The organic matter rich soils promote the activities of these organisms [6, 7]. Also, free-living nitrogen-fixing bacteria *eg Azotobacter chroococcum* and *Azospirillum lipoferum*, were found to have not only the ability to fix nitrogen but also the ability to release phytohormones similar to gibberellic acid and indole acetic acid, which could stimulate plant growth, absorption of nutrients, and photosynthesis [8].

Many authors have shown the positive effect inoculation of wheat with Azotobacter or/and *Azospirillum* [9, 10, 11, 12]. Tilak [13] reported positive effects of double-inoculation of *Azotobacter* and *Azospirillum* on dry matter of maize and sorghum. Rai and Caur [14] studied *Azotobacter* and *Azospirillum* and double-inoculation and alone inoculation effects on wheat growth and yield. Double-inoculation of *Azotobacter* and *Azospirillum* had positive effects on plant height, spike length, grain yield, biological yield and harvest index in various wheat genotypes.

Present study aims to evaluate the importance of bio-fertilization in the improvement growth and productivity of bread wheat crop as well as the expansion of bio-agriculture to reduce agriculture costs and environmental pollution via lowering mineral fertilizers application.

2. Materials and methods

The field experiments were conducted at the Experimental Farm, Faculty of Agriculture, South Valley University (latitude 26°10′ N, longitude 32°43′ E, Altitude 79 m above sea level), Qena, Egypt during 2010-11 and 2011-12 seasons. The soil of the experimental site is sandy throughout its profile (73.7% coarse sand, 16.8% fine sand, 5.8% silt and 3.7% clay). Its pH value of 7.62, 1.75 EC (dSm^{-1}), 0.45% organic matter content, 0.25% total N, and available P and K of 7.42 and 170 ppm, respectively. The weather is very hot and dry from May to October where temperatures can reach up to 40 °C. On the other hand, the weather is usually warm during winter months and rainfall is rare.

The dose of nitrogen (230 kg N ha^{-1}) was manipulated at various levels in combination with different biofertilizers as per the treatment schedule. The different treatment combination as follows:

T_1- 100% mineral N (MN), T_2- *Azotobacter* (AZB) alone, T_3- *Azospirillum* (AZS) alone, T_4- AZB + AZS, T_5- 75 % MN + AZB, T_6- 50 % MN + AZB, T_7- 25 % MN + AZB, T_8- 75 % MN + AZS, T_9- 50 % MN + AZS, T_{10}- 25 % MN + AZS, T_{11}- 75 % MN + AZB + AZS, T_{12}- 50 % MN + AZB + AZS, T_{13}- 25 % MN + AZB + AZS, T_{14}- Control (without nitrogen and uninoculated). The seeds were inoculated by liquid culture of locally isolated strains of *Azotobacter lipoferum* and *Azospirillum chroococcum* (≈10^7 CFU/ml) which obtained from Biofertilizers Production Unit of Faculty of Agriculture, South Valley University. 1% of carboxy methyl cellulose (CMC) was added to the culture to increase its viscosity to gel form to act as adhesive biostabilizer, the addition of CMC was made just before using.

The experiment was carried out in a randomized block design with three replications. Experimental unit measured 3.0 m in width and 4 m in length. Bread wheat (Giza 168 cv.) was sown on the 10th of November in each season. The other cultural practices were carried out as recommended for the crop.

At harvest time, ten fertile stems were taken at random from each plot for measuring plant height, spike length, number of spikelets/spike and kernel weight /spike. Also, 1000-kerenl weight was estimated for each plot. Meanwhile, grain and straw yields were estimated at plot basis. Harvest index (%) of each plot was calculated by using the following formula:

$$Harvest\ Index\ (\%) = \frac{Grain\ \ yield}{Biolo\ gical\ \ yield}\ x\,100$$

The data were analyzed by analysis of variance (ANOVA) using MSTAT-C statistical software. Treatment means were compared using Duncan's multiple tests [15]. Since data followed the homogeneity test, pooling was carried out over the seasons and mean data are given.

3. Results and discussion

3.1. Yield attributes

Data presented in Table 1 indicated that various studied treatments had a significant effect ($P < 0.01$) on plant height, spike length, number of spikelets/spike, kernel weight /spike and 1000-kernel weight. Table 2 shows that greatest values of such traits were from treatment T_{11} (75% mineral N + biofertilizer with *Azotobacter* and *Azospirillum*). Also, T_{11} significantly increased plant height, spike length, number of spikelets/spike, kernel weight /spike and 1000-kernel weight by 4.1, 13.2, 10.2, 9.6 and 12.0 %, respectively, compared to T_1 (100% mineral nitrogen and uninoculated) and by 30.9, 67.2, 53.5, 100, 76.3%, respectively, compared to T_{14} (without nitrogen and uninoculated). However, T_{12} (50% mineral N + biofertilizer with *Azotobacter* and *Azospirillum*) resulted also higher values for the above mentioned yield components comparing with T_1 (100% nitrogen and uninoculated) but the differences among the two treatments almost did not attain the statistical differences. Meanwhile, T_1 resulted yield and yield components almost significantly higher in their values than those in T_{14} (without nitrogen and uninoculated) treatment. These findings are in agreement with those of Sharief *et al* [16], Elsayed *et al* [17] and El-Garhi *et al* [18].

Inoculation with *Azospirillum* alone (T_3) produced significantly higher plant height (98.4 cm), kernel weight per spike (2.035 g) and 1000-kernel weight (33.33 g) than those of inoculation with *Azotobacter* alone (T_2).

3.2. Grain and straw yields

The effects of studied treatments on the grain and straw yields were significant at 1 % level (Table 1). Means in Table 2 indicates that superiority of grain and straw yields were achieved by application of double-inoculation of *Azotobacter* and *Azospirillum* plus 75% mineral N (T_{11}) with a grain and straw yields of 5.046 and 6.470 tons ha^{-1}, respectively. Meanwhile, double-inoculation of *Azotobacter* and *Azospirillum* plus 50% mineral N (T_{12}) resulted higher value for the studied grain yield (4.684 t ha^{-1}) comparing with T_1 (4.486 t ha^{-1}) but the differences among the two treatments did not attain the statistical differences. Also, T_{12} treatment did not differ significantly with application with 100% mineral N (T_1) concerning the effect of straw yield as its values attained 6.059 and 6.058 t ha^{-1} for the two treatments, respectively.

Source of variance	d.f	Plant height	Spike length	No. of spiklet/ spike	Kernel weight/ spike	1000-seed weight	Grain yield /ha	Straw yield /ha	Harvest index
Year (Y)	1	112.0	5.054	9.11	0.342	58.73	0.238	0.323	0.0082
Rep./Y(Ea)	6	38.04	2.317	8.38	0.048	12.66	0.167	0.218	0.0551
Treatment (T)	13	493.8**	17.15**	48.35**	1.656**	255.9**	4.518**	6.594**	0.176**
Y x T	13	2.288	0.408	6.42	0.057	0.082	0.020	0.161	0.041
Error (Eb)	78	16.311	1.973	3.86	0.0421	9.78	0.102	0.167	0.039

** significant at $P < 0.01$ level

Table 1. Analysis of variance of measured parameters

Treatments	Plant height (cm)	Spike length (cm)	No. of spike-lets/ spike (g)	Kernel weight/ spike (g)	1000-kernel weight (g)	Grain yield (t/ha)	Straw yield (t/ha)	Harvest index (%)
T_1- 100% mineral N (MN)	112.7bc	12.63 b	22.18 b	2.800 b	41.72 b	4.486 b	6.058 b	42.5 ab
T_2- *Azotobacter* (AZB)	94.3 g	10.33 c	17.93 d	1.799 f	29.78 f	2.993 f	4.389 e	40.5 ce
T_3- *Azospirillum* (AZS)	98.4 f	11.48bc	18.13 d	2.035 e	33.33 e	3.362 e	4.829 d	41.0 bc
T_4- AZB + AZS	100.2 f	11.38bc	20.65bc	2.275 d	37.50cd	3.708cd	5.411 c	40.7 ce
T_5- 75 % MN + AZB	110.8bc	12.00 b	18.75cd	2.747 b	36.89 d	4.422 b	6.059 b	42.2 ab
T_6- 50 % MN + AZB	104.3 e	12.00 b	20.50bc	2.423cd	37.63cd	3.870 c	5.575 c	41.0 bc
T_7- 25 % MN + AZB	100.1 f	11.43bc	18.50cd	2.013 e	33.08 e	3.385de	4.977 d	40.5 ce
T_8- 75 % MN + AZS	111.3bc	12.60 b	21.90 b	2.758 b	41.63 b	4.521b	6.060 b	42.7 ab
T_9- 50 % MN + AZS	108.9cd	12.50 b	21.80 b	2.553 c	40.63bc	3.877 c	5.647 c	40.7 ce
T_{10}- 25 % MN + AZS	105.3de	11.75bc	20.44bc	2.050 e	35.83de	3.481de	5.000 d	41.0 bc
T_{11}- 75 % MN + AZB + AZS	117.8 a	14.30 a	24.45 a	3.070 a	46.73 a	5.046 a	6.470 a	43.8 a
T_{12}- 50 % MN + AZB + AZS	113.6 b	12.80 b	22.25 b	2.850 b	41.75 b	4.684 b	6.059 b	43.6 a
T_{13}- 25 % MN + AZB + AZS	109.5 c	12.25 b	19.25cd	2.288 d	37.50cd	3.955 c	5.557 c	41.6 b
T14- Control (without)	90.0 h	8.55 d	15.93 e	1.520 g	26.50 g	2.545 g	3.978 f	39.0 e

The same letters within columns means not significant differences at 5% level.

Table 2. The associative influence of biofertilizers and reduced doses of mineral nitrogen on wheat yield parameters (data over two seasons).

Application of T_{11} had significantly higher grain and straw yields by 12.5 and 6.8 % relative to T_1 and by 98.3 and 62.4%, respectively relative to T_{14}. Also, T_{12} had significantly higher grain and straw yields by 84.1 and 52.3%, respectively relative to T_{14}. Also it is showed in Table 2 that *Azospirillum* is more effective than *Azotobacter* on grain yield due to more role of *Azospirillum* in up taking nitrogen produced by biological fixing by *Azospirillum* bacteria that finally will cause to more grain yield of plant. The lower values of grain and straw yields (2.545 and 3.978 t ha^{-1}, respectively) were obtained from T_{14} (without nitrogen and uninoculated). It is evident from the data in Table 2 that combined application of mineral and biofertilizers were favorable in enhancing yield than using mineral or biofertilizer alone.

Such increase in yields (grain and straw) and grain yield attributes, due to application of T_{11} or T_{12}, might be due to the role of biofertilizer (*Azotobacter* and *Azospirillum*) in enhancing soil biological activity, which improved nutrient mobilization from organic and chemical sources. Also, the biofertilizer plays a significant role in regulating the dynamics of organic matter decomposition and the availability of plant nutrients and in increasing nitrogen fixer. In this case, Radwan and Hussein [19], Sharief *et al* [16], Elsayed *et al* [17], El-Garhi *et al* [18], Badr *et al* [11] and Bahrani *et al* [12] found positive effect on yield and yield attributes of wheat when inoculated with biofertilizer. In controlled field trials in Iran, Khavazi *et al* [20] found that yield improvements of more than 20% have been observed for wheat as a result of application of *Azotobacter* and *Azospirillum* inoculums.

3.3. Harvest index

Variance analyzing of harvest index, data showed that harvest index was significant influenced by various studied treatments at 1% probability level (Table 1). Application of T_{11} resulted highest value of harvest index (43.8%) and it was followed by T_{12} (43.6%), T_8 (42.7%), T_1 (42.5%) and T_5 (42.2%) without any differences significant among them (Table 2). Meanwhile, the lower value of harvest index (39.0%) was obtained from T_{14}. Thus it is indicated that using bio-fertilizers caused to increasing harvest index due to effect on dry weight and allocating more photosynthetic matters to grain.

The interaction effect of fertilization and year was not significant for all yield attributes traits and grain yield as well as harvest index (Table 1). Such results indicated that fertilization treatments showed similar effects from season to season.

4. Conclusion

In conclusion, the use of biofertilizers became ineludible to minimize the environmental pollution, caused by the chemical ones, and to improve the yield quality of various crops needed at the time being. Although 25 or 50 % of mineral N was replaced by biofertilizers (double-inoculation of *Azotobacter* and *Azospirillum*), the yield and its components of wheat increased compared to that obtained with the recommended dose of mineral nitrogen. Finally, the biofertilizers of efficient strains could save 25 or 50 % of the recommended dose of mineral nitrogen.

5. References

[1] De Datta, S.K. and R.J. Buresh, 1989. Integrated nitrogen management in irrigated rice. Advances in Soil Science 10, 143–169.

[2] Choudhury, A.T.M.A. and I.R. Kennedy, 2005. Nitrogen fertiliser losses from rice soils and control of environmental pollution problems. Communications in Soil Science and Plant Analysis 36, 1625–1639.

[3] Choudhury, A.T.M.A. and I.R. Kennedy, 2004. Prospects and potentials for systems of biological nitrogen fixation in sustainable rice production. Biology and Fertility of Soils 39, 219–227.

[4] Kennedy, I.R., A.T.M.A. Choudhury and M.L. Kecskés, 2004. Non-symbiotic bacterial diazotrophs in cropfarming systems: can their potential for plant growth promotion be better exploited? Soil Biology and Biochemistry 36, 1229–1244.

[5] Hafeez, F.Y., S. Hameed, A.H. Zaidi, K.A. Malik, 2002. Biofertilizers for Sustainable Agriculture. In: Techniques for Sustainable Agriculture (F. Azam, M. M. Iqbal, C. Inayatullah and K. A. Malik, Editors), pp. 67-73. ISBN, NIAB, Faisalabad, Pakistan

[6] Lakshami, K., N.S. Subba, K.V. Tilak and C.S. Singh, 1972. *Azospirillum*, a new bacterial fertilizer for tropical crops. Sci. Reporter, Council of Scientific and Industrial Research (India), 16:690-692.

[7] Dobereiner, J. and J.M. Day, 1975. Nitrogen fixation in the rhizosphere of tropical grasses. In: Nitrogen fixation by free living microorganisms. W.D.P. Stewart (ed.). Cambridge Univ. Press, London. pp: 39-56.

[8] biofertilizers in Iran. Ministry of Jihad-e- Agriculture, Agricultural Research and Education Organization (AREO) and Soil and Water Research Institute (SWRI), Tehran, Iran, pp: 419.

[9] Fayez, M., N.F. Emam, and H.E. Makboul, 1985. The possible use of nitrogen fixing *Azospirilum* as biofertilizer for wheat plants. Egypt. J. Microbiol., 20(2), 199-206.

[10] Tawfik, M.M. and A.M. Gomaa, 2005. Effect of organic and biofertilizers on the growth and yield of wheat plants Egypt .J. Agrci. Res., 2(2): 711-725.

[11] Abbasdokht, H., 2008. The study of Azotobacter chroococcum inoculation on yield and post harvest quality of wheat. International Meeting on Soil Land Management and Agroclimatology, Turkey, 885-889.

[12] Badr, A. Elham, O.M. Ibrahim and M.F. El-Kramany, 2009. Interaction effect of biological and organic fertilizers on yield and yield components of two wheat cultivars. Egypt .J. Agron., 31(1): 17-27.

[13] Bahrani, A.J. Pourreza and M. Hagh Joo, 2010. Response of Winter Wheat to Co-Inoculation with Azotobacter and Arbescular Mycorrhizal Fungi (AMF) under Different Sources of Nitrogen Fertilizer. American-Eurasian J. Agric. & Environ. Sci., 8 (1): 95-103

[14] Tilak, K.V.B.R., 1992. *Azospirillum* brasilense and *Azotobacter* chrooccocum inoculum effect of mayze and sorghum. Soil Bio. Biochem., 14: 417-418.

[15] Rai, S.N. and A.C. Caur, 1998. Characterization of *Azotobacter* Spp. And effect of *Azospirilum lipoferum* on the yield and N-Uptake of wheat crop, Plant and Soil, 109: 131-134.

[16] Steel, R.G.D. and J.H. Torrie, 1980. Principles and procedures of statistics, 2nd ed. McGraw-Hill, New York.

[17] Sharief, A.E., S.E. El-Kalla, A.A. Leilah and H.E.M. Mostafa, 1998. Response of wheat cultivars to nitrogen fertilizer levels and biological fertilization. J.Agric.Sci. 23(12):5807-5816. Mansoura Univ., Egypt.

[18] El-Sayed, M.Z., A.E. Abd El-Sattar, H.A. Basha and I.M. Abd El-Hammeed, 2005. Improvement of wheat productivity in newly reclaimed soil in Egypt. Annals UMCS, Sec. E.60:113-121.

[19] El-Garhi, A.S., N.A. Atia and Sara E.E. Fouda, 2007. Effect of inoculating N-fixing bacteria (Cerealine) on wheat (Triticum aestivum, L.) growth and nutrient content. Zagazig J. Agric. Res., 34(2): 249-273.

[20] Radwan, S.M.A. and H.F. Hussein, 1996. Effect of bio. and organic fertilization on wheat yield under different weed control treatments. Egyptian J. Appl. Sci., 11:267-281.

[21] Khavazi, K., H. Asadi-Rahmani and M.J. Malakouti, (eds.), 2005. Necessity for the production of biofertilizers in Iran. Ministry of Jihad-e- Agriculture, Agricultural Research and Education Organization (AREO) and Soil and Water Research Institute (SWRI), Tehran, Iran, pp: 419.

Permissions

The contributors of this book come from diverse backgrounds, making this book a truly international effort. This book will bring forth new frontiers with its revolutionizing research information and detailed analysis of the nascent developments around the world.

We would like to thank the Organising Committee, for lending their expertise to make the book truly unique. They have played a crucial role in the development of this book. Without their invaluable contribution this book wouldn't have been possible. They have made vital efforts to compile up to date information on the varied aspects of this subject to make this book a valuable addition to the collection of many professionals and students.

This book was conceptualized with the vision of imparting up-to-date information and advanced data in this field. To ensure the same, a matchless editorial board was set up. Every individual on the board went through rigorous rounds of assessment to prove their worth. After which they invested a large part of their time researching and compiling the most relevant data for our readers. Conferences and sessions were held from time to time between the editorial board and the contributing authors to present the data in the most comprehensible form. The editorial team has worked tirelessly to provide valuable and valid information to help people across the globe.

Every chapter published in this book has been scrutinized by our experts. Their significance has been extensively debated. The topics covered herein carry significant findings which will fuel the growth of the discipline. They may even be implemented as practical applications or may be referred to as a beginning point for another development. Chapters in this book were first published by InTech; hereby published with permission under the Creative Commons Attribution License or equivalent.

The editorial board has been involved in producing this book since its inception. They have spent rigorous hours researching and exploring the diverse topics which have resulted in the successful publishing of this book. They have passed on their knowledge of decades through this book. To expedite this challenging task, the publisher supported the team at every step. A small team of assistant editors was also appointed to further simplify the editing procedure and attain best results for the readers.

Our editorial team has been hand-picked from every corner of the world. Their multi-ethnicity adds dynamic inputs to the discussions which result in innovative outcomes. These outcomes are then further discussed with the researchers and contributors who give their valuable feedback and opinion regarding the same. The feedback is then collaborated with the researches and they are edited in a comprehensive manner to aid the understanding of the subject.

Apart from the editorial board, the designing team has also invested a significant amount of their time in understanding the subject and creating the most relevant covers. They scrutinized every image to scout for the most suitable representation of the subject and create an appropriate cover for the book.

The publishing team has been involved in this book since its early stages. They were actively engaged in every process, be it collecting the data, connecting with the contributors or procuring relevant information. The team has been an ardent support to the editorial, designing and production team. Their endless efforts to recruit the best for this project, has resulted in the accomplishment of this book. They are a veteran in the field of academics and their pool of knowledge is as vast as their experience in printing. Their expertise and guidance has proved useful at every step. Their uncompromising quality standards have made this book an exceptional effort. Their encouragement from time to time has been an inspiration for everyone.

The publisher and the editorial board hope that this book will prove to be a valuable piece of knowledge for researchers, students, practitioners and scholars across the globe.

List of Contributors

Sukru Dursun, Serpil Onder, Ramazan Acar, Mithat Direk and Osman Mucehver
Selcuk University, Konya, Turkey

Ali A Torahi
Department of Forestry, Behbehan University of Technology, Behbehan, Iran

Vrushal T. Ghoble
Jai Hind College, India

H. Zarei Mahmoodabady
Department of Environmental Engineering, Islamic Azad University, Maybod Branch, Yazd, Iran

M. Yazdi
Department of Agricultural Engineering, Payam E Noor University, Yazd Center, Yazd, Iran

Safar Ghaedrahmati and Mohammad Reza Rezaei
University of Yazd, Iran

Mahdi Paktinat, Bagher Mojazi Amiri and Hamid Farahmand
Department of Fisheries, Faculty of Natural Resources, University of Tehran, Karaj, Iran

Mahvash Khodabandeh
Industrial and Environmental Biotechnology Department, National Institute of Genetic Engineering and Biotechnology (NIGEB), Tehran, Iran

Amirsasha Bnanankhah and Farhad Nejadkoorki
Department of Environmental Engineering, Yazd University, Yazd, Iran

M.T. Dastorani and S. Poormohammadi
Faculty of Natural Resources, Yazd University, Iran

Elham Yousefi and Farhad Nejadkoorki
Department of Environmental Engineering, Yazd University, Yazd, Iran

El-Tatawy Nashwa
Economics and Agribusiness Dept., Faculty of Agriculture, Alexandria University, Egypt

Majid Loghmanpour
Yerevan State Agrinean University, Armenia and Expert of Forest, Range & Watershed Management Organization, Armenia

Zhirair Vardanian
Dept. of Yerevan State Agrinean University, Armenia

Hadi Kiadaliri
Science & Research Unit of Azad University, Tehran, Iran

Mohsen Elahi
Tarbiat Modares University of Noor-Mazandarn, Iran

Kheira Hammadi, Mahjouba Aznouz, Miloud Halbouche
Laboratory of PAA Faculty SNV, Abdelhamid Ben Badis Mostaganem University, Algeria

Fatemeh Bateni, Sima Fakheran and Alireza Soffianian
Department of Natural Resources, Isfahan University of Technology, Isfahan, Iran

Shima Ziajahromi and Ali Daryabeigi Zand
Environmental Sciences Research Institute, Shahid Beheshti University, G.C., Tehran, Iran

Meysam Khanizadeh
Faculty of Nuclear Engineering and Physics, Amirkabir University, Tehran, Iran

Hakimeh Khalifipour
Department of Natural Resources, Isfahan University of Technology, Isfahan, Iran

Ali Talebi
Faculty of Natural Resources, Yazd University, Yazd, Iran

Bongumusa M. Zuma
Institute for Environmental Biotechnology, Rhodes University, Grahamstown 6140, South Africa
Environmental Health and Biotechnology Research Group, Division of Pharmaceutical Chemistry, Faculty of Pharmacy, Rhodes University, Grahamstown 6140, South Africa

Catherine D. Luyt, Tatenda Chirenda and Roman Tandlich
Environmental Health and Biotechnology Research Group, Division of Pharmaceutical Chemistry, Faculty of Pharmacy, Rhodes University, Grahamstown 6140, South Africa

Neda Bihamta Toosi
Department of Natural Resources, Isfahan University of Technology, Isfahan, Iran

Fardad Edalatkhah
Municipality of Tehran, Tehran, Iran

Mahdi Shafaghati, Akram Norouzi and Somaye Motaghi
Payame Noor University, Taleghan, Iran

Maryam Mirbahaei
Payame Noor University, Varamin, Iran

A. Niaei, D. Salari, P. Nakhostin Panahi and S. M. Mousavi
Department of Applied Chemistry & Chem. Eng., Faculty of Chemistry, University of Tabriz, Tabriz, Iran

Maliheh Alsadat Madanian
Department of Natural Resources, Isfahan University of Technology, Isfahan, Iran

Korous Khoshbakht and Hadi Veisi
Environmental Sciences Research Institute, Shahid Beheshti University, G.C., Evin, Tehran, Iran

Mohammad Ebrahim Rezai
Agronomy and Plant Breeding Department, University of Mohaghegh, Ardabili, Iran

Meena Al-Mansoori, Aydin Basarir, and Sherin Sherif
Agribusiness and Consumer Sciences Department, United Arab Emirates University, UAE

Abdurrahman Dinc
Department of Tourism Guide, Tourism Faculty, Konya University, Konya

Nurhan Kocan
Department of Landscape Architecture, Agriculture Faculty, Ege University, Izmir

Said Sawy, Ali Abdel-Hameed and Kais A. Sultan
Faculty of Agriculture, Cairo University, Egypt

Fadila Khaldi
Department of Biology,University of Mohammed Cherif Messaadia, Souk-Ahras, Algeria
Laboratory of Cell Toxicology, Department of Biology, Faculty of Sciences, Algeria

Houria Berrebbah
Laboratory of Cell Toxicology, Department of Biology, Faculty of Sciences, Algeria

Med. Réda. Djebar
University of Badji Mokhtar, Annaba, Algeria

Jan M. Mari
Department of Plant Protection Sindh Agriculture University, Tandojam, Pakistan

Faris N. A. Alhady Alibawi and Ali G. Alhuweizi
College of Science, University of Babylon, Babylon, Iraq

Sahib Y. Al-Morshidy
University of Kufa, Najaf, Iraq

Ali Akbar Karimian
Faculty of Natural Resources, Yazd University, Yazd, Iran

Rabha Bennama
Department of Biotechnology, University of Mostaganem, Mostaganem, Algeria

Victor Ladero, Miguel A. Alvarez and María Fernández
Instituto de Productos Lácteos de Asturias (IPLA-CSIC), Villaviciosa, Spain

Ahmed Bensoltane
Department of Biology, University of Oran, Algeria

M. Nabiuni, L. Karimzadeh and Z. Nazari
Department of Cell and Molecular Biology, Faculty of Biological Sciences, Kharazmi University, Tehran, Iran

S. Nasri and F. Poyanmanesh
Department of Biology, Faculty of Agriculture and Basic Science, Payamnoor University, Tehran, Iran

Elham Azimi and Abdolhosein Shiravi
Department of Biology, Faculty of Science, Islamic Azad University, Damghan, Iran

Saad Merza Al-Araji, Ala H. Abbase, Zainab F. Hassan
Department of Medical Physiology, College Of Medicine, University Of Babylon, Iraq

Nedjoud Grara and Fadila Khaldi
Laboratory of Sciences and Technology of Water and the Environment, University of Mohammed cherif Messaadia, Souk-Ahras, Algeria

Amira Atailia, Mounir Boucenna, Houria Berrebbah and Mohamed Reda Djebar
Laboratory of Cellular Toxicology, University of Annaba, Algeria

Samiullah
Department of Plant Sciences, Quaid-i-Azam University, Islamabad
College of Pharmacy, University of Hawaii at Hilo, USA

Asghari Bano
Department of Plant Sciences, Quaid-i-Azam University, Islamabad

Sisay Girmay and Ghee Tan
College of Pharmacy, University of Hawaii at Hilo, USA

Farida Becir
SNV Institute, Department of Biology, El-Tarf University, Algeria
Laboratory of Aquatic and Terrestrial Ecosystems (Ecostaq), Badji-Mokhtar University, Algeria

Zihad Bouslama
Laboratory of Aquatic and Terrestrial Ecosystems (Ecostaq), Badji-Mokhtar University, Algeria

Idir Bitam
Vectorial Systems Ecology Service, Pasteur Institute of Algeria, Algeria

Hadjira Hannachi and M'Barek Chetoui
Laboratory of Animal Ecology, Faculty of Sciences, El-Manar University, Tunis

Mohammad S. Jawarneh, Riad Muhidat1, Jamil N. Lahham and Ahmad Ali El-Oqlah
Department of Biological Sciences, Faculty of Science, Yarmouk University, Saudi Arabia

Mohammad H. Brake
Science Department, Faculty of Science, Jerash University, Saudi Arabia

Hussein M. Migdadi
Plant Production Department, Collage of Food and Agricultural Sciences, King Saud University, Saudi Arabia

Darlina Md. Naim
School of Biological Sciences, Universiti Sains Malaysia, Penang, Malaysia

Hurul Adila-Aida Mohamad Rosly
Centre for Marine and Coastal Studies, Universiti Sains Malaysia, Penang, Malaysia

Siti Azizah Mohd. Nor
School of Biological Sciences, Universiti Sains Malaysia, Penang, Malaysia
Centre for Marine and Coastal Studies, Universiti Sains Malaysia, Penang, Malaysia

Sabrina Bouchelaghem and Djebar M. R
Laboratory of Toxicology and Cellular Bioenergetics, University of Annaba, Algeria

Shafariatul A.I, Hasidah M. Sidek and Salmijah Surif
Department of Biomedical Science, Faculty of Allied Health Sciences, Universiti Kebangsaan, Kuala Lumpur, Malaysia

Noraziah Mohamad Zin and Mohamed Kamel A Ghani
School of Diagnostic and Applied Health, Faculty of Health Sciences, Universiti Kebangsaan Malaysia, Kuala Lumpur

Nor Farha Hussain
Pathology Unit, Hospital of Tengku Anis, Pasir Puteh, Kelantan

Rahizan Isa
Bacteriology Unit, Infectious Disease Research Centre, Institute of Medical Research, Kuala Lumpur

Nik Marzuki Sidik
School of Biosciences and Biotechnology, Faculty of Science and Technology, University Kebangsaan Malaysia, Bangi Selangor

Alireza Hosseini, Mohammad-Saber Fallah Nezhad and Yahya Zare Mehrjardi
Department of Industrial Engineering, Yazd University, Iran

Reza Hosseini
Division of Biostatistics, University of Southern California, USA

Farzad Eskandari
Department of Agricultural Economics, Faculty of Agriculture, Kurdistan University, Sanandaj, Iran

Reza Kamrani
Islamic Azad University, Bam Branch, Iran